Edited by
Richard P. Mildren
and James R. Rabeau

Optical Engineering of Diamond

Related Titles

Rafailov, E. U., Cataluna, M. A., Avrutin, E. A.

Ultrafast Lasers Based on Quantum Dot Structures

Physics and Devices

2011
ISBN: 978-3-527-40928-0

Quinten, M.

Optical Properties of Nanoparticle Systems

Mie and Beyond

2011
ISBN: 978-3-527-41043-9

Bovensiepen, U., Petek, H., Wolf, M. (eds.)

Dynamics at Solid State Surfaces and Interfaces

Volume 1: Current Developments

2010
ISBN: 978-3-527-40937-2

Santori, C., Fattal, D., Yamamoto, Y.

Single-photon Devices and Applications

2010
ISBN: 978-3-527-40807-8

Mandel, P.

Nonlinear Optics

An Analytical Approach

2010
ISBN: 978-3-527-40923-5

Brillson, L. J.

Surfaces and Interfaces of Electronic Materials

2010
ISBN: 978-3-527-40915-0

Ohtsu, M. (ed.)

Nanophotonics and Nanofabrication

2009
ISBN: 978-3-527-32121-6

Koizumi, S., Nebel, C., Nesladek, M. (eds.)

Physics and Applications of CVD Diamond

2008
ISBN: 978-3-527-40801-6

Wehrspohn, R. B., Kitzerow, H.-S., Busch, K. (eds.)

Nanophotonic Materials

Photonic Crystals, Plasmonics, and Metamaterials

2008
ISBN: 978-3-527-40858-0

Edited by Richard P. Mildren and James R. Rabeau

Optical Engineering of Diamond

WILEY-VCH

WILEY-VCH Verlag GmbH & Co. KGaA

The Editors

Assoc. Prof. Richard P. Mildren
Macquarie University,
MQ Photonics Research Centre
Department of Physics and Astronomy
Sydney, NSW 2109
Australia

Assoc. Prof. James R. Rabeau
Macquarie University (Honorary Fellow)
Department of Physics and Astronomy
Sydney, NSW 2109
Australia
(currently at Deloitte Touche Tohmatsu, Australia)

Cover Picture
Concept image of diamond photonic nanowire arrays with embedded nitrogen-vacancy color centers. By applying large-scale semiconductor manufacturing techniques to single-crystal diamond, we may utilize its unique and exceptional material properties in diverse areas of quantum science, optics and photonics, and nanotechnology. The work was performed by researchers at the Harvard University and Texas A&M. Created by Jay Penni, used with permission of Marko Loncar and Thomas Babinec.

■ All books published by **Wiley-VCH** are carefully produced. Nevertheless, authors, editors, and publisher do not warrant the information contained in these books, including this book, to be free of errors. Readers are advised to keep in mind that statements, data, illustrations, procedural details or other items may inadvertently be inaccurate.

Every effort has been made to trace copyright holders and secure permission prior to publication. If notified, the publisher will rectify any error or omission at the earliest opportunity.

Library of Congress Card No.: applied for

British Library Cataloguing-in-Publication Data
A catalogue record for this book is available from the British Library.

Bibliographic information published by the Deutsche Nationalbibliothek
The Deutsche Nationalbibliothek lists this publication in the Deutsche Nationalbibliografie; detailed bibliographic data are available on the Internet at <http://dnb.d-nb.de>.

© 2013 Wiley-VCH Verlag GmbH & Co. KGaA, Boschstr. 12, 69469 Weinheim, Germany

All rights reserved (including those of translation into other languages). No part of this book may be reproduced in any form – by photoprinting, microfilm, or any other means – nor transmitted or translated into a machine language without written permission from the publishers. Registered names, trademarks, etc. used in this book, even when not specifically marked as such, are not to be considered unprotected by law.

Cover Design Adam-Design, Weinheim
Typesetting Toppan Best-set Premedia Limited, Hong Kong
Printing and Binding Markono Print Media Pte Ltd, Singapore

Print ISBN: 978-3-527-41102-3
ePDF ISBN: 978-3-527-64863-4
ePub ISBN: 978-3-527-64862-7
mobi ISBN: 978-3-527-64861-0
oBook ISBN: 978-3-527-64860-3

Contents

Foreword XV
Preface XVII
List of Contributors XXI

1	**Intrinsic Optical Properties of Diamond** 1	
	Richard P. Mildren	
1.1	Transmission 2	
1.2	Lattice Absorption 3	
1.2.1	The Two-Phonon Region 4	
1.2.2	Absorption at Wavelengths Longer than 5 μm 7	
1.2.3	Temperature Dependence 8	
1.2.4	Isotopic Content 9	
1.3	UV Edge Absorption 11	
1.4	Refractive Index 13	
1.4.1	Temperature Dependence of the Refractive Index 14	
1.5	Verdet Constant 16	
1.6	First-Order Raman Scattering 16	
1.6.1	Wavelength Dependence 20	
1.6.2	Raman Linewidth 21	
1.6.3	Temperature Dependence 22	
1.6.4	Isotopic Content 23	
1.7	Stimulated Raman Scattering 24	
1.8	Brillouin Scattering 25	
1.9	Electronic Nonlinearity 27	
1.9.1	Nonlinear Refractive Index 29	
1.9.2	Two-Photon Absorption 29	
	Acknowledgments 30	
	References 31	
2	**Optical Quality Diamond Grown by Chemical Vapor Deposition** 35	
	Ian Friel	
2.1	Introduction 35	
2.2	CVD Diamond Growth Principles 36	

2.2.1	Fundamentals of Growth 36
2.2.2	Morphology and Texture 42
2.3	Properties of Optical Quality CVD Diamond 43
2.3.1	Absorption 44
2.3.2	Nonoptical Wavelengths 47
2.3.3	Isotopic Purity 49
2.3.4	Strain-Induced Birefringence 49
2.3.5	Scatter 52
2.3.6	Other Properties of CVD Diamond 54
2.3.6.1	Thermal Properties 54
2.3.6.2	Strength 56
2.4	Optical Applications of CVD Diamond 60
2.4.1	Applications of Polycrystalline Diamond 60
2.4.2	Applications of Single-Crystal Diamond 61
2.5	Summary 63
2.6	Acknowledgments 64
	References 64

3	**Polishing and Shaping of Monocrystalline Diamond** 71
	Jonathan R. Hird
3.1	Introduction: Background and Historical Overview 71
3.2	Shaping Diamond: Cleaving, Bruting, and Sawing 73
3.3	Practical Aspects of Diamond Polishing 74
3.3.1	Apparatus and Preparation 74
3.3.2	Directional Dependence of Polishing: Wear Anisotropy 75
3.4	The Science of Mechanical Polishing 77
3.4.1	Wear Anisotropy 79
3.4.2	Velocity Dependence 79
3.4.3	Diamond Polishing Wear Debris 81
3.4.4	The Polished Diamond Surface 82
3.4.5	Subsurface Damage 84
3.4.6	The Scaife: Its Surface and Preparation 84
3.4.7	Atmosphere Dependence 88
3.4.8	Triboluminescence 88
3.4.9	Wear Mechanism 89
3.5	Tribological Behavior of Diamond 92
3.5.1	Slow-Speed Sliding of Diamond against Diamond 92
3.5.2	Sliding of Diamond against Other Materials 95
3.6	Other Polishing Methods 96
3.6.1	Wear of Diamond by Other Materials 96
3.6.2	Hot Metal Polishing 97
3.6.3	Chemical–Mechanical Planarization 97
3.7	Producing High-Quality Planar Surfaces on Diamond 98
3.7.1	Cleaving 98
3.7.2	Post-Mechanical Polishing Treatment 99

3.7.3	Dry Chemical Etching	99
3.8	Nonplanar and Structured Geometries	101
3.9	Summary	102
	References	103

4 Refractive and Diffractive Diamond Optics 109
Fredrik Nikolajeff and Mikael Karlsson

4.1	Introduction	109
4.2	Windows and Domes	110
4.3	Refractive Devices	112
4.3.1	Lenses	112
4.3.2	Prisms	118
4.3.3	Other Devices	118
4.4	Diffractive Components	119
4.5	Polishing	126
4.6	Micromachining	128
4.7	Coatings	133
4.8	Applications	134
4.9	Conclusions and Outlook	137
	References	138

5 Nitrogen-Vacancy Color Centers in Diamond: Properties, Synthesis, and Applications 143
Carlo Bradac, Torsten Gaebel, and James R. Rabeau

5.1	Introduction	143
5.2	Defects in Diamond	144
5.2.1	Intrinsic and Extrinsic Defects	144
5.2.2	Nitrogen-Related Defects in Diamond	145
5.2.2.1	C-Center	146
5.2.2.2	A-Center	147
5.2.2.3	B-Center	147
5.2.2.4	Platelets	147
5.2.2.5	Other Nitrogen-Vacancy Complexes	147
5.2.3	Nitrogen-Vacancy (NV) Center	148
5.2.3.1	Structure	149
5.2.3.2	Charge State	150
5.2.3.3	Energy Level Scheme	152
5.2.4	Optical and Spin Properties of NV^- Centers in Diamond	153
5.2.4.1	Polarization	153
5.2.4.2	Control	155
5.2.4.3	Measurement	156
5.2.4.4	Relaxation Time	156
5.3	Synthesis of Diamond	159
5.3.1	High-Pressure, High-Temperature (HPHT) Synthesis	159
5.3.2	Chemical Vapor Deposition (CVD) Synthesis	160

5.3.3	Detonation *161*	
5.3.4	Enhancement of Color Center Concentration *162*	
5.3.4.1	Irradiation *162*	
5.3.4.2	Ion Implantation *163*	
5.3.4.3	Incorporation during CVD Growth *163*	
5.4	Applications of Color Centers in Diamond *163*	
5.4.1	Quantum Information Technology *164*	
5.4.2	Life Sciences *165*	
5.4.3	High-Resolution Magnetometry *165*	
5.5	Feasibility of NV Center-Based Nanotechnologies *165*	
5.5.1	Fabricating Ultrasmall (<10 nm) Fluorescent Nanodiamonds *166*	
5.5.2	Some NV Luminescence "Anomalies" at the Nanoscale *167*	
5.6	Conclusions and Outlook *168*	
	References *169*	

6 n-Type Diamond Growth and Homoepitaxial Diamond Junction Devices *177*
Satoshi Koizumi and Toshiharu Makino

6.1	n-Type Diamond Growth and Semiconducting Characteristics *177*	
6.1.1	Background *177*	
6.1.2	Chemical Vapor Deposition of n-Type Diamond Thin Films *178*	
6.1.3	Electrical Properties of n-Type Diamond [20] *182*	
6.2	Electrical Properties of Diamond pn Junctions *185*	
6.3	Diamond Deep-UV LEDs *187*	
6.3.1	Diamond pn Junction LEDs *188*	
6.3.1.1	{111} Diamond pn Junction LEDs *188*	
6.3.1.2	{100} Diamond pn Junction LEDs *190*	
6.3.2	Diamond pin Junction LEDs *192*	
6.3.2.1	{100} Diamond pin Junction LEDs *192*	
6.3.2.2	{111} Diamond pin Junction LEDs *194*	
6.3.2.3	{111} Diamond pin Junction LED with Thick i-Layer *197*	
6.3.3	Light-Emission Mechanisms *200*	
6.4	Recent Progress of Diamond Junction Devices *202*	
6.5	Summary *204*	
6.6	Acknowledgments *205*	
	References *206*	

7 Surface Doping of Diamond and Induced Optical Effects *209*
Vladimira Petrakova, Miroslav Ledvina, and Milos Nesladek

7.1	Introduction *209*	
7.2	NV Centers in Diamond *210*	
7.3	Theoretical Considerations of Surface Manipulation with Optical Defects by Transfer Doping *212*	
7.3.1	Band Bending Model Calculations *212*	

7.3.2	Implication of Surface Functional Changes on Luminescent Properties of NV Centers Implanted to Different Depths	216
7.4	Formation of Variously Charged NV Centers in Diamond	217
7.4.1	Introduction of Vacancies to the Diamond Lattice: Different Irradiation Strategies	220
7.4.1.1	Creation of Vacancies: GR1 versus ND1	220
7.4.1.2	Resulting NV PL of Variously Irradiated NDs	221
7.4.2	Formation of NV Centers: An Annealing Study	223
7.5	Transfer Doping Effects: Luminescent Properties of NV Centers in Variously Terminated Nanodiamonds	225
7.5.1	Quenching of NV-Luminescence on ND Particles	226
7.5.2	Surface and Size Effects in the Variously Terminated Nanodiamond	228
7.5.2.1	Fluorinated NDs	228
7.5.2.2	Size Dependence of NV Luminescence on Variously Terminated ND	229
7.5.3	Fluorescent Nanodiamond Particles as a Sensor of Charged Molecules	231
7.5.3.1	Interactions with Charged Polymers	231
7.5.4	NDs for Biology: Monitoring of Transfection by ND Carriers	232
7.6	Conclusions	235
	References	236

8 Diamond Raman Laser Design and Performance 239
Richard P. Mildren, Alexander Sabella, Ondrej Kitzler, David J. Spence, and Aaron M. McKay

8.1	Introduction and Background	239
8.1.1	Crystalline Raman Laser Principles	242
8.1.1.1	Basic SRS Theory	243
8.1.1.2	Raman Laser Design and Modeling	245
8.2	Optical, Thermal, and Physical Properties of Diamond	246
8.2.1	Raman Gain Coefficient	248
8.2.1.1	Dependence on Polarization	250
8.2.1.2	Dependence on Pump Linewidth	253
8.2.2	Thermal Properties	254
8.2.2.1	Thermal Lens Strength	255
8.2.2.2	Thermally Induced Stress Birefringence	255
8.2.2.3	Thermal Stress Fracture	256
8.2.3	Laser Damage Threshold	256
8.2.4	Design Implications for CVD-Grown Material	257
8.3	Diamond Raman Laser Development	258
8.3.1	External Cavity Wavelength Conversion	258
8.3.2	Intracavity Diamond Raman Lasers	261
8.3.3	Synchronously Pumped Mode-Locked Lasers	262
8.4	Extending the Capability of Raman Lasers Using Diamond	264

8.4.1	Long-Wavelength Generation 264
8.4.2	Deep Ultraviolet Generation 265
8.4.3	High Average Power 267
8.5	Conclusions and Outlook 270
	Acknowledgments 271
	References 272

9 Quantum Optical Diamond Technologies 277
Philipp Neumann and Jörg Wrachtrup

9.1	Introduction 277
9.1.1	Single Quantum Systems 277
9.1.2	The Nitrogen-Vacancy (NV) Center in Diamond 278
9.1.3	The Present Situation 278
9.2	The NV Center's Electron Spin as a Master Qubit 279
9.3	Nuclear Spins as a Qubit Resource 280
9.3.1	Interaction of a Single Electron Spin with nearby Nuclear Spins 282
9.3.1.1	Nuclear spin Hamiltonian 282
9.3.1.2	Secular Approximation and Nonsecular Terms 283
9.3.1.3	Examples of Nearby Nuclear Spins 284
9.3.1.4	Quantum Gates Using Nuclear Spins 287
9.3.1.5	The Nuclear Spin Bath 288
9.3.2	Nonlocal States: The Heart of a Quantum Processor 293
9.3.2.1	Two Nearest-Neighbor ^{13}C Nuclear Spins 293
9.3.2.2	Characterization of the Qubit System 295
9.3.2.3	Generation and Detection of Entanglement 296
9.4	Summary and Outlook 303
	References 305

10 Diamond-Based Optical Waveguides, Cavities, and Other Microstructures 311
Snjezana Tomljenovic-Hanic, Timothy J. Karle, Andrew D. Greentree, Brant C. Gibson, Barbara A. Fairchild, Alastair Stacey, and Steven Prawer

10.1	Introduction 311
10.1.1	Motivation: Applications of Fluorescent Diamond Devices 311
10.1.1.1	Sensing 312
10.1.1.2	Single-Photon Sources for Quantum Key Distribution and Quantum Metrology 313
10.1.1.3	Quantum Information Processing with NV Diamond 315
10.2	Optical Properties 318
10.3	Design of Diamond-Based Optical Structures 319
10.4	Single-Crystal Diamond 321
10.4.1	Lift-Off of Thin and Ultra-Thin Single-Crystal Diamond Films 322
10.4.1.1	Lift-Off 322
10.4.1.2	Lift-Out 322
10.4.2	Fabrication: Lithography and Etching 323

10.4.2.1	Photolithography	325
10.4.2.2	E-Beam Lithography	325
10.4.2.3	FIB Milling	325
10.4.2.4	Ga Hard Mask	326
10.4.2.5	Reactive Ion Etching: A Scalable Process	326
10.4.2.6	Etching of Bulk Diamond: Microlenses	327
10.4.3	Optical Waveguiding in Single-Crystal Diamond	327
10.4.3.1	Waveguides: FIB Lithography	328
10.4.3.2	RIE-Fabricated Waveguides	330
10.4.3.3	Light Guiding in Ion-Implanted Diamond	330
10.4.4	Surface Emission: Solid Immersion Lenses and Nanopillars	330
10.4.4.1	Solid Immersion Lenses	331
10.4.4.2	Nanopillars	332
10.4.5	Cavities	333
10.4.5.1	Nanobeams	333
10.4.5.2	Two-Dimensional Photonic Crystals	333
10.4.5.3	Ring Resonators	334
10.5	Polycrystalline Thin Films	335
10.5.1	Properties	335
10.5.2	Optical Structures in Polycrystalline Films	337
10.5.2.1	Ring Resonators	337
10.5.2.2	Two-Dimensional Photonic Crystals	337
10.6	Hybrid Approaches	338
10.6.1	Nanomanipulation of Nanodiamond	338
10.7	Conclusions and Outlook	340
	Acknowledgments	343
	References	343
11	**Thermal Management of Lasers and LEDs Using Diamond**	**353**
	Alan J. Kemp, John-Mark Hopkins, Jennifer E. Hastie, Stephane Calvez, Yanfeng Zhang, Erdan Gu, Martin D. Dawson, and David Burns	
11.1	Introduction	353
11.1.1	The Requirement for Thermal Management of Lasers and LEDs	353
11.1.2	The Advantages of Diamond: Thermal Conductivity and Rigidity	354
11.2	The Use of Diamond in Lasers: A Brief Review	355
11.2.1	Diamond as a Sub-Mount for Lasers	355
11.2.2	Intracavity Applications of Diamond for Thermal Management	355
11.2.3	Direct Exploitation of Diamond as a Laser Gain Material	356
11.3	Exploiting the Extreme Properties of Diamond	357
11.3.1	Intracavity versus Extracavity Use of Diamond	357
11.3.2	Material Requirements for Intracavity Applications	360
11.4	Current Uses of Diamond: Semiconductor Disk Lasers	360
11.4.1	Semiconductor Disk Lasers: Basic Principles	360
11.4.2	Thermal Management Strategies for Semiconductor Disk Lasers	362
11.4.2.1	Approaches to the Use of Diamond	362

11.4.2.2	Intracavity and Extracavity Approaches to the Use of Diamond *363*
11.4.2.3	Intracavity Diamond Heat Spreaders for Wavelength Diversity *367*
11.4.3	A Review of Progress to Date, and Future Prospects *368*
11.5	Current Uses of Diamond: Doped-Dielectric Disk Lasers *369*
11.5.1	Doped-Dielectric Disk Lasers: Diamond Sub-Mounting for Mechanical Rigidity *369*
11.5.2	Intracavity Use of Diamond in Doped-Dielectric Disk Lasers *371*
11.5.3	Future Prospects *374*
11.6	Current Uses of Diamond: Light-Emitting Diodes *375*
11.6.1	Introduction *375*
11.6.2	Diamond as a Heat Spreader in LEDs *375*
11.6.3	Monolithic Structures and the Epitaxy of Gallium Nitride on Diamond *376*
11.6.4	Diamond LEDs *376*
11.7	Conclusions and Future Directions *376*
	Acknowledgments *378*
	References *378*

12 Laser Micro- and Nanoprocessing of Diamond Materials *385*
Vitaly I. Konov, Taras V. Kononenko, and Vitali V. Kononenko

12.1	Introduction *385*
12.2	Laser-Induced Surface Graphitization *388*
12.2.1	Mechanisms of Diamond Surface Graphitization *388*
12.2.2	Experimental Data *389*
12.3	Laser Ablation *394*
12.3.1	Vaporization Ablation *395*
12.3.2	Nanoablation *402*
12.3.3	Photoionization of Diamond *406*
12.4	Bulk Graphitization of Diamond *410*
12.4.1	Threshold Conditions *410*
12.4.2	Laser-Induced Graphitization Waves *412*
12.4.3	Three-Dimensional (3-D) Laser Writing in Diamond *414*
12.5	Diamond Laser Processing Techniques *419*
12.5.1	Laser Polishing *420*
12.5.2	Formation of Conductive Structures in Diamond *424*
12.5.3	Surface Structuring *428*
12.5.4	Diamond Optics *431*
12.6	Conclusions *438*
	Acknowledgments *438*
	References *438*

13 Fluorescent Nanodiamonds and Their Prospects in Bioimaging *445*
Nitin Mohan and Huan-Cheng Chang

| 13.1 | Introduction *445* |
| 13.2 | Color Centers *446* |

13.3	Red Fluorescent Nanodiamonds	448
13.3.1	Mass Production	448
13.3.2	Fluorescence Spectra	451
13.3.3	Photostability	451
13.3.4	Fluorescence Lifetimes	453
13.4	Smaller FNDs	454
13.5	Biological Applications	458
13.5.1	Fluorescence Resonance Energy Transfer	458
13.5.2	Cellular Uptake and Fluorescence Imaging	459
13.5.3	Two-Photon Excited Fluorescence Imaging	460
13.5.4	Fluorescence Lifetime Imaging	461
13.5.5	Super Resolution Imaging	463
13.5.6	Single Particle Tracking	464
13.5.7	*In Vivo* Imaging in *Caenorhabditis elegans*	465
13.5.8	Other Imaging Techniques	467
13.6	Conclusion	469
	References	469

Index *473*

Color Plates *489*

Foreword

Practically every publication on research involving diamond materials begins by touting the extreme and unique physical and chemical properties of diamond. It is for this reason that research and applications involving diamond materials has expanded tremendously during the past 25 years. This expansion has been driven by improvements in the synthesis of diamond materials, achieving purity, quality, uniformity, and morphologies that are unattainable in geologically mined diamonds, and thus enabling many new technological applications.

Today, engineered diamond materials are available in many forms, ranging from single-crystal gems and plates, through a range of polycrystalline plates and shapes with varying grain sizes and morphologies, to films and coatings with nano- to micro-crystalline grain sizes, and nanocrystalline powders. This diversity in diamond materials has attracted research and applications in many fields, including optics and lasers, quantum computing and communication, biology, high-energy and high-pressure physics, thermal management, tribology, electrochemistry, electronics, micro-electromechanical systems (MEMS), chemical sensing, and corrosion resistance.

The primary driver behind this blossoming of diamond materials is the rapid improvements and expansion in diamond synthesis achieved via by chemical vapor deposition (CVD). Improvements in diamond synthesis by high-pressure, high-temperature (HPHT) processes, and of nanopowders by detonation processes, have also contributed to the diverse research and applications of diamond.

Our expanding knowledge base on diamond materials, and their properties and processing, requires that we expand upon the existing array of books and reviews in the field. Given the intense interest in the applications of the many forms of diamond to optics and optical applications, this new book – *Optical Engineering of Diamond* – is a welcome and valuable addition to the field. The book will immediately assist many of the exciting and important developments employing diamond materials and their optical properties that are about to happen. Examples of the fields which will be impacted are quantum devices, lasers, infrared sensing, radiation detection, synchrotron and accelerator technologies, fusion research, biological research and drug delivery, jewelry and gems, and high-power and high-voltage electronics.

This book contains 13 chapters written by the leading experts in their subfields. The chapters compile and review a knowledge base that is not available anywhere else, and provide guidance for the processing, forming, shaping, and building of devices and structures from diamond materials. As such, *Optical Engineering of Diamond* will become a valuable reference work for researchers and technologists working with diamond materials.

U.S. Naval Research Laboratory, Washington, DC *J.E. Butler*
Cubic Carbon Ceramics, Huntingtown, MD *Retired*

Preface

If one were to carry out a survey on the most important materials in optics that come to mind, the responses might include silica, yttrium aluminum garnet, and gallium arsenide. Over and above core optical properties such as transparency and luminescence, other important considerations come into play including durability, ease of manufacture, and thermal conductivity. But what if a material existed that displayed *all* of these properties on a vastly superior scale? A material of exceptional hardness, thermal conductivity, and transparency, that was also cheap, and easy to fabricate and modify? It is well known that diamond exhibits some of these properties and is indeed often touted as a being a "super material"; however, it also has a reputation for being expensive and difficult to synthesize, and still maintains the perception of being reserved for the wealthy. Nevertheless, the synthesis of diamond has advanced enormously in recent years, with diamonds of sizes spanning from several nanometers to multiple centimeters now routinely produced with qualities that often exceed that of natural diamond. This capability has come after innumerable failed attempts to make diamond, and we can now consider ourselves privileged to witness and participate in this "golden" age of diamond optics and photonics.

Today, considerable research efforts are underway in diverse areas such as diamond optics, photonics, lasers, quantum computing, and biomedicine. In quantum technologies for example, it is the room-temperature rigidity of the crystal lattice that is being harnessed to study quantum effects in an ideal isolated environment. The possibilities are multiplied with ground-breaking parallel developments in nanoscale imaging and manipulation, enabling the use of ultra-small diamonds in unique bioimaging technologies. Science magazines, general media, and blogs continuously report new developments, and some have already led to commercially available products. Just as the unique aesthetic properties of diamond have fascinated over millennia, so the fascination continues with renewed intensity for properties which have always resided just below the surface.

By capturing the state of the art in diamond optical engineering, this book aims to provide a resource in a convenient and accessible form to support further research and development. Written by 39 experts from 12 leading research groups, the book includes chapters on detailed optical properties, fabrication, and engineering techniques of diamond, and reviews several of its topical optical applications. By way of introduction, Chapter 1 provides an in-depth review of the linear and nonlinear optical properties intrinsic to diamond. Six subsequent chapters deal with growth and shaping methods, and with engineering the content of diamond. In Chapter 2, *I. Friel* reviews diamond growth by chemical vapor deposition, and includes descriptions of the growth principles and of properties and applications of high-optical-quality material. In Chapter 3, *J.R. Hird* describes the fascinating history of diamond polishing, and reviews in detail modern polishing techniques and the associated wear mechanisms and surface properties. This chapter also includes a description of techniques used to achieve surface roughnesses approaching atomic dimensions. In Chapter 4, written by *F. Nikolajeff* and *M. Karlsson*, techniques used to create refractive elements such as domes, diffractive optics, and lens arrays are reviewed. Recipes for plasma based-techniques for producing micro-structures are a valuable feature of this chapter. In Chapter 5, *C. Bradac, T. Gaebel*, and *J.R. Rabeau*, review the properties, synthesis, and applications of nitrogen-vacancy color centers in diamond, a vitally important topic that underpins a large focus of present-day diamond optical research. The possibility of creating diamond light-emitting diodes and laser diodes was one of the early drivers for diamond synthesis, and progress in this challenging task is reviewed in Chapter 6 by *S. Koizumi* and *T. Makino*, who also discuss in detail the electro-optical properties of n-doped diamond. Surface optics is a topic of growing importance; hence, in Chapter 7, *V. Petrakova, M. Ledvina*, and *M. Nesladek* report experimental and modeling results on the effects of the surface proximity and the lattice termination on the properties of color centers. The applications for this phenomenon in areas such as biosensing are also discussed in this chapter.

The chapters in the final section of the book provide a comprehensive review of the major and topical optical applications. In Chapter 8, *R.P. Mildren, A. Sabella, O. Kitzler, D.J. Spence,* and *A.M. McKay* review the recent progress of diamond Raman lasers, and discuss the highly promising outlook for creating high-power devices of broad wavelength reach. In Chapter 9, *P. Neumann* and *J. Wrachtrup* provide a detailed description of the optical and spin properties of the nitrogen vacancy center and discuss the important application area in quantum information processing. Subsequently, in Chapter 10, *S. Tomljenovic-Hanic, T.J. Karle, A.D. Greentree, B.C. Gibson, B.A. Fairchild, A. Stacey,* and *S. Prawer* describe applications for fluorescent diamond in quantum key distribution and metrology, and comprehensively review the methods used to create the enabling photonic structures such as waveguides and optical cavities. The use of diamond as heat-spreaders in optical systems such as high-average-power lasers and light-emitting diodes is reviewed in Chapter 11 by *A.J. Kemp, J.-M. Hopkins, J.E. Hastie, S. Calvez, Y. Zhang, E. Gu, M.D. Dawson,* and *D. Burns*. This chapter includes model results that highlight design considerations and optical applications in which diamond provides striking advantages. Direct write laser fabrication is a highly practical method for manipulating the surface and bulk of diamond, and in Chapter 12, *V.I. Konov, T.V. Kononenko,* and *V.V. Kononenko* report extensively on the principles of laser processing; these authors also discuss in detail a large range of applications such as laser polishing, the fabrication of diffractive optics, and direct-write of conductive structures in the bulk. Finally, in Chapter 13, *N. Mohan* and *H.-C. Chang* review the applications of fluorescent nanodiamond in biomedicine. This chapter describes the properties of nanoprobes, the challenges involved in their mass production, and the numerous ways that they are being used *in vivo* and *in vitro* for particle tracking and bioimaging.

This book constitutes important milestones in the fields of optical device engineering and diamond science. We are very grateful to the contributing authors, who all responded enthusiastically to our invitation and produced extremely valuable chapters. We also acknowledge the contributions of numerous colleagues who have shared our enthusiasm for this project, and the editorial staff at Wiley for their assistance, in particular Valerie Moliere and Anja Tschörtner. Many thanks to Andy Edmonds, Stefania Castelletto, and Torston Gaebel for their assistance in proofing manuscripts. We are grateful to Jim Butler for supporting the book concept from the outset, providing expert advice along the way, and for contributing the Foreword. The future of diamond optics is bright, and we look forward to the exciting developments ahead.

Sydney, August 2012 *Richard P. Mildren and James R. Rabeau*

List of Contributors

Carlo Bradac
Macquarie University
Department of Physics and Astronomy
Sydney, NSW 2109
Australia

David Burns
University of Strathclyde
Institute of Photonics
106 Rottenrow
Glasgow G4 0NW
UK

Stephane Calvez
University of Strathclyde
Institute of Photonics
106 Rottenrow
Glasgow G4 0NW
UK

Huan-Cheng Chang
Academia Sinica
Institute of Atomic and Molecular Sciences
Taipei 106
Taiwan

Martin D. Dawson
University of Strathclyde
Institute of Photonics
106 Rottenrow
Glasgow G4 0NW
UK

Barbara A. Fairchild
University of Melbourne
School of Physics
Parkville, Vic 3010
Australia

Ian Friel
Element Six
King's Ride Park
Ascot SL5 8BP
UK

Torsten Gaebel
Macquarie University
Department of Physics and Astronomy
Sydney, NSW 2109
Australia

Brant C. Gibson
University of Melbourne
School of Physics
Parkville, Vic 3010
Australia

Andrew D. Greentree
University of Melbourne
School of Physics
Parkville, Vic 3010
Australia

Erdan Gu
University of Strathclyde
Institute of Photonics
106 Rottenrow
Glasgow G4 0NW
UK

Jennifer E. Hastie
University of Strathclyde
Institute of Photonics
106 Rottenrow
Glasgow G4 0NW
UK

Jonathan R. Hird
University of California
Los Angeles
Department of Physics and Astronomy
Los Angeles, CA 90095-1547
USA

John-Mark Hopkins
University of Strathclyde
Institute of Photonics
106 Rottenrow
Glasgow G4 0NW
UK

Timothy J. Karle
University of Melbourne
School of Physics
Parkville, Vic 3010
Australia

Mikael Karlsson
Uppsala University
The Ångström Laboratory
P.O. Box 534
751 21 Uppsala
Sweden

Alan J. Kemp
University of Strathclyde
Institute of Photonics
106 Rottenrow
Glasgow G4 0NW
UK

Ondrej Kitzler
Macquarie University
MQ Photonics Research Centre
Sydney, NSW 2109
Australia

Satoshi Koizumi
National Institute for Materials Science
Wide Bandgap Materials Group
1-1 Namiki
Tsukuba 305-0044
Japan

Taras V. Kononenko
General Physics Institute
Natural Sciences Center
Vavilova Street, 38
Moscow 119991
Russia

Vitali V. Kononenko
General Physics Institute
Natural Sciences Center
Vavilova Street, 38
Moscow 119991
Russia

Vitaly I. Konov
General Physics Institute
Natural Sciences Center
Vavilova Street, 38
Moscow 119991
Russia

Miroslav Ledvina
Institute of Organic Chemistry and
Biochemistry
Academy of Sciences of the Czech
Republic
v.v.i., Flemingovo n. 2
166 10 Prague 6
Czech Republic

Toshiharu Makino
National Institute of Advanced
Industrial Science and Technology
Energy Technology Research Institute
1-1-1 Umezono
Tsukuba 305-8568
Japan

Aaron M. McKay
Macquarie University
MQ Photonics Research Centre
Sydney, NSW 2109
Australia

Richard P. Mildren
Macquarie University
MQ Photonics Research Centre
Sydney, NSW 2109
Australia

Nitin Mohan
Academia Sinica
Institute of Atomic and Molecular
Sciences
Taipei 106, Taiwan

Milos Nesladek
University Hasselt
Institute for Materials Research
IMEC
IMOMEC Division
Wetenschapspark 1
3590 Diepenbeek
Belgium

Philipp Neumann
Universität Stuttgart
Physikalisches Institut
70569 Stuttgart
Germany

Fredrik Nikolajeff
Uppsala University
The Ångström Laboratory
P.O. Box 534
751 21 Uppsala
Sweden

Vladimira Petrakova
Institute of Physics
Academy of Sciences of the Czech
Republic
v.v.i., Na Slovance 2
182 21 Prague 8
Czech Republic
and
Czech Technical University in Prague
Faculty of Biomedical Engineering
Sítná sq. 3105
272 01 Kladno
Czech Republic

Steven Prawer
University of Melbourne
School of Physics
Parkville, Vic 3010
Australia

James R. Rabeau
Macquarie University
Department of Physics and Astronomy
Sydney, NSW 2109
Australia

Alexander Sabella
Macquarie University
MQ Photonics Research Centre
Sydney, NSW 2109
Australia
and
Defence Science and Technology Organisation
Edinburgh, South Australia 5111
Australia

David J. Spence
Macquarie University
MQ Photonics Research Centre
Sydney, NSW 2109
Australia

Alastair Stacey
University of Melbourne
School of Physics
Parkville, Vic 3010
Australia

Snjezana Tomljenovic-Hanic
University of Melbourne
School of Physics
Parkville, Vic 3010
Australia

Jörg Wrachtrup
Universität Stuttgart
Physikalisches Institut
70569 Stuttgart
Germany

Yanfeng Zhang
University of Strathclyde
Institute of Photonics
106 Rottenrow
Glasgow G4 0NW
UK

1
Intrinsic Optical Properties of Diamond
Richard P. Mildren

Diamond comprises the lowest mass element that can form a stable covalently bonded crystal lattice, and this lattice is highly symmetric and tightly bound. Its resulting extreme properties, along with the recent developments in its synthesis, have led to an explosion of interest in the material for a diverse range of optical technologies including sensors, sources, and light manipulators. The optical properties in many respects sit well apart from those of other materials, and therefore offer the tantalizing prospect of greatly enhanced capability. A detailed knowledge base of the interaction of electromagnetic radiation with the bulk and the surface of diamond is of fundamental importance in assisting optical design.

For any material, the dataset characterizing optical performance is large and diamond is no exception despite its inherent lattice simplicity. The properties of interest extend over a large range of optical frequencies, intensities and environmental parameters, and for many variants of the diamond form including defect and impurity levels, crystal size, and isotopic composition. Over and above the fascination held for this ancient material, its highly symmetric structure and pure natural isotopic content (98.9% ^{12}C) provides an outstanding example for underpinning solid-state theory. As a result, diamond has been extensively studied and its optical properties are better known than most other materials.

Many excellent reviews of optical properties have been reported previously (see e.g. Refs [1–3]). These concentrate mainly on linear optical properties, often focus on extrinsic phenomena, and are written from perspectives outside of the field of optics, such as electronics and solid-state physics. Consequently, there is a need to consolidate the data from the perspective of optical design. Furthermore, the nonlinear optical properties of diamond have not to date been comprehensively reviewed. The aim of this chapter is to do this, with emphasis placed on the intrinsic properties of single-crystal diamond (i.e., pure Type IIa diamond[1]). The chapter

---

1) Type IIa represents the most pure form; other categories (Types Ia, Ib, and IIb) have substantial levels of nitrogen (Type Ia and Ib) and/or boron impurity (Type IIb). Note that the delineations between types are not well defined. Type IIa are rarely found as large homogeneous crystals in nature as nitrogen aids the formation process. It is thus for historical reasons that nitrogen-doped diamonds, which provide the major source of natural gemstones, are categorized as Type I.

Optical Engineering of Diamond, First Edition. Edited by Richard P. Mildren and James R. Rabeau.
© 2013 Wiley-VCH Verlag GmbH & Co. KGaA. Published 2013 by Wiley-VCH Verlag GmbH & Co. KGaA.

also includes the dependence of optical properties on basic variables such as wavelength, temperature, and isotopic composition. Although the scope is limited to bulk intrinsic properties, the intention is to stimulate a further expansion of the knowledge base as the limits of measurement resolution and performance are extended, and as more detailed investigations emerge into areas such as surface optics, crystal variants, and nano-optical effects.

The chapter focuses on the optical properties spanning from ultraviolet (UV) to infrared (IR). It should be noted that, throughout the chapter, Système Internationale (SI) units have been used, apart from some exceptions to stay with conventions. The data provided refer to diamond with the naturally occurring isotopic ratio, unless specifically stated otherwise.

1.1
Transmission

Diamond has a wide bandgap and lacks first-order infrared absorption, which makes it one of the most broadly transmitting of all solids. As shown in Figure 1.1, the transmission spectrum for a diamond window is featureless for wavelengths longer than approximately 225 nm ($\alpha < 1\,\text{cm}^{-1}$ for $\lambda > 235\,\text{nm}$), apart from a moderate absorption in the range 2.6 to 6.2 μm and extending weakly outside each side. Indeed, there is no absorption in the long-wavelength limit, which is a characteristic of the Group IV elements (e.g., Si and Ge) that share the diamond

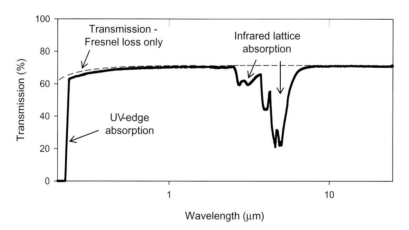

Figure 1.1 Transmission spectrum for a Type IIa diamond window ("Type IIIa," Element 6) of 1 mm thickness. The spectrum was measured using a Cary 5000 spectrometer (UV-near IR) and Bruker Zertex 80 (>2 μm; resolution 4 cm^{-1}). The transmission for Fresnel loss only (dashed curve) was calculated using the relation described in the text and in Equation (1.6). The small difference between the dashed and measured curves in the regions away from the UV-edge and lattice absorption is largely attributed to the combination of spectrometer calibration error and scatter in the sample.

lattice symmetry. UV-edge absorption, infrared lattice absorption and Fresnel reflection dominate the wavelength dependence for transmission. The Fresnel reflection at each diamond–air interface is approximately 17% in the visible ($R = 17.1\%$ at 632 nm), and when accounting for multiple reflections between each surface this leads to a maximum transmission of $(1 - R)^2/(1 - R^2) = 70.8\%$. Using dispersion data for the refractive index (see Section 1.4), the transmission upper limit (no absorption) is shown as a function of wavelength (dashed line in Figure 1.1).

1.2
Lattice Absorption

The absorption in the mid-IR, which is most prominent in the range 2.6 to 6.2 μm, arises due to the coupling of radiation with the movement of nuclei, and is often referred to as "lattice" or "multiphonon" absorption. The magnitude and shape of the absorption spectrum is a consequence of the vibrational properties of the crystal lattice, which are governed by the forces between neighboring atoms and the symmetry of collective vibrations. The theoretical framework that most successfully describes the spectrum has been developed since the 1940s, stimulated by the pioneering work of Sir C.V. Raman on diamond's optical properties and Max Born on the quantum theory of crystals. It is interesting to note that, although diamond's lattice is one of the most simple, there have been substantial controversies in explaining the spectrum (see e.g., Ref. [4]) and there are on-going challenges to thoroughly explain some of the features.

A brief and qualitative summary of the theory of lattice absorption is provided here to assist in an understanding of the IR spectrum's dependence on important material and environmental parameters such as impurity levels, isotopic content, and temperature. A greatly simplifying and important aspect is that there is no one-phonon absorption in pure, defect-free diamond (which would appear most strongly near 7.5 μm for diamond), as also for other monatomic crystals with inversion symmetry such as Si and Ge. The movement of nuclei in vibrational modes of the lattice are countered by equal and opposite movement of neighbors, so that no dipole moment for coupling with radiation is induced. One-phonon absorption may proceed by spoiling the local symmetry through, for example, lattice imperfections (impurities and defects) or by the application of electric field. Dipole moments may also be induced in the crystal via interaction of the incident photon with more than one phonon, although with reduced oscillator strength; this is the origin of lattice absorption in pure diamond. From a classical viewpoint, the absorption mechanism can be qualitatively understood as one phonon inducing a net charge on atoms, and a second phonon (or more) vibrating the induced charge to create a dipole moment [5]. The maximum phonon frequency that can be transmitted by the lattice is 1332 cm^{-1} (which corresponds to the zero-momentum optical phonon and the Raman frequency), and integer multiples of this frequency at 3.75 μm (2665 cm^{-1}) and 2.50 μm (3997 cm^{-1}) mark

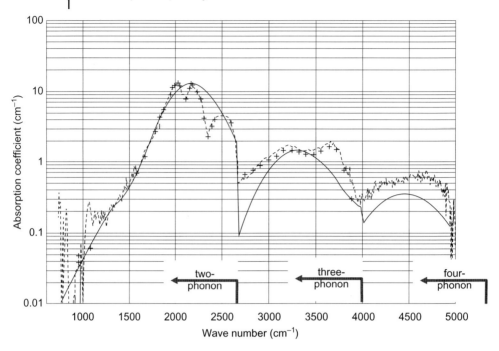

Figure 1.2 Two-, three-, and four-phonon lattice absorption bands. The underlying figure showing the calculated (smooth solid curve) and measured absorption spectra is reprinted with permission from Ref. [6]; © 1994, SPIE. The measurements were collated from several sources, as detailed in the reference.

the short-wavelength limits for two- and three-phonon absorption regions. The demarcations between two- and three-phonon absorption are clearly evident in the transmission spectrum of Figure 1.1 and the logarithmic plot of lattice absorption in Figure 1.2 [6]. Between wavelengths 3.75 and 6 μm, the lattice absorption at room temperature is strongest with a peak of approximately $10\,\text{cm}^{-1}$ at 4.63 μm, and is primarily attributable to two-phonon absorption.

1.2.1
The Two-Phonon Region

Absorption may involve the creation and destruction of phonons, which are constrained to certain energies and wavevectors as a result of the symmetry and interatomic forces. For two-phonon creation, the absorption at a given frequency is proportional to the number of pairs of modes of the created phonons and a transition probability that takes into consideration allowed phonon combinations (e.g., longitudinal or transverse) and the transition oscillator strength. The number of allowed combinations of a given energy is usually highest for phonon wavevectors along directions of high symmetry in the lattice, and with momenta that correspond to phonon wavelengths of the order of the atomic spacing in that direction (i.e., at the edge of the Brillouin zone). Along with the generally higher density of modes at the Brillouin zone edge, the transition probability is also higher as the largest charge deformations are induced. For diamond, there is also a peak in the density of modes in one symmetry direction (<110>) for momenta corresponding to wavevectors at approximately 70% of the Brillouin zone. These peaks in the density of states are the so-called "critical points" in the lattice phonon dispersion curves.

The primary directions of symmetry for the diamond lattice, along with the frequency of critical point phonons, are listed in Table 1.1. Also listed are the critical points derived from the relatively recent studies of Vogelsegang et al. [7] and Klein et al. [8], for data derived using a combination of neutron-scattering data, second-order Raman spectra[2] [9] and, in the case of Ref. [8], by using impurity spectra to access single phonon information. There remains significant uncertainty in the frequency of many critical points, however, and as a result there is disagreement between some phonon assignments.

For temperatures below 1000 K, the populations of critical-point phonons are small (due to their characteristically high energies in diamond), and only phonon summations appear strongly in the spectrum. Momentum–energy conservation and symmetry impose selection rules for the type of phonons created. As the photon momentum is negligible compared to the Brillouin zone-edge phonons and to conserve crystal momentum, the wavevector of each phonon is equal in magnitude and opposite in direction. The character of the phonons must also be different; that is, they should correspond to different dispersion branches (either optical or acoustic phonon), or have a different polarization (longitudinal or transverse direction). The resulting absorption features are referred to as "combination lines." Pairs of phonons of the same type (overtones) are absent or weak.

Due to the large number of possible phonon modes, the spectrum appears as a fairly smooth continuum extending to wavelengths that extend beyond 10 μm. A joint-density-of-states calculation [10] was successful in reproducing the gross features of the two-phonon spectrum, as shown in Figure 1.3, including the broad peak near 2500 cm^{-1}, the region of highest absorption from 1800–2300 cm^{-1}, the local minimum at 2100 cm^{-1}, and the tail at frequencies less than 1750 cm^{-1}. An improved agreement would require a better knowledge of the dispersion curves and transition probabilities. Unfortunately, however, there are large uncertainties in the phonon dispersion data obtained by neutron scattering data, due to the lack of test samples of sufficient size. The more recently published critical point values [7, 8], obtained with the assistance of optical spectra of variable impurity and isotopic content, have enabled some features to be assigned to critical point

---

2) The second-order Raman scattering involves two phonons. The second-order spectrum contains a peak at twice the Raman frequency that is more than two orders of magnitude weaker than the first-order peak, and a broad feature extending to lower frequencies (see e.g., Ref. [9]).

Table 1.1 Directions of high symmetry, critical point phonons, the corresponding frequencies and two-phonon combinations identified in the IR absorption spectrum. Combinations corresponding to the major peaks are shown in bold. L = longitudinal, T = transverse, O = optical, and A = acoustic. Note that the use of L and T labels for the K symmetry points is not conventional and should not be taken as indicating branches of purely transverse or longitudinal character.

Crystal direction (as viewed in perspective through the 4 × 4 unit cell)	Crystal direction, K-space symmetry label	Critical point frequencies (cm^{-1})		Observed spectrum features in the two-phonon region (see Figure 1.4) with assigned phonon combinations (cm^{-1})
		Ref. [7]	Ref. [8]	
	<100>, X	1170 (L)[a] 1088 (TO) 786 (TA)	1191 ± 3 (L)[a] 1072 ± 2 (TO) 829 ± 2 (TA)	2260 ± 6 L + TO (X) 1895 ± 6 TO + TA (X)[c]
	<110>, K	1236 (LO) 1112 (TO1) 1051 (TO2)[b] 986 (LA) 982 (TA1)[b] 748 (TA2)	1239 ± 2 (LO) 1111 ± 1 (TO1) 1042 ± 2 (TO2)[b] 992 ± 3 (LA) 978 ± 1 (TA1)[b] 764 ± 4 (TA2)	1977 ± 2 **LA + TA1** (Σ) 2005.5 ± 2 LO + TA2 (Σ) **2029 ± 2 TO2 + (?)[d] (Σ)** 2097 ± 2 TO1 + LA (Σ) **2160 ± 2 TO1 + TO2 (Σ)** 2293 ± 8 LO + TO2 (Σ)
	<111>, L	1245 (LO) 1208 (TO) 1009 (LA) 572 (TA)	1256 ± 4 (LO) 1220 ± 2 (TO) 1033 ± 2 (LA) 553 ± 2 (TA)	1777.5 ± 4 TO + TA (L) 1816.5 ± 4 TA + LO (L) 2260 ± 6 LA + (?)[e] (L) 2293 ± 6 LO + LA (L)
	<210>, W	1164 (L)[a] 1012 (TO) 915 (TA)	1146 ± 1 (L)[a] 1019 ± 3 (TO) 918 ± 12 (TA)	2175 ± 5 L + TO (W)

a) The LO and LA modes are degenerate.
b) "Accidental" critical points occur for the TO2- and TA1-labeled phonons for wavevectors nearby the symmetry point.
c) Assignment agrees with Ref. [8] data only.
d) Ambiguous assignment – Ref. [8] suggests LA, whereas Ref. [7] suggests TA1.
e) Ambiguous assignment – Ref. [8] suggests LO, whereas Ref. [7] suggests TO.

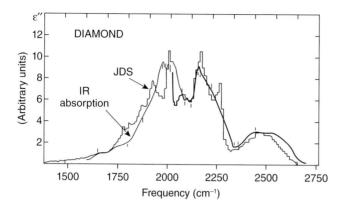

Figure 1.3 Comparison of two-phonon absorption band with the joint density of states (JDS) calculation of Wehner et al. [10], showing qualitative agreement with several of the main features in the measured spectrum. The stepped appearance of the calculated spectrum is a consequence of the digitizing procedure used to sample the phonon dispersion curves. Reproduced with permission from Ref. [10]; © 1967, Elsevier.

combinations with greater confidence. The last column in Table 1.1 lists the frequencies of features that can be readily seen in the spectrum of Figure 1.4, along with their suggested assignments of the likely critical point phonons. The major peaks at 4.63 μm (2160 cm^{-1}), 4.93 μm (2030 cm^{-1}) and 5.06 μm (1975 cm^{-1}) all correspond to phonons in the <110> symmetry direction, for which there is an "accidental" critical point in the dispersion curves for one or both of the phonons involved in the assigned combinations. A similar type of analysis can be performed, at least in principle, for the third- and higher-order phonon bands at wavelengths <3.75 μm (>2665 cm^{-1}); however, such a procedure is very difficult due to the greatly increased number of phonon combinations, the lack of detailed knowledge on transition probabilities, and the poor visibility of critical point locations.

1.2.2
Absorption at Wavelengths Longer than 5 μm

In the range 5 to 10 μm, lattice absorption decreases approximately exponentially from 10 cm^{-1} to approximately 0.05 cm^{-1}. The weaker longer-wavelength absorption results primarily from combination pairs of low-energy acoustic phonons away from the phonon dispersion critical points. A calculation for multiphonon absorption using polynomial fits to the acoustic phonon densities of states [6], reproduces this trend satisfactorily (as seen Figure 1.2). At wavelengths longer than 8 μm (<1250 cm^{-1}), a significant departure of experiment from theory is observed as the weaker absorption approaches the level of impurity absorption and the resolution of the measurement. Due to interest in diamond as a window

1 Intrinsic Optical Properties of Diamond

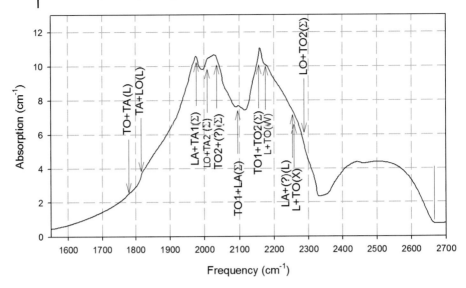

Figure 1.4 Detail of the two-phonon absorption region (using the data of Figure 1.1) with identification of major features in the spectrum with suggested critical point phonon summations (refer to Table 1.1). The vertical line at 2665 cm^{-1} corresponds to twice the Raman frequency, and indicates the upper limit for two-phonon absorption.

material for missile domes and high-power CO_2 lasers, absorption in the long-wavelength atmospheric window at 8–12 μm and at 10.6 μm has been studied in depth (see e.g., Refs [6, 11, 12]). Absorptions as low as 0.03 cm^{-1} at 10.6 μm have been measured for single-crystal and polycrystalline material [13]. Intrinsic absorption is expected to decrease monotonically at longer wavelengths *ad infinitum* due to the diminishing number of phonon modes and, indeed, low-loss material has been observed up to and beyond 500 μm [13, 14].

1.2.3
Temperature Dependence

Temperature affects lattice absorption via changes in the phonon ambient population density and shifts in phonon mode frequency. The effect of the thermal phonons in the material is to stimulate absorption events coinciding with the incident photon. As described by Lax [15], the relationship between two-phonon absorption coefficient α and temperature is given by

$$\alpha \propto (n_1 + 1)(n_2 + 1) \tag{1.1}$$

where n_i are the occupation numbers at thermal equilibrium of the final state phonons (also called Bose–Einstein factors)

$$n_i(\omega, T) = (\exp[\hbar\omega/k_B T] - 1)^{-1} \tag{1.2}$$

where k_B is Boltzmann's constant and $\hbar$ is Planck's constant divided by 2π. A further consideration noted by Lax was that of induced emission, which involves the annihilation of thermal phonons and the creation of an IR photon at the frequency sum. This is proportional to $n_1 n_2$ for two phonons, so that the net absorption becomes

$$\alpha \propto (n_1+1)(n_2+1) - n_1 n_2 = n_1 + n_2 + 1 \qquad (1.3)$$

However, as $n_1 n_2$ can be neglected for temperatures less than approximately 1000 K, Equation (1.1) holds for most temperatures of interest. At room temperature and below, the density of thermal phonons is small so that absorption is essentially constant and reflects the spontaneous component (i.e., the component contribution caused by quantum fluctuations). At elevated temperatures, however, absorption increases notably (as shown in Figure 1.5a [16]) for temperatures up to 800 K and for wavelengths spanning the two- and three-phonon bands (from 2.5 μm to beyond 20 μm). The temperature dependence of a given feature in the absorption band varies slightly according to the thermal populations of the responsible phonons of the feature combination. Although the agreement with Equations (1.1) and (1.2) above is quite good for some spectral features [17], for others – such as those near the 4.9 μm peak – the increase in absorption exceeds the prediction. A more thorough treatment would need to consider contributions from higher-order multiphonon processes and also two-phonon difference bands (of frequency $\omega_i - \omega_j$, where the phonon ω_i is absorbed and ω_j phonon is emitted) which play an increasingly important role at higher temperatures and longer wavelengths [18].

Lattice absorption spectra between 2 and 20 μm for several temperatures up to 500 K are shown in Figure 1.5c. In addition to increased absorption by thermal phonons, the absorption spectrum is altered via small decreases in the phonon frequencies, which occur as a result of the change in interatomic forces and average atomic spacing. The shift varies according to the branches of the phonons involved, and is thus a complex function of wavelength. By using measurements of numerous absorption features, Picarillo et al. [18] determined that for the two-phonon region at temperatures between 14 and 825 K, the shifts of absorption features $\Delta \nu$ are within a maximum deviation of ±13% with the average fit given by

$$\Delta \nu / \nu = -0.027 \times n(\omega_e, T) \qquad (1.4)$$

(where $\omega_e = 860 \, \text{cm}^{-1}$) and which corresponds to approximately $0.013 \, \text{cm}^{-1} \text{K}^{-1}$ at room temperature near 4 μm. The calculated temperature shift at the two-phonon peak ($2160 \, \text{cm}^{-1}$) and experimental measurements for the nearby $2286.5 \, \text{cm}^{-1}$ shoulder feature are shown in Figure 1.5b.

1.2.4
Isotopic Content

The effect of the isotopic ^{12}C to ^{13}C ratio on the spectrum on the two-phonon spectrum has been investigated [7]. As the ^{13}C content is increased from the

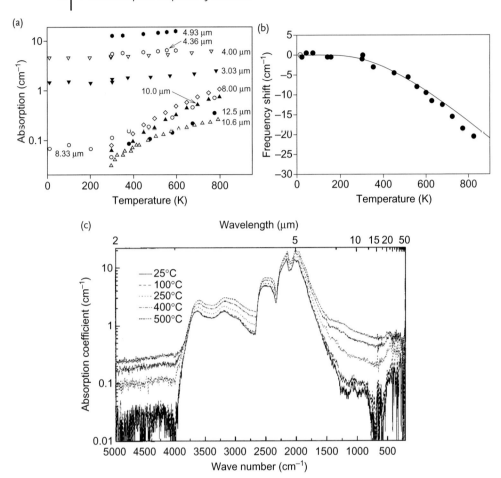

Figure 1.5 (a) Temperature dependence of absorption at selected IR wavelengths [11, 16–18]; (b) Shift of the 2286.5 cm^{-1} shoulder as a function of temperature in the two-phonon spectrum (data points [18]). The solid line is the calculated shift using Equation (1.4); (c) Lattice absorption spectrum as a function of temperature. Reproduced with permission from Ref. [11]; © 2003, Institute of Physics.

natural ratio (^{12}C:^{13}C = 1 − x:x, where x = 0.011), to the almost pure ^{13}C (x = 0.987), the spectrum shifts towards longer wavelengths (see Figure 1.6) in good agreement with the expected $M^{-0.5}$ frequency dependence on the reduced mass M. For an approximately equal mix of the two isotopes, the authors reported that the features seen in the above two-phonon spectra were either broadened or unresolvable.

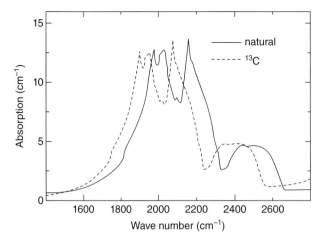

Figure 1.6 Two-phonon absorption spectra for enriched ^{13}C and natural isotopic fractions. Reproduced with permission from Ref. [7]; © 1998, American Physical Society.

1.3
UV Edge Absorption

The UV absorption edge begins for wavelengths slightly longer than the lowest energy bandgap at 227 nm (5.47 eV). The gap is indirect, and requires the excited electron to gain momentum in one of the <100> crystal directions. As a result, absorption near the UV edge is a three-body interaction, involving either the absorption or emission of a lattice phonon, and is weaker than would be the case for a direct bandgap. Close to the gap absorption is also influenced by interaction of the created electron and hole, which are weakly attractive (forming an exciton of binding energy $E_x = 0.07$ eV at room temperature), and which acts to slightly reduce the energy required to otherwise span the bandgap. A schematic diagram (Figure 1.7a) shows the indirect and direct bandgaps of diamond, along with the transitions involved in the near bandgap absorption.

A detailed description of the absorption edge (as shown in detail in Figure 1.7b) is given by Clark et al. [20]. At room temperature, the main onset of absorption occurs at 236 nm (labeled E_2 at 5.26 eV, 0.21 eV below the indirect bandgap energy E_g), and coincides with the excitation of an outer electron from the top of the valence band to the excitonic state just below the conduction band minimum and assisted by the absorption of a highest energy phonon. The conduction band minimum resides at a momentum value 76% of the Brillouin zone [21], where the highest energy phonon has energy of approximately 0.15 eV. For shorter wavelengths, absorption increases due to the increased density of states (which scales with $[\hbar c/\lambda - E_{th}]^{1/2}$, where E_{th} is the threshold energy for the phonon-assisted transition), and due to added contributions from the lower-energy transverse optical and

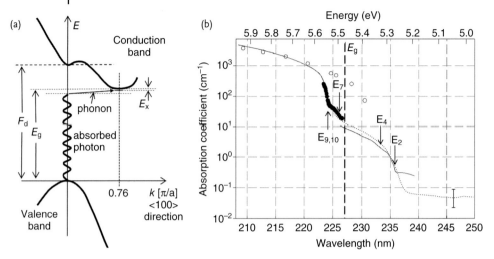

Figure 1.7 (a) Electronic band diagram highlighting the absorption mechanism for photon energies just below the bandgap. E_d and E_g are the direct and indirect band gaps, and E_x is the exciton binding energy; (b) The UV absorption edge at 295 K (solid lines, solid circles) [20]. The labels E_n, which retain the system of Ref. [20], show the features in the spectrum described in the text. The data of Philipp and Taft [19] (open circles) were determined from reflectance measurements. The dotted trace indicates the absorption coefficient determined from the transmission data for the Type IIa CVD sample of Figure 1.1.

acoustic phonon modes as the incident photon energy increases above their respective thresholds at 235.5 nm (5.27 eV) and 233.5 nm (E_4; 5.31 eV). For wavelengths shorter than 230.8 nm, the photon and phonon have sufficient energy to create an unbound electron–hole pair, and for such bound–unbound transitions the absorption increases according to $(\hbar c/\lambda - E_{th})^{3/2}$. Due to the low ambient density of high-energy phonons in diamond at room temperature, the absorption remains moderate until the photon energy exceeds the threshold for phonon emission. Evidence for phonon emission thresholds appears at 226 nm for the TA phonon (E_7; 5.482 eV), and at 224 nm for the closely spaced TO and longitudinal phonons (E_9; 5.531 eV and E_{10}; 5.544 eV, respectively). Transmission may be readily measured through thin samples up to the first direct band gap at $E_d = 7.3$ eV (170 nm), although at shorter wavelengths the absorption depth falls sharply and is limited to only a few microns of material. Absorption at wavelengths shorter than 200 nm has also been studied extensively (see e.g., Refs [19, 22]).

With increasing temperature, the absorption edge is influenced by the increased role of thermally excited phonons and the downshift in the phonon energy due to lattice expansion. The absorption spectra for a range of temperatures between 126 and 661 K are shown in Figure 1.8 [20]. The frequency downshift is small compared to the photon energy, and corresponds to less than 0.1 eV over the entire temperature range. The increasing absorption at wavelengths longer than 236 nm (<5.26 eV) for temperatures above 300 K is attributed to the increased density

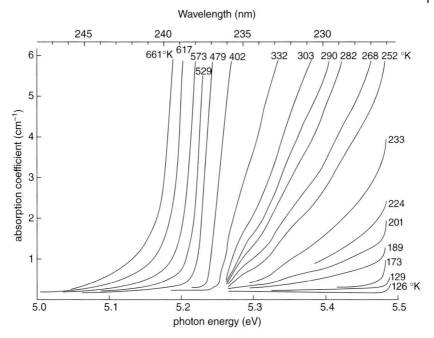

Figure 1.8 UV absorption edge as a function of temperature. The top wavelength scale has been added for convenience. Reproduced with permission from Ref. [20]; © 1964, Royal Society Publishing.

of thermal phonons and the onset of absorption events involving two or more phonons. Below room temperature, absorption relying on single-phonon absorption diminishes so that the absorption edge retreats progressively towards the 226 nm threshold (E_7; 5.482 eV) corresponding to phonon emission.

To date, the effect of isotopic composition on UV absorption edge does not appear to have been studied in great detail. For higher concentrations of ^{13}C, the lower phonon frequencies shift the E_i thresholds towards shorter wavelengths in the case of phonon absorption, and towards longer wavelengths in the case of phonon emission. In addition, the bandgap wavelength shifts due to the change in electron–phonon coupling for the heavier atomic constituents and a minor contribution (of approximately one-fourth) from the change in molar volume. For pure ^{13}C, the indirect bandgap increases by 13.6 meV, corresponding to a blue-shift of 0.56 nm [23].

1.4
Refractive Index

The refractive index values of Peter [24] and Edwards and Ochea [25] are plotted in Figure 1.9a (for an extended list of values, see Ref. [3] and page 670 in Ref. [2]).

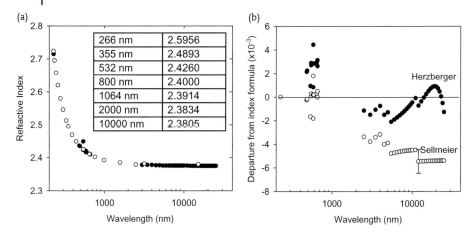

Figure 1.9 (a) Refractive index as a function of wavelength (hollow–[24], filled–[25] and references therein). The inset table contains Equation (1.6) values for several key wavelengths listed; (b) The departure of measured index values from the index formula.

Two common fits to the data are:

$$\text{Sellmeier [24]} \quad n^2 = 1 + \frac{0.3306\lambda^2}{\lambda^2 - (175.0)^2} + \frac{4.3356\lambda^2}{\lambda^2 - (106.0)^2} \quad (1.5)$$

$$\text{Herzberger [2]} \quad n = A + \frac{B}{\frac{\lambda}{1000} - 0.028} + \frac{C}{\left(\frac{\lambda}{1000} - 0.028\right)^2} + D\lambda^2 + E\lambda^4 \quad (1.6)$$

where $A = 2.37837$, $B = 1.18897$, $C = -1.0083 \times 10^{-4}$, $D = -2.3676 \times 10^{-5}$, $E = 3.24263 \times 10^{-8}$, and where λ is in nanometers. Note that the coefficients for the Herzberger formula published earlier in Ref. [25] are in error. The two formulae provide slightly different fits to the data, with improved agreement for the visible and IR data respectively, as shown in Figure 1.9b. The shift in index with isotopic ratio has been estimated to be +0.0004 for ^{13}C in the long-wavelength limit ([26]; see also Refs [19] and [20]).

1.4.1
Temperature Dependence of the Refractive Index

The temperature dependence of n has been obtained using a number of methods, including electronic measurements of the dielectric coefficient and from the interference fringes in Fourier transform spectra for thin heated windows. Some results are summarized for measurements at several wavelengths in Figure 1.10a. By using the latter method, Ruf et al. [26] have measured the dependence at long wavelengths (100 μm), and have shown a good agreement with the empirical relationship

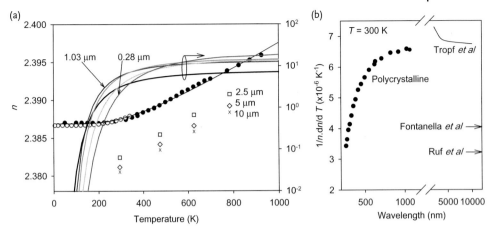

Figure 1.10 (a) Refractive index and thermo-optic coefficient versus temperature. Index data includes $\lambda = 100\,\mu m$ (solid circles [26]), $\lambda \to \infty$ (open circles [27]) and $\lambda = 2.5, 5,$ and $10\,\mu m$ [28]. The solid black curves for n and $1/n \times dn/dT$ are obtained using Equation (1.7). The colored curves correspond to Equation (1.7) using values ω_{eff} and K obtained for polycrystalline material at wavelengths 1.03, 0.62, 0.41, and 0.28 μm (brown, orange, light green, and dark green) [29]; (b) Values of $1/n \times dn/dT$ at 300 K as a function of wavelength [26–28]. The solid circles correspond to polycrystalline material [29]. For a color version of this figure, please see the color plate at the end of this book.

$$n(T) = n_0 + K\left[n_{BE}(\hbar\omega_{eff}, T) + \frac{1}{2}\right] \quad (1.7)$$

where $n_{BE}(\hbar\omega, T)$ is the Bose–Einstein factor of Equation (1.2) for an effective phonon frequency $\hbar\omega_{eff} = 711\,cm^{-1}$, and where $K = 0.019 \pm 0.007$. The index is found to remain constant at temperatures up to approximately 200 K, and increases linearly at temperatures above 450 K. The unusually high Debye temperature for diamond extends the range of constant index to notably higher temperatures compared to other materials. As a further consequence, the room temperature thermo-optic coefficient $(1/n) \times dn/dT$ is a steep function of temperature for diamond. Equation (1.7) describes the increase in vibrational degrees of freedom with temperature. An analysis of how the quantum and thermal motion affects the temperature dependence is presented in Refs [26, 30], although as with many other materials a comprehensive theory is yet to be developed.

The thermo-optic coefficient is also dependent on wavelength. The coefficient in the long-wavelength limit was determined from capacitive measurements for $T = 5.5$–$340\,K$ by Fontanella et al. [27] (open circles in Figure 1.10a), and at $100\,\mu m$ by Ruf et al. (solid symbols), both of which are in good agreement. Fontanella et al. reported a slightly higher room-temperature value ($4.04 \times 10^{-6}\,K^{-1}$; cf. $3.2 \times 10^{-6}\,K^{-1}$ for Ruf et al.). Tropf et al. [28] have also reported a value of $6.7 \times 10^{-6}\,K^{-1}$

near 10 μm, although this was likely to be overvalued as it most likely presumed a linear dependence near room temperature [26].

Knowledge of the thermo-optic coefficient at shorter wavelengths is relatively poor, especially for single-crystal materials. Hu and Hess [29] used ellipsometry to study the refractive index at wavelengths spanning the near-IR to UV for nanocrystalline films grown by chemical vapor deposition. The results in Figure 1.10b show that the coefficient $(1/n) \times dn/dT$ decreases by as much as half of the long wavelength value for wavelengths approaching the bandgap. Comparisons of the Hu and Hess data with the aforementioned long-wavelength results of Ruf et al. and Fontanella et al. suggest that the Hu and Hess data are systematically overvalued; however, if the trend can be applied to bulk single crystal, then $(1/n) \times dn/dT$ values less than $2 \times 10^{-6}\,K^{-1}$ would be expected at wavelengths less than 400 nm.

1.5
Verdet Constant

The Faraday rotation of polarization was investigated for diamond by Ramaswasan [31]. The Verdet constant V, which is a function of material dispersion and thus wavelength, is related to the magnetic anomaly γ via the relationship:

$$V = \gamma \frac{e}{2m_e c^2} \lambda \frac{dn}{d\lambda} \tag{1.8}$$

where m_e and e are the electron mass and charge, respectively. Ramaswasan found that γ was constant at 27.8% in the visible, giving, for example, a value of V of approximately $6.8\,rad\,T^{-1}\,m^{-1}$ at 589 nm.

1.6
First-Order Raman Scattering

First-order Raman scattering results from the interaction of an incident electromagnetic photon with a near zero-wavevector optical phonon, which is the vibrational mode involving the relative movement of the two face-centered cubic lattices that comprise the diamond lattice in the direction of the linking carbon–carbon bond (see Figure 1.11a). At this point in the phonon dispersion curves, the longitudinal and two transverse vibrational branches converge and the Raman mode is triply degenerate. The Raman frequency ω_r is $1332.3\,cm^{-1}$ (or 39.99 THz) at room temperature.

The quantum mechanical theory of spontaneous Raman scattering in crystals has been reviewed by Loudon [32], in which diamond is used as a main example. The coupling between incident and output photons is mediated by a photon–electron interaction in which the incident photon deforms the periodic electron potential and induces a lattice phonon. For a polarized beam entering a crystal of

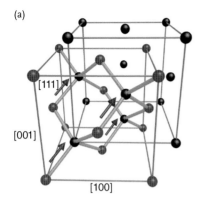
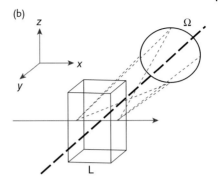

Figure 1.11 (a) The diamond lattice showing the two interpenetrating face-centered cubic lattices and the direction of relative movement for the zero wave-vector optical phonon involved in first-order Raman scattering; (b) Configuration for perpendicular Raman scattering. For a color version of this figure, please see the color plate at the end of this book.

length L, the fraction of photons scattered perpendicular to the incident beam into a detector of collection solid angle Ω is given by (see e.g., Refs [33, 34])

$$S = \frac{\hbar(2\pi)^4 N(n_{BE}+1)\Omega L}{2\lambda_s^4 M \omega_r} \sum_j |e_s R_j e_i|^2 \qquad (1.9)$$

where n_{BE} is the Bose–Einstein occupation number at ω_r, $N = 4/a_0^3$ is the number of unit cells, a_0 the lattice constant (8.81×10^{28} cells m^{-3} and 3.567 Å respectively at room temperature), $M = \rho/4N$ is the reduced mass of the carbon–carbon scattering center, and ρ is the crystal density (3.52 g cm^{-3}). Diamond's combination of high density, small mass and large deformation potential is responsible for a high scattering efficiency compared to other crystals. The efficiency for anti-Stokes scattering can be obtained using the same relation by replacing $(n_{BE} + 1)$ with n_{BE}, and is typically 600 times weaker at room temperature. The Raman tensor R_j describes the strength of the interaction as a function of the incident and Stokes scattered polarizations e_i and e_s and with respect to each of the three degenerate Raman modes, $j = 1, 2$, and 3. The symmetry of the zone-center optical phonon ($\Gamma_{25}+$ in the notation of Birman, or equivalently F_{2g}) imposes the following form of R_j in the frame of the cubic axes $x = [100]$, $y = [010]$, and $z = [001]$:

$$R_1 = \begin{bmatrix} 0 & d & 0 \\ d & 0 & 0 \\ 0 & 0 & 0 \end{bmatrix} R_2 = \begin{bmatrix} 0 & 0 & 0 \\ 0 & 0 & d \\ 0 & d & 0 \end{bmatrix} R_3 = \begin{bmatrix} 0 & 0 & d \\ 0 & 0 & 0 \\ d & 0 & 0 \end{bmatrix} \qquad (1.10)$$

where $d = d\alpha/dq$ is the change in crystal polarizability α with movement of the two sub-lattices along q, a characteristic of the material related to the lattice deformation potentials. The Raman tensors for the common crystal orientations of $x' = [110]$, $y' = [1\bar{1}0]$, and $z' = [001]$, obtained by rotation of the coordinate system about 45° around the z-axis, are:

$$R_1 = \begin{bmatrix} 0 & 0 & \frac{d}{\sqrt{2}} \\ 0 & 0 & \frac{d}{\sqrt{2}} \\ \frac{d}{\sqrt{2}} & \frac{d}{\sqrt{2}} & 0 \end{bmatrix} \quad R_2 = \begin{bmatrix} 0 & 0 & \frac{d}{\sqrt{2}} \\ 0 & 0 & -\frac{d}{\sqrt{2}} \\ \frac{d}{\sqrt{2}} & -\frac{d}{\sqrt{2}} & 0 \end{bmatrix} \quad R_3 = \begin{bmatrix} d & 0 & 0 \\ 0 & -d & 0 \\ 0 & 0 & 0 \end{bmatrix} \quad (1.11)$$

The Raman tensors for the common crystal orientation of $x'' = [110]$, $y'' = [\bar{1}\bar{1}2]$, and $z'' = [1\bar{1}1]$, obtained by a rotation of arcos $[1/\sqrt{3}](\approx 54.7\,°)$ around the x'-axis, are

$$R_1 = \begin{bmatrix} 0 & \frac{d}{\sqrt{3}} & \frac{d}{\sqrt{6}} \\ \frac{d}{\sqrt{3}} & \frac{2d}{3} & -\frac{d}{3\sqrt{2}} \\ \frac{d}{\sqrt{6}} & -\frac{d}{3\sqrt{2}} & \frac{2d}{3} \end{bmatrix} \quad R_2 = \begin{bmatrix} 0 & \frac{d}{\sqrt{3}} & \frac{d}{\sqrt{6}} \\ \frac{d}{\sqrt{3}} & -\frac{2d}{3} & \frac{d}{3\sqrt{2}} \\ \frac{d}{\sqrt{6}} & \frac{d}{3\sqrt{2}} & \frac{2d}{3} \end{bmatrix} \quad R_3 = \begin{bmatrix} d & 0 & 0 \\ 0 & -\frac{d}{3} & \frac{\sqrt{2}d}{3} \\ 0 & \frac{\sqrt{2}d}{3} & -\frac{2d}{3} \end{bmatrix}$$
(1.12)

It should be noted that the symmetric nature of R is only strictly valid for photon frequencies much larger than ω_r.

The absolute value of d, which is a function of wavelength, has been determined using a variety of methods, and ranges from 3.4 to $5.8 \times 10^{-16}\,cm^2$ [35, 36] in the visible range. Owing again to the fact that diamond is often used as a simple and representative example for developing theory, the knowledge of the absolute value of the Raman tensor is better known compared to other materials, and often represents a reference for scattering intensity. Perhaps the most certain value is $|d| = 4.4 \pm 0.3 \times 10^{-16}\,cm^2$ at 514.5 nm of Grimsditch and Ramdas [37], who developed a method based on the ratio of the Brillouin and Raman scattered intensities that avoids uncertainties introduced by absolute photometric measurements. Using this value of d, $S = 6.1 \times 10^{-7}$ for the case of y-polarized incident light for a cubic cut crystal in the configuration of Figure 1.11b (i.e., polarized along [010], in the plane including the incident beam and detector) and with $L = 1\,cm$ and $\Omega = 1\,srad$. The scattered light is unpolarized, whereas for a z-polarized incident beam the scattering efficiency is half this value and polarized in the x-direction. The scattering efficiency as a function of incident polarization has been calculated using Equation (1.9) for the standard cases of perpendicular and axially directed scattering for a linearly polarized incident beam for input beams directed along the major crystallographic axes <100>, <110>, and <111>. The results, which represent the scattering behavior of any material with an F_{2g} Raman mode, are plotted in Figure 1.12 as functions of the incident and scattered polarization directions. The two scattering geometries discussed above correspond to Figure 1.12b with

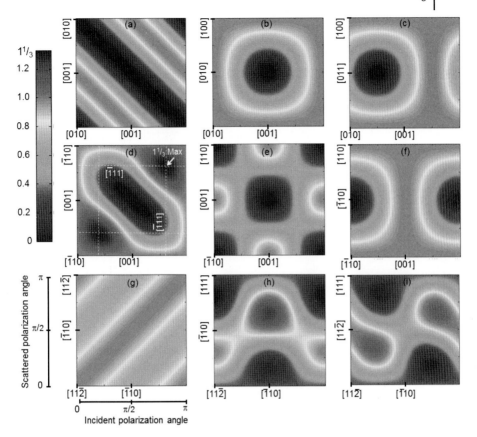

Figure 1.12 Scattering efficiency $\left(\sum_j |e_s R_j e_i|^2\right) / d^2$ as a function of input and Stokes beam polarizations for several beam directions. (a–c) For an incident beam along a <100> axis for axially directed scattering, perpendicular scattering in direction <100>, and perpendicular scattering in the direction <110>, respectively; (d–f) For an incident beam along a <110> axis for axially directed scattering, perpendicular scattering in direction <110>, and perpendicular scattering in the direction <100>, respectively; (g–i) For an incident beam along a <111> axis for axially directed scattering, perpendicular scattering in direction <112>, and perpendicular scattering in the direction <110>, respectively. For a color version of this figure, please see the color plate at the end of this book.

input polarizations [010] and [001]. The highest value for the scattering matrix element is $4d^2/3$, which is obtained for light directed along the <110> direction for incident and scattered light polarized along the <111> direction (see Figure 1.12d). Polarized scattering data can be a useful aid for orienting diamond samples [38]. The effects of the surface, sample size and crystal imperfections have been reviewed in Ref. [39].

1.6.1
Wavelength Dependence

Efficiency increases at shorter wavelengths due to the characteristic λ^{-4} dependence for scattering, but is also enhanced by resonances between the sum and difference combinations of the pump, Stokes and Raman frequencies with transitions across the electronic bandgap. Figure 1.13a shows a comparison of $|d|$ as a function of the pump wavelength with other materials. The increase in $|d|$ at shorter wavelengths is attributed to resonance with the first direct bandgap [33]. By calculating the resonant contribution using a parabolic two-band model [33, 36],[3)] a relation for the dispersion of gain was derived as:

$$|d| = 6.5 \times 10^{-16} \, g\,(\lambda_g/\lambda) \quad (1.13)$$

where d is in units of cm² and where

$$g(x) = x^{-2}[2 - (1-x)^{-1/2} - (1+x)^{-1/2}] \quad (1.14)$$

takes into account the increase in density of states at shorter wavelengths. Here, λ is the pump wavelength and λ_g is taken as 207 nm (6 eV). Although measurements at short wavelengths are scarce, theoretical predictions suggest that the

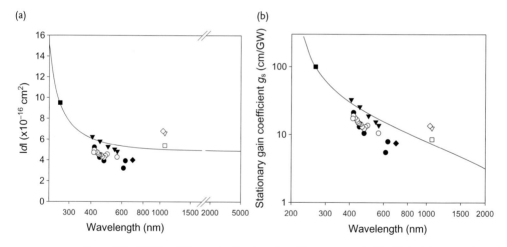

Figure 1.13 (a) $|d|$ and (b) g_s as a function of pump wavelength. Conversion between $|d|$ and g_s was performed where needed using Equation (1.20) with the Raman linewidth taken to be 1.5 cm⁻¹ and $\sum_j |e_s R_j e_i|^2 = d^2$. The data points are obtained from Refs [40] (closed squares), [36, 41] (solid triangles), [33] (solid and open circles), [42] (open triangle; see Section 8.2.1, open square), [43] (solid diamond), and [44] (open diamond). The curve in (a) was obtained using Equations (1.13) and (1.14), and in (b) using Equations (1.13), (1.14), and (1.20).

3) See the note added in proof in Ref. [33].

resonance for wavelengths approaching the bandgap is not as strong as for the Group IV analogs Si and Ge [35, 45].

1.6.2
Raman Linewidth

The finite linewidth of the Stokes shift, which is observed in the broadened frequency spectrum of the scattered light, results from anharmonic forces on the zone-center phonon oscillator and its resultant damping (with rate Γ) [46]. The full-width at half-maximum of the Lorentzian line shape $\Delta v = \Gamma/\pi c$ (in units of cm^{-1}) has been measured by several groups, with values ranging from 1 to 4 cm^{-1} at room temperature (see the recent study in Ref. [47] and references therein), most of which are based on direct measurements of the Raman spectrum. Determining accurate widths is not straightforward due to the compounding effects of instrument resolution and sample purity. The linewidth is of the order of the resolution of most conventional spectroscopic instruments (i.e., diffraction grating and Fourier transform IR spectrometers); thus, accurate measurements require detailed consideration of the instrument function. Liu et al. [48] aimed at carefully considering the spectrometer resolution and reported a room temperature value of approximately 1.2 cm^{-1}. By using a Fabry–Perot interferometer to analyze the broadened lineshape, McQuillan et al. [43] measured widths between 2.04 and 2.22 cm^{-1} for three Type IIa samples, with a quoted uncertainty of 0.04 cm^{-1}. Solin and Ramdas measured 1.65 ± 0.02 cm^{-1} [9]. Alternative methods based on ultrafast coherent phonon spectroscopy [47, 49], which measure the phonon damping rate using a pair of ultrafast pulses, yielded 1.5 ± 0.07 cm^{-1} and 1.54 ± 0.39 cm^{-1}. Levenson et al. [50] used a value of 1.02 cm^{-1} in order to fit the observed wavelength dependence of intensity of the coherent anti-Stokes Raman signal. As the damping rate measurements by ultrafast phonon spectroscopy are less affected by instrument functions, these values are expected to be more accurate. It is presently unclear how sample impurities and lattice defects may affect the Raman linewidth, however. Highly nitrogen-doped samples have been shown to introduce a small amount of broadening (from 1.54 to 1.81 cm^{-1} with up to 100 ppm of nitrogen) [49], variations in the nitrogen content may be responsible for the inconsistencies seen in measured values.

A linewidth of 1.5 cm^{-1} corresponds to a phonon dephasing time ($T_2 = 1/\Gamma$) of 7.1 ps and an oscillator quality factor ($T_2\omega_r$) of 1800. The main damping mechanism has been proposed to result from a resonant coupling with pairs of acoustic phonons of opposite momentum at the Raman frequency [51]. The dephasing time is shorter compared to the other Group IV crystals sharing a diamond lattice symmetry (Si, Ge, and Sn), and reflects a slightly higher density of states for the allowed two-phonon decay products [52]. It is also notably shorter than the strong 1047 cm^{-1} Raman mode of barium nitrate ($T_2 = 20$ ps at room temperature [53]), which is an outstanding example of a molecular crystal containing a phonon oscillator with very small anharmonic forces.

1.6.3
Temperature Dependence

Temperature influences the scattering efficiency, linewidth, and center frequency of Raman scattering. Observed ratios of the Stokes and anti-Stokes signals are consistent with the predicted temperature dependence resulting from n_{BE} in Equation (1.9) [54]. It is also found that as temperature is raised, the center frequency decreases and the linewidth broadens. The temperature dependence of the center frequency has been investigated by numerous groups, including a study conducted more than 60 years ago by Krishnan at temperatures up to 1000 K [55]. Data obtained from several of the more recent studies are shown in Figure 1.14a. Accurate determinations have been problematic due to difficulties in accurately recording the sample temperature and deconvolving the instrument function [58], and the absence of satisfactory *ab initio* theory to underpin the data [59]. The decrease in center frequency with temperature, which occurs primarily due to the change in force constants as the lattice expands, is described well by the semi-empirical relationship [58]:

$$\omega_r(T) = \omega_r(0) - A \cdot n_{BE}(B \cdot \omega_r(T=0)) \tag{1.15}$$

where the Raman frequency at absolute zero is taken as $\omega_r(0) = 1332.7 \pm 0.2$ [48], and $A = 56 \pm 2$ and $B = 0.75 \pm 0.02$ are determined by fitting to data in Figure 1.14a. The results of the study in Ref. [48] suggest that B should be equal to 0.5, in accordance with a damped oscillator model based on two acoustic phonons of opposite momentum from the same branch (the Klemens model [51]); however, the agreement with experiment is less satisfactory when using this value.

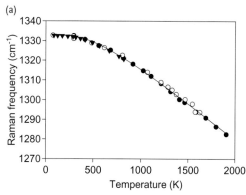
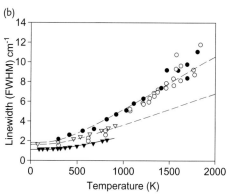

Figure 1.14 (a) Dependence of the first-order Stokes center frequency on temperature; (b) Dependence of Raman linewidth on temperature. The data are from Refs [56] (filled circles), [54] (open circles), [48] (filled triangles), and [57] (open triangles). In (b), the solid line corresponds to Equation (1.16) with $A' = 2$ and $B' = 0.5$ (Klemens model) and with $\Gamma(0) = 1.103\,\text{cm}^{-1}$. The dashed lines show how the parameters A' and B' may be varied in order to fit the data. The lower and upper curves $A' = 2$ and 3 respectively, both with $B' = 0.5$.

1.6 First-Order Raman Scattering

The linewidth is known with much less certainty, as highlighted by the spread of the selected data compiled in Figure 1.14b. The Klemens model has been proposed in many studies [48, 51, 56] as a good approximation in which the linewidth dependence is given by

$$\Delta n(T) = \Delta n(0) \cdot (1 + A' \cdot n_{BE}(B'w_r[T = 0])) \tag{1.16}$$

where $A' = 2$ and $B' = 0.5$. Good agreement was obtained by Liu et al. [48], Borer et al. [57], and Herchen and Capelli [54] for T less than 1000 K. In order to fit the higher-temperature data of Zouboulis and Grimsditch [56] and Herchen and Capelli [54], a value of $A' \approx 3.0$ is required. The dashed curves in Figure 1.14b show Equation (1.16) for these two choices of A'. It is clear that accurate determinations of linewidth temperature dependence are challenging, and further investigations on this topic are required to fully test the validity of the Klemens model.

1.6.4
Isotopic Content

The dependence of the center frequency and lineshape with isotopic content has been investigated in several studies [60–63]. The Raman spectrum for extreme and intermediate concentrations of ^{12}C and ^{13}C is shown in Figure 1.15a. From $\omega_r = 1332.4 \, \text{cm}^{-1}$ for natural diamond (^{13}C fraction $x = 0.011$), the measured room-temperature center frequency increases to $1332.7 \, \text{cm}^{-1}$ for $x = 0.001$, and decreases to $1282 \, \text{cm}^{-1}$ for a highly enriched ^{13}C ($x = 0.99$) [63]. Based on these measure-

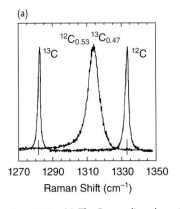
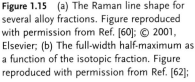
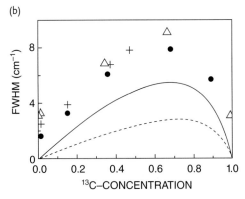

Figure 1.15 (a) The Raman line shape for several alloy fractions. Figure reproduced with permission from Ref. [60]; © 2001, Elsevier; (b) The full-width half-maximum as a function of the isotopic fraction. Figure reproduced with permission from Ref. [62]; © 1993, Elsevier. As detailed in the reference, the symbols correspond to data compiled from several studies, and the two curves show model results obtained using differing assumptions for the density of states.

ments and those of Ref. [61], the following polynomial was used to fit the room-temperature data:

$$\omega_r(x) = 1332.82 - 34.77x - 16.98x^2 \tag{1.17}$$

The frequency shift is explained by two contributions affecting the Raman frequency: one involving the change in average mass proportional to $M_{av}^{-0.5}$ where $M_{av} = (1-x)M_{12} + xM_{13}$; and the other involving the random distribution of isotopes in the bulk referred to as "isotopic disorder," which is responsible for a small frequency increase. A thorough theoretical treatment involves consideration of the disorder-induced anharmonic terms, as well as contributions arising from a relaxation in the requirement for wavevector conservation due to breakdown in the lattice translational invariance [60–63].

The shape and width are also found to vary significantly with x. The width as a function of isotopic ratio is shown in Figure 1.15b [62]. The notably larger values for the natural isotope ratio compared to the measurements of others (see Figure 1.14b) is not explained by the authors, but may be a result of a lower instrument resolution or by sample impurities. The ratio of ^{12}C and ^{13}C widths is slightly greater than unity consistent with the expected ratio of $M_{13}/M_{12} = 1.08$ [60]. For alloys, the width increases notably and the shape is no longer purely Lorentzian. This is highlighted by the spectrum for $x = 0.47$ in Figure 1.15a, which shows an extended shoulder on the low mass side and a linewidth of approximately $6\,cm^{-1}$. Explanation of the observed change in shape requires consideration of disorder-induced additional damping of the zone center phonon (which is sensitive to the precise form of the phonon dispersion curves near the Raman frequency) and the relaxation of wave-vector conservation [60–62]. The maximum width of approximately $8\,cm^{-1}$ is otained near $x = 0.7$.

1.7
Stimulated Raman Scattering

Stimulated Raman scattering, in which growth in a Stokes field results from parametric coupling with the pump and phonon fields, was first observed in solids in a study involving diamond [64]. For the simple case of a plane wave pump pulse of duration much longer than T_2 and intensity I_p, the growth in the Stokes intensity in the z-direction is (refer also to Section 8.1.1.1)

$$dI_s/I_s = g_s I_p(z) \cdot dz \tag{1.18}$$

The gain coefficient g_s factors in the accumulated density of phonons (which decay on the order of T_2), and can be derived from the spontaneous Raman scattering values of d and Γ described above, using the relationship for the steady-state gain coefficient [65]

$$g_s(\omega_s) = \frac{4\pi^2 \omega_s N}{n_s n_l c^2 M \omega_r} \frac{\Gamma}{(\omega_s - \omega_l + \omega_r)^2 + \Gamma^2} \sum_j |e_s R_j e_i|^2 \tag{1.19}$$

which, at the Stokes frequency line center, reduces to

$$g_s = \frac{4\pi^2 \omega_s N}{n_s n_i c^2 M \omega_r \Gamma} \sum_j |e_s R_j e_i|^2 \tag{1.20}$$

where ω_i and ω_s are the pump and Stokes frequencies, respectively, and n_i and n_s are the corresponding refractive indices. The gain coefficient is related to the imaginary part of the Raman susceptibility χ_R through

$$\chi_R = -n_s n_i g_s / 2\mu_0 \omega_s \tag{1.21}$$

Direct measurements of g_s have been made from the threshold for stimulated Raman scattering or by using pump–probe measurements. For single-pass stimulated Raman scattering of an intense pump beam, the pump intensity I_{th} threshold is conventionally obtained using the relationship

$$g_s = 25 / L \cdot I_{th} \tag{1.22}$$

where it is assumed that the pump intensity is maintained over L and that a gain factor of e^{25} marks the onset of observable stimulated Raman scattering. In the presence of a Stokes resonator, the gain can be determined by equating the steady-state gain with the resonator losses at threshold. Since gain depends on intensity, a good knowledge of the beam diameters and beam shape is required. A more detailed review of g_s values is given in the context diamond Raman lasers in Chapter 8.

Gain coefficient values have been determined for pump wavelengths between 266 nm and 1064 nm, as shown in Figure 1.13b. For pumping at 1064 nm, the gain varies from more than 12.5 cm GW^{-1} for propagation along a <100> axis as determined from a measurement of SRS threshold [42], to 8.5 cm GW^{-1} using pump-probe techniques (see Section 8.2.1). The theoretical curve shows the expected wavelength dependence using Equation (1.20) and the semi-empirical expression for d (Equations 1.13 and 1.14) under the assumption that the linewidth is 1.5 cm^{-1} ($T_2 = 7.1$ ps). It is also assumed that a configuration is used in which $\sum_j |e_s R_j e_i|^2 = d^2$, which is the case, for example, for co-propagating pump and Stokes beams along the <100> direction. Thus, a 33% enhancement may be obtained for the pump and Stokes polarization parallel to <111> axes (refer to Section 1.6). At long wavelengths, the gain is linear with Stokes frequency. It should be noted, however, that the above theory is only valid for Stokes frequencies much greater than ω_r which, for diamond, corresponds to $\lambda_s \ll 7.5\,\mu m$.

1.8
Brillouin Scattering

The wavelength shift and intensity of Brillouin scattering is determined by the elastic strain and photoelastic tensors, which for the cubic symmetry of diamond

Table 1.2 Elastic strain c_{ij} and photoelastic tensor p_{ij} elements [37]. For a discussion on the accuracy of the p_{ij}, see values in Ref. [66]. For reference, the dispersion of the p_{ij} has been considered in Ref. [67] (see also Ref. [3]), and the temperature and isotopic variation in the elastic strain is considered in Refs [68] and [69].

c_{11} (10^{11} Pa)	c_{12} (10^{11} Pa)	c_{44} (10^{11} Pa)	p_{11}	p_{12}	p_{44}
10.764	1.252	5.744	−0.249	0.043	−0.172

is characterized by the three pairs of parameters c_{11}, c_{12}, and c_{44}, and p_{11}, p_{12}, and p_{44}. Experimentally derived values for these are listed in Table 1.2.

The frequency shift is given by

$$\Delta\omega = \pm 2\omega_i n(v_s/c)\sin(\theta/2) \tag{1.23}$$

where θ is the angle between the incident and scattered radiation from the acoustic wave directed at the intermediate angle and n is the refractive index at ω. The speed of the longitudinal or transverse acoustic phonon $v_s = (X_q/\rho)^{0.5}$ can be determined from the appropriate combination of elastic strain tensor elements X_q. For phonons propagating along a cubic axis, for example, $X_q = c_{11}$ or c_{44} for longitudinal or transverse polarization respectively, giving $v_s = 1.8 \times 10^4$ ms^{-1} and 1.3×10^4 ms^{-1}. For Brillouin scattering at visible wavelengths, the typical shift for perpendicular scattering from longitudinal acoustic phonons is 3.5–4.5 cm^{-1} and 2–3 cm^{-1} for transverse phonons.

The scattering efficiency at the Stokes wavelength can be calculated using [37, 70]

$$S = \frac{k_B T \omega_s^4 n^8 \Omega L}{128\pi^2 c^4} \frac{\sum_j (e_s T_j e_i)^2}{X_q} \tag{1.24}$$

where the sum is over the longitudinal and transverse acoustic waves and the T_j are the scattering tensors. For $k_B T$ much greater than acoustic phonon energy $\hbar\Delta\omega$, which is the case at room temperature, the intensity of Stokes and anti-Stokes are equal. For incident and scattered beams directed along perpendicular <110> axes, the relevant scattering tensors for the <100> directed longitudinal phonon and two degenerate transverse phonons are

$$T_1 = 2\begin{bmatrix} p_{11} & 0 & 0 \\ 0 & p_{12} & 0 \\ 0 & 0 & p_{12} \end{bmatrix} \quad T_2 = 2\begin{bmatrix} 0 & p_{44} & 0 \\ p_{44} & 0 & 0 \\ 0 & 0 & 0 \end{bmatrix} \quad T_3 = 2\begin{bmatrix} 0 & 0 & p_{44} \\ 0 & 0 & 0 \\ p_{44} & 0 & 0 \end{bmatrix} \tag{1.25}$$

Further, if the light beams are polarized normal to the scattering plane, $\sum_j (e_s T_j e_i)^2 = 4p_{12}^2$ and $X_q = c_{11}$ so that the Brillouin scattering efficiency per unit solid angle S/Ω is approximately 10^{-8} srad^{-1} at 500 nm for $L = 1$ cm. In this case, generated phonons are longitudinally polarized in the <100> direction. Scattering tensors and X_q values are listed in Ref. [37] for phonons directed along the symmetry directions <110> and <111>.

For stimulated Brillouin scattering, the steady-state gain coefficient at line center is given by [71]

$$g_B = n^7 \omega_s^2 p^2 / \rho v_s c^3 \Gamma_B \qquad (1.26)$$

where p is the appropriate photoelastic tensor component. The damping of the acoustic wave Γ_B, has not been investigated in detail for diamond. According to Boyd [71], the damping rate of the scattering phonons or Brillouin scattering linewidth for materials is

$$\Gamma_B = q^2/\rho[4/3\eta_s + \eta_b + k(C_p/C_v - 1)/C_p] \qquad (1.27)$$

where q is the acoustic phonon wave-vector, η_s is the shear viscosity, η_b is the bulk viscosity, κ is the thermal conductivity and C_p and C_v are the specific heats at constant pressure and volume. The author is unaware of published η_s and η_v values, or measurements of acoustic phonon lifetime values. If, instead, the damping rate for Ge of $\Gamma_B = 1\,\text{ns}^{-1}$ (from Ref. [72] and scaled for visible wavelengths $\omega_s \approx 4 \times 15\,\text{rad s}^{-1}$) is used, then $g_B \approx 1.3\,\text{cm GW}^{-1}$. Thus, pulse intensities of approximately $1\,\text{GW cm}^{-2}$ with temporal coherence longer than 1 ns are expected to be necessary to achieve significant gain in crystals of length 1 cm long. To the present author's knowledge, no details of stimulated Brillouin scattering in diamond have yet been reported.

1.9
Electronic Nonlinearity

The third-order electronic nonlinearity $\chi^{(3)}$ results from a deformation of the lattice electron cloud at high incident electric field intensities, and is responsible for a wide range of fast (10^{-16} s) optical effects in the transparent region. These include four-wave mixing, third-harmonic generation, self-focusing (intensity-dependent refractive index), DC and optical Kerr effects, and multi-photon absorption. Direct measurements of the $\chi^{(3)}$ have been determined from experiments in four-wave mixing [41, 50], self-focusing ("Z-scan technique") [73], and the DC Kerr effect [74]. Values for the three independent tensor coefficients that characterize cubic crystals are listed in Table 1.3, and plotted as a function of wavelength in Figure 1.16 [75]. The fact that $\chi^{(3)}_{1111} = 3\chi^{(3)}_{1221} = 3\chi^{(3)}_{1122}$ holds approximately for diamond indicates that the electronic orbitals responsible for the nonlinearity are spherical on average, and that the electronic linearity is thus approximately directionally independent [41]. The magnitude is in general agreement with theoretical calculations based on the band structure [76], and is approximately one-twentieth of the real part of the resonant Raman nonlinearity. The magnitude is similar to lead silicate glasses, but notably lower compared to other crystals of similar refractive index. The wavelength dependence has not been modeled in detail for wide indirect bandgap materials such as diamond; however, general models for the nonlinear absorption [77] combined with a Kramers–Kronig analysis [73] seem to be useful in qualitatively predicting the wavelength dependence.

Table 1.3 The measured third-order susceptibility tensor components χ_{ijkl} and the nonlinear refractive index n_2. $\chi^{(3)}$ is related to the nonlinear polarization $P_i^{(3)}$ through $P_i^{(3)} = \varepsilon_0 \chi_{ijkl}^{(3)} E_j \cdot E_k \cdot E_l$. Note that the values reported in Ref. [41] have been increased by a factor of 4 to conform to the definition of the susceptibility used here, and in the convention of most other studies. The $\chi^{(3)}$ values in italics are calculated from n_2 measurements (and vice versa), as described in Section 1.9.1.

Wavelength (nm)	$\chi_{1111}^{(3)}$	$\chi_{1221}^{(3)}$	$\chi_{1122}^{(3)}$	n_2 (10^{-20} m^2 W^{-1})	Reference
	(10^{-21} m^2 V^{-2})				
530[a]	3.2			15.6	[50]
545[a]	2.6	0.96	1.02	12.6	[41]
565[a]		0.88		12.8[b]	[41]
450[a]		1.22		17.8[b]	[41]
407[a]		1.46		21.3[b]	[41]
650		0.217		3.17	[74]
532	1.44			7.0	[73]
355	−2.76			−13.4	[73]
1064	0.864			4.2	[73]
532	1.44			7.0	[73]
355	−2.47			−12	[73]
266	−6.99			−34	[73]

a) The wavelengths listed for the four-wave mixing experiments represent the pump beam wavelength.
b) Calculated using Equation (1.28) and assuming $\chi_{1122}^{(3)} = 3\chi_{1221}^{(3)}$.

The data indicate that $\chi_{1111}^{(3)}$ is approximately 1–3 × 10^{-21} m^2 V^{-2} in the visible range, and increases slightly for shorter wavelengths approaching the threshold for two-photon absorption ($\lambda < 450$ nm). Values determined using "Z-scan" measurements of the nonlinear refractive index (see Section 1.9.1) indicate a change in sign at a wavelength approximately 70% of the bandgap (ca. 400 nm) and a significant increase in magnitude as the bandgap is approached. This behavior is characteristic of many other semiconductors. A model including the various contributions to the nonlinearity was developed in Ref. [78], which assumed a simple two-band structure (valence and conduction bands) and neglected the role of free carriers generated by two-photon absorption. This model was successful in reproducing the behavior of a large number of semiconductors, and suggests that the effects of two-photon absorption dominates the $\chi_{ijkl}^{(3)}$ wavelength dependence as the bandgap is approached. For photon energies much smaller than the bandgap, no significant dispersion is expected, as is confirmed by the data for wavelengths longer than 600 nm.

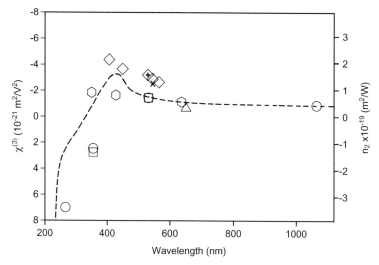

Figure 1.16 The dependence of susceptibility $\chi^{(3)}_{1111}$ on wavelength. Data are from Refs [50] (diamond with center cross), [73] (circles and squares), [74] (upward triangle), and [75] (hexagons). The hollow diamonds and (x) correspond to $3\chi^{(3)}_{1221}$ and $\chi^{(3)}_{1111}$ of Ref. [41]. The right-hand axis shows the nonlinear refractive index n_2 calculated from $\chi^{(3)}_{1111}$ using Equation (1.28). The dashed line is the theoretical wavelength dependence for n_2 obtained by a Kramers–Kronig analysis of the nonlinear absorption, as described in Ref. [73].

1.9.1
Nonlinear Refractive Index

The nonlinear refractive index arising from the nonresonant electronic nonlinearity can be measured directly using the Z-scan technique [73], and is related to $\chi^{(3)}_{ijkl}$ by [71]

$$n_2 = 3\chi^{(3)}_{1111}/4cn_0^2\varepsilon_0 \qquad (1.28)$$

where n_2 is defined as the index change n_2I from the linearly polarized incident beam of intensity I. The results are included in Table 1.3 and Figure 1.16. In the visible range, n_2 is approximately $1-2 \times 10^{-19}\,\text{m}^2\,\text{W}^{-1}$. As for $\chi^{(3)}_{1111}$, n_2 changes sign at a wavelength of approximately 400 nm.

1.9.2
Two-Photon Absorption

Two-photon absorption coefficients β_2 for materials are known to scale reasonably well with the inverse fourth power of the band gap [77, 79]. Consequently, the two-photon absorption coeffiecient in diamond is relatively small compared to

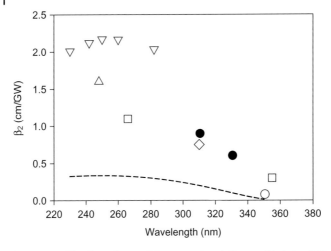

Figure 1.17 The dependence of β_2 on wavelength from solid circles [75], downward triangles [81], upward triangle [80], and squares [73]. The dashed line is the theoretical wavelength dependence of Ref. [79], assuming a direct bandgap energy of 7.0 eV.

that of most other semiconductors. The β_2 values are determined by measurement of the decrease in transmission as a function of the incident intensity. A summary of reported measurements is shown in Figure 1.17; also shown is the theoretical scaling formula of Van Stryland et al. [79], which has been used with good success to predict β_2 dispersion for direct bandgap semiconductors. Clearly, the measured β_2 values for indirect-bandgap diamond are notably larger than predicted by theory, although the general shape as a function of wavelength is in reasonable agreement. Although other models for two-photon absorption in diamond have been suggested [75, 81], to the present author's knowledge a comprehensive theory for the observed magnitudes and dispersion is yet to be reported. Three-photon absorption measurements have been recently reported in the range of 350 to 430 nm [75].

Acknowledgments

The author is grateful to Alex Sabella for his advice on Raman gain measurements, and to Andrew Lehmann and Ondrej Kitzler for assistance in preparing some of the figures. Thanks also to Brian Orr, David Coutts, Gerald Bonner, David Spence, and Marcos Grimsditch for useful comments on the manuscript. This material is partly based on research sponsored by the Australian Research Council Future Fellowship Scheme (FT0990622), and the Air Force Research Laboratory under agreement numbers AOARD-10-4078 and 12-4055.

References

1 Walker, J. (1979) Optical absorption and luminescence in diamond. *Rep. Prog. Phys.*, **42** (10), 1605.
2 Edwards, D. and Philipp, H.R. (1985) Cubic carbon (diamond), in *Handbook of Optical Constants of Solids* (ed. E.D. Palik), Academic Press, Orlando, FL, pp. 665–673.
3 Zaitsev, A.M. (2001) *Optical Properties of Diamond: A Data Handbook*, Springer, Berlin.
4 Smith, H.M.J. (1948) The theory of the vibrations and the Raman spectrum of the diamond lattice. *Philos. Trans. R. Soc. Lond. Ser. A Math. Phys. Sci.*, **241** (829), 105–145.
5 Lax, M. and Burstein, E. (1955) Infrared lattice absorption in ionic and homopolar crystals. *Phys. Rev.*, **97** (1), 39–52.
6 Thomas, M. (1994) Multiphonon model for absorption in diamond. *Proc. SPIE*, **2286**, 152.
7 Vogelgesang, R., Alvarenga, A.D., Hyunjung, K., Ramdas, A.K., Rodriguez, S., Grimsditch, M., and Anthony, T.R. (1998) Multiphonon Raman and infrared spectra of isotopically controlled diamond. *Phys. Rev. B*, **58** (9), 5408–5416.
8 Klein, C.A., Hartnett, T.M., and Robinson, J. (1992) Critical point phonon frequencies of diamond. *Phys. Rev. B*, **45** (22), 854–863.
9 Solin, S. and Ramdas, A. (1970) Raman spectrum of diamond. *Phys. Rev. B*, **1** (4), 1687–1698.
10 Wehner, R., Borik, H., Kress, W., Goodwin, A., and Smith, S. (1967) Lattice dynamics and infra-red absorption of diamond. *Solid State Commun.*, **5** (4), 307–309.
11 Mollart, T.P., Lewis, K.L., Pickles, C.S.J., and Wort, C.J.H. (2003) Factors affecting the optical performance of CVD diamond infrared optics. *Semicond. Sci. Technol.*, **18** (3), S117.
12 Harris, D. (1994) Properties of diamond for window and dome applications. *Proc. SPIE*, **2286**, 218–228.
13 Sussmann, R.S., Brandon, J.R., Coe, S.E., Pickles, C.S.J., Sweeney, C.G., Wasenczuk, A., Wort, C.J.H., and Dodge, C.N. (1998) CVD diamond: a new engineering material for thermal, dielectric and optical applications. *Ind. Diam. Rev.*, **3**, 69–77.
14 Sussmann, R.S., Pickles, C.S.J., Brandon, J.R., Wort, C.J.H., Coe, S.E., Wasenczuk, A., Dodge, C.N., Beale, A.C., Krehan, A.J., Dore, P., Nucara, A., and Calvani, P. (1998) CVD diamond windows for infrared synchrotron applications. *Il Nuovo Cimento D*, **20** (4), 503–525.
15 Lax, M. (1958) Optical properties of diamond type crystals. *Phys. Rev. Lett.*, **1** (4), 131–132.
16 Hahn, D.V., Thomas, M.E., Blodgett, D.W., and Kaplan, S.G. (2003) Characterization and modeling of the infrared properties of diamond and SiC. *Proc. SPIE*, **5078**, 148–158.
17 Piccirillo, C., Davies, G., Mainwood, A., and Penchina, C. (2001) The variation of optical absorption of CVD diamond as a function of temperature. *Phys. B Condens. Matter*, **308–310**, 581–584.
18 Piccirillo, C., Davies, G., Mainwood, A., Scarle, S., Penchina, C.M., Mollart, T.P., Lewis, K.L., Nesládek, M., Remes, Z., and Pickles, C.S.J. (2002) Temperature dependence of intrinsic infrared absorption in natural and chemical-vapor deposited diamond. *J. Appl. Phys.*, **92** (2), 756.
19 Philipp, H. and Taft, E. (1964) Kramers–Kronig analysis of reflectance data for diamond. *Phys. Rev.*, **136** (5A), 1445–1448.
20 Clark, C.D., Dean, P.J., and Harris, P.V. (1964) Intrinsic edge absorption in diamond. *Proc. R. Soc. A Math. Phys. Eng. Sci.*, **277** (1370), 312–329.
21 Dean, P., Lightowlers, E., and Wight, D. (1965) Intrinsic and extrinsic recombination radiation from natural and synthetic aluminum-doped diamond. *Phys. Rev.*, **140** (1), A352–A386.
22 Roberts, R. and Walker, W. (1967) Optical study of the electronic structure of diamond. *Phys. Rev.*, **161** (3), 730–735.
23 Collins, A.T., Lawson, S.C., Davies, G., and Kanda, H. (1990) Indirect energy gap

of ^{13}C diamond. *Phys. Rev. Lett.*, **65** (7), 891–894.
24 Peter, F. (1923) Uber brechungsindizes und absorptionkonstanten des Diamanten zwischen 644 und 226 mµ. *Z. Phys. A Hadrons Nucl.*, **15** (1), 358–368.
25 Edwards, D.F. and Ochoa, E. (1981) Infrared refractive index of diamond. *J. Opt. Soc. Am.*, **71** (5), 607–608.
26 Ruf, T. and Cardona, M. (2000) Temperature dependence of the refractive index of diamond up to 925 K. *Phys. Rev. B*, **62** (24), 16578–16581.
27 Fontanella, J., Johnston, R.L., Colwell, J.H., and Andeen, C. (1977) Temperature and pressure variation of the refractive index of diamond. *Appl. Opt.*, **16** (11), 2949–2951.
28 Tropf, W.J., Thomas, M.E., and Linevsky, M.J. (1998) Infrared refractive indices and thermo-optic coefficients for several materials. *Proc. SPIE*, **3425**, 160.
29 Hu, Z.G. and Hess, P. (2006) Optical constants and thermo-optic coefficients of nanocrystalline diamond films at 30–500 °C. *Appl. Phys. Lett.*, **89** (8), 1906–1906.
30 Karch, K., Dietrich, T., Windl, W., Pavone, P., Mayer, A., and Strauch, D. (1996) Contribution of quantum and thermal fluctuations to the elastic moduli and dielectric constants of covalent semiconductors. *Phys. Rev. B*, **53** (11), 7259–7266.
31 Ramaseshan, S. (1946) The Faraday effect in diamond. *Proc. Math. Sci.*, **24** (1), 104–113.
32 Loudon, R. (1964) The Raman effect in crystals. *Adv. Phys.*, **13** (52), 423–482.
33 Grimsditch, M., Cardona, M., Calleja, J.M., and Meseguer, F. (2005) Resonance in the Raman scattering of CaF$_2$, SrF$_2$, BaF$_2$ and diamond. *J. Raman Spectrosc.*, **10** (1), 77–81.
34 Anastassakis, E., Iwasa, S., and Burstein, E. (1966) Electric-field-induced infrared absorption in diamond. *Phys. Rev. Lett.*, **17** (20), 1051–1054.
35 Calleja, J., Kuhl, J., and Cardona, M. (1978) Resonant Raman scattering in diamond. *Phys. Rev. B*, **17** (2), 876–883.
36 Cardona, M. (1982) Resonance phenomena, in *Light Scattering Solids II* (eds M. Cardona and G. Guntherodt), Springer, Berlin, pp. 19–178.
37 Grimsditch, M. and Ramdas, A.K. (1975) Brillouin scattering in diamond. *Phys. Rev. B*, **11** (8), 3139–3148.
38 Mossbrucker, J. and Grotjohn, T.A. (1996) Determination of local crystal orientation of diamond using polarized Raman spectra. *Diamond Relat. Mater.*, **5** (11), 1333–1343.
39 (a) Prawer, S., Nugent, K.W., Jamieson, D.N., Orwa, J.O., Bursill, L.A., and Peng, J.L. (2000) The Raman spectrum of nanocrystalline diamond. *Chem. Phys. Lett.*, **332** (1-2), 93–97; (b) Prawer, S. and Nemanich, R.J. (2004) Raman spectroscopy of diamond and doped diamond. *Philos. Trans. Ser. A Math. Phys. Eng. Sci.*, **362** (1824), 2537–2565.
40 Granados, E., Spence, D.J., and Mildren, R.P. (2011) Deep ultraviolet diamond Raman laser. *Opt. Express*, **19** (11), 10857–10863.
41 Levenson, M. and Bloembergen, N. (1974) Dispersion of the nonlinear optical susceptibility tensor in centrosymmetric media. *Phys. Rev. B*, **10** (10), 4447–4463.
42 Kaminskii, A.A., Hemley, R.J., Lai, J., Yan, C.S., Mao, H.K., Ralchenko, V.G., Eichler, H.J., and Rhee, H. (2007) High-order stimulated Raman scattering in CVD single crystal diamond. *Laser Phys. Lett.*, **4** (5), 350–353.
43 McQuillan, A., Clements, W., and Stoicheff, B. (1970) Stimulated Raman emission in diamond: spectrum, gain, and angular distribution of intensity. *Phys. Rev. A*, **1** (3), 628–635.
44 Feve, J.-P.M., Shortoff, K.E., Bohn, M.J., and Brasseur, J.K. (2011) High average power diamond Raman laser. *Opt. Express*, **19** (2), 913–922.
45 Swanson, L. and Maradudin, A. (1970) Pseudopotential calculation of the Raman tensor for homopolar semiconductors. *Solid State Commun.*, **8** (11), 859–865.
46 Loudon, R. (1963) Theory of the first-order Raman effect in crystals. *Proc. R. Soc. Lond. Ser. A Math. Phys. Sci.*, **275** (1361), 218–232.
47 Lee, K.C., Sussman, B.J., Nunn, J., Lorenz, V.O., Reim, K., Jaksch, D., Walmsley, I.A., Spizzirri, P., and Prawer, S. (2010) Comparing phonon dephasing

lifetimes in diamond using Transient Coherent Ultrafast Phonon Spectroscopy. *Diamond Relat. Mater.*, **19** (10), 1289–1295.

48 Liu, M., Bursill, L., Prawer, S., and Beserman, R. (2000) Temperature dependence of the first-order Raman phonon line of diamond. *Phys. Rev. B*, **61** (5), 3391–3395.

49 Ishioka, K., Hase, M., Kitajima, M., and Petek, H. (2006) Coherent optical phonons in diamond. *Appl. Phys. Lett.*, **89**, 231916.

50 Levenson, M., Flytzanis, C., and Bloembergen, N. (1972) Interference of resonant and nonresonant three-wave mixing in diamond. *Phys. Rev. B*, **6** (10), 3962–3965.

51 Klemens, P. (1966) Anharmonic decay of optical phonons. *Phys. Rev.*, **148** (2), 845–848.

52 Menendez, J. and Cardona, M. (1984) Temperature dependence of the first-order Raman scattering by phonons in Si, Ge, and α-Sn: anharmonic effects. *Phys. Rev. B*, **29** (4), 2051–2059.

53 Zverev, P.G., Liu, H., and Basiev, T.T. (1995) Vibrational dynamic of the Raman-active mode in barium nitrate crystal. *Opt. Lett.*, **20** (23), 2378–2380.

54 Herchen, H. and Cappelli, M. (1994) Second-order Raman scattering in diamond up to 1900 K. *Phys. Rev. B*, **49** (5), 3213–3216.

55 Krishnan, R. (1946) Temperature variations of the Raman frequencies in diamond. *Proc. Math. Sci.*, **24** (1), 45–57.

56 Zouboulis, E. and Grimsditch, M. (1991) Raman scattering in diamond up to 1900 K. *Phys. Rev. B*, **43** (15), 12490–12493.

57 Borer, W., Mitra, S., and Namjoshi, K. (1971) Line shape and temperature dependence of the first-order Raman spectrum of diamond. *Solid State Commun.*, **9** (16), 1377–1381.

58 Cui, J.B., Amtmann, K., Ristein, J., and Ley, L. (1998) Noncontact temperature measurements of diamond by Raman scattering spectroscopy. *J. Appl. Phys.*, **83** (12), 7929–7033.

59 Lang, G., Karch, K., Schmitt, M., Pavone, P., Mayer, A.P., Wehner, R.K., and Strauch, D. (1999) Anharmonic line shift and linewidth of the Raman mode in covalent semiconductors. *Phys. Rev. B*, **59** (9), 6182–6188.

60 Cardona, M. and Ruf, T. (2001) Phonon self-energies in semiconductors: anharmonic and isotopic contributions. *Solid State Commun.*, **117** (3), 201–212.

61 Hass, K., Tamor, M., Anthony, T., and Banholzer, W. (1992) Lattice dynamics and Raman spectra of isotopically mixed diamond. *Phys. Rev. B*, **45** (13), 7171–7182.

62 Spitzer, J., Etchegoin, P., Cardona, M., Anthony, T., and Banholzer, W. (1993) Isotopic-disorder induced Raman scattering in diamond. *Solid State Commun.*, **88** (7), 509–514.

63 Vogelgesang, R., Ramdas, A., Rodriguez, S., Grimsditch, M., and Anthony, T. (1996) Brillouin and Raman scattering in natural and isotopically controlled diamond. *Phys. Rev. B*, **54** (6), 3989–3999.

64 Eckhardt, G., Bortfeld, D., and Geller, M. (1963) Stimulated emission of Stokes and anti-Stokes Raman lines from diamond, calcite, and α-sulfur single crystals. *Appl. Phys. Lett.*, **3** (8), 137–138.

65 Penzkofer, A., Laubereau, A., and Kaiser, W. (1978) High intensity Raman interactions. *Prog. Quantum Electron.*, **6**, 55–140.

66 Lang, A.R. (2009) The strain-optical constants of diamond: a brief history of measurements. *Diamond Relat. Mater.*, **18** (1), 1–5.

67 Grimsditch, M., Anastassakis, E., and Cardona, M. (1979) Piezobirefringence in diamond. *Phys. Rev. B*, **19** (6), 3240–3243.

68 Zouboulis, E.S. and Grimsditch, M. (1998) Temperature dependence of the elastic moduli of diamond: a Brillouin-scattering study. *Phys. Rev. B*, **57** (5), 2889–2896.

69 Ramdas, A., Rodriguez, S., and Grimsditch, M. (1993) Effect of isotopic constitution of diamond on its elastic constants: ^{13}C diamond, the hardest known material. *Phys. Rev. Lett.*, **71** (1), 189–192.

70 Loudon, R. (1964) The Raman effect in crystals. *Adv. Phys.*, **13** (7), 423–482.

71 Boyd, R.W. (2003) *Nonlinear Optics*, Academic Press, London.

72 Kovalev, V., Musaev, M., and Faizullov, F. (1984) Stimulated Brillouin scattering gains and decay times of hypersonic waves in optical crystals at the 10.6 μ wavelength. *Sov. J. Quantum Electron.*, **14** (1), 110–112.

73 (a) Sheik-Bahae, M., DeSalvo, R., Said, A., Hagan, D., Soileau, M., and Van Stryland, E. (1992) Nonlinear refraction in UV transmitting materials. *Proc. SPIE*, **1624**, 25–30; (b) Sheik-Bahae, M., DeSalvo, R., Said, A., Hagan, D., Soileau, M., and Van Stryland, E. (1995) Optical nonlinearities in diamond. *Proc. SPIE*, **2428**, 605–609.

74 Zhao, J., Jia, G., Liu, X., Chen, Z., Tang, J., and Wang, S. (2010) Measurement of third-order nonlinear optical susceptibility of synthetic diamonds. *Chin. Opt. Lett.*, **8** (7), 685–688.

75 Kozák, M., Trojánek, F., Dzurňák, B., and Maly, P. (2012) Two-and three-photon absorption in chemical vapor deposition diamond. *J. Opt. Soc. Am. B*, **142** (1), 1141–1145.

76 Arya, K. and Jha, S.S. (1979) Tight-binding bonding orbital model for third-order nonlinear optical susceptibilities in group-IV crystals. *Phys. Rev. B*, **20** (4), 1611–1616.

77 Wherrett, B.S. (1984) Scaling rules for multiphoton interband absorption in semiconductors. *J. Opt. Soc. Am. B*, **1** (1), 62–72.

78 Sheik-Bahae, M., Hutchings, D., Hagan, D.J., and Van Stryland, E.W. (1991) Dispersion of bound electron nonlinear refraction in solids. *IEEE J. Quantum Electron.*, **27** (6), 1296–1309.

79 Van Stryland, E.W., Guha, S., Vanherzeele, H., Woodall, M., Soileau, M., and Wherrett, B. (1986) Verification of the scaling rule for two-photon absorption in semiconductors. *J. Mod. Opt.*, **33** (4), 381–386.

80 Preuss, S. and Stuke, M. (1995) Subpicosecond ultraviolet laser ablation of diamond: nonlinear properties at 248 nm and time-resolved characterization of ablation dynamics. *Appl. Phys. Lett.*, **67** (3), 338–340.

81 Gagarskii, S. and Prikhodko, K. (2008) Measuring the parameters of femtosecond pulses in a wide spectral range on the basis of the multiphoton-absorption effect in a natural diamond crystal. *J. Opt. Technol.*, **75** (3), 139–143.

2
Optical Quality Diamond Grown by Chemical Vapor Deposition
Ian Friel

2.1
Introduction

Diamond is a remarkable substance; its extreme physical and chemical properties are unmatched by any other material. It has the highest thermal conductivity and highest stiffness, possesses extreme hardness and wear resistance, and is optically transparent almost continually from the ultraviolet (UV) to the terahertz (THz) range. Diamond is also chemically inert, radiation hard, and possesses excellent electrochemical properties.

Despite diamond's clear potential as an engineering material, during the early 1990s its high technology applications were limited to just a few niche examples based either on natural diamond or high-pressure, high-temperature (HPHT) synthetic. Optical applications were restricted by the lack of availability of consistent, high-purity diamond, by crystal size and – particularly in the case of natural diamond – by cost. The first commercial products based on chemical vapor deposition (CVD) diamond were released at about this time [1], and today the situation is transformed. A raft of high-technology CVD diamond-based applications that harness diamond's extreme properties are now available [2], achieved through a continuous process of innovation and by leveraging the key advantage of the CVD synthesis process, namely the ability to exercise an unprecedented level of control over diamond's physical properties. As a result, a range of CVD diamond grades – which have been engineered for specific applications such as thermal, mechanical, electrochemical, optical and electronic – are now available.

For the optical engineer, CVD diamond provides a platform for developing technologies that require extreme performance, beyond the capability of standard optical materials. This includes the ability to manage significant heat loads (such as in high-power lasers or X-ray beams) without detriment to performance, or to interface with a chemically or mechanically hostile environment.

CVD diamond is available in both single-crystal and polycrystalline forms. Most commercial CVD diamond technologies developed during the 1990s were based on polycrystalline diamond [3]. Optical quality polycrystalline diamond has the advantage of being available in large areas: for example, low-loss windows in excess

Optical Engineering of Diamond, First Edition. Edited by Richard P. Mildren and James R. Rabeau.
© 2013 Wiley-VCH Verlag GmbH & Co. KGaA. Published 2013 by Wiley-VCH Verlag GmbH & Co. KGaA.

of ⌀100 mm and 1.5 mm thickness are employed in kilowatt (kW) CO_2 lasers and high-power gyrotrons. Large-area polycrystalline hemispherical domes for infrared (IR) imaging applications were also developed during this period [4]. However, for applications in the UV to near-IR range of the spectrum, where low birefringence and low scatter is required, single-crystal diamond is preferred. The past ten years have witnessed several key breakthroughs in the development of high-quality, single-crystal CVD diamond for optical and quantum optical applications [5–8], achieved through careful control of the point, extended, and isotopic defects. In the wake of these breakthroughs, new technologies are emerging [9–15] which, from an industrial perspective, show significant potential. Some of these are discussed in greater depth in other chapters of this book.

The purpose of the present chapter is to review the synthesis and properties of polycrystalline and single-crystal CVD diamond in the context of the most relevant issues for the optical engineer. An introductory section on CVD growth principles is aimed at providing the reader with a basis for understanding how synthesis feeds into microstructure, material properties, and performance. Key optical properties such as absorption, scatter and birefringence are reviewed and related to the most common defects in CVD diamond. The technological applications of diamond usually combine two or more of its extreme properties. Hence, from an optical engineering perspective it is important to understand CVD diamond's thermal and mechanical properties, and how they differ by commercial grade and between polycrystalline and single-crystal diamond; these differences are summarized. Finally, several established commercial optical applications of CVD diamond are described, which demonstrate how diamond's properties can be combined to enable a variety of high-performance optical technologies.

2.2
CVD Diamond Growth Principles

The gas-phase synthesis of diamond at low pressures was first demonstrated in 1952 by William Eversole [16], predating HPHT synthesis by a few years [17, 18]. This early CVD process yielded relatively low growth rates, whereas the HPHT method led to diamond synthesis on an industrial scale. It was not until the 1980s that a breakthrough in CVD diamond growth was made in Japan [19], leading to respectable growth rates and stimulating renewed interest. The current understanding of state-of-the-art diamond growth from gas-phase precursors is briefly summarized in the following section.

2.2.1
Fundamentals of Growth

Most methods of CVD diamond synthesis utilize a hydrocarbon source gas, typically methane, diluted in hydrogen. Growth takes place at pressures ranging from a few tens to several hundred Torr, and at substrate temperatures around 700 to

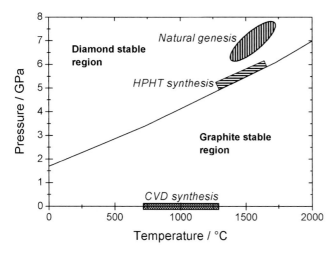

Figure 2.1 Carbon phase diagram. The regions of pressure–temperature space corresponding to metastable CVD synthesis, HPHT synthesis, and natural genesis are indicated schematically [2].

1200 °C. Under these conditions, while graphite is the thermodynamically stable allotrope of carbon (Figure 2.1), diamond synthesis is possible based on the kinetics of the growth surface.

Atomic hydrogen plays the key role in facilitating the preferential deposition of diamond over graphite by terminating the diamond surface during growth, thus preventing sp^2-type surface reconstruction [20, 21]. In addition, atomic hydrogen efficiently etches away any sp^2-bonded phases at typical CVD growth temperatures.

In order to provide the reactive species needed for high-quality diamond growth, the feed gas mixture must first be activated. In fact, it is the activation method used that differentiates between the various CVD growth techniques and reactor technologies. These include hot filament, arc-jet, microwave plasma, combustion flame and others (a comprehensive review is available in Ref. [22]). In an industrial context, CVD diamond deposition requires a versatile and robust reactor technology that is capable of growing high-quality material at economic growth rates over large areas. Currently, microwave plasma chemical vapor deposition (MPCVD) provides the most effective route to achieving these requirements for the majority of applications.

In MPCVD, a microwave generator (typically capable of delivering kW of power) provides the power source to drive the gas-phase chemistry via the production of a plasma. The reactor system must be carefully designed to provide optimum performance in terms of the efficient coupling of microwave energy, to achieve the appropriate shape and position of the plasma with respect to the deposition area, and to attain the high gas temperatures necessary to create the key species for high-quality growth [23].

Electromagnetic energy from the generator is transmitted via a waveguide to the reactor vessel, where it is coupled in by either electric or magnetic coupling schemes. The reactor geometry itself is chosen to provide a resonant cavity for a given electromagnetic mode which, when excited at its resonant frequency, provides the high electric fields necessary to ignite the plasma. Thus, plasma ignition is achieved and sustained without the use of electrodes, which might otherwise act as a source of material contamination. In order to maximize the flux of growth species and to attain a uniform deposition, the plasma should be close to the deposition surface. Consequently, the chamber is designed to provide an electric field antinode above the substrate, and with a microwave-transparent dielectric window that confines the chamber gases to a volume above the deposition region. In this way, parasitic plasma ignition at other electric field antinodes is avoided.

Although a detailed description of the gas-phase and surface chemical reactions involved in CVD diamond growth is beyond the scope of this chapter (see Refs [24–26]), the key processes are outlined here.

The importance of atomic hydrogen in preventing graphitic deposition, and therefore in growing diamond of high crystalline quality, has already been noted. Atomic hydrogen also facilitates all of the critical gas-phase and growth surface reactions. The mechanism of producing atomic hydrogen in the plasma varies according to the gas pressure; at lower pressures the dissociation of molecular hydrogen occurs primarily via collisions with energetic electrons, whilst at higher pressures thermal dissociation dominates. Atomic hydrogen reacts with the hydrocarbon source gas in a series of hydrogen abstraction or addition reactions of the form:

$$\text{H-abstraction: } CH_x + H \rightleftharpoons CH_{x-1} + H_2 \; (x = 4-1) \tag{2.1}$$

$$\text{H-addition: } CH_{x-1} + H + M \rightleftharpoons CH_x + M \; (x = 4-1) \tag{2.2}$$

where M is a third body. (The formation of higher hydrocarbons via the combination of methyl radicals and further hydrogen abstraction/addition reactions also occurs but is not considered here.) The plasma ball is hottest in the middle, where it reaches temperatures in excess of 2500 K in a typical high-performance MPCVD reactor, and decreases in temperature from the center to the edge. It has been calculated [23, 27] that the concentration of methyl (CH_3) radicals, which generally are accepted as being the key hydrocarbon species responsible for carbon addition [28, 29], are highest in the region where the gas temperature varies between around 1300 and 2200 K, typically a few millimeters above the deposition surface.

To date, the CVD growth of single-crystal diamond of any significant size has only been demonstrated by homoepitaxy on single-crystal diamond substrates. Growth on nondiamond substrates leads to polycrystalline diamond which can, through a careful choice of nucleation and growth conditions, be engineered such that large, well-oriented grains are produced (textured, highly oriented polycrystalline diamond [30–32]). However, due to a lack of reported optical characterization data for this material, it will not be discussed further.

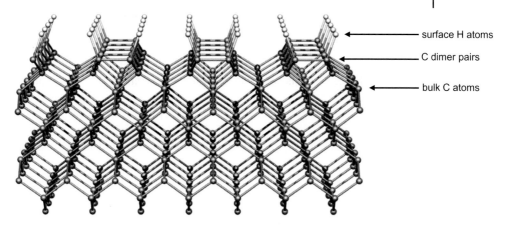

Figure 2.2 C{100}:H 2 × 1 surface viewed along a <110> direction. Courtesy of A.K. Tiwari (private communication).

For commercial, optical quality, single-crystal CVD diamond, homoepitaxial growth is usually performed on {100}-oriented substrates. During growth, the high flux of atomic hydrogen leads to a hydrogenated diamond surface. Steric effects result in a C{100}:H 2 × 1 surface reconstruction [33, 34], in which surface carbon atoms form rows of dimer pairs, resulting in an uppermost five-membered ring, as shown in Figure 2.2.

Each C–H surface site undergoes many collisions per second with H radicals, which leads to a constant exchange of hydrogen between the gas phase and the surface. As such, about 1 to 10% of the surface consists of unterminated surface radical sites, into which new carbon atoms can be inserted. In a recent theoretical study [35], the energetics associated with the insertion of a CH_2 growth unit into a C–C dimer bond on the C{100}:H 2 × 1 surface were calculated; the reaction pathway for this process is shown in Figure 2.3. The same group also calculated that the surface migration of CH_2 groups, either along dimer rows or between adjacent dimer rows, is relatively facile in either direction. Earlier calculations also predicted the surface migration of CH_2 [36]. These surface migration mechanisms are mediated through processes of bond breaking and reformation. The surface diffusion of physisorbed adatoms is deemed unlikely, due to the relatively high growth temperatures involved. Thermally activated surface migration of growth units should allow for a degree of control over the surface roughness of as-grown diamond films by varying both the flux of carbon and the surface temperature. This has indeed been observed experimentally both by the present author's group (as illustrated in Figure 2.4) and by others [37, 38].

The CVD growth of polycrystalline diamond has been demonstrated on numerous nondiamond substrate materials; those that form refractory carbides (silicon, molybdenum, tungsten, tantalum, and others) are generally preferred. The process

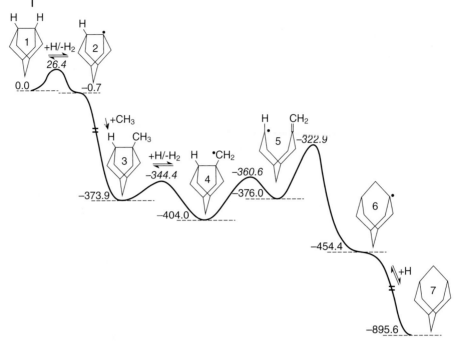

Figure 2.3 Reaction path for incorporating a CH$_2$ group into a C–C dimer bond. Energies (B3LYP QM/MM, 6-311G** basis set) are quoted in units of kJ mol^{-1}, relative to that of structure **1**. Reprinted with permission from Ref. [35]; © 2008, American Chemical Society.

of surface carburization upon exposure to the plasma is thought to facilitate the nucleation of diamond [39–41]. Once formed, the individual nuclei increase in size and coalesce; growth then proceeds according to the van der Drift model of competitive grain growth [42]. The result is a fully dense diamond layer with broadly columnar, well inter-grown grains (Figure 2.5).

During growth, dopant impurities (such as boron) can be deliberately introduced in a controlled manner. Impurities can also be incorporated in an uncontrolled manner, such as those entering through air leaks due to poor vacuum integrity or as impurities in the source gases. Even without these sources of uncontrolled impurities, some level of defect formation is unavoidable in CVD diamond growth. A simplified surface reaction model for a standard carbon/hydrogen chemistry [43] predicts that the concentration of point defects forming in the diamond lattice during growth, such as sp^2 species, is scaled according to the ratio $G/[H]^2$, where G is the growth rate and $[H]$ the atomic hydrogen flux. As well as quantifying the relationship between quality and atomic hydrogen, this ratio illustrates that in order to grow high-quality material, a trade-off in growth rate is generally required.

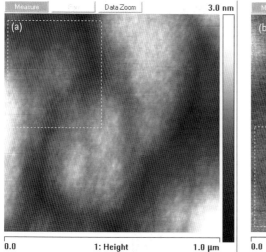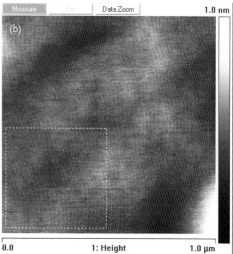

Figure 2.4 Atomic force microscopy images over a $1 \times 1\,\mu m^2$ area of a {100} as-grown CVD diamond film grown under conditions of (a) 2.0% CH_4 in H_2 and substrate temperature <800 °C and (b) 0.8% CH_4 in H_2 and substrate temperature >900 °C. Values of the rms roughness measured were 0.43 and 0.08 nm, respectively. For a color version of this figure, please see the color plate at the end of this book.

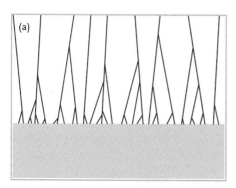

Figure 2.5 (a) Schematic illustration of competitive, columnar grain growth; (b) Scanning electron microscopy image of the surface of an unpolished, optical quality CVD layer.

2.2.2
Morphology and Texture

Crystal morphology evolution during single-crystal diamond growth, and texture evolution during polycrystalline growth, is determined by the relative growth rates of the most commonly occurring low-index crystallographic planes, namely {100} and {111} (and to a lesser extent {110} and {113}) [44–47]. In essence, the slowest growing crystal planes tend to dominate the final morphology after long growth times. For the purposes of morphology classification and modeling, it is useful to define a set of linear growth rate ratios. By convention, the growth rate of each plane, G_{hkl}, is normalized by the modulus of the corresponding direction vector (e.g., $G_{111} \rightarrow G_{111}/\sqrt{3}$, and so on), and the ratios are taken with respect to the {100} growth rate:

$$\alpha \equiv \frac{\sqrt{3}G_{100}}{G_{111}}, \quad \beta \equiv \frac{\sqrt{2}G_{100}}{G_{110}}, \quad \gamma \equiv \frac{\sqrt{11}G_{100}}{G_{113}}. \quad (2.3)$$

In practice, these parameters are dependent on CVD growth conditions such as the CH_4 mole fraction and growth temperature. For instance, material grown under conditions corresponding to high α ($\alpha \geq 3$) and low β and γ produces a morphology in which {111} planes eventually dominate, irrespective of the initial morphology. The results of theoretically modeling the morphology development of single-crystal growth under these conditions are illustrated in Figure 2.6.

For single-crystal CVD diamond growth at a fixed substrate temperature, it was reported [48] that α, β, and γ were all increased as the methane concentration was increased from 4% to 7%. Conversely, under conditions of fixed methane concentration, these parameters all decreased as the substrate temperature was increased from 750 to 950 °C. The γ parameter was reported to be the most sensitive function

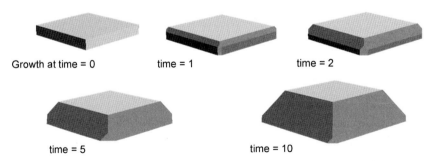

Figure 2.6 Theoretical morphology development under conditions of $\alpha = 4$, with low β and γ from a {100}-oriented, <110>-sided substrate (time = 0). At time > 0 {100} facet growth is shown in white, {111} in green, and {110} in pink. Morphology model courtesy of DTC Research Center, Maidenhead [2]. For a color version of this figure, please see the color plate at the end of this book.

Figure 2.7 Morphology development and surface area increase during growth of {100} single-crystal diamond.

of conditions. Knowledge of how these growth rate ratios depend on reactor conditions can allow a significant degree of control over morphology development during growth, such as choosing conditions that lead to an increase in the surface area, as illustrated in Figure 2.7.

Texture development during the CVD growth of polycrystalline diamond is also governed by relative values of α, β, and γ. Initially, crystallites of different orientations nucleate on the substrate, growing both laterally and vertically. Subsequently, those crystallites oriented such that the fastest growth direction points along the vertical direction will overtake and overgrow other crystallites that are less favorably oriented. Polycrystalline diamond films grown under $\alpha = 1$ conditions will possess predominantly {100} facets with a <111> texture, while those grown under $\alpha = 3$ will have mainly {111} facets with a <100> texture. The average lateral grain size increases monotonically, from sub-micron at the nucleation face to many hundreds of microns at the growth face for a 1 mm-thick layer. This anisotropy – which is due to both the columnar nature of the grains and the increase in the lateral grain size with thickness – leads to anisotropies in several key physical properties of polycrystalline diamond layers, such as lateral versus vertical thermal conductivity, or nucleation versus growth face strength. These characteristics are discussed in more detail in Section 2.3.6.

2.3
Properties of Optical Quality CVD Diamond

For optical applications of CVD diamond in which the diamond is an optical element of the system, material of high crystalline quality is required. Although the detailed requirements may be application-specific, diamond with low optical losses is generally desirable. For example, even if the application is harnessing a combination of diamond's optical, thermal, and mechanical properties (as is often the case), the use of a mechanical or thermal grade of diamond would usually be insufficient, due to the higher optical losses associated with these lower-cost diamond grades. On the other hand, higher-cost electronic-grade material, which is characterized by the very lowest point defect densities (such as a single substitutional nitrogen concentration <5 ppb) may be overspecified for many optical applications. It is vitally important to understand the key optical properties of CVD diamond – and the factors that affect them – in order to develop strategies to control defect formation and produce material that is fit for purpose.

In the following section, consideration is given (in the context of CVD synthesis) to the key linear optical properties of diamond. Several nonoptical properties of CVD diamond (e.g., mechanical and thermal properties) are also described, due to their relevance to most optical applications. In all cases these properties are strongly influenced by one or more types of defect.

2.3.1
Absorption

The absorption spectrum of intrinsic diamond, from the UV to far IR, is illustrated in Figure 2.8. The absorption edge at 225 nm corresponds to the indirect band gap of diamond, while an IR absorption band between around 2.5 to 6.5 μm exists due to two- and three-phonon excitation. The strength of absorption in this region has been shown to be temperature-dependent [50, 51]. Single-phonon excitation is forbidden in intrinsic diamond by symmetry considerations. Beyond the phonon absorption band, transmittance measurements on optical polycrystalline CVD diamond windows found the material to be transparent up to at least 500 μm [52].

Point defects and defect complexes lead to additional absorption features throughout the UV-visible-IR range, and also lead to a lifting of the restriction of the one-phonon absorption band, centered at around 7.5 μm (1332 cm^{-1}). A vast array of point defects in diamond and their optical properties have been cataloged [53]. In the context of the present chapter, however, many of these can be ignored

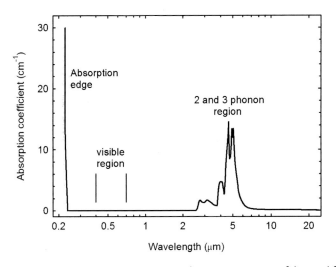

Figure 2.8 Room temperature intrinsic absorption spectrum of diamond from the UV to long-wave IR. Reproduced with permission from Ref. [50]; © 2001, Gemmological Association of Great Britain.

as they are specific to HPHT synthetics or to natural diamond. Also excluded are defects that are added intentionally either to alter the electrical properties (e.g., boron or phosphorus) or to introduce color centers of interest for particular quantum optics applications (see Chapters 5, 6, and 9).

For the purposes of the present discussion, the extrinsic absorption characteristics of optical quality CVD diamond are determined predominantly by the effects of nitrogen. It is well known that nitrogen added to the gas phase can increase the growth rate [54, 55]; however, nitrogen also influences the optical properties, and accordingly a balance must be struck. In CVD diamond, even a few hundred parts per billion of nitrogen in the solid can lead to effects in the growth process which result in broad band absorption peaks centered at around 270, 365, and 520 nm [56], with an underlying broad "ramp" of increasing absorption with decreasing wavelength. The feature at 270 nm is associated with single substitutional nitrogen in its neutral charge state (N_s^0), while the origins of the 365 and 520 nm features are currently unclear. The absorption ramp has been attributed to the formation of vacancy clusters [57, 58]. Taken together, these latter absorption features lead to an overall brown coloration in CVD diamond, the strength of which can (depending on the growth conditions) be correlated with the amount of added nitrogen. However, the precise growth mechanisms that lead to the brown-related defect (or defects) are not well understood.

The relationship between low levels of nitrogen incorporation and absorption in the UV-visible range is illustrated in Figure 2.9. It can be seen that, in the sample with a lower nitrogen concentration, the absorption coefficient across this spectral range is significantly reduced. More accurate measurements of the absorption coefficient can be made using laser calorimetry [59, 60], and this method was

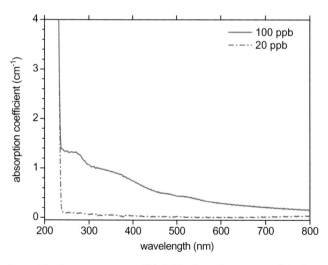

Figure 2.9 Room temperature UV-visible absorption spectra of single-crystal CVD diamond containing approximately 100 ppb and 20 ppb of nitrogen.

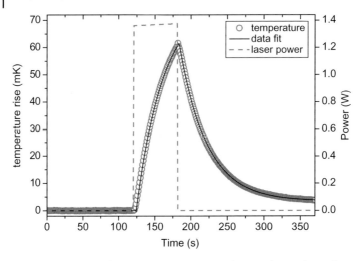

Figure 2.10 Typical calorimetry temperature rise of a CVD diamond sample (as reported in Ref. [7]).

applied at the key laser wavelengths of 532 and 1064 nm on a series of as-grown single crystal CVD samples with nitrogen concentrations of approximately 20, 50, and 100 ppb [7]. A typical temperature trace from a calorimetry measurement made at 1064 nm using a Nd:YVO$_4$ laser at a fixed power for 60 s is shown in Figure 2.10. When the absorption coefficients extracted from the calorimetry data were plotted (see Figure 2.11), it was clear that at both wavelengths there was a strong correlation between the absorption coefficient and the incorporated nitrogen under the conditions of growth investigated. Absorption was observed to be stronger at 532 nm, most likely due to the nearby 520 nm band, while at 1064 nm there is no known discrete absorption feature in CVD diamond. However, due to the correlation with nitrogen, it might be speculated that absorption at this wavelength could be ascribed (at least in part) to the tail of the broad absorption ramp. Alternatively, it might be related to a hydrogen-induced absorption at 1356 nm. For the samples with the lowest concentration of nitrogen, the absorption coefficient was measured to be at or below 10^{-3} cm^{-1} at 1064 nm, and $<5 \times 10^{-3}$ cm^{-1} at 532 nm. This represented a significant improvement on previous values measured on single-crystal diamond, and would be sufficiently low for many optical applications.

The absorption coefficient of optical quality diamond has been well characterized at 10.6 μm due to the importance of diamond as a material for CO$_2$ laser optics. The room temperature absorption coefficient of natural Type IIa and CVD single-crystal diamond has been measured at about 0.02 to 0.03 cm^{-1} [61–63]. These values probably represent an intrinsic lower limit given by the tail of the two-phonon absorption band. In optical polycrystalline CVD diamond, the presence of "black spot" microfeatures increases this to around 0.05 cm^{-1}, a value

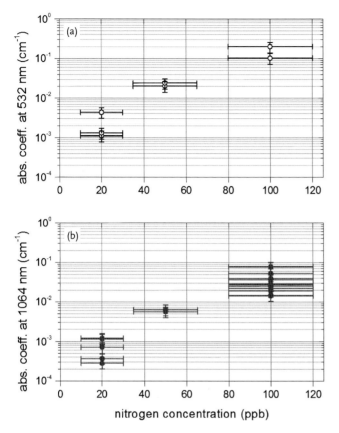

Figure 2.11 Absorption coefficients determined at (a) 532 nm and (b) 1064 nm using laser calorimetry for single-crystal CVD diamond samples as a function of nitrogen concentration.

that has been proven to be perfectly acceptable for commercial CO_2 laser systems operating at kW output powers, and which is routinely achievable. This is demonstrated in Figure 2.12, which shows the distribution of absorption coefficients measured at 10.6 μm on 57 optical polycrystalline diamond output couplers, manufactured sequentially by Element Six. The mean absorption coefficient of these samples was found to be 0.055 cm^{-1} with a standard deviation of 0.010 cm^{-1}.

2.3.2
Nonoptical Wavelengths

Synthetic diamond is of increasing interest for optical applications at X-ray wavelengths [64, 65]. For synchrotrons in particular – where the trend is towards higher brilliance X-ray beams – the high thermal conductivity of diamond, the low thermal expansion coefficient and low X-ray absorption of diamond make it an attractive

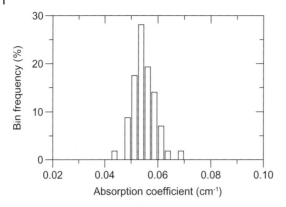

Figure 2.12 Distribution of absorption coefficient at 10.6 μm measured on 57 optical polycrystalline samples of size Ø25 mm × 1 mm.

Table 2.1 Absorption coefficient (α) of diamond at several common laboratory and synchrotron X-ray wavelengths (λ).

Source	Ag Kα	Mo Kα	Synchrotron	Cu Kα	Synchrotron	Cr Kα
λ (nm)	0.056	0.071	0.100	0.154	0.200	0.229
α (cm^{-1})	1.315	2.025	4.640	15.85	34.66	52.73

material for X-ray optics such as windows, polarizers, filters, beam splitters, and monochromators. High-quality HPHT IIa crystals have been demonstrated by Element Six with regions free of dislocations and stacking faults of up to around 4 × 4 mm [66]. The absorption coefficient of diamond has been tabulated at various X-ray wavelengths corresponding to popular laboratory sources and two common synchrotron wavelengths [67]; these data are listed in Table 2.1.

Due to its transparency and other extreme properties, CVD diamond is of interest as a high-power terahertz material. Current terahertz free electron lasers can produce continuous-wave output powers of up to several hundred watts, and diamond could potentially be used as a window or beam splitter material [68]. Calorimetric absorption measured at 130.3 μm on an optical quality polycrystalline CVD diamond disc of diameter 55 mm yielded an absorption coefficient of 0.067 ± 0.003 cm^{-1}.

For transmission at microwave frequencies (millimeter wavelengths), the key parameters are the dielectric constant ($\varepsilon_r \approx 5.7$ between 10 and 200 GHz) and the dielectric loss factor tan δ, as these determine the power reflected and absorbed by the window. The reflected power of a multi-passband window can be minimized by ensuring that the window thickness is an integer times $\lambda/(2\varepsilon_r^{1/2})$ [69], where λ

is the free space wavelength. As an alternative ellipsoidal Brewster windows can be used. At microwave frequencies, dielectric loss factors less than 10^{-5} at 145 GHz have been reported [70]. Good homogeneity was demonstrated on a polycrystalline CVD diamond disc of size Ø100 × 1.6 mm, while temperature-dependent measurements performed between 70 and 370 K showed no significant variations.

2.3.3
Isotopic Purity

As described in detail elsewhere in this book, several point defects in diamond – such as the nitrogen-vacancy center – are, by virtue of their spin properties, of great interest for quantum optical applications. For several practical applications, the spin dephasing time (T_2) should be maximized, ideally at room temperature, and this can be facilitated by growing diamond with a low concentration of paramagnetic defects. Even in single-crystal diamond with very high chemical purity, spin decoherence of the nitrogen-vacancy center occurs via coupling to the nuclear spin of the ^{13}C isotope, which has a naturally occurring abundance of around 1.1%. Recently, the growth of both chemically *and* isotopically pure diamond has been achieved [8]. Through the use of a ^{12}C isotopically enriched CH_4 source gas and optimized CVD growth conditions, single-crystal diamond with only 0.3% ^{13}C was attained simultaneously with a total nitrogen level of the order of 0.05 ppb. This material led to the longest room temperature spin dephasing time ever measured in a solid-state system ($T_2 = 1.8$ ms), and represents a key breakthrough in the realization of future diamond-based quantum optical technologies.

2.3.4
Strain-Induced Birefringence

Diamond is a cubic material and should therefore be optically isotropic. In practice, however, the presence of strain – typically from grown-in defects – leads to birefringence [71, 72]. In polycrystalline CVD diamond, the stress between individual grains resulting from competitive grain growth is relatively high, and consequently polycrystalline diamond is not suitable for applications that require low levels of birefringence.

In single-crystal CVD diamond, several sources of strain can be present. Lattice mismatch between the substrate and epilayer causes bulk strain, but this is easily addressed by removing the substrate. The differential uptake of point defects within the CVD layer may produce strain due to spatial variations in the lattice parameter; however, in optical quality diamond the point defect densities are usually low enough that this does not play any significant role. On the other hand, strain due to extended defects can lead to significant birefringence. Of these extended defects, crystal twins (a planar defect) and dislocations (a line defect) are the most relevant in as-grown single-crystal CVD diamond [73, 74].

In diamond, twinning occurs on {111} planes, whereby the twinned crystal is rotated about the <111> direction by 180° (or equivalently by 60°) with respect to the parent crystal. The relatively low energy for twin formation (around $0.14\,\mathrm{J\,m^{-2}}$ based on estimates of half the single stacking fault energy) [75] means that twins form readily at typical CVD growth temperatures on any developed {111} planes, leading to incoherent lattice boundaries between neighboring twinned and untwinned regions [76] and subsequent defect formation. The apparent twinning on {100} surfaces (penetration twins) is attributed to the formation of {111}-facetted etch pits during growth, which subsequently twin. Once formed, twinned sectors can grow into macroscopic defects that exert a strain on the surrounding lattice; thus, their formation must be controlled during the synthesis of low birefringence material. Any twinned lateral {111} growth sectors can be removed post-growth by laser sawing.

For homoepitaxial CVD growth on (001)-oriented single crystal substrates, dislocations tend to have a line direction close to the [001] growth direction, although local inclinations in the growth surface can produce deviations toward a [101] line direction [77, 78]. Dislocations in CVD diamond either originate at the epitaxial interface as a result of lattice errors caused by substrate processing damage or surface contamination, or are caused by dislocations in the substrate that subsequently propagate into the CVD growth layer. X-ray topography studies of CVD diamond grown on {100} Type Ib HPHT substrates have shown that two dislocation types predominate: pure edge, and 45° mixed:

- Pure edge dislocations were reported to nucleate in isolated clusters at the substrate interface, although the precise origin of these types was unclear. Such dislocations produce a characteristic petal-shape birefringence pattern that, depending on the dislocation cluster size, can extend over several hundred microns [79].

- The 45° mixed dislocations were found to nucleate in pairs at the substrate–epilayer interface, and were associated with mechanical processing damage of the substrate.

Dislocations present in the substrate continue through into the epitaxial layer. By contrast, stacking faults, which occur commonly in HPHT substrates, were found not to continue propagating into the CVD growth layer. Rather, the line of intersection of the {111} stacking fault plane with the (001) growth interface leads to a line of [001] dislocations in the CVD layer [6, 80].

For light of wavelength λ passing through a crystal of thickness d and birefringence Δn, the phase shift δ between the components of light polarized along the local fast and slow axes is given by:

$$\delta = \frac{2\pi}{\lambda}(\Delta n)d. \tag{2.4}$$

Single-crystal CVD diamond typically exhibits a nonuniform birefringence pattern with $\Delta n > 10^{-4}$ possible in regions containing dense clusters of dislocations. The distribution of dislocations in CVD diamond can be such that the principal strain axes – and hence the slow axis – varies spatially. This is illustrated in Figure 2.13a,

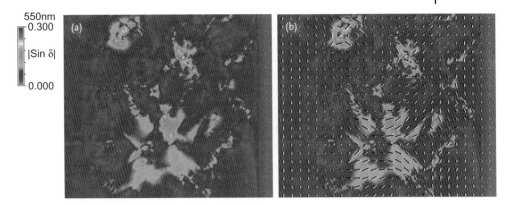

Figure 2.13 (a) |sin δ| false color image of a 2 × 2 × 2 mm³ free-standing {100} CVD diamond plate illuminated along the <100> growth direction with λ = 550 nm light; (b) The same image overlaid with a grid of vectors representing the relative magnitude and direction of the local slow axis. For a color version of this figure, please see the color plate at the end of this book.

b, which show birefringence micrographs (|sin δ| false color images [81]) of a free-standing {100} CVD crystal viewed along the growth direction. The effects of a birefringence distribution of this nature cannot be completely negated by rotationally aligning the crystal. In any case, for some optical applications it may be desirable to fix polarization along a given crystallographic direction (such as Raman lasers; see Chapter 8).

It has been clearly demonstrated that birefringence in single-crystal CVD diamond can be reduced significantly through a combination of:

- Performing CVD growth on substrates that contain low densities of extended defects such as dislocations and stacking faults.
- Growing on substrates with reduced levels of surface and sub-surface damage.
- Processing the finished CVD diamond optic with care to avoid generating strain post-growth.
- Designing the optic such that beam propagation is perpendicular to the CVD growth direction, and hence the dislocation line direction.

The effect of reducing the dislocation density on birefringence is demonstrated in Figure 2.14, which shows {111} X-ray projection topographs and |sin δ| maps imaged parallel and perpendicular to the CVD growth direction, of three {100}-oriented, free-standing CVD blocks of size ~3 × 3 × 2 mm³. As shown in the X-ray topographs, samples A, B, and C exhibit a decreasing density of dislocations, which clearly results in reduced levels of retardation. The anisotropy in |sin δ| is also apparent. Sample C, when viewed along the growth direction, exhibits |sin δ| of up to 0.85 within the strain field of a large isolated cluster of dislocations; this is equivalent to a birefringence averaged over the path length of 4×10^{-5}. The same

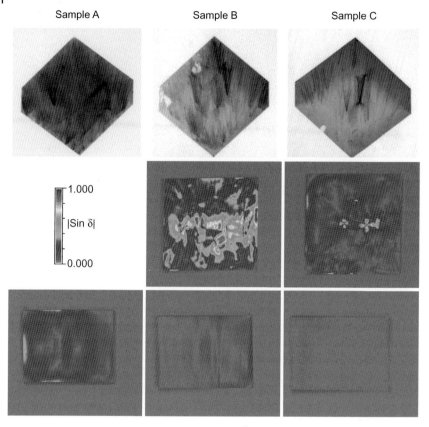

Figure 2.14 Free-standing, {100} CVD blocks imaged using {111} X-ray projection topography (top row) and birefringence microscopy as viewed parallel (middle row) and perpendicular (bottom row) to the direction of growth. |sin δ| > 1 parallel to growth direction for sample A (not shown). For a color version of this figure, please see the color plate at the end of this book.

sample, when viewed perpendicular to the growth direction, exhibits a birefringence averaged over the path length of $<1 \times 10^{-6}$.

The ability to control extended crystal defects such as twins and dislocations has enabled the production of a consistent material that is suitable for many applications that require low birefringence.

2.3.5
Scatter

For optical applications of CVD diamond at shorter wavelengths (UV to near-IR), the level of scatter leads to a clear differentiation between polycrystalline and

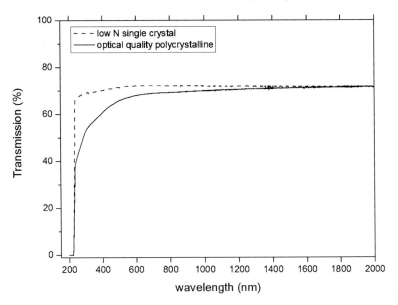

Figure 2.15 UV-visible-near IR transmission spectrum of optical grade polycrystalline and low N (~20 ppb) single-crystal CVD diamond. The samples measured were uncoated, with thicknesses of 1.1 and 1.7 mm, respectively. The accuracy of the measurement was about ±1%.

single-crystal material. This is illustrated in the transmission spectra shown in Figure 2.15. At wavelengths greater than around 2 μm, the transmission in these samples is very similar, and is essentially limited by total reflection, which puts the maximum transmission at approximately 71%. However, below a wavelength of about 1 μm, transmission through the polycrystalline sample falls off relatively rapidly due to an increase in scattering loss. This characteristic of polycrystalline CVD diamond was quantified in an early study [82]. Measurements reported on a wide range of CVD polycrystalline samples of different grades showed that high-angle scatter was correlated with the density of micro-crack features (as discussed in Section 2.3.6.2), while low-angle scatter was attributed to strain [83]. At both 1064 nm and 10.6 μm, the contribution to optical loss from scatter was found to exceed the loss due to absorption. In an optical grade sample of thickness 0.7 mm, the total forward scatter was found to be 3% at 1064 nm, and 0.5% at 10.6 μm, in broad agreement with data reported elsewhere [82].

In single-crystal diamond free from macroscopic defects, much lower levels of scatter have been reported. A total forward scatter at 1064 nm, ranging from 0.04 to 0.63%, was reported for a set of 15 single-crystal samples of different sizes and geometries that were supplied by different sources. In a study undertaken by Element Six [63, 84] on several CVD single-crystal and natural plates polished into windows, total forward scatter at 1064 nm ranging from 0.02 to 0.67% was obtained. In these studies, scatter was determined by measuring the total power in the forward hemisphere at angles greater than 2.5° to the direction of the beam. No clear correlation of scatter with sample thickness, depolarization loss or surface roughness was apparent, and the origin and observed variation of scatter in single-crystal diamond remains unclear. Nevertheless, scatter in single-crystal diamond is clearly much lower than in polycrystalline diamond at shorter wavelengths. To the present author's knowledge, scatter has not been reported to be a limiting factor in optical applications.

In summary, the lower scattering and depolarization losses measured in single-crystal diamond at shorter wavelengths mean that it is generally preferred to polycrystalline diamond for UV, visible, and near-IR systems. For longer wavelength applications such as CO_2 laser optics, or mid- to long-wave IR imaging, the scattering losses are sufficiently low in optical quality polycrystalline diamond that the availability of large-area wafers (flat and domed) is generally advantageous.

2.3.6
Other Properties of CVD Diamond

The relevant thermal and mechanical properties of CVD diamond are reviewed in the following subsections, and related to the microstructure of single-crystal and polycrystalline materials.

2.3.6.1 Thermal Properties

Diamond has the highest room temperature thermal conductivity of any material, with a value around fivefold that of copper, and typically orders of magnitude greater than that of other common optical materials such as sapphire, ZnSe, or Nd:YAG. A high thermal conductivity can lead to reduced temperature gradients, either in the diamond itself or in another optical component bonded to diamond. This is critical in high-power laser applications, in which thermally induced variations in the refractive index can cause detrimental effects such as thermal lensing in laser crystals or wave-front distortion in windows. Temperature gradients may also lead to mechanical strain, which can produce form-shape aberrations and, where component cooling is insufficient, mechanical failure.

Diamond's thermal properties have been extensively reviewed [2, 85] and are considered, in the context of optical engineering, in Chapter 11 of this book. At this point, the basic material factors that influence the thermal conductivity of optical polycrystalline diamond will be briefly reviewed and contrasted with other grades of CVD diamond, and also with high-quality Type IIa natural diamond.

As an electrical insulator, thermal conductivity in diamond is governed by phonon – not electron – transport processes. The high Debye temperature ensures extremely high values of room temperature thermal conductivity in high-quality diamond, while any lattice impurities reduce thermal conductivity due to phonon scattering. The naturally occurring ^{13}C isotopic impurity reduces thermal conductivity by around 30% [86] compared to isotopically pure (^{12}C) diamond. In polycrystalline diamond, additional phonon scattering occurs at grain boundaries, such that the grain size can significantly affect low-temperature thermal conductiv-

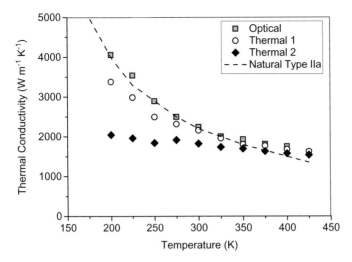

Figure 2.16 Thermal conductivity (measured "through plane") of optical and thermal grades of polycrystalline CVD diamond as a function of temperature. The dashed line provides a comparison to Type IIa natural diamond.

ity; typically, a smaller grain size will lead to a lower thermal conductivity due to increased phonon scattering [87]. At higher temperatures, however, phonon–phonon scattering is the dominant scattering mechanism, leading to a convergence in the conductivity of all diamond grades with increasing temperature.

The thermal conductivity of optical and several lower-cost thermal grades of polycrystalline CVD diamond was measured as a function of temperature and compared to high-quality Type IIa natural diamond [88]. These measurements were taken "through-plane" (i.e., parallel to the CVD growth direction) using a laser flash technique. The results (see Figure 2.16) showed that, for optical quality polycrystalline CVD diamond, the thermal conductivity was essentially the same as that of high-quality natural Type IIa diamond over the temperature range investigated.

The thermal conductivity of polycrystalline diamond measured in-plane was found to be reduced compared to the through-plane value [89]. Furthermore, the conductivity in both directions was found to be a function of film thickness. These observations were explained by the columnar grain structure, leading to anisotropic and thickness-dependent grain boundary scattering of phonons. However, in another study [90] on optical and near-optical quality samples, in-plane versus through-plane thermal conductivities measured at room temperature were found to be in agreement, within the accuracy of the measurements (8%).

It should be borne in mind that diamond's excellent thermal characteristics can only be fully harnessed if sufficient heat-sinking is incorporated. The ability, therefore, to form low thermal resistance, robust mechanical bonds between diamond and other materials (such as metals, alloys and ceramics) is paramount.

Achieving this is never a trivial undertaking, however, and the requirements differ according to the application. Diamond intracavity heat spreaders for disk lasers require a bond to the laser disk that must be not only optically transparent but also thermomechanically robust. Diamond windows for ultrahigh-vacuum systems require mounting and bonding into a metal vacuum flange, and the entire assembly must be capable of withstanding vacuum system baking to several hundred degrees Celsius. For high-power diamond CO_2 laser optics, direct or indirect water cooling is required, and the mount must incorporate this facility. A comprehensive review of the bonding and mounting of diamond is beyond the scope of this chapter (moreover, the methodologies employed are often proprietary), although a brief overview of some of the important considerations is provided in the following subsections.

It is ironic that the extreme properties which make diamond so useful in engineering applications tend to count against it in the context of bonding (chemical inertness, low thermal expansion coefficient, high Young's modulus, high yield stress, low toughness). Furthermore, diamond generally graphitizes or decomposes far below its melting temperature. Diamond does, however, form carbides with several elements such as titanium, silicon, chromium, tungsten, and molybdenum, and this can provide a route for the brazing or soldering of other materials to diamond, by using a judicious combination of materials. Surface preparation is also an important consideration, and the ability to bond can be highly dependent on surface flatness, roughness, and cleanliness. In addition, the diamond surface can be terminated with different elements such as hydrogen and oxygen, which may also influence bonding.

A diamond optic and its mounting configuration may be under stress as a result of the mounting process. It may be desirable to minimize this or, conversely, to engineer stress in to counter any stresses developed under operation. An accurate prediction of the coupled thermal–mechanical–optical response usually requires finite element modeling. For extremely high-power applications where a significant temperature rise is expected, the temperature dependence of key parameters must be considered. In extreme cases, thermal runaway effects are possible. Consider the case of a CVD diamond output coupler for a multi-kW CO_2 laser operating at 10.6 µm. The absorbed power from the laser produces a temperature rise that (for noncryogenic operation) reduces diamond's thermal conductivity, thus further increasing the temperature. This rise in temperature leads in turn to an increase in absorption in the two- to three-phonon band (see Figure 2.8), the long wavelength tail of which produces an increase in absorption coefficient at 10.6 µm, and therefore an increase in absorbed power. A thermal runaway of this nature can lead to mechanical failure of the window, but this scenario is avoidable if the initial heat load can be adequately managed.

2.3.6.2 Strength

The strength of any ceramic depends on the distribution of pre-existing flaws, and is statistical in nature. A calculation of diamond's theoretical strength, based on the Young's modulus, fracture surface energy and equilibrium separation between atoms, gives a value of around 200 GPa. Experimentally, fracture in bulk ceramics always occurs at much lower values due to flaws such as surface or bulk cracks, which act as stress concentrators. According to the Griffiths model for fracture of brittle materials [91], the presence of surface cracks of length c will in practice reduce the fracture strength by around a factor of $\sqrt{a_0/c}$ times the theoretical strength, where a_0 is the atomic equilibrium separation. Thus, if this picture is applicable to diamond, a measured fracture strength of 1 GPa implies a critical flaw size of a few microns, while a measured strength of 100 MPa implies a flaw size of several hundred microns. In diamond, once the applied stress exceeds the strength of the critical flaw, failure occurs by cleavage along {111} planes, as these are the lowest energy planes for fracture [92].

Since a size and spatial distribution of flaws is to be expected, the strength of diamond is not single-valued and must be treated using Weibull statistics [93]. This analysis yields a Weibull modulus, which should always be quoted alongside the average measured strength of a set of test samples, and provides an indication of the consistency of the material's strength (a higher Weibull modulus implies a more consistent strength).

As already discussed, diamond can be synthesized to produce different grades for different applications, while strength can be measured by using different techniques on test samples of various sizes and geometries. Each of these factors plays a role in determining the measured strength of a specimen. As the single-crystal and polycrystalline forms have different microstructures, their strengths must be considered separately. For polycrystalline diamond, the picture is further complicated by the anisotropic grain structure, in which the lateral grain size increases by several orders of magnitude from the nucleation to the growth face. Thus, variations in fracture strength reported by different research groups on materials grown and tested under different conditions are to be expected.

Due to the limited availability and prohibitive cost of large samples, the strength testing of single-crystal diamond on a statistically significant sample set entails the use of samples that are only a few millimeters in size, which makes the measurements more technically challenging. In a study carried out by Element Six [2, 94], the strength of a set of 30 CVD single-crystal samples of nominally the same bulk characteristics was measured using a cantilever beam test. Rectangular specimens of thickness 0.2 to 0.3 mm and lateral size 5 × 3 mm were prepared and mechanically polished to an R_q of around 1 nm. A mean strength of 2860 MPa was determined with a standard deviation of 1200 MPa, and this variation was reflected in a low Weibull modulus of 2.5. The maximum value measured was 5100 MPa. While these results clearly showed that single-crystal diamond can be a very strong material, further investigations will be required to understand and minimize the variations observed.

For polycrystalline CVD diamond, the fracture strength has been reported for a range of bulk materials of different thicknesses (>50 μm) tested under different configurations. In all instances the average fracture strength was found to be markedly reduced with respect to both natural and synthetic single-crystal diamond.

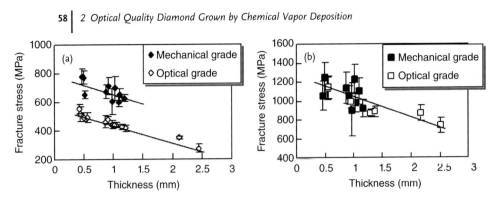

Figure 2.17 Fracture strength of CVD polycrystalline diamond measured as a function of thickness with (a) growth face and (b) nucleation face in tension.

For optical grade polycrystalline diamond, different values of strength have been reported on samples originating from different sources. In a study employing a ring-on-ring flexural strength test [95], the nucleation face was found to have a strength of 300 MPa and a Weibull modulus equal to 3, independent of the sample thickness. In contrast, the growth face was found to have a strength that decreased with sample thickness, falling to less than 100 MPa for samples greater than 0.5 mm thickness.

In further studies conducted by Element Six [96, 97], a series of as-grown, CVD polycrystalline specimens of size 2×18 mm were strength-tested as a function of thickness by using a three-point bend configuration. Optical and mechanical material grades were included in the study, in which both the growth and nucleation faces were placed in tension. The data obtained are plotted in Figure 2.17. The strength of samples with the nucleation face placed in tension was determined to be consistently greater than when the growth face was placed in tension. It was also found that the strength decreased with increasing thickness for both faces. These observations were attributed to an increase in the critical flaw size with thickness due to the increase in lateral grain size from nucleation to growth faces. In other words, flaws within the bulk of individual grains were found to play a key role in the fracture strength. A simple model based on critical flaw size being of the order of the grain size was found to describe the data satisfactorily. The observed increase in the growth face strength of mechanical samples was attributed to the smaller lateral grain sizes, and hence smaller flaw sizes, of this material grade. No clear difference in the nucleation face strength was observed between the mechanical and optical materials, and further studies appeared to indicate that the nucleation face strength is insensitive to the nucleation density [98]. The value of the Weibull moduli (accounting for the thickness variation) for samples ranging from 0.4 to 1.4 mm thick was calculated from the fracture strength data; these data are summarized in Table 2.2.

Evidence supporting the existence of flaws within individual grains was found by examining the fracture surfaces of test specimens, in which both intergranular

Table 2.2 Weibull moduli calculated for data from Figure 2.17a, b for samples of between 0.4 and 1.4 mm thickness.

Polycrystalline CVD diamond grade	Growth face under tension	Nucleation face under tension
Mechanical	11.6	6.5
Optical	23.1	11

Figure 2.18 Micro-fractures ("black spots") visible in a polished cross-section of optical polycrystalline CVD diamond.

and transgranular fracture were observed. Optical microscopy of polycrystalline diamond samples in general reveals the presence of microscopic crack-like features (termed "black spots") that are formed during synthesis within individual grains, and which are most probably a result of within-grain stress (Figure 2.18).

For a polycrystalline CVD diamond window application, it is important to design the system to ensure that the side with greater strength (i.e., the nucleation side) is the side under tension. The requirements on the thickness-to-diameter ratio for a window subject to a pressure differential can be estimated using simple analytical expressions [63, 99].

2.4
Optical Applications of CVD Diamond

Some of the key commercial applications of both polycrystalline and single-crystal CVD diamond are briefly reviewed in the following subsections.

2.4.1
Applications of Polycrystalline Diamond

Polycrystalline diamond windows have been used in commercial, high-power CO_2 laser systems since the mid-1990s [100]. Kilowatt-class CO_2 lasers are generally employed in industrial manufacturing processes for high-speed welding and cutting applications, and the range of diamond optics in these systems includes output couplers (Figure 2.19a), beam splitters, and exit windows. Diamond's superior thermal and mechanical properties allow relatively thin windows to be employed with minimal temperature gradients under operation, which in turn ensures that any thermally induced beam distortion is kept to a minimum. The advantage of using diamond over ZnSe for high-power CO_2 laser optics has been clearly illustrated in finite element analysis simulations [3, 88]. For a 5 kW laser beam at 10.6 µm, the thermo-optical response of a 1 mm-thick diamond window was calculated and compared to that of a 6 mm-thick ZnSe window (the increased thickness is required due to the lower mechanical strength of ZnSe). Both windows were edge-cooled and anti-reflective-coated. ZnSe has a much lower absorption coefficient than diamond at 10.6 µm (~5×10^{-4} cm^{-1}), and the absorbed power was found to be dominated by absorption in the coatings. Calculations indicated that the radial variation in optical path length due to temperature gradients across the ZnSe window was over 240-fold larger than that of the diamond window. Hence, a much lower thermal lensing effect is predicted in diamond compared to ZnSe for high-power CO_2 laser applications. Form shape deviations from perfect flatness may also lead to beam distortion, which is particularly important for laser output couplers.

Figure 2.19 Examples of large-area polycrystalline diamond windows for high-power applications. (a) CO_2 laser output coupler; (b) Double flange mounted transmission window for a gyrotron tube assembly.

The combination of a low dielectric loss, high thermal stability, insensitivity to radiation damage, and excellent mechanical properties ensures that CVD diamond is the material of choice for high-power microwave transmission. Megawatt power gyrotrons operating at between 70 and 170 GHz and containing diamond output windows have been extensively deployed in nuclear fusion research reactors based on electron cyclotron heating. For such applications, the diamond window can be brazed into a vacuum flange assembly for simple edge cooling by water (see Figure 2.19b). Other possible window materials used for high-power gyrotron tubes, such as sapphire- or gold-doped silicon, either require sophisticated or impractical cooling systems, or may suffer from detrimental thermal effects due to variations in the dielectric loss factor and dielectric constant with temperature [69, 101].

CVD diamond has long been of interest for military thermal imaging applications in harsh environments, where delicate IR detectors require protection by a robust, IR-transparent window. For example, heat-seeking missiles fitted with IR-transparent domes must be able to withstand erosion under the impact of high-velocity particles such as water droplets, dust and even insects, and also be able to survive the aerothermal heating that occurs upon launch. Compared to other materials transparent in the 8–14 µm band (such as ZnS or germanium), much lower levels of sand erosion damage are measured in polycrystalline CVD diamond. Single-crystal and polycrystalline diamond also exhibit superior damage threshold velocities under waterjet impact. Polycrystalline CVD diamond dome segments and full hemispherical domes up to 70 mm in diameter have been successfully fabricated, with good thickness and crystal morphology uniformity. For further information on this subject, the reader is referred to Refs [3, 4, 99, 102] and references therein.

2.4.2
Applications of Single-Crystal Diamond

The availability of optical quality, single-crystal CVD diamond as a consistent, engineered material, and the ability to process this material into a range of geometric shapes (Figure 2.20) is stimulating the emergence of a range of new

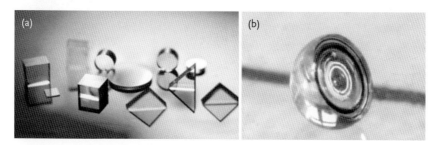

Figure 2.20 (a) A selection of single-crystal diamond prisms, windows and etalons; (b) A single-crystal diamond solid immersion lens.

technologies that have great potential for commercial application. In addition, optical components which historically have been created from natural diamond are now being replaced by the more consistent CVD material.

An example of an established commercial application of single-crystal CVD diamond which previously was based on natural diamond, is that of attenuated total reflectance (ATR) crystals for IR spectroscopy [103, 104]. In conventional IR spectroscopy, substances are analyzed in transmission mode; for solids, this often requires the preparation of a thin sample or for the sample to be ground into a powder and dispersed in a matrix material, while in the case of liquids a thin film can be prepared within a transparent sample cell. The prepared sample is then mounted in the spectrometer for analysis, usually in the laboratory. By contrast, in the ATR method, the IR probe beam is totally internally reflected off the surface of the ATR crystal, which in turn is placed in direct contact with the material to be analyzed. The interaction of the evanescent wave with the sample as a function of wavelength is detected and an IR spectrum produced. In many situations the ATR method is preferred as it does not require sample preparation, which can be both impractical and inconsistent. CVD diamond ATR crystals have significant advantages over other IR materials due to diamond's superior scratch resistance, chemical inertness and wide IR transparency, which enables the measurement of hard surfaces or chemically hostile environments, without degradation. The ATR crystals can also be formed in different geometries and mounted in holders or probes, which in turn permits the development of remote, portable devices or systems that can be used to analyze processes *in situ*.

The creation of a bulk diamond solid immersion lens (SIL) has also been demonstrated [105]. By using a high-refractive index material (such as diamond), SILs permit image resolution beyond the diffraction limit. At present, both hemispherical and super-hemispherical diamond SILs are commercially available, and may find future application in diamond-based quantum optical systems or in optical data storage involving UV lasers [106].

Single-crystal CVD diamond can be used as a solid etalon for wavelength locking in wavelength division multiplexing, in combination with a tunable laser. The combination of diamond's high thermal conductivity, low thermo-optic coefficient and low coefficient of thermal expansion provides a high level of thermal stability compared to conventional etalon materials, such as fused silica [107, 108]. Such benefits were demonstrated on an uncoated 50 GHz diamond etalon, built into a butterfly package and mounted on a thermoelectric cooler held at 25 °C. When the thermal stability was evaluated by varying the package case temperature from around −5 to +75 °C, the lock frequency of the diamond etalon was found to shift by only 0.15 GHz. In contrast, the shift measured in a fused silica etalon under the same conditions was found to be 1.6 GHz (see Figure 2.21).

New optical and quantum optical applications of single-crystal CVD diamond have emerged during recent years, and have included Raman lasers, intracavity heat spreaders for semiconductor and doped dielectric disk lasers, and diamond-based magnetometers (examples of these applications are described in detail elsewhere in this book).

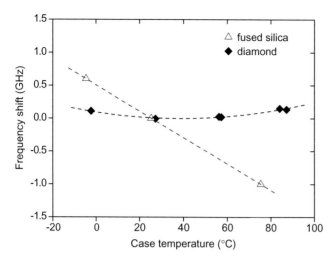

Figure 2.21 Shift in lock frequency measured as a function of the package case temperature for uncoated diamond and fused silica etalons.

2.5 Summary

Today, CVD represents a well-established and robust method of diamond synthesis, with a versatility that enables a high degree of control over the formation of point and extended defects in diamond. This in turn allows a range of single-crystal and polycrystalline diamond grades to be engineered for specific applications.

As an optical material, CVD diamond can provide solutions for extremely demanding applications, with a wide range of CVD diamond optical products – such as kW CO_2 laser optics, high-power microwave transmission windows and ATR prisms – being commercially available. Together, these products harness CVD diamond's outstanding optical properties in combination with its mechanical and thermal properties to enable a step change in performance.

Recent breakthroughs in the industrial synthesis of single-crystal CVD diamond have resulted in an availability of materials with improved birefringence and absorption characteristics, and high chemical and isotopic purities. These improvements in material qualities are in turn driving the development of new optical technologies with potential for significant new commercial applications. Yet, as ever, the success of these new technologies will to a great extent continue to depend on the creativity of engineers and scientists to integrate diamond with other materials by developing new bonding and coating techniques, and processing diamond in new and innovative ways. Moreover, from the perspective of synthesis, continued development is required to increase the size of high-quality single-crystal materials, to reduce production costs, and to control defects to even higher levels of accuracy, thus releasing further diamond's potential as an optical engineering material.

2.6 Acknowledgments

The author would like to acknowledge the contributions and support of the Element Six Research and Development team (Ascot, UK) and, in particular, useful discussions with Joe Dodson, Daniel Twitchen, Tim Mollart, Chris Wort, and Geoff Scarsbrook. The use of X-ray topographs from the Diamond Trading Company (Maidenhead, UK) and support from Diamond Microwave Devices Limited is gratefully acknowledged.

References

1 Sussmann, R.S. (1993) DIAFILM, a new material for optics and electronics. *Ind. Diamond Rev.*, **53**, 63–72.

2 Balmer, R.S., Brandon, J.R., Clewes, S.L., Dhillon, H.K., Dodson, J.M., Friel, I., Inglis, P.N., Madgwick, T.D., Markham, M.L., Mollart, T.P., Perkins, N., Scarsbrook, G.A., Twitchen, D.J., Whitehead, A.J., Wilman, J.J., and Woollard, S.M. (2009) Chemical vapour deposition synthetic diamond: materials, technology and applications. *J. Phys. Condens. Matter*, **21**, 364221-1–364221-23.

3 Sussmann, R.S. (2000) *Handbook of Ceramic Hard Materials* (ed. R. Riedel), Wiley-VCH Verlag GmbH, Weinheim, pp. 573–622.

4 Wort, C.J.H., Brandon, J.R., Dorn, B.S.C., Savage, J.A., Sussmann, R.S., and Whitehead, A.J. (1995) *Applications of Diamond Films and Related Materials: Third International Conference* (eds A. Feldman, Y. Tzeng, W. Yarbrough, M. Yoshikawa, and M. Murokawo), NIST Special Publications 885, pp. 569–572.

5 Isberg, J., Hammersberg, J., Johansson, E., Wikstrom, T., Twitchen, D.J., Whitehead, A.J., Coe, S.E., and Scarsbrook, G.A. (2002) High carrier mobility in single-crystal plasma-deposited diamond. *Science*, **297**, 1670–1672.

6 Friel, I., Clewes, S.L., Dhillon, H.K., Perkins, N., Twitchen, D.J., and Scarsbrook, G.A. (2009) Control of surface and bulk crystalline quality in single crystal diamond grown by chemical vapour deposition. *Diamond Relat. Mater.*, **18**, 808–815.

7 Friel, I., Geoghegan, S.L., Twitchen, D.J., and Scarsbrook, G.A. (2010) Development of high quality single crystal diamond for novel laser applications. *Proc. SPIE*, **7838**, 783819-1–783819-8.

8 Balasubramanian, G., Neumann, P., Twitchen, D., Markham, M., Kolesov, R., Mizuoschi, N., Isoya, J., Achard, J., Beck, J., Tissler, J., Jacques, V., Hemmer, P.R., Jelezko, F., and Wrachtrup, J. (2009) Ultralong spin coherence time in isotopically engineered diamond. *Nat. Mater.*, **8**, 383–387.

9 Millar, P., Kemp, A.J., and Burns, D. (2009) Power scaling of Nd:YVO$_4$ and Nd:GdVO$_4$ disk lasers using synthetic diamond as a heat spreader. *Opt. Lett.*, **34**, 782–784.

10 Hopkins, J.M., Hempler, N., Rösener, B., Schulz, N., Rattunde, M., Manz, C., Köhler, K., Wagner, J., and Burns, D. (2008) High-power, (AlGaIn)(AsSb) semiconductor disk laser at 2.0 µm. *Opt. Lett.*, **33**, 201–203.

11 Mildren, R.P. and Sabella, A. (2009) Highly efficient diamond Raman laser. *Opt. Lett.*, **34**, 2811–2813.

12 Sabella, A., Piper, J.A., and Mildren, R.P. (2010) 1240 nm diamond Raman laser operating near the quantum limit. *Opt. Lett.*, **35**, 3874–3876.

13 Lubeigt, W., Savitski, V.G., Bonner, G.M., Geoghegan, S.L., Friel, I., Hastie,

J.E., Dawson, M.D., Burns, D., and Kemp, A.J. (2011) 1.6 W continuous-wave Raman laser using low-loss synthetic diamond. *Opt. Express*, **19**, 6938–6944.

14 Toyli, D.M., Weis, C.D., Fuchs, G.D., Schenkel, T., and Awschalom, D.D. (2010) Chip-scale nanofabrication of single spins and spin arrays in diamond. *Nano Lett.*, **10**, 3168–3172.

15 Ladd, T.D., Jelezko, F., Laflamme, R., Nakamura, Y., Monroe, C., and O'Brien, J.L. (2010) Quantum computers. *Nature*, **464**, 45–53.

16 Kiffler, A.D. (1956) *Report, Tonowanda Laboratories, Linde Air Products Co., Synthesis of diamond from carbon monoxide*.

17 Bundy, F.P., Hall, H.T., Strong, H.M., and Wentorf, R.H. (1955) Man-made diamonds. *Nature*, **176**, 51–55.

18 Liander, H. and Lundblad, E. (1960) Some observations on the synthesis of diamonds. *Ark. Kemi.*, **16**, 139–149.

19 Matsumoto, S., Sato, Y., Tsutsumi, M., and Setaka, N. (1982) Growth of diamond particles from methane-hydrogen gas. *J. Mater. Sci.*, **17**, 3106–3112.

20 Lander, J.J. and Morrison, J. (1966) Low energy electron diffraction study of the (111) diamond surface. *Surf. Sci.*, **4**, 241–246.

21 Tsuno, T., Tomikawa, T., Shikata, S., Imai, T., and Fujimori, N. (1994) Diamond(001) single-domain 2×1 surface grown by chemical vapor deposition. *Appl. Phys. Lett.*, **64**, 572–574.

22 Asmussen, J. and Reinhard, D.K. (eds) (2002) *Diamond Films Handbook*, Marcel Dekker.

23 Silva, F., Hassouni, K., Bonnin, X., and Gicquel, A. (2009) Microwave engineering of plasma-assisted CVD reactors for diamond deposition. *J. Phys. Condens. Matter*, **21**, 36402-1–36402-16.

24 Butler, J.E. and Woodin, R.L. (1993) Thin-film diamond growth mechanisms. *Philos. Trans. R. Soc. A*, **342**, 209–224.

25 Butler, J.E., Mankelevich, Y.A., Cheeseman, A., Ma, J., and Ashfold, M.N.R. (2009) Understanding the chemical vapour deposition of diamond: recent progress. *J. Phys. Condens. Matter*, **21**, 364201-1–364201-20.

26 Hassouni, K., Silva, F., and Gicquel, A. (2010) Modelling of diamond deposition microwave cavity generated plasmas. *J. Phys. D Appl. Phys.*, **43**, 153001-1–153001-43.

27 Mankelevich, Y.A. and May, P.W. (2008) New insights into the mechanism of CVD diamond growth: single crystal diamond in MW PECVD reactors. *Diamond Relat. Mater.*, **17**, 1021–1028.

28 Chu, C.J., D'Evelyn, M.P., Hauge, R.H., and Margrave, J.L. (1991) Mechanism of diamond growth by chemical vapor deposition on diamond (100), (111), and (110) surfaces: carbon-13 studies. *J. Appl. Phys.*, **70**, 1695–1705.

29 Harris, S.J., Weiner, A.M., and Perry, T.A. (1991) Filament-assisted diamond growth kinetics. *J. Appl. Phys.*, **70**, 1385–1391.

30 Jiang, X. and Klages, C.P. (1993) Heteroepitaxial diamond growth on (100) silicon. *Diamond Relat. Mater.*, **2**, 1112–1113.

31 Wolter, S.D., Stoner, B.R., Glass, J.T., Ellis, P.J., Buhaenko, D.S., Jenkins, C.E., and Southworth, P. (1993) Textured growth of diamond on silicon via in situ carburization and bias-enhanced nucleation. *Appl. Phys. Lett.*, **62**, 1215–1217.

32 Hessmer, R., Schreck, M., Geier, S., Rauschenbach, B., and Stritzker, B. (1995) The influence of the growth process on the film texture of epitaxially nucleated diamond on silicon (001). *Diamond Relat. Mater.*, **4**, 410–415.

33 Bobrov, K., Mayne, A., Comtet, G., Dujardin, G., Hellner, L., and Hoffman, A. (2003) Atomic-scale visualization and surface electronic structure of the hydrogenated diamond C(100)-(2×1):H surface. *Phys. Rev. B*, **68**, 195416-1–195416-8.

34 Ristein, J. (2006) Diamond surfaces: familiar and amazing. *Appl. Phys. A*, **82**, 377–384.

35 Cheeseman, A., Harvey, J.N., and Ashfold, M.N.R. (2008) Studies of carbon incorporation on the diamond {100} surface during chemical vapor deposition using density functional

theory. *J. Phys. Chem. A*, **112**, 11436–11448.

36 Skokov, S., Weiner, B., and Frenklach, M. (1994) Elementary reaction mechanism for growth of diamond (100) surfaces from methyl radicals. *J. Phys. Chem.*, **98**, 7073–7082.

37 Lee, N. and Badzien, A. (1995) Effect of methane concentrations on surface morphologies and surface structures of (001) homoepitaxial diamond thin films. *Appl. Phys. Lett.*, **67**, 2011–2013.

38 Okushi, H., Watanabe, H., Ri, S., Yamanaka, S., and Takeuchi, D. (2002) Device grade homoepitaxial diamond film growth. *J. Cryst. Growth*, **237-239**, 1269–1276.

39 Spitsyn, B.V., Bouilov, L.L., and Derjaguin, B.V. (1981) Vapor growth of diamond on diamond and other surfaces. *J. Cryst. Growth*, **52**, 219–226.

40 Williams, B.E. and Glass, J.T. (1989) Characterization of diamond thin films: diamond phase identification, surface morphology, and defect structures. *J. Mater. Res.*, **4**, 373–384.

41 Haubner, R., Lindlbauer, A., and Lux, B. (1996) Diamond nucleation and growth on refractory metals using microwave plasma deposition. *Int. J. Refract. Metals Hard Mater.*, **14**, 119–125.

42 van der Drift, A. (1967) Evolutionary selection, a principle governing growth orientation in vapour-deposited layers. *Philips Res. Rep.*, **22**, 267–288.

43 Goodwin, D.G. (1993) Scaling laws for diamond chemical-vapor deposition. I. Diamond surface chemistry. *J. Appl. Phys.*, **74**, 6888–6894.

44 Wild, C., Herres, N., and Koidl, P. (1990) Texture formation in polycrystalline diamond films. *J. Appl. Phys.*, **68**, 973–978.

45 Clausing, R.E., Heatherly, L., Horton, L.L., Specht, E.D., Begun, G.M., and Wang, Z.L. (1992) Textures and morphologies of chemical vapor deposited (CVD) diamond. *Diamond Relat. Mater.*, **1**, 411–415.

46 Wild, C., Koidl, P., Müller-Sebert, W., Walcher, H., Kohl, R., Herres, N., Locher, R., Samlenski, R., and Brenn, R. (1993) Chemical vapour deposition and characterization of smooth {100}-faceted diamond films. *Diamond Relat. Mater.*, **2**, 158–168.

47 Silva, F., Bonnin, X., Achard, J., Brinza, O., Michau, A., and Gicquel, A. (2008) Geometric modeling of homoepitaxial CVD diamond growth: I. The {100}{111}{110}{113} system. *J. Cryst. Growth*, **310**, 187–203.

48 Silva, F., Achard, J., Brinza, O., Bonnin, X., Hassouni, K., Anthonis, A., De Corte, K., and Barjon, J. (2009) High quality, large surface area, homoepitaxial MPACVD diamond growth. *Diamond Relat. Mater.*, **18**, 683–697.

49 Collins, A.T. (2001) The colour of diamond and how it may be changed. *J. Gemmol.*, **27**, 341–359.

50 Mollart, T.P. and Lewis, K.L. (2001) The infrared optical properties of CVD diamond at elevated temperatures. *Phys. Status Solidi A*, **186**, 309–318.

51 Piccirillo, C., Davies, C.G., Mainwood, A., Scarle, S., Penchina, C.M., Mollart, T.P., Lewis, K.L., Nesládek, M., Remes, Z., and Pickles, C.S.J. (2002) Temperature dependence of intrinsic infrared absorption in natural and chemical-vapor deposited diamond. *J. Appl. Phys.*, **92**, 756–763.

52 Dore, P., Nucara, A., Cannavò, D., De Marzi, G., Calvani, P., Marcelli, A., Sussmann, R.S., Whitehead, A.J., Dodge, C.N., Krehan, A.J., and Peters, H.J. (1998) Infrared properties of chemical-vapor deposition polycrystalline diamond windows. *Appl. Opt.*, **37**, 5731–5736.

53 Zaitsev, A.M. (2001) *Optical Properties of Diamond: A Data Handbook*, Springer-Verlag.

54 Jin, S. and Moustakas, T.D. (1994) Effect of nitrogen on the growth of diamond films. *Appl. Phys. Lett.*, **65**, 403–405.

55 Achard, J., Silva, F., Brinza, O., Tallaire, A., and Gicquel, A. (2007) Coupled effect of nitrogen addition and surface temperature on the morphology and the kinetics of thick CVD diamond single crystals. *Diamond Relat. Mater.*, **16**, 685–689.

56 Martineau, P.M., Lawson, S.C., Taylor, A.J., Quinn, S.J., Evans, D.J.F., and Crowder, M.J. (2004) Identification of synthetic diamond grown using

chemical vapor deposition. *Gems Gemol.*, **40**, 2–25.

57 Maki, J.M., Tuomisto, F., Kelly, C., Fisher, D., and Martineau, P.M. (2007) Effects of thermal treatment on optically active vacancy defects in CVD diamonds. *Phys. B*, **401/402**, 613–616.

58 Jones, R., Hounsome, L.S., Fujita, N., Öberg, S., and Briddon, P.R. (2007) Electrical and optical properties of multivacancy centres in diamond. *Phys. Status Solidi*, **204**, 3059–3064.

59 Turri, G., Chen, Y., Bass, M.A., Orchard, D.A., Butler, J.E., Magana, S., Feygelson, T., Thiel, D., Fourspring, K., Dewees, R., Bennett, J.M., Pentony, J.M., Hawkins, S., Baronowski, M., Guenthner, A.J., Seltzer, M.D., Harris, D.C., and Stickley, C.M. (2007) Optical absorption, depolarization, and scatter of epitaxial single-crystal chemical-vapor-deposited diamond at 1.064 μm. *Opt. Eng.*, **46**, 064002-1–064002-10.

60 ISO (2003) *Test method for absorptance of optical laser components*, ISO 11551:2003.

61 McGeogh, S.P., Gibson, D.R., and Savage, J.A. (1992) Assessment of type IIA diamond as an optical material for use in severe environments. *Proc. SPIE*, **1760**, 122–142.

62 Remes, Z., Nesladek, M., and Pickles, C.S.J. (2001) Local variations and temperature dependence of optical absorption coefficient in natural IIa type and CVD diamond optical windows. *Phys. Status Solidi A*, **186**, 297–301.

63 Dodson, J.M., Brandon, J.R., Dhillon, H.K., Friel, I., Geoghegan, S.L., Mollart, T.P., Santini, P., Scarsbrook, G.A., Twitchen, D.J., Whitehead, A.J., Wilman, J.J., and de Wit, H. (2011) Single crystal and polycrystalline CVD diamond for demanding optical applications. *Proc. SPIE*, **8016**, 80160L-1–80160L-11.

64 Burns, R.C., Chumakov, A., Carbone, G., Connell, S.H., Dube, D., Godfried, H.P., Hansen, J.O., Härtwig, J., Masiello, F., Rebak, M., Rommeveaux, A., Setshedi, R., Van Vaerenbergh, P., and Gibaud, A. (2007) Diamonds for X-ray optical applications at 3rd and 4th generation X-ray sources. *Proc. SPIE*, **6705**, 67050K-1–67050K-6.

65 Goto, S., Takahashi, S., Kudo, T., Yabashi, M., Tamasaku, K., Nishino, Y., and Ishikawa, T. (2007) Characterization of beryllium and CVD diamond for synchrotron radiation beamline windows and x-ray beam monitor. *Proc. SPIE*, **6705**, 67050H-1–67050K-8.

66 Härtwig, J. (2006) X-ray characterisation of diamonds for X-ray optical applications. 2nd International Conference on Diamonds for Modern Light Sources, July 23–26, 2006, Pilanesberg, South Africa.

67 Moore, M. (2009) Imaging diamond with X-rays. *J. Phys. Condens. Matter*, **21**, 364217-1–364217-15.

68 Kuberev, V.V. (2009) Optical properties of CVD-diamond in terahertz and infrared ranges. *Nucl. Instrum. Methods Phys. Res. A*, **603**, 22–24.

69 Thumm, M. (1998) Development of output windows for high-power long-pulse gyrotrons and EC wave applications. *Int. J. Infrared Millimeter Waves*, **19**, 3–14.

70 Heidinger, R., Spörl, R., Thumm, M., Brandon, J.R., Sussmann, R.S., and Dodge, C.N. (1998) CVD diamond windows for high power gyrotrons, Programme Committee, 23rd International Conference on Infrared and Millimetre Waves, Colchester, UK, pp. 223–225.

71 Lang, A.R. (1967) Causes of birefringence in diamond. *Nature*, **213**, 248–251.

72 Howell, D. (2009) *Quantifying Stress and Strain in Diamond*, PhD Thesis, University College London.

73 Blank, E. (2003) *Thin-Film Diamond I* (eds C.E. Nebel and J. Ristein), Semiconductors and Semimetals, vol. 76, Elsevier, pp. 49–144.

74 Gaukroger, M.P., Martineau, P.M., Crowder, M.J., Friel, I., Williams, S.D., and Twitchen, D.J. (2008) X-ray topography studies of dislocations in single crystal CVD diamond. *Diamond Relat. Mater.*, **17**, 262–269.

75 Suzuki, K., Ichihara, M., Takeuchi, S., Ohtake, N., Yoshikawa, M., Hirabayashi, K., and Kurihara, N. (1992) Electron

microscopy studies of dislocations in diamond synthesized by a CVD method. *Philos. Mag. A*, **65**, 657–664.

76 Butler, J.E. and Oleynik, I. (2008) A mechanism for crystal twinning in the growth of diamond by chemical vapour deposition. *Philos. Trans. R. Soc. A*, **366**, 295–311.

77 Martineau, P.M., Gaukroger, M.P., Khan, R.U.A., and Evans, D.J.F. (2009) Effect of steps on dislocations in CVD diamond grown on {001} substrates. *Phys. Status Solidi C*, **6**, 1953–1957.

78 Davies, N., Khan, R.U.A., Martineau, P.M., Gaukroger, M.P., Twitchen, D.J., and Dhillon, H.K. (2011) Effect of off-axis growth on dislocations in CVD diamond grown on {001} substrates. *J. Phys. Conf. Ser.*, **281**, 012026, 1–012026, 7.

79 Pinto, H. and Jones, R. (2009) Theory of the birefringence due to dislocations in single crystal CVD diamond. *J. Phys. Condens. Matter*, **21**, 364220, 1–364220, 7.

80 Martineau, P.M., Gaukroger, M.P., Guy, K.B., Lawson, S.C., Twitchen, D.J., Friel, I., Hansen, J.O., Summerton, G.C., Addison, T.P.G., and Burns, R. (2009) High crystalline quality single crystal chemical vapour deposition diamond. *J. Phys. Condens. Matter*, **21**, 364205, 1–364205, 8.

81 Glazer, A.M., Lewis, J.G., and Kaminsky, W. (1996) *Proc. R. Soc. Lond. A*, **452**, 2751–2765.

82 Harris, D.C. (1994) Properties of diamond for window and dome applications. *Proc. SPIE*, **2286**, 218–228.

83 Mollart, T.P., Lewis, K.L., Pickles, C.S.J., and Wort, C.J.H. (2003) Factors affecting the optical performance of CVD diamond infrared optics. *Semicond. Sci. Technol.*, **18**, S117–S124.

84 Whitehead, A.J. (2003) *Internal report*, DeBeers Industrial Diamond.

85 Berman, R. (1992) *The Properties of Natural and Synthetic Diamond* (ed. J.E. Field), Cambridge Academic, pp. 473–513.

86 Bray, J.W. and Anthony, T.R. (1991) On the thermal conductivity of diamond under changes to its isotopic character. *Z. Phys. B Condens. Matter*, **84** (1), 51–57.

87 Robinson, C.J., Hartnett, T.M., Miller, R.P., Willingham, C.B., Graebner, J.E., and Morelli, D.T. (1993) Diamond for high heat flux applications. *Proc. SPIE*, **1793**, 146–156.

88 Godfried, H.P., Coe, S.E., Hall, C.E., Pickles, C.S.J., Sussmann, R.S., Tang, X., and van der Voorden, W.K. (2000) Use of CVD diamond in high-power CO_2 lasers and laser diode arrays. *Proc. SPIE*, **3889**, 553–563.

89 Graebner, J.E. (1998) Thermal conductivity of diamond films: $0.5\,\mu m$ to $0.5\,mm$. *Isr. J. Chem.*, **38**, 1–15.

90 Twitchen, D.J., Pickles, C.S.J., Coe, S.E., Sussmann, R.S., and Hall, C.E. (2001) Thermal conductivity measurements on CVD diamond. *Diamond Relat. Mater.*, **10**, 731–735.

91 Griffiths, A.A. (1920) The phenomena of rupture and flow in solids. *Philos. Trans. R. Soc.*, **A221**, 163–198.

92 Telling, R.H., Pickard, C.J., Payne, M.C., and Field, J.E. (2000) Theoretical strength and cleavage of diamond. *Phys. Rev. Lett.*, **84**, 5160–5163.

93 Weibull, W. (1951) A statistical distribution function of wide applicability. *J. Appl. Mech.*, **13**, 293–297.

94 Whitehead, A.J. (1995) *Internal report*, DeBeers Industrial Diamond.

95 Klein, C.A. (2002) Diamond windows and domes: flexural strength and thermal shock. *Diamond Relat. Mater.*, **11**, 218–227.

96 Pickles, C.S.J. (2002) The fracture stress of chemical vapour deposited diamond. *Diamond Relat. Mater.*, **11**, 1913–1922.

97 Pickles, C.S.J., Brandon, J.R., Coe, S.E., and Sussmann, R.S. (1999) Factors influencing the strength of chemical vapour deposited diamond, in *9th Cimtec-World Forum on New Materials Symposium IV – Diamond Films* (ed. P. Vincenzini), Techna Srl., pp. 435–454.

98 Davies, A.R., Field, J.E., and Pickles, C.S.J. (2003) Strength of free-standing chemically vapour-deposited diamond measured by a range of techniques. *Philos. Mag.*, **83**, 4059–4070.

99 Harris, D.C. (1999) *Materials for Infrared Windows and Domes*, SPIE, Washington, DC.

100 Rofin Sinar Technologies (1997) *Annual Report*, pp. 6–7.

101 Thumm, M. (2004) State-of-the-art of high power gyro-devices and free electron maser. Update 2003. *Wissenchaftliche Berichte*, FZKA 6957, 1–88. ISSN 0947-8620.

102 Seward, C.R., Field, J.E., and Coad, E.J. (1994) Liquid impact erosion of bulk diamond, diamond composites and diamond coatings. *J. Hard Mater.*, **5**, 49–62.

103 Perkin-Elmer (2005) *FT-IR Spectroscopy – Attenuated Total Reflectance*, technical note. Available at: http://www.perkinelmer.com.

104 Specac. *Reflectance Spectroscopy*, technical note. Available at: http://www.specac.com.

105 Schaich, T.J., van Oerle, B.M., Godfried, H.P., Kriele, P.A., Houwman, E.P., Nelissen, W.G., Pels, G.J., and Spaaij, P.G. (2005) High NA diamond lenses for near-field optical storage, in *International Symposium on Optical Memory and Optical Data Storage*, OSA Technical Digest Series, paper WD6.

106 Siyushev, P., Kaiser, F., Jacques, V., Gerhardt, I., Bischof, S., Fedder, H., Dodson, J., Markham, M., Twitchen, D., Jelezko, F., and Wrachtrup, J. (2010) Monolithic diamond optics for single photon detection. *Appl. Phys. Lett.*, **97**, 241902–241904.

107 Godfried, H.P. (2006) US patent application US 2006/0209380 A1.

108 Godfried, H.P., Scarsbrook, G.A., Twitchen, D.J., Houwman, E.P., Nelissen, W.G.M., Whitehead, A.J., Hall, C.E., and Martineau, P.M. (2010) US patent application 2010/0116197 A1.

3
Polishing and Shaping of Monocrystalline Diamond
Jonathan R. Hird

3.1
Introduction: Background and Historical Overview

The task of cutting and polishing diamond for scientific purposes has always been marked by the difficulty of overcoming the very same physical and material properties that make the material scientifically interesting. Diamond is highly resistant to plastic deformation, possesses a high thermal conductivity, is highly anisotropic in its wear properties, and has a propensity for fracture along its octahedral plane. All of these factors make it very difficult to polish. For scientific research, the quality of the surface finish must be excellent, the subsurface must be damage-free, and control over morphology must be exact. Attempts to conquer these difficulties have a remarkable historical precedent: for example, diamond microscope lenses were first manufactured about 180 years ago by the Scottish optician Andrew Pritchard [1], and the first precision curved surfaces on diamond were probably made by John Harrison 260 years ago (Figure 3.1) [2].

Both pioneers met with considerable – though not insurmountable – difficulties during polishing. Pritchard succeeded after many months of trial and error, writing that this was ". . . contrary to the expectations of many, whose judgment in these matters was thought of much weight, who predicted that the crystalline structure of the diamond would not permit it to receive a spherical shape" [3]. Ultimately, and almost immediately upon their completion, this seemingly impossible and expensive feat of optical engineering proved to be a scientific cul-de-sac due to more practical advances in optics [4]. Almost 200 years later, however, diamond lenses are routinely being fabricated and are useful in a variety of applications.

John Harrison, regarded today as the father of modern chronometry, also found that shaping diamond for use in his timekeeper escapements was a difficult task, remarking that ". . . diamond is a hard thing to wrestle with." Although Harrison's timekeepers still function today, the use of diamond in chronometry was confined to a handful of watches owing to the prohibitive expense in purchasing diamond. Harrison's diamond pallets, like Pritchard's diamond lenses, were ahead of the time, and some 250 years would pass before diamond was again used in the escapements of watches to great effect [5].

Optical Engineering of Diamond, First Edition. Edited by Richard P. Mildren and James R. Rabeau.
© 2013 Wiley-VCH Verlag GmbH & Co. KGaA. Published 2013 by Wiley-VCH Verlag GmbH & Co. KGaA.

Figure 3.1 A diamond pallet mounted in a brass collet made by master horologist Derek Pratt, after John Harrison (1693–1776). The diamonds are unusual in that they have a curved facet. The major dimensions of the diamond are ~2 mm × 0.35 mm × 1 mm, of which 1 mm of the length resides in the collet. This escapement was designed and made by John Harrison during the 1750s, and is a critical component of his timekeeper for the longitude–H4. Image courtesy of Charles Frodsham & Co. Ltd.

With the advent of readily available synthetic diamond materials the considerable expense is no longer relevant. Diamond, however, remains a difficult material to work, and this is an important consideration when the decision is made to use diamond in an experiment or application. Diamond polishing has been carried out at least since the end of the Middle Ages [6], yet it is remarkable that, although refinements have been made, the polishing process used most commonly today would be recognizable to an early diamond lapidary. The science of how diamond undergoes wear and polishing has been studied for nearly a century, yet debate continues as to the exact mechanism of how material is removed from the diamond. Diamond polishing continues to be a time-consuming and labor-intensive exercise; moreover, the process may introduce subsurface damage, and contaminate or change the surface chemistry. Fortunately, from a diamond device standpoint, a range of techniques are emerging today which promise to overcome these factors. The aim of this chapter is to introduce the reader to the science and art of traditional diamond polishing, and also to some more of the exotic techniques that are emerging.

3.2 Shaping Diamond: Cleaving, Bruting, and Sawing

Traditionally, there are four methods by which diamonds are worked into the desired shape: cleaving, bruting, sawing, and polishing:

Cleaving: This is perhaps the oldest method of shaping diamond, as the propensity of the diamond to cleave along its octahedral {111} plane must have been noted since the most ancient times. Practical accounts of cleaving can be found elsewhere [7], but the basics are as follows. First, a V-shaped notch (kerf) of millimeter dimensions is scribed into the appropriate plane using a sharp-pointed diamond. Second, the diamond to be cleaved is inserted into a wooden stick and held in place with cement. The cleaver's blade (a knife of length 10 cm, height 5 cm and 0.3 cm thickness, which is tapered to blunt edge) is inserted into the kerf such that the blade does not touch the bottom of the notch. A sharp tap is given to the top of the blade by means of a wooden or steel rod of 12 mm diameter. The key to cleavage is to keep the fracture propagating in a straight line without the crack branching on to other cleavage planes. This is a task that requires a high degree of manual skill, but the surface that it produces exhibits macroscopic step patterns. The cleaving of planes other than {111} is possible, and examples of (221), (110), (322), (331), (211) and (332) have been found in Nature [8]. By using an *ab initio* approach, Telling *et al.* [9] have demonstrated that the contribution of bond bending in addition to the energy taken to break a bond must be taken into account for the natural dominance of {111} cleavage to be observed.

Sawing: This is carried out using a rotating thin phosphor bronze disc charged with a mixture of diamond grit and oil. The saw blades are of 7.5 cm diameter, and have a thickness ranging between 75 and 160 μm. The thickness selected depends on the dimensions of the stone being cut. Sawing is carried out either along the dodecahedral plane or the cube plane and the diamond is gravity-fed onto the saw blade. The rotational speed of the saw is set according to the size of the stone. Smaller stones (<1 carat) are sawn at higher speeds (7000–7500 rpm) while larger stones (>1 carat) are sawn using lower speeds (4000–4500 rpm) [7].

Bruting: This is essentially the turning of diamond in a lathe using an inferior quality diamond to produce a circular section via a process of abrasion and micro-chipping [7].

Polishing: Although, eventually, laser cutting will render the bruting, sawing, and cleaving process superfluous, the mechanical polishing of diamond is still widely used both in the gem industry and for scientific sample preparation.

3.3
Practical Aspects of Diamond Polishing

3.3.1
Apparatus and Preparation

In its traditional form – and today, by far the most common manifestation – diamond polishing is carried out using a rotating horizontal disk of about 30 cm diameter known as a *skive* or *scaife* (Figure 3.2). The scaife plate spins around a vertical axis with a speed of around 3000 rpm, which corresponds to a tangential linear velocity at its periphery of about $30\,\mathrm{m\,s^{-1}}$. The scaife plate is manufactured from a high-grade cast iron and, when new, is either sandblasted to give a matt finish, or more commonly, "striped" by using a machine which scribes curved radial arcs into the surface of the cast iron. Scaife plates can be purchased either with these radial grooves or blank. Any damage to the surface will require the grooves to be refinished. The possibility of microstructural defects from the casting process, or those introduced through damage, means that it is prudent to subject scaife plates to X-ray analysis prior to their use. Whether the surface finish is sandblasted or striped, the texture is intended to facilitate the application and fixture of a mixture of diamond grit and oil (typically olive oil) that is worked into the surface prior to its use. This is especially important when the surface is new, but the process is also carried out before starting work; diamond lapidaries will perform this procedure daily. The initial "running in" of the scaife is essential to the function, performance and lifetime of the scaife and, ultimately, to the degree of smoothness

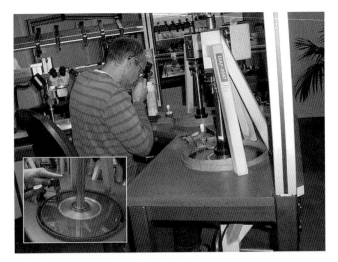

Figure 3.2 State-of-the-art diamond polishing bench and workshop. Inset: A scaife used for obtaining very smooth surfaces; note the mirror-quality finish of the lap. Images courtesy of Element 6.

in the surface finish that can be achieved on the diamond. To prepare a scaife for use, it is first cleaned using a soft cloth and a solvent (typically denatured alcohol). When the alcohol has evaporated, 1–2 carats (0.2–0.4 g) of diamond powder (having a size distribution <40 µm) are sprinkled evenly around the surface of the scaife and rubbed in with the fingertips, using a circular motion. A few drops of olive oil are applied around the circumference of the scaife and a similar rubbing in with the fingertips is performed. An even, very thin coating that has a slightly dull appearance to the eye is what is required here. Next, an inferior quality diamond (traditionally boart or carbanado) or, more recently, a sintered diamond material (such as Syndax), is used to work this in while the scaife is spinning at speed. The diamond is held in a small cup or collet known as a *dop*, which is in turn fastened to a *tang*; this is a hand-held tool which allows the diamond to be oriented and moved along the radius of scaife along a guide mounted on the scaife bench. Using hand pressure only, the diamond (oriented in a "soft" direction; see below) or Syndax (no orientation required) is worked back and forth until the lap gradually undergoes a change in appearance, from the initial dull light gray color to an almost glossy dark gray color. Chemically, this is caused by the evaporation and burning of the olive oil in addition to the movement, fracture, orientation, and polishing of the diamond grit particles. This glossy, dark gray color is an indication that the scaife is ready for use. When stationary, the scaife surface will feel waxy, but not oily, to the touch. The procedure may be repeated as required and is typically carried out several times a day if it is being used continuously.

When the scaife is ready for use, the diamond to be polished is mounted into a dop and hand tang. Traditionally, molten lead, or a mixture of shellac and powdered pumice stone was used to fix the diamond into the dop. A friendlier alternative to lead is a mixture of wood glue, gypsum plaster (plaster of Paris) and water, which is allowed to dry when the diamond is mounted in position. If the diamond is substantial and/or regular in size and shape, then metal claws or a collet can be used to clamp it, taking precautions to avoid fracture from excessive heating while it is being polished.

3.3.2
Directional Dependence of Polishing: Wear Anisotropy

There are many surprises to be encountered surrounding the art and science of diamond polishing, and these have provided an interesting source of scientific questions. Perhaps the most surprising is the extreme anisotropy in tribological properties that is encountered when the diamond is placed on the scaife. Diamond is very resilient to wear unless it is oriented correctly with respect to the direction of scaife rotation. Indeed, an incorrectly oriented diamond will score the scaife and render it unusable, requiring it to be skimmed flat, striped, and redressed. In the gem-cutting industry, lapidaries will prepare the surface of the scaife into annular sectors having different lap qualities. A diamond is first "tried" on the innermost circumference of the scaife; this is known appropriately as the "trying ring," and is where orientation and alignment can be adjusted as required. It is

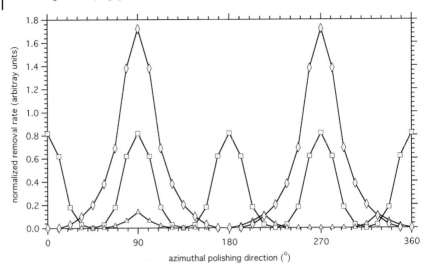

Figure 3.3 Wear anisotropy of diamond for the three basal planes. The cube {100} plane (represented above by squares) has 4-fold symmetry; {100}<100> are soft while {100}<110> are hard. The dodecahedral {110} (represented by parallelograms) has 2-fold symmetry and can be polished relatively quickly in <100> directions while {110}<110> are hard. Although difficult to polish, the octahedral plane {111} has 3-fold symmetry, with <112> being a viable polishing direction. The curves shown here were re-plotted using Tolkowsky's original data points [10].

sometimes useful to use a permanent marker pen or similar to mark the diamond surface to check where the scaife is in contact with the diamond. Gem cutters – like all highly skilled artisans – use a combination of sensory perception and experience with the geometry of the diamond to form it. The vibration detected through the tang and the sound emitted from the diamond as it rubs against the rotating scaife, indicate to the lapidary how well the diamond is cutting. The generation of heat serves as a useful indication that a soft direction has been found, although the diamond should never be allowed to glow to red heat.

In diamond polishing parlance, diamond has "hard" and "soft" directions of wear on each crystallographic plane, with the structure of the diamond sometimes being referred to as the grain; this is in direct analogy to the ways of working wood. This anisotropy in wear properties is shown in Figure 3.3., where it can be seen that the cube plane {001} of diamond has four soft directions, the dodecahedral has two {110}, while the octahedral {111} is the hardest to polish but has three polishing directions. It can be seen that slight deviations from the "grain" cause rapid reductions in the amount of material that can be worn away.

While there is no substitute for the skill of a diamond cutter, X-ray diffraction in the Laue geometry is useful to locate the best polishing direction and crystal plane and is, of course, essential if a precise crystallographic plane is required for scientific purposes. If an X-ray apparatus is available, then it is relatively simple to set up an apparatus that allows the diamond, already mounted in a dop, to be aligned perpendicular to the source, and transferred directly back in to the tang where it can be polished in this direction. The Laue patterns from the basal planes of diamond and their polishing directions, together with a stereogram showing the relative ease of polishing when moving from one crystallographic plane to another, are shown in Figure 3.4.

3.4
The Science of Mechanical Polishing

For most of the last century, a multitude of scientific techniques and methodologies have been directed at diamond polishing in attempts to understand the mechanisms which allow the hardest known material to be facetted and polished with mirror-like surfaces. Ostensibly, this has been carried out for industrial purposes with the aim of speeding up the process, since diamond faceting and polishing is still a slow and relatively laborious process (tens of $\mu m\,min^{-1}$ are typical). However, even a cursory inspection of a table of physical properties of diamond [12] reveals that there is interesting science lurking behind the vast material property parameter space.

From a fundamental perspective, the question is: What is the scientific process by which carbon atoms are plucked individually or in clusters from the diamond surface during mechanical polishing?

Unfortunately, the tests and analysis that can be readily applied to investigate the wear behavior of other materials – even brittle solids – are not so easily applied to diamond, which has a combination of properties that do not encourage wear. Furthermore, the traditional diamond polishing apparatus is not generally suitable for adoption in the laboratory, and it is difficult to change parameters in a scientific fashion. It is perhaps for these reasons that the technique developed for polishing diamonds in antiquity is largely unchanged today.

It is reasonable to assume that the wear rate – usually defined as the removal or transfer of material per unit area in unit time – is some complex function of a particular combination of material properties, interfacial pressure, atmosphere, geometry, and velocity. However, for many materials there is a range of processes that can result in material removal or transfer. For brittle materials, these could occur via a mechanism which invokes one or more of the following: seizure; melt wear (a thermal process which may include cracking); spalling; or structural change (a tribochemical process or a fracture-based process such as indentation cracking or brittle spalling) [13].

Mapping the parameter space by varying the polishing speed, interfacial pressure, and geometry is a logical start to unveiling the fundamental wear process. It was Tolkowsky [10] who put the subject on a scientific footing. Tolkowsky, famous for the invention of the brilliant cut [14], is also noted for reproducing the anisotropic wear behavior, noted by lapidaries since the earliest times, under laboratory

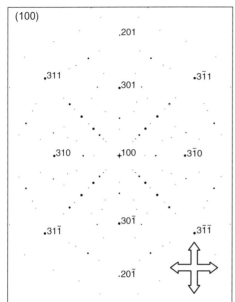
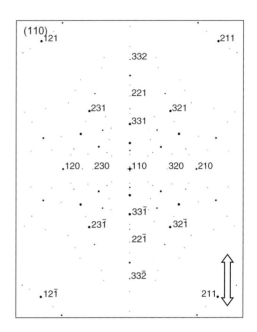
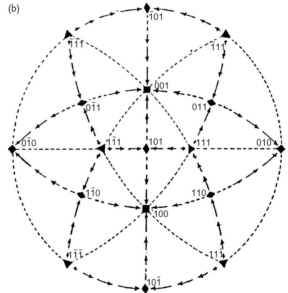

Figure 3.4 (a) Simulated Laue patterns for cube (100) and dodecahedral surfaces (110) using 37.5 kV X-rays at a sample distance of 35 mm. The ideal polishing directions are indicated; for these planes these are <100> type directions; (b) Kraus and Slawson's stereogram, which shows the relative ease of opening facets when moving from one plane on to another. The longer arrows represent easier polishing planes and directions [11].

conditions. In doing so, he caused a debate that continues to this day on the precise mechanisms governing diamond polishing. Tolkowsky's results are shown in Figure 3.3, re-plotted from his thesis.

3.4.1
Wear Anisotropy

In the parlance of the industry, the directions of polish <d> on any diamond surface {s} in which it may be worn with no damage to the scaife are known as "soft" directions of polish. If the diamond is rotated from these directions a resistance to wear will be noted, and damage to the scaife may occur; these directions are dubbed "hard" directions. For the cube plane, the directions of polish {100}<100> are soft, while {100}<110> are hard; that is, there is fourfold symmetry. On the dodecahedral plane there are two directions of soft polish, <100>, while <110> is hard. The octahedral plane {111} is almost uniformly difficult to polish, but it may be polished with difficulty in <112> directions, of which there are three. In practice, however, it is usual to tilt the octahedral surface slightly towards the dodecahedral {110} plane for a good surface finish. There are, of course, an infinite number of directions and planes, and a skilled lapidary will navigate these using prior knowledge of the stone's orientation, combined with sensory information. From a crystallographic perspective, Kraus and Slawson [11] have made use of a stereogram (see Figure 3.4b) that greatly simplifies an understanding of this parameter space. It will be noted that a tilt towards {110} used for polishing {111} enables stones to run quicker. Experiments demonstrating the anisotropic wear behavior described by Tolkowsky have since been reproduced in different ways, and the consensus is that while it is difficult to arrive at the results, these are essentially correct [11, 15–29].

3.4.2
Velocity Dependence

The agreement on the general form of the anisotropic wear behavior does not, however, extend to the relationship between the wear rate and the sliding speed (rotational speed of the scaife), the radius of the lap, the contact pressure, or the geometry–the very parameters which are useful in mapping wear processes.

The anisotropy of wear and the difficulty in orientating a diamond is partly responsible for this discord, susceptible as it is to slight variations in orientation and tilt. The scaife will become warm as polishing progresses, causing changes in the geometry and alignment as the scaife expands, and also perhaps changing the lap condition. This latter variable is a major concern when carrying out wear measurements, as it inevitably changes during the course of an experiment– disastrously so if a slight misorientation of the stone occurs. Furthermore, geometry changes occurring when monitoring the wear of the diamond may contribute to erroneous results if equilibrium has not been reached, or has not been accounted for.

A lapidary artisan will correct for lap changes and misalignments throughout the course of the day, but this is not something that is easily reproduced in the laboratory. In an attempt to simplify the situation, others have sought to replace the traditional scaife used by Tolkowsky with a miniature rotating grinding wheel which is either cast iron and charged with diamond grit (like a mini-scaife) or with diamond grit embedded in a cobalt matrix, much like a true grinding wheel [21–23, 26–28]. Approximating the action of the scaife in this way, however, provides a number of results which are not observed during polishing on a scaife. This is perhaps not surprising given that the rotation speeds are higher, the grit particles are reduced in number, and the geometry of the situation is dissimilar. The comparison of results obtained in this manner has generally led to an obfuscation of the true diamond wear mechanism.

The complications inherent to diamond polishing are matched by a number of experimental difficulties which include a difficult environment to make sensitive wear measurements. Vibration, electrical noise, and a harsh environment of debris combined with a lack of access to the small diamond being polished, all place limitations on what measurements can be made.

Hukao [20] and Wilks and Wilks [25–28] used an optical technique for measuring the wear, while Tolkowsky [10] had a static tang manufactured with a depth gage attached so that the wear could be monitored while polishing proceeded. The Grodzinski–Stern investigation [22] utilized a chemical balance to measure mass loss of the diamond at intervals after it was removed from the experiment. The results from these experiments are not consistent with each other.

With the benefit of X-ray orientation and relatively large high-pressure, high-temperature (HPHT) monocrystalline diamonds of constant cross-section, Hird and Field [30, 31] have employed a variety of techniques to revisit this problematic area, with the aim of achieving reliable and reproducible results. A mass-loss technique was first employed [32] aided by a experimental tang. This was rigidly mounted to the bench and could be loaded with lead masses to vary the interfacial contact pressure between the scaife and the diamond. The tang and dop were designed so that the diamond could be transferred to a Laue X-ray backscattering alignment apparatus, where it could be positioned and returned to the polishing rig so that a specific plane and direction of polish could be experimented upon. The dop had jaws that rigidly held the diamond during wear, but which also facilitated removal of the diamond whereupon it was sonicated in acetone before being weighed on a chemical balance. A second method monitored displacement directly through the use of an electronic depth gage. This was mounted over the diamond on a kinetic mount when the experiment was stopped at intervals and allowed to cool to 300 K [30, 31]. Both methods circumvent the issue of thermal expansion as noted above and most importantly, the results from both experiments corroborated each other.

In a third method, the same experimental tang was employed as in the second study, but this time fitted with an integrated linear variable displacement transducer (LVDT). This considerably simplified the data collection, allowed for unprec-

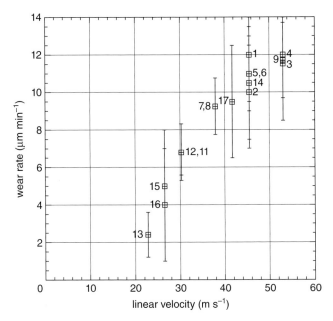

Figure 3.5 The wear rate of diamond plotted as a function of the tangential velocity of the scaife. The numbers by the data points indicate the numerical order in which the measurements were taken. It can be seen that a linear fit to the data does not extrapolate through the origin. The experimental parameters were a load of 1.52 kg, an interfacial area of ~40 mm² (i.e., a contact area of ~0.4 MPa), and the circumference of track was 0.91 m [31].

edented accuracy and for wear maps to be produced [30]. The data obtained from all the experiments shown in Figure 3.5 confirmed diamond polishing as a velocity-dependent process that consists of a number of components which must be considered as a whole to understand the process.

3.4.3
Diamond Polishing Wear Debris

The debris produced by frictional sliding on diamond, and particularly during polishing, has been the subject of analysis by many authors. Feng and Field [33] used infrared spectroscopy, electron and X-ray diffraction for this purpose, while another approach has been to use electron energy loss spectroscopy (EELS) [34–36]. All of the evidence obtained has indicated that the wear debris consists of a low-density form of carbon (non-diamond) which is predominantly sp^2-bonded. The debris produced by low-speed frictional sliding of a diamond stylus on a diamond surface also consists of sp^2-bonded and/or amorphous carbon.

3.4.4
The Polished Diamond Surface

The nature of the polished diamond surface has received considerable attention, both in terms of its surface topology and its unusual surface chemistry [37]. The as-polished diamond surface is found to be hydrogen-bonded [38], and a curious observation is that a diamond taken from the scaife will often appear very clean with a mirror-like facet. A study of a diamond surface using Raman spectroscopy did not find any evidence of a thin layer of non-diamond carbon [39].

Differential interference contrast microscopy (also referred to as Nomarski microscopy) offers the most rapid means of making qualitative observations on the topology of diamond surfaces, and is used extensively in research and industry. Diamonds viewed in this way usually exhibit polishing lines (also termed "running lines") along the polishing direction which, depending on the quality of the surface finish, may sometimes be visible to the unaided eye.

Prior to the invention of the scanning probe microscope, several studies of the topology of diamond surface were conducted using reflection electron microscopy [40–43], scanning electron microscopy [44], and transmission electron microscopy [45]. In all cases, the authors reported a polishing structure on a scale smaller than the "optical" grooves, and observations of the diamond surface included microcracks, nanometer-sized grooving [40], and (sub)surface damage [46]. Thornton and Wilks [45] employed a tungsten replica technique and reported a structure which they claimed was consistent with microcleavage. The propensity of insulating diamond surfaces to be charged electrically under the electron beam means that it is difficult to obtain a sharp image, however, and the interpretation of contrast can be difficult. Ennos [47] reported that carbon structures tended to form under the electron beam, and Seal and Mentor [40] attributed this to a worse-than-expected image quality when attempting to obtain higher-magnification images. Electrical charging can, of course, be eliminated by studying Type IIb diamonds, or by coating the diamond prior to examination with a thin conducting film. Unfortunately, however, it appears that this latter method may obscure some of the finer topography on the diamond surface [44, 48].

Scanning probe microscopy offers a rapid and accurate method for obtaining quantitative topological information on atomic length scales, with the pioneering studies on polished diamond surfaces being carried out by Couto *et al.* [49–51] at a relatively early stage in the development of these microscopes. These authors studied a variety of HPHT and chemical vapor-deposited diamond surfaces that were polished along various crystallographic directions. It was quickly apparent that diamond surfaces that had been polished in both "soft" and "hard" directions had very different topologies on the micro- and nano-scales, despite the facets having a mirrored appearance to the unaided eye.

Couto *et al.* showed that diamond surfaces polished in soft directions have a topology which appeared to be a superposition of undulating, smooth, parallel grooves, but on a much smaller scale than the running lines which are visible optically (Figure 3.6). Notably, it was observed that the cross-section of the grooves

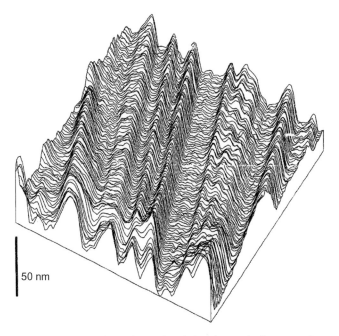

Figure 3.6 The topology of a scaife-polished diamond. The image shows a 500 × 500 nm area of a cube face of diamond that has been polished in a "soft" [001] direction. Well-polished diamond surfaces exhibit smooth grooves that are tens of nanometers wide [49].

did not change in shape over distances exceeding several microns. For a (100)<001> surface, the dimensions of the grooving ranged between 20 nm and 100 nm in width, while depths between 4 and 12 nm were measured. A dodecahedral surface (110)<001> was found to be similarly grooved, but here the dimensions were found to be deeper (32 nm) and wider (200 nm). These grooves are several orders of magnitude smaller than the polishing grits typically used in polishing, and different in nature to those seen using microscopy (which are several orders of magnitude larger in width). The smooth nature of the tracks extended to the angles of inclination of their walls, which were shown to have a Gaussian-type distribution [51]; that is, they had no preferred crystallographic orientation. This observation, together with the smooth nature of the grooves and lack of ring cracks, instantly ruled out any theory involving microcleavage along {111} planes for "soft" direction polishing. A similarity should be noted here between these observations and those of the tracks made by low-speed sliding of sharp indenters. In these studies grooving was also found that was suggestive of plastic deformation [40, 42, 43].

Another important observation, pertinent to obtaining a high-quality finish on diamond surfaces, was that nanostructure was observed on surfaces to which *zoeten*—a radial and lateral movement of the diamond across the scaife during polishing—had been applied. On these surfaces, nanometer-sized grooving was still apparent with a 30 nm width, but with an average height of 3 nm.

Diamonds polished in hard directions, however, show a different surface topology. On a (100) surface polished in a direction 15° from <011>, Couto et al. reported a "hill and valley" topology together with a second type of structure. The dimensions of the undulations had an average width of 38 nm, a height of 4 nm, and a 12° inclination. The second structure consisted of elongated microstructures in the direction of polishing. Again, the sidewalls of this morphology were found to be inclined at 12° but the width was 76 nm and the depth 8 nm. An octahedral face (111) polished along <110> – another hard direction – showed a different structure altogether, namely parallel polishing grooves with 12° sidewall inclinations together with perpendicular grooves.

It is difficult to reconcile the differences between soft and hard direction polishing with a single wear mechanism, and the authors suggested that the structures apparent in hard directions are all indicative of material removal by microfracture. The origin of the smooth nanogrooving is, however, debatable and will be discussed later in this chapter.

3.4.5
Subsurface Damage

As will be seen in Section 3.5, it is relatively easy – despite the great hardness of diamond – to induce fracture on a diamond surface either via indentation or scratching with a sharp tip. During the mechanical polishing of diamond on a scaife, the surface and subsurface of the diamond may be damaged by fracture in this way. Generally, the damage can be kept to a minimum by using small grit particles and an experienced lapidary [52]; however, such damage is difficult to detect and quantify, with depths ranging from about 1 µm upwards. Larger extensive features may be detected optically using, for example, Nomarski microscopy. Samuels and Wilks [53] have suggested that the viewing of such damage may be enhanced by smearing grease onto the surface of the diamond. Damage can also be observed using cathodoluminescence, where the surface is irradiated by electrons which generates luminescence of the diamond. Dislocations tend to increase luminescence, while surface cracking decreases it [54]. Friel et al. [52] have noted that dry etching in oxygen leads to the preferential etching of such damage, so that it is enhanced and readily visible. Other methods of quantifying surface quality have been explored using techniques such as X-ray reflectivity and grazing incidence in-plane diffraction [55].

3.4.6
The Scaife: Its Surface and Preparation

Two complementary studies of the scaife have been made [31, 51]. Prior to these investigations – perhaps owing to the difficulty of either sectioning a scaife into manageable specimens, or of placing the scaife plate under the microscope – the scaife was largely assumed to be a matrix of randomly oriented diamond grit particles, whereby only those particles oriented in a hard direction relative to the diamond being polished were responsible for material removal. Consequently, it was surprising to discover that the reality was quite different. Couto *et al.* [51] used optical microscopy to reveal that the scaife surface had been plastically deformed, or grooved, by the diamond being polished. They also discovered that the larger diamond grit particles (in this case >2 μm) had flat tops, indicating that they too had been polished. The dimensions and structure of the grooves seen on the scaife surface through the optical microscope appeared to match the optical grooves found on the polished diamond surface. Hird and Field [31] extended this line of inquiry, using optical and electron microscopies to view scaife sections that had been etched using a gas torch. The etching removed a layer of surface debris, so as to expose additional detail; moreover, further evidence was found of reciprocal polishing between the grit particle and the workpiece, with the larger grit particles being able to move and rotate to orient themselves (Figure 3.7).

Importantly, the nanodimension grooves found on diamond surfaces are also present on the polished diamond grit particles, which is indicative of soft direction polishing. This observation immediately rules out any suggestion that the grooves are somehow formed by the plowing action of an asperity on a macroscale diamond grit particle. The nanogrooves found on polished diamond surfaces are, therefore, either a direct consequence of diamond–diamond sliding contact, or they are due to the effect of a third body acting on the diamond under polish and the diamond grit constrained on the scaife. Hird and Field [31] have taken the latter view, and suggest that the answer to the existence of nanogrooves and the reciprocal polishing of the grit particles lies in the scaife preparation process. As noted in Section 3.3, great importance is ascribed to preparing the scaife for use and for keeping it in good condition during its lifetime. The salient features of the scaife preparation process are that traditionally, a diamond oriented in the soft direction with respect to the tangential direction of the rotation is used to work the diamond grit particles into the scaife. During this process, the friction and heat generated is observed to be high and a chemical change is induced on the scaife plate, thus generating the ubiquitous black layer associated with a well-prepared scaife. Couto *et al.* [51] employed a variety of techniques to study the debris found on scaife surfaces, and identified amorphous carbon, oxides and carbides of iron and diamond particles, in addition to minute quantities of Na, Cl, Si, and Ca which, presumably, were present as contaminants.

The process of scaife preparation has been studied in detail by Hird *et al.* [39], who used Raman spectroscopy (which is well suited to distinguishing between the many allotropes of carbon [56–58]) to analyze the chemical evolution of wear debris as it developed during the procedure. When comparing the use of silicone oil and olive oil during scaife preparation, oil and diamond particles were found in the debris during the early stages. After further working in, the Raman signatures of both oils disappeared. When a scaife was prepared with silicone oil, the oil component disappeared without trace, leaving amorphous carbon and small diamond particles, Raman spectroscopy of which showed a peak at ~1326 cm^{-1} that was associated with hexagonal diamond or lonsdaleite. The scaife surface was also

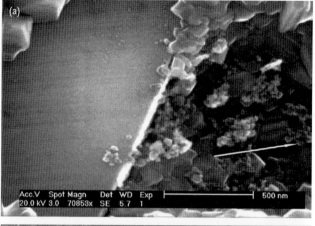

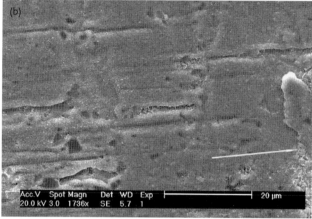

Figure 3.7 Electron micrographs of a scaife surface. (a) A flat-topped grit particle on a scaife plate, with its sharp edge aligned to interact with the oncoming diamond workpiece. Nanometer-sized grooves can be seen on the surface of this, which is an indication that reciprocal polishing has occurred; (b) Comet-like trails produced in the cast iron by the grit particles as they undergo alignment into high-friction directions. In both images the directions of motion of the scaife are indicated by the arrow [31, 32].

covered in an unconstrained fine powder, which suggested that a comminuting action had taken place on the grits during preparation. In contrast, the scaife prepared with olive oil showed debris with large quantities of amorphous carbon but very few diamond particles, appeared black and shiny, and was slightly sticky to the touch. Subsequent results obtained from a simple experiment in which both oils were heated on a glass plate and their residues analyzed were consistent with the observations made during the scaife preparation, namely that olive oil transforms leaving the signature of the D and G bands of amorphous carbon while silicone oil simply evaporates. Consistent with the findings of Couto et al. [51], iron compounds were also found. Additionally, it was found that increasing the power of the excitation laser induced chemical changes in these compounds.

There are two salient results from this work. First, amorphous carbon is found during the scaife preparation process, and its occurrence is not dependent on the presence of either silicone oil or olive oil. Second, olive oil degrades and forms a layer that entrains the fine powder, which is clearly visible during preparation when using silicone oil. The former point indicates that diamond is undergoing the same phase transformation as it does during polishing, while the latter point indicates that a comminution of diamond grit particles is also occurring. Indeed, the comminution of diamond grit particles was suggested as a possible cause of nanogrooving by Hird and Field [31]. In order for a brittle material to fail by fracture, sufficient surface energy must be provided in order to create the new surface. This has an interesting consequence during comminution, in that a brittle-to-ductile transition will be observed when insufficient surface energy is available for fracture [59–61].

Using the formulation adopted by Hagan [61], whereby it is assumed that the threshold of brittle fracture, c, is related to the hardness, H, and critical stress intensifier, K_{IC}, by

$$c \cong 30\left(\frac{K_{IC}}{H}\right)^2 \tag{3.1}$$

an order-of-magnitude approximation of the size at which an initially brittle particle becomes ductile upon repeated comminution can be made. For diamond, K_{IC} lies between $3.4\,\mathrm{MN\,m^{-3/2}}$ and $5\,\mathrm{MN\,m^{-3/2}}$, and H between 57 GPa and 104 GPa. The ranges of either value reflect conditions of loading, indenter type, and crystallographic orientation [12]. Substitution into this equation yields a range of values between 32 nm and 230 nm. Importantly, from a consideration of the fracture mechanics of the process of comminution, it can be shown that the minimum size of particle at which it is no longer considered useful as a sliding indenter (as it yields plastically at this point) is the same order of magnitude as the smallest grooves observed on polished diamond surfaces.

If all of the evidence so far is integrated, a detailed picture emerges as to the rôle and nature of the scaife. During preparation with a diamond oriented in the soft direction, or when using a sintered product such as SYNDAX, diamond grit particles contained in a slurry with olive oil will undergo comminution and/or embed themselves into the soft cast iron, or will be contained in grooves. The grit particles become oriented in a soft direction and are polished by the same process that transforms the diamond to a lesser form of carbon. The fracture of embedded grits along cleavage planes on the scaife ensures that a keen edge is presented to the oncoming diamond while, at the same time, the heat generated via frictional sliding causes a transformation of the olive oil to amorphous carbon (this is the black layer that serves to act as a transport and entrainment layer for small diamond particles). When such particles become trapped between a flat-topped diamond grit part and the diamond surface, they are able to act as sliding indenters and cause nanogrooves on the diamond surface. A grit particle on the scaife is thus seen to act like a chisel and an anvil.

If the scaife is to be improved, then one critical consideration must be the size of the diamond grit particles formed via comminution, as this ultimately dictates the topology of a diamond surface. It has already been noted that, in general terms, the smaller the grit used on the scaife, the less subsurface damage [52]. Comminution also goes some way to explaining the general finding that a large decrease in grit polishing size does not produce a correspondingly large decrease in surface roughness [41].

3.4.7
Atmosphere Dependence

A few studies have been reported on the environmental dependence of the wear rate during diamond polishing. Among these, Hitchiner et al. [62] noted that, during "hard" direction <011> polishing of a {011} surface, the wear rate could be decreased by lowering the pressure to 0.1 Torr. On readmitting air to 760 Torr, however, the wear rate was found to recover. Hard direction polishing in argon produced a similar result, with distinct differences being seen in the surface topology (examined using Nomarski microscopy) produced in vacuum and in air. The presence of isopropanol and water on a bonded scaife were found to increase the wear rate for {011}<100> [63].

More recently, Yarnitsky and coworkers have conducted a series of (unpublished) experiments in which a variety of gases (nitrogen, oxygen, carbon dioxide, argon, and helium) were used as the polishing atmosphere. With the exception of argon and helium, all were reported to increase the wear rate found under normal atmospheric conditions, and clearly there is room for further investigation of this topic.

3.4.8
Triboluminescence

One of the more curious observations made during diamond polishing is the possible emission of light at the interface between the scaife and a diamond as it is being polished. The intensity of the emitted light has been shown to exhibit an anisotropy [64]. Using a traditional polishing scaife, Hird et al. [64] showed that the maximum intensity is emitted in "hard" directions, whereas J. Chapman (unpublished results), using an impregnated scaife, noted that "soft" directions produce the most intense light emission. Chapman also noted that the light emission bore a resemblance to the UV-induced fluorescence of the stone. Hird et al. [64] showed that the spectrum had the resemblance of electroluminescence, and found that spectral change could be induced by slight changes in orientation of the diamond on the scaife. It was postulated that this was due to changes in frictional heating, a result which was consistent with observations of temperature dependence in diamond electroluminescence.

Light may also be emitted during the scratching of diamond by a sharp diamond-tipped stylus, though this has a different spectrum [64]. Irrespective of the cause of light emission during these processes, the appearance of triboluminescence during polishing warrants further attention, as it may provide direct information on the behavior of electrons during bond stretching and breaking, and during frictional electrification.

A further luminescent effect that occurs during polishing – albeit much dimmer and requiring significant dark adaption – involves the appearance of an annulus of light around the entire circumference of the track on which the diamond is being polished. Spectral analysis has shown this light to be broad band and to have a decay time of seconds. It is thought that this may be a chemiluminescence associated with iron oxide [64].

3.4.9
Wear Mechanism

The propensity of diamond to cleave along its octahedral plane [8, 9, 65] was the basis for early theories which attempted to describe the anisotropy in wear, and how material might be removed from the diamond during polishing. The first such theory was proposed by Tolkowsky [10], expanded by Bergheimer [15], given a modern scientific footing by Wilks and Wilks [28, 63] and Jeynes [66], but had its roots in the early ideas of crystal structure [67]. Essentially, this model depended on the relative ease of removing the minute octahedral and tetrahedral diamond fragments that had become separated from the diamond work piece. While such a mechanism would exhibit anisotropy in the ease of material removal, it could not account for the polishing debris, which has the appearance of soot rather than crystalline diamond. Neither could it explain the observations from friction experiments conducted with sharp styli [33, 42], where smooth grooves could be found on a nanometer scale. The nanogrooves found on polished diamond surfaces (as described by Couto *et al.* [49–51]) mean that fracture-based models are considered redundant for all but the wear induced in hard directions. As fracture conditions tend to be avoided due to the possibility of damaging the scaife plate and/or the valuable work piece, attention will be focused on soft direction polishing at this point.

It is possible on the basis of evidence acquired to date to rule out one possibility, namely that the wear mechanism involving or including iron carbide formation can (on the basis of studies of diamond polishing debris using Raman spectroscopy [39] and EELS [35, 36]) be neglected, as no significant quantities of iron or iron compounds have been found. Although scaife plates do wear down over time, cast iron has become the material of choice based on its material properties [32] rather than its chemistry. However, it is notable that diamond can be worn by metals during high-speed rubbing, and the diffusion of carbon from the diamond through metal films at elevated temperatures forms the basis of a novel processing technique (see Section 3.6.2).

Owing to the high-speed nature of the polishing process, a thermally induced mechanism of wear has always remained attractive, as diamond is readily graphitized in air above 900 K and the temperature at asperity hot spots (minute areas of localized heating) must be very high [42]. Further, non-diamond debris is found during diamond polishing. Such a process must be independent of any process that produces frictional anisotropy (as is present during slow-speed sliding). The thermal conductivity of diamond is very high at low temperatures but decreases rapidly as the diamond is heated [68]. Nonetheless, it has been observed that when being well polished, the diamond does not candesce even from the scaife–diamond interface; indeed, the bulk temperature of a diamond while being polished has generally been reported as between 500 and 700 K. It should also be noted that, whilst it is possible to make a diamond glow to red heat through overloading, under these circumstances damage to the scaife occurs, the possibility of fracture of the workpiece is increased, the surface finish is poor, and the wear rate appears to slow, possibly as a result of damage to the scaife. While there appears to be an activation energy for diamond polishing [30], there is however no evidence for a wear mechanism based solely on the diamond being graphitized or burning to form carbon dioxide during traditional diamond polishing.

Any theory of how a diamond wears must explain three key observations:

- Reciprocal polishing of the workpiece and of the scaife grit particles occur simultaneously.
- The wear debris from diamond polishing is non-diamond carbon.
- Nanogrooving is observed on the surfaces of grit particles and workpiece. The widths of these grooves match the calculated sizes of small diamond particles formed via comminution through theoretical arguments of surface energy requirements.

It will be seen in Section 3.5 that sliding indenters of small radii produce tensile stresses which can exceed the yield strength of diamond, and whose tracks have been observed to produce nanometer-sized grooving. It seems likely, therefore, that the nanogrooving observed on diamond surfaces and the grit particles is a direct consequence of the action of the comminuted diamond grit particles trapped between the flat grit particles and the diamond workpiece. These are contained within an interfacial transport layer composed of non-diamond carbon. The presence of these small diamond particles imposes a limit on the degree of surface smoothness that can be achieved using a diamond scaife.

Overcoming the yield stress of diamond is not a material removal mechanism in itself; rather, a breaking of bonds must occur for the material to be removed. Clearly, this must be either sp^3-bonded, or after some phase transformation, sp^2-bonded material. Motivated by insights provided by numerous research groups [34, 36, 49, 50, 69, 70], Jarvis *et al.* [69] found that a sp^3–sp^2 transformation can

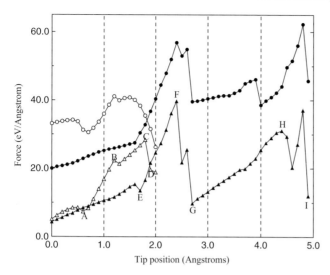

Figure 3.8 Forces experienced by a rigid tip during the deformation of an asperity on a (110) surface as a function of tip position. Triangles refer to the force that opposes the advance of the tip, and circles refer to the normal force. The soft direction is denoted by open symbols, while the hard direction is denoted by filled symbols. See main text for an explanation of the feature indicated by A–D (soft direction) [69].

occur by solely mechanical means. Using *ab initio* quantum mechanical simulations, it was possible to model the anisotropy in polishing rates, nanogroove formation, and post-wear debris by the microscopic action of a realistic diamond asperity which was allowed to slide in distinct crystallographic directions on the (110) diamond surface. In this way, by monitoring the forces on the asperity tip as a function of tip position, regimes could be demonstrated that corresponded to various events in the wear process. The salient features of these experiment are shown in Figure 3.8; the "soft" direction results indicated by positions A to D are discussed in the following subsections.

At point A in Figure 3.8, a slight decrease in the retarding force is experienced as the removal of a hydrogen atom occurs; this is followed by adhesion between the tip and workpiece, due to the formation of a C–C bond. Subsequent progression of this asperity along the surface causes a failure of one such carbon–carbon bond at point B. A period of compression causes the bonds between asperity and surface to become highly deformed, until a reconstruction of the surface occurs between C and D. At this point, the asperity loses its remaining structural strength, which allows it to be removed with relative ease from the surface. The structure of the asperity after these events is rendered amorphous, being neither graphitic nor diamond-like. These theoretical investigations fit many of the experimentally observable results, and have been extended and applied by others [29–31, 34–36, 39].

Direct evidence of a phase transformation has come about through combining indentation studies with micro-Raman spectroscopy [71, 72]. Indentation provides shear stresses in addition to hydrostatic stresses, and is attractive as an experimental technique since it is possible to drive and monitor phase transformations simultaneously. Amorphization and graphitization of the diamond, and the presence of other carbon polytypes in the impression, were detected in the indentations. Significantly, the transformations occurred during unloading of the stress.

More recently, Pastewka et al. [73] have provided the most intimate details of the phase transformation. This is of particular interest with regards to the future of the subject, namely the optimization of the diamond-polishing process. By using classical molecular dynamics, Pastewka et al. have successfully modeled experimental wear rates, anisotropy and the chemical nature of the wear debris by modeling the realistic interaction of two diamond surfaces sliding at $30\,\mathrm{m\,s^{-1}}$ with a force of 10 GPa between them (the experimental parameters for diamond polishing). The two diamond surfaces are shown to degrade mechanically and to result in an amorphous interface layer consisting of 60% sp^2-bonded and 20% sp^3-bonded carbon. This layer may be removed via two mechanisms:

- By chemical means, whereby an unstable thin layer of sp-bonded carbon chain 2 nm thick degrades to form carbon monoxide or carbon dioxide through a reaction with the surrounding atmosphere.

- By mechanical planarization, since grit particles are known to exhibit sharp edges [31]. This process is illustrated schematically in Figure 3.9.

With regards to the first mechanism, there is some evidence that atmospheric pressure may reduce the wear rate, but these experiments have only been conducted in hard polishing directions [62]. Further experiments on the atmospheric dependence of the wear rate may be able to corroborate this further. Experiments to detect carbon monoxide or carbon dioxide are probably futile, as diamond polishing debris – being of a carbonaceous nature – would also be expected to produce such gases upon thermal degradation.

3.5
Tribological Behavior of Diamond

3.5.1
Slow-Speed Sliding of Diamond against Diamond

In its most deconstructed form, the diamond–scaife interaction has often been considered as the interaction of a single, infinitely hard and sharp diamond grit particle sliding on a diamond surface [69]. Thus, it is interesting to compare the results obtained during slow-speed sliding diamond–diamond friction, where the

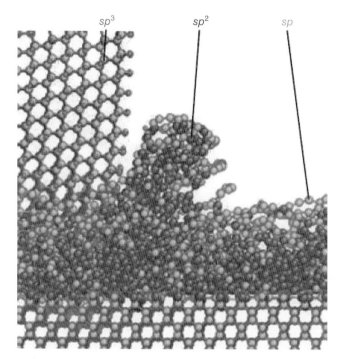

Figure 3.9 An illustration from a molecular dynamics study of diamond polishing by Pastewka et al. [73]. This image shows the planing action of a sharp diamond grit particle on the amorphous sp^2 (green), sp (beige) layer on a diamond surface as it is polished on a scaife. Such simulations can now be carried out using realistic speeds and loads. For a color version of this figure, please see the color plate at the end of this book.

action of a diamond stylus sliding over a diamond surface can be likened to a grit particle on a scaife, with a diamond surface undergoing high-speed sliding on a scaife. Such comparisons should be made cautiously due to the five-order of magnitude different in sliding speeds (friction experiments <0.5 mm s^{-1}, where frictional heating may be all but neglected; tangential scaife speed used in polishing ~50 000 mm s^{-1}, where heating is considerable). Further, the environmental conditions found in the scaife environment, which include iron (and other elements present in cast iron, such as cobalt), carbonaceous species (which might include oil residues and iron carbides, graphite, diamond and amorphous carbon) are in stark contrast to the rather more austere environment used in the environments where diamond–diamond friction experiments are carried out. Despite these seemingly irreconcilable experimental conditions, a number of striking similarities between the two processes have emerged. A complete review of the friction of diamond with a discussion of its many intricacies may be found in Ref. [54].

However, for the purpose of this chapter, the observed frictional anisotropy, the wear tracks, and the debris produced during sliding will be discussed.

Although the various beneficial tribological properties of diamond had been noted and used for several hundred years previously in horological escapements [2], it was not until 1951 [74] that the friction of diamond was re-studied. It did not take long before Kenyon [75] noticed that the friction – much like the wear behavior during polishing – was also anisotropic, and this discovery was further investigated by Seal [42], who established that the frictional anisotropy on cube {100} faces varied by a factor of between two and three, but was absent with a very low coefficient on octahedral surfaces. In the case of the cube plane, a low friction was observed in directions of polishing considered "hard" <110>, while a high friction was found in directions <100> considered "soft". The frictional behavior of other planes are similarly correlated with polishing anisotropy. Others have found that the frictional anisotropy becomes more marked with the cone angle of stylus [76] and of the stylus area of contact [46, 53], and have shown that the polishing lines (i.e., surface topography) contributed to the marked anisotropy. Seal [42] also showed that the degree of loading and area of contact produced different and distinctive wear tracks that tended towards ring cracks at higher loads (this is characteristic of Hertzian failure under tensile stress when the maximum stresses are directed radially out from the edge of the circle of contact). The cracks generally appeared parallel to where the octahedral planes intersected the plane of interest. Most importantly, it was shown that Auerbach's law (which states that for a brittle solid, the critical load p_c required to form a crack is directly proportional to the radius r of the indenter, or stylus) was upheld for diamond in this loading range. For simplicity of calculation, Seal considered the static case. However during the sliding of a stylus additional tensile stresses appear at the rear of the contact, and the tensile stress must be multiplied by an additional factor $(1+\mu A)$ [77], where μ is the coefficient of friction and A is proportional to the Poisson ratio σ and equal to

$$A = \frac{3\pi(4+\sigma)}{8(1-2\sigma)} \quad (3.2)$$

Although the linearity of Auerbach's law is, of course, unaffected by this factor, it is significant that the critical load for crack formation can be reduced by a factor of almost two if this is taken into account. Further, in addition to the frictional anisotropy, the elastic moduli of diamond are also highly anisotropic [12]; hence, a variety of tensile strengths are produced for each crystal surface, applied load, and sliding direction. It is important to note that the ring cracks will penetrate into the bulk so as to form distinctive cones if static pressure is further applied.

The main consequence of the Auerbach relation is that it implies that an indenter with a small radius can produce tensile stresses which exceed the yield strength of the material in question, and suggests that a failure mechanism other than cracking may occur. In the case of diamond, Seal [42] found that when diamond styli with radii between 15 and 40 µm (the smallest used) were traversed across a diamond wear tracks with heavy damage were produced. These exhibited smooth grooving features that were orders of magnitude less than the styli dimensions when imaged using reflection electron microscopy. The origin of these grooves was attributed to asperities present on the stylus. Many subsequent studies have been conducted on the friction of diamond, the results of which have upheld those of Seal. More recent improvements in the various microscopies have established that frictional sliding on diamond always produces some type of wear track [43, 78], and that the smooth grooving produced when sliding in high-friction directions ("soft" polishing directions) is of nanometer dimensions.

It was rapidly established during early studies of diamond friction that environmental conditions play a significant role in the frictional behavior, with heating in air, water, or paraffin all affecting the friction in some manner [33, 79, 80]. In particular, the presence of water has a dramatic effect on the coefficient of friction, by reducing it to about ~0.025 from an initial value of ~0.11 after approximately 1000 traversals of the track. There is also a slight pH dependence, with a minimum of friction at between pH 6 and pH 8 [81]. Under ultrahigh-vacuum conditions, the friction coefficient tends to become very high (>1.2) after a number of traversals. If the sliding surfaces are separated and left so that a few monolayers of residual gas in the chamber are allowed to settle, when sliding is continued the friction decreases before rising again after a few traversals, which clearly demonstrates the role of a surface film [82]. Partial pressures of various gases also affect friction; typically, friction was slightly reduced by oxygen but unchanged by nitrogen and hydrogen. Marked decreases in friction were seen, however, when oxygen and hydrogen were ionized by a hot tungsten cathode [80].

Clearly, the parameter space involved with the tribology of diamond surfaces is both large and varied, with fracture, adhesion, plastic deformation, surface topology, phononic vibrations [83] and the nature of the surface chemistry all having been shown to contribute to energy dissipation during frictional sliding.

3.5.2
Sliding of Diamond against Other Materials

The friction between monocrystalline diamond and other materials is of interest here as the scaife is made from cast iron. To date, however, there is only limited amount of published work in this research area. Using a tungsten carbide stylus sliding on a diamond {100} plane, Enomoto and Tabor [46] showed that frictional anisotropy is not observed over a range of loads, and further that no damage was produced on the diamond. Hird *et al.* (2008) found that synthetic oil could stabilize and reduce the friction of diamond sliding against hardened steel. In a vacuum environment, a correlation between the relative chemical activity of various transition metals and diamond has been noted [84], while a strong adhesion between diamond and copper, causing high friction, has also been described by Pepper [85], though only at elevated temperatures (1200 K).

3.6
Other Polishing Methods

3.6.1
Wear of Diamond by Other Materials

The wear of single-crystal diamond by materials other than by a scaife has been noted for some time, and is most often encountered during the machining of materials by diamond tools (for a review, see Ref. [63]). Bowden and Freitag [86] conducted an elegant experiment in which a metallic ball of copper, chromium, or steel was rotated electromagnetically at very high speeds (up to $700\,\mathrm{m\,s^{-1}}$) and allowed to decelerate freely when a diamond surface was brought into contact. Distinct transitions in the coefficient of friction were noted, and a dependence on the melting point of the metal was observed which led the authors to conclude that the formation of hotspots was important in the friction and wear of diamond by metals. The wear debris was shown to be carbon-based, and the possibility of a thermochemical-aided wear mechanism was discussed and not discounted. In the same year, Bowden and Scott [87] investigated the wear of diamond by glass, with the surprising result that significant wear of the diamond could be produced. Notably, the wear was significantly greater in a nitrogen atmosphere than in an oxygen atmosphere, and showed wear rates that exceeded those produced on a traditional diamond scaife under laboratory conditions. The wear debris was found to be carbon-based. An oxidative wear mechanism was ruled out, due to the fact that the wear would have been expected to be greater in an oxygen-rich environment rather than in a nitrogen environment. A high level of humidity was found to reduce the wear rate. Interestingly, the presence of a glow discharge was noted. This had also been observed by Valov [88], who had experimented with the use of flint and diamond some years earlier. Bowden and Scott noted that the discharge was green-colored in oxygen but violet-colored in nitrogen, which was in agreement with the appearance of glow discharges in those gases. The electrical discharge nature of this triboluminescence has more recently been confirmed [89], and under normal atmospheric conditions consists mainly of the second positive band of nitrogen. These results suggest that tribochemical reactions are indeed taking place, although a conclusive link to a wear mechanism has not yet been established. Gaissmaier and Weis [90] have also investigated the effects of discs made from a variety of metals and polymers on diamond surfaces. By using a geometry reminiscent of the diamond sawing operation, they showed that a circular blade made from tin of 0.6 mm width, and with a tangential speed of $6.2\,\mathrm{m\,s^{-1}}$ (diameter 125 mm), could produce very smooth wear scars on diamond. Subsequently, both steel and copper blades produced wear scars reminiscent of those produced by micro-abrasion testing [63]. In the case of the tin blade, an atmosphere of nitrogen suppressed the wear, whereas nitrogen appeared to increase the wear of diamond against steel.

3.6.2
Hot Metal Polishing

The polycrystalline nature of some chemical vapor-deposited diamond material does not exhibit the wear anisotropy demonstrated by monocrystalline diamond. A consequence of this is that mechanical lapping by the traditional scaife is impossible, and will result in permanent damage to the scaife. Consequently, abrasive lapping processes using bonded abrasive wheels and liquid cooling have been commonly adopted. Owing to the modest surface finish obtainable via this method, a variety of novel ways have been sought to effect wear and macroscale smoothing, such that today there exists a vast literature describing a myriad of techniques and experimental parameters. These take advantage of chemical reactions to attack diamond and degrade the diamond surface to graphitic or amorphous carbon which can be readily removed by an acid wash or via planarization [91].

Iron films [92] and manganese powders [93] heated at ~900 °C in contact with diamond have been noted to cause wear through catalysis and diffusion processes. The solubility of carbon into the metal determines the wear rate, and a degree of success has been achieved by using molten rare earth metals such as cerium [94] and lanthanum–nickel eutectic alloys [95] in enacting wear and smoothing of the chemical vapor-deposited diamond. As with the wet etching of diamond, wear proceeds faster on some planes than others [96], and a saturation of the metallic film is reached which limits further diffusion. Consequently, there is a tail-off in the wear rate.

In other experiments by Yoshikawa [97], diamond has been worn by contact with a rotating hot (~1000–1200 K) metallic disc of iron or nickel, but discs of cast iron and molybdenum failed to produce any appreciable wear on polycrystalline diamond. This is an interesting result since the catalytic conversion of diamond to graphitic or amorphous carbon by the cast iron scaife has often been mooted to be a contributory factor in the mechanism behind traditional diamond polishing. Yoshikawa also identified an atmospheric dependence and showed that the wear rate was greatest under vacuum conditions.

In more recent experiments, Bussone *et al.* [55] used an iron plate heated to 1100 K to polish a single-crystal diamond {111} surface. Such surfaces are known to be particularly problematic to polish on-plane.

Experiments conducted with a heated traditional scaife plate have also shown a temperature-dependent wear rate [98]. However, it was also found that direct heating of the diamond that was being polished had no effect on the wear rate.

3.6.3
Chemical–Mechanical Planarization

The most practical of all new polishing technologies is chemical–mechanical planarization (for a review, see Ref. [91]), which combines wet chemical etching with mechanical planarization and is used extensively in the semiconductor industry to polish materials other than diamond. The process has the advantage of being inexpensive and simple, and is currently used commercially to polish diamond. The basic principle of the scheme involves the application of oxidizing agents such as KOH or KNO_3 to a rotating disc of alumina, against which the diamond is loaded. Through a process of brittle fracture, micro- or nano-cracks are generated in the diamond, and these react readily with the oxidizing chemical to produce wear and polishing. If desired, diamond grit may be added to increase the polishing rate.

3.7
Producing High-Quality Planar Surfaces on Diamond

3.7.1
Cleaving

The propensity of diamond to cleave along its octahedral planes might, in principle, appear to be an ideal method for producing very flat surfaces on diamond. In practice, however, the cleavage planes are relatively rough, exhibiting steps that may vary in height between 10 and 200 nm, and have been described as river-like owing to their appearance (Figure 3.10). From a physical perspective, however, the presence of these steps is not entirely unexpected.

Cracks in brittle isotropic solids, such as glasses, bifurcate when v_F/c_R (v_F is the velocity of the propagating crack, c_R is the Rayleigh wave velocity) exceeds

Figure 3.10 The octahedral surface of a Type II diamond after cleaving, as viewed using Nomarski microscopy. These surfaces exhibit Wallner lines rather than being atomically smooth [99].

~0.6. In solids with well-defined cleavage planes (including diamonds) this ratio approaches unity [65, 100]; however, the longitudinal elastic wave velocity c_L and transverse elastic wave velocity c_S are very high in diamond: 17 300–19 500 m s^{-1} and 10 500–13 000 m s^{-1}, respectively [101]. The velocity of the growing cleavage crack during cleavage is significantly lower, and has been estimated as between 1000 m s^{-1} and 3000 m s^{-1} [102]. The fastest crack in diamond, as recorded by Field, was 7200 m s^{-1}. The stress wave travels ahead of the growing crack, and will be reflected at any unbound interface. This will in turn cause crack instabilities when the reflected wave interacts with the more slowly propagating crack tip. Wavelets are also generated by the growing crack tip, and these can also be reflected and subsequently add a tensile component to the propagating crack. The result of these interactions between the reflected waves and the crack tip are the steps that characterize the {111} cleavage plane. They are an example of "Wallner lines" that are readily observed on fracture surfaces on other materials [103]. Field [65] has pointed out that acoustic impedance matching the diamond with its clamp during cleavage would allow the stress wave to exit the diamond, thereby reducing crack tip instabilities; however, the ideal material for this is gold, which probably limits its widespread use. Another idea to circumvent the cleavage steps might be to use waveguide geometry to attenuate the velocity of the stress wave to match the crack during cleavage (J.R. Hird *et al.*, unpublished results). This might allow very flat, damage-free, high-aspect ratio diamond plates to be produced.

3.7.2
Post-Mechanical Polishing Treatment

A well-polished mechanical diamond surface might have a typical rms surface roughness of about 1 nm. Whilst decreasing the grit size increases the smoothness of the surfaces, the effect is not in direct proportion to the grit size [41]. The use of smaller grit sizes is known to reduce the chance of subsurface damage, however [52]. *Zeoten* – that is, the radial motion of the diamond over the scaife – does not produce a flat surface on the nanoscale [50, 51]. One technique used widely in the gemstone industry to produce mirror-like finishes is to mis-orient the stone into a "hard" direction – that is, a few degrees from the ideal polishing direction. This results in a smoother, but not necessarily damage-free, surface. This technique has been used in conjunction with one of several post-polishing etching techniques to uniformly remove the top several hundred nanometers of the polished surface, allowing the production of very smooth, damage-free diamond surfaces for scientific research [104].

3.7.3
Dry Chemical Etching

It has been recognized for some time [105] that atomic hydrogen may promote diamond growth during synthesis by assisting sp^3 bond formation, and can also

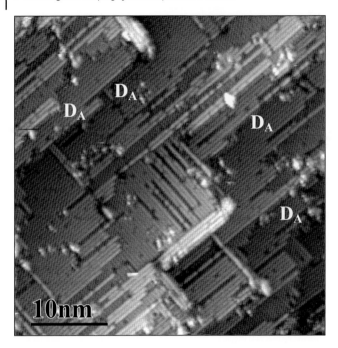

Figure 3.11 A smooth, hydrogen-free diamond (100)-(2 × 1) surface (double steps denoted by D_A). This surface has been etched by atomic hydrogen at 1273 K for 5 min [108].

be used as an etchant of graphite and of diamond, on which it preferentially etches (100) facets. Ravi *et al.* [106] described the details of macroscale atomic smoothing using a hydrogen etch on polycrystalline diamonds films, while Thoms *et al.* [107] extended these studies to include single-crystal diamond and were able to demonstrate smooth and well-ordered (100) surfaces that were stable in air. The experimental parameters required to produce such surfaces were similar to those employed during diamond growth, namely a hydrogen atmosphere of 10 Torr, with the substrate held at a temperature of ~1100 K for 10–90 min. Subsequent atomic force microscopy measurements showed that the treated surfaces had height variations of only 0.4 nm over a 2 × 2 μm area.

Stallcup and Perez [108] obtained atomic-resolution images of (100) diamond films, again by etching with atomic hydrogen (Figure 3.11), and showed that the surface quality obtained was a function of the substrate temperature. In this case, the experimental parameters were 30 Torr hydrogen with a flow rate of 200 sccm (standard cubic centimeter per minute), and a filament temperature of 2473 K. The substrate temperature found to produce the smoothest films was 1273 K.

Recently, Schwitters *et al.* [104] have reported atomically smooth surfaces produced by polishing cube facets in soft directions with 6–12 μm grit and a second polish along the "hard" direction <110> with a grit size distribution <1 μm), using a low load. When applying this technique, the authors reported that no polishing grooves could be observed using Nomarski microscopy. Following thorough cleaning and rinsing, the diamond was subjected to a hydrogen etch using a 600 W chemical vapor deposition (CVD) reactor in a 10 Torr hydrogen atmosphere for 30 min. The substrate was first heated to 1100 K and then cooled in the plasma to 500 K in order to maximize hydrogen termination. Ultimately, the rms roughness of the surface was found to be only ~0.1 nm.

For many applications, it is not only the surface roughness that must be removed but also any subsurface damage resulting from prior mechanical polishing. For this, the state-of-art is to use an Ar/Cl_2 based inductively coupled plasma (ICP) which shows no preferential etching of damage [109]. In this approach, the experimental parameters used were a platen power of 300 W, a coil power of 400 W, 5 mTorr chamber pressure, and flow rates of 25 sccm Ar and 40 sccm Cl_2. The authors reported a macroscale smoothing of polished diamond surfaces to an rms roughness of just 0.18 nm over a 1 µm^2 area for an etch duration of 10 min. The etch rate at a coil power of 400 W was found to be ~0.06 µm min^{-1}, and the as-received diamond had a roughness of 1.1 nm over a 5 × 5 µm area.

3.8
Nonplanar and Structured Geometries

It is possible to overcome the anisotropy in wear to form curved, even spherical, highly polished surfaces on single-crystal diamond via mechanical means. Bruting—that is, the turning of a gemstone diamond on a lathe using another diamond as the tool—has been used in the gemstone industry for many centuries to form the girdle (the dividing circular cross-section where the crown facets meet the base facets) of a round-cut gemstone [7]. The procedure proceeds via a microchipping or abrasive wear process, and the resultant surface is relatively rough. Ramaseshan [8] has reported that some natural diamond fractures may even exhibit curved shapes.

Although Andrew Pritchard's diamond microscope lenses of the early nineteenth century are perhaps the best example of a precision spherical surface on diamond produced by mechanical means, they remain relatively unknown to those outside the field of the history of science. Pritchard, an optician working in Edinburgh, recorded the precise means by which he made the lenses [3]. The first stage was to grind the diamond parallel in cross-section, after which a bruting-type process was employed in which a diamond was used to create, by abrasion, a spherical form on one of the faces. A concave tool of cast iron of the required curvature and radius was then used which has been "paved" with diamond grit hammered into it, using a steel punch, and rotated in the lathe at about 60 revolutions per second. The diamond workpiece was then mounted in a short handle and held against the concave tool while it was rotating. Having roughly formed the initial convex surface, a second concave tool containing a very fine diamond grit and oil mixture was then used to finish the surface. The sizes of lenses that

Pritchard created were minute at between 1 and 2 mm, with foci varying between 0.75 and 1.3 mm.

Another famous technological example of curved diamond surfaces occurred in the escapement of John Harrison's watches that were produced in the mid-eighteenth century [2]. These were similarly minute but had roughly cylindrical facets (see Figure 3.1). It is not known exactly how Harrison polished these lenses, but a method similar to that used by Pritchard must have been applied, albeit using the inside of a cylinder for the form or a concave ring cut into a scaife plate. The latter method has recently been used successfully by Derek Pratt to recreate Harrison's diamond pallets (as seen in Figure 3.1). Provided that the diamond part is small with respect to the diameter of the scaife, any nonlinearity in form that arises when using this technique will be small.

Whereas Pritchard's jewel microscope lenses were obsolete almost immediately after he had perfected them, diamond lenses have recently re-emerged owing to other attractive properties of diamond including its wide optical bandwidth and ability to manage high optical power densities. Recently, diamond microlenses with excellent sphericity have been fabricated using top-down fabrication techniques (see Chapter 6 for a full description of this technique).

3.9
Summary

Despite the considerable progress that has been made in the area of diamond polishing during recent years, key scientific and technological challenges remain and will need to be overcome if the true benefits of the industrial processing of diamond are to be fully realized.

Currently, the polishing and shaping process remains relatively slow and expensive, with a refined version of the diamond polishing apparatus – which bears a great likeness to that used in the Renaissance – still being the preferred method for the rapid placement of high-quality planar surfaces on diamond. Unfortunately, this process is heavily dependent on the skill of the operator, and subsurface damage can easily be introduced which must then be removed by other means. Moreover, the polished surface is not atomically planar, but rather consists of smooth grooves of nanometer dimensions. These grooves are caused by the comminution of diamond particles, and their size represents a fundamental limit of achievable surface roughness. The use of a post-polishing process, or perhaps a new polishing technique, is necessary if higher degrees of flatness are to be achieved.

The present understanding of the polishing mechanism is also far from complete, with many interesting scientific questions remaining unanswered. For example, the dependence of diamond polishing on atmosphere has never been adequately explored. While such studies might increase the wear rate and/or elucidate the mechanism of polishing, further investigations of triboluminescence during polishing may provide an interesting handle on the precise mechanism of wear, as the emitted light must be a result of electronic transitions from within the diamond. These effects will be dependent on temperature and stress. The ability to measure the interfacial temperature between the diamond and the scaife, using non-contact IR techniques or perhaps even using the junction of the scaife (iron) and diamond (n- or p-type) as a thermocouple, would provide further insight into the mechanism.

The past two decades have witnessed major advances in the scientific understanding of diamond polishing processes, which have been shown to be both subtle and complex. Ever-more sophisticated apparatus and ingenuity will be required to take this topic forward: diamond does not give up its secrets without a fight!

References

1 Pritchard, A. (1832) *The Microscopic Cabinet of Select Animated Objects; with A Description of the Jewel and Doublet Microscope, Test Objects, &C*, Whittaker, Treacher, and Arnot, London.
2 Hird, J.R., Betts, J.D., and Pratt, D.F. (2008) The diamond pallets of John Harrison's fourth longitude timekeeper-H4. *Ann. Sci.*, **65** (2), 171–200.
3 Pritchard, A. (1829) On the art of forming diamonds into single lenses for microscopes. *Edinburgh J. Sci.*, **1**, 147–155.
4 Turner, G.L. (1968) The rise and fall of the jewel microscope, 1824–1837. *Microscopy*, **31**, 85–94.
5 Delfs, T. (2004) Diamond balance springs. *Horol. J.*, **49**, 50.
6 Lenzen, G. (1970) *[Produktions- Und Handelsgeschichte Des Diamanten.] The History of Diamond Production and the Diamond Trade*, Barrie & Jenkins, London.
7 Watermeyer, B. (1982) *Diamond Cutting: A Complete Guide to Diamond Processing*, 2nd edn, Centaur, Johannesburg.
8 Ramaseshan, S. (1946) The cleavage properties of diamond. *Proc. Math. Sci.*, **24** (1), 114–121.
9 Telling, R.H., *et al.* (2000) Theoretical strength and cleavage of diamond. *Phys. Rev. Lett.*, **84** (22), 5160–5163.
10 Tolkowsky, M. (1920) Research on the Abrading, Grinding or Polishing of Diamond, DSc thesis, City and Guilds College, University of London.
11 Kraus, E.H. and Slawson, C.B. (1939) Variation of hardness in the diamond. *Am. Mineral.*, **24**, 661–676.
12 Field, J.E. (1992) Appendix: table of properties, in *The Properties of Natural and Synthetic Diamond* (ed. J.E. Field), Academic Press, pp. 667–699.
13 Ashby, M.F. and Lim, S.C. (1990) Wear-mechanism maps. *Scripta Metall. Mater.*, **24** (5), 805–810.
14 Tolkowsky, M. (1919) *Diamond Design: A Study of the Reflection and Refraction of Light in A Diamond. With 37 Illustrations*, E. & F. N. Spon, London.
15 Bergheimer, H.N. (1938) Die schleifhärte des diamanten und seine struktur. *Neues Jahrbuch für Miner. Geol. und Palaeont. Beilagen Bd.*, **74**, 318–332.
16 Peters, C.G., Nefflen, K.F., and Harris, F.K. (1945) Diamond cutting accelerated by an electric arc. *J. Res. Natl Bur. Stand.*, **34** (6), 587–593.
17 Whittaker, H. and Slawson, C.B. (1946) Vector hardness in diamond tools. *Am. Mineral.*, **31**, 143–149.
18 Winchell, H. (1946) Observations on orientation and hardness variations. *Am. Mineral.*, **31**, 149–152.
19 Denning, R.M. (1953) Directional grinding hardness in diamond. *Am. Mineral.*, **38**, 108–117.
20 Hukao, Y. (1955) Abrading tests of diamonds. *Ind. Diamond Rev.*, **15**, 45–48, 107–109.

21 Grodzinski, P. (1949) Diamond and gemstone industrial production methods. *Ind. Diamond Rev.*, **9**, 118–124, 152–156.

22 Grodzinski, P. and Stern, W. (1949) Abrasion tests on diamond. *Nature*, **164**, 193–194.

23 Grodzinski, P. (1953) *Diamond Technology: Production Methods for Diamond and Gem Stones*, 2nd edn, N.A.G. Press, London.

24 Slawson, C.B. and Kohn, J.A. (1950) Maximum hardness vectors in the diamond. *Ind. Diamond Rev.*, **10**, 168–172.

25 Wilks, E.M. (1952) An interferometric investigation of the abrasion hardness properties of diamond. *Philos. Mag.*, **43**, 1140–1145.

26 Wilks, E.M. and Wilks, J. (1954) The hardness properties of cube faces of diamond. *Philos. Mag.*, **45**, 844–850.

27 Wilks, E.M. and Wilks, J. (1959) The resistance of diamond to abrasion. *Philos. Mag.*, **4**, 158–170.

28 Wilks, E.M. and Wilks, J. (1972) The resistance of diamond to abrasion. *J. Phys. D Appl. Phys.*, **5** (10), 1902.

29 Grillo, S.E., Field, J.E., and van Bouwelen, F.M. (2000) Diamond polishing: the dependency of friction and wear on load and crystal orientation. *J. Phys. D Appl. Phys.*, **33** (8), 985–990.

30 Hird, J.R. and Field, J.E. (2005) A wear mechanism map for the diamond polishing process. *Wear*, **258** (1-4), 18–25.

31 Hird, J.R. and Field, J.E. (2004) Diamond polishing. *Proc. R. Soc. A Math. Phys. Eng. Sci.*, **460** (2052), 3547–3568.

32 Hird, J.R. (2002) *The Polishing of Diamond*, Cambridge University Press, Cambridge.

33 Feng, Z. and Field, J.E. (1992) The friction and wear of diamond sliding on diamond. *J. Phys. D Appl. Phys.*, **25** (1A), A33–A37.

34 Grillo, S.E. and Field, J.E. (1997) The polishing of diamond. *J. Phys. D Appl. Phys.*, **30** (2), 202–209.

35 van Bouwelen, F.M., Field, J.E., and Brown, L.M. (2003) Electron microscopy analysis of debris produced during diamond polishing. *Philos. Mag.*, **83** (7), 839–855.

36 van Bouwelen, F.M., et al. (1996) Wear by friction between diamonds studied by electron microscopical techniques. *Diamond Relat. Mater.*, **5** (6-8), 654–657.

37 Chakrapani, V., et al. (2007) Charge transfer equilibria between diamond and an aqueous oxygen electrochemical redox couple. *Science*, **318** (5855), 1424–1430.

38 Pate, B.B. (1986) The diamond surface: atomic and electronic structure. *Surf. Sci.*, **165** (1), 83–142.

39 Hird, J.R., Bloomfield, M., and Hayward, I.P. (2007) Investigating the mechanisms of diamond polishing using Raman spectroscopy. *Philos. Mag.*, **87** (2), 267–280.

40 Seal, M. and Menter, J.W. (1953) Crystallographic slip in diamond. *Philos. Mag. Ser. 7*, **44** (359), 1408–1410.

41 Bailey, A.I. and Seal, M. (1956) The surface topography of a polished diamond. *Ind. Diamond Rev.*, **16**, 145–148.

42 Seal, M. (1958) The abrasion of diamond. *Proc. R. Soc. Lond. Ser. A Math. Phys. Sci.*, **248** (1254), 379–393.

43 Wang, Z.L., Feng, Z., and Field, J.E. (1991) Imaging friction tracks on diamond surfaces using Reflection Electron-Microscopy (REM). *Philos. Mag. A Phys. Condens. Matter Struct. Defects Mech. Prop.*, **63** (6), 1275–1289.

44 Postek, M.T. and Evans, C.J. (1989) Inspection of single-point diamond turning tools at low accelerating voltage in a scanning electron microscope. *Scanning Microsc.*, **3** (3), 435–442.

45 Thornton, A.G. and Wilks, J. (1976) An examination of polished diamond surfaces in the electron microscope. *J. Phys. D Appl. Phys.*, **9** (1), 27.

46 Enomoto, Y. and Tabor, D. (1981) The frictional anisotropy of diamond. *Proc. R. Soc. Lond. A Math. Phys. Sci.*, **373** (1755), 405–417.

47 Ennos, A.E. (1953) The origin of specimen contamination in the electron microscope. *Br. J. Appl. Phys.*, **4**, 101–106.

48 Decker, D.L., et al. (1984) Preselection of diamond single-point tools. *SPIE Proc.*, **508**, 132–139.
49 Couto, M., et al. (1992) Scanning tunneling microscopy of polished diamond surfaces. *Appl. Surf. Sci.*, **62** (4), 263–268.
50 Couto, M.S., van Enckevort, W.J.P., and Seal, M. (1994) On the mechanism of diamond polishing in the soft directions. *J. Hard Mater.*, **5** (2), 31–47.
51 Couto, M.S., van Enckevort, W.J.P., and Seal, M. (1994) Diamond polishing mechanisms: an investigation by scanning tunnelling microscopy. *Philos. Mag. B*, **69** (4), 621–641.
52 Friel, I., et al. (2009) Control of surface and bulk crystalline quality in single crystal diamond grown by chemical vapour deposition. *Diamond Relat. Mater.*, **18** (5–8), 808–815.
53 Samuels, B. and Wilks, J. (1988) The friction of diamond sliding on diamond. *J. Mater. Sci.*, **23** (8), 2846–2864.
54 Tabor, D. and Field, J.E. (1992) Friction of diamond, in *The Properties of Natural and Synthetic Diamond* (ed. J.E. Field), Academic Press, London, pp. 547–571.
55 Bussone, G., et al. (2011) Investigation of surface and sub-surface damage in high quality synthetic diamonds by X-ray reflectivity and grazing incidence in-plane diffraction. *Phys. Status Solidi A Appl. Mater. Sci.*, **208** (11), 2612–2618.
56 Knight, D.S. and White, W.B. (1989) Characterization of diamond films by Raman spectroscopy. *J. Mater. Res.*, **4**, 385–393.
57 Ferrari, A.C. and Robertson, J. (2004) Raman spectroscopy of amorphous, nanostructured, diamond-like carbon, and nanodiamond. *Philos. Trans. R. Soc. A Math. Phys. Eng. Sci.*, **362** (1824), 2477–2512.
58 Prawer, S. and Nemanich, R.J. (2004) Raman spectroscopy of diamond and doped diamond. *Philos. Trans. R. Soc. Lond. Ser. A Math. Phys. Eng. Sci.*, **362** (1824), 2537–2565.
59 Steir, K. and Schönert, K. (1972) Verformung und bruchphänomene unter druckbeanspruchung von sehr kleinen körnern aus kalkstein, quartz, und polystyrol. *Dechema-Monographien*, **69**, 167.
60 Kendall, K. (1978) The impossibility of comminuting small particles by compression. *Nature*, **272** (5655), 710–711.
61 Hagan, J.T. (1981) Impossibility of fragmenting small particles: brittle–ductile transition. *J. Mater. Sci.*, **16** (10), 2909–2911.
62 Hitchiner, M.P., Wilks, E.M., and Wilks, J. (1984) The polishing of diamond and diamond composite materials. *Wear*, **94** (1), 103–120.
63 Wilks, J. and Wilks, E. (1991) *Properties and Applications of Diamond*, Butterworth-Heinemann.
64 Hird, J.R., Chakravarty, A., and Walton, A.J. (2007) Triboluminescence from diamond. *J. Phys. D Appl. Phys.*, **40** (5), 1464–1472.
65 Field, J.E. (1971) Brittle fracture; its study and application. *Contemp. Phys.*, **12**, 1–31.
66 Jeynes, C. (1983) A proposed diamond polishing process. *Philos. Mag. A*, **48** (2), 169–197.
67 Bragg, W.L.S. (1962) *The Crystalline State: A General Survey*, 2nd edn, Bell, London.
68 Berman, B. (1979) Thermal properties, in *The Properties of Diamond* (ed. J.E. Field), Academic Press, pp. 3–22.
69 Jarvis, M.R., et al. (1998) Microscopic mechanism for mechanical polishing of diamond (110) surfaces. *Phys. Rev. Lett.*, **80** (16), 3428–3431.
70 van Bouwelen, F.M., et al. (1996) Wear by friction between diamonds studied by electron microscopical techniques. *Diamond Relat. Mater.*, **5** (6–8), 654–657.
71 Gogotsi, Y.G., Kailer, A., and Nickel, K.G. (1998) Pressure-induced phase transformations in diamond. *J. Appl. Phys.*, **84** (3), 1299–1304.
72 Gogotsi, Y.G., Kailer, A., and Nickel, K.G. (1999) Materials–transformation of diamond to graphite. *Nature*, **401** (6754), 663–664.
73 Pastewka, L., et al. (2011) Anisotropic mechanical amorphization drives wear in diamond. *Nat. Mater.*, **10** (1), 34–38.
74 Bowden, F.P. and Young, J.E. (1951) Friction of diamond, graphite, and

carbon and the influence of surface films. *Proc. R. Soc. Lond. Ser. A. Math. Phys. Sci.*, **208** (1095), 444–455.

75 Bowden, F.P. (1954) Redwood Lecture: the friction of non-metallic solids. *J. Inst. Petrol.*, **40**, 89–103.

76 Bowden, F.P. and Brookes, C.A. (1966) Frictional anisotropy in nonmetallic crystals. *Proc. R. Soc. Lond. Ser. A Math. Phys. Sci.*, **295** (1442), 244–258.

77 Hamilton, G.M. and Goodman, L.E. (1966) The stress field created by a circular sliding contact. *J. Appl. Mech.*, **33** (2), 371–376.

78 Couto, M.S., van Enckevort, W.J.P., and Seal, M. (1997) Friction tracks on diamond surfaces imaged by atomic force microscopy. *Diamond Relat. Mater.*, **6** (8), 975–982.

79 Feng, Z. and Field, J.E. (1991) Friction of diamond on diamond and chemical vapor-deposition diamond coatings. *Surf. Coat. Technol.*, **47** (1–3), 631–645.

80 Feng, Z., Tzeng, Y., and Field, J.E. (1992) Friction of diamond on diamond in ultra-high vacuum and low-pressure environments. *J. Phys. D Appl. Phys.*, **25** (10), 1418–1424.

81 Grillo, S.E. and Field, J.E. (2000) Very low friction for natural diamond in water of different pH values. *Eur. Phys. J. B*, **13** (3), 405–408.

82 Bowden, F.P. and Hanwell, A.E. (1966) The friction of clean crystal surfaces. *Proc. R. Soc. Lond. Ser. A Math. Phys. Sci.*, **295** (1442), 233–243.

83 Cannara, R.J., et al. (2007) Nanoscale friction varied by isotopic shifting of surface vibrational frequencies. *Science*, **318** (5851), 780–783.

84 Miyoshi, K. and Buckley, D.H. (1980) Adhesion and friction of single-crystal diamond in contact with transition metals. *Appl. Surf. Sci.*, **6** (2), 161–172.

85 Pepper, S.V. (1982) Diamond (111) studied by electron energy loss spectroscopy in the characteristic loss region. *Surf. Sci.*, **123** (1), 47–60.

86 Bowden, F.P. and Freitag, E.H. (1958) The friction of solids at very high speeds. I. Metal on metal; II. Metal on diamond. *Proc. R. Soc. Lond. Ser. A Math. Phys. Sci.*, **248** (1254), 350–367.

87 Bowden, F.P. and Scott, H.G. (1958) The polishing, surface flow and wear of diamond and glass. *Proc. R. Soc. Lond. Ser. A Math. Phys. Sci.*, **248** (1254), 368–378.

88 Valov, A.A. (1948) *Ind. Diamond Rev.*, **8**, 22.

89 Miura, T. and Nakayama, K. (2000) Spectral analysis of photons emitted during scratching of an insulator surface by a diamond in air. *J. Appl. Phys.*, **88** (9), 5444–5447.

90 Gaissmaier, K. and Weis, O. (1993) Superpolishing of diamond. *Diamond Relat. Mater.*, **2** (5–7), 943–948.

91 Malshe, A.P., et al. (1999) A review of techniques for polishing and planarizing chemically vapor-deposited (CVD) diamond films and substrates. *Diamond Relat. Mater.*, **8** (7), 1198–1213.

92 Jin, S., et al. (1992) Massive thinning of diamond films by a diffusion process. *Appl. Phys. Lett.*, **60** (16), 1948–1950.

93 Jin, S., et al. (1992) Polishing of CVD diamond by diffusional reaction with manganese powder. *Diamond Relat. Mater.*, **1** (9), 949–953.

94 Jin, S., et al. (1993) Shaping of diamond films by etching with molten rare-earth-metals. *Nature*, **362** (6423), 822–824.

95 Johnson, C.E. (1994) Chemical polishing of diamond. *Surf. Coat. Technol.*, **68–69**, 374–377.

96 Jin, S., et al. (1994) Anisotropy in diamond etching with molten cerium. *Appl. Phys. Lett.*, **65** (21), 2675–2677.

97 Yoshikawa, M. (1990) Development and performance of a diamond-film polishing apparatus with hot metals. *Proc. SPIE*, **1325**, 210–221.

98 Yarnitsky, Y., et al. (1988) The dynamometer as a qualitative gauge in diamond polishing. *Mater. Sci. Eng. A*, **105–106**, Part 2 (0), 565–569.

99 Wilks, J. and Wilks, E.M. (1992) Wear and polishing of diamond, in *The Properties of Natural and Synthetic Diamond* (ed. J.E. Field), Academic Press, pp. 573–604.

100 Field, J.E. (1979) Strength and fracture properties of diamond, in *The Properties of Diamond* (ed. J.E. Field), Academic Press, London, pp. 281–324.

101 Willmott, G.R. and Field, J.E. (2006) A high-speed photographic study of fast cracks in shocked diamond. *Philos. Mag.*, **86** (27), 4305–4318.

102 Field, J.E. (1992) Strength, fracture and erosion properties of diamond, in *The Properties of Natural and Synthetic Diamond* (ed. J.E. Field), Academic, pp. 473–513.

103 Wallner, H. (1939) Linenstrukturen an Bruchflächen. *Z. Phys.*, **114**, 368–378.

104 Schwitters, M., *et al.* (2011) Contribution of steps to optical properties of vicinal diamond (100):H surfaces. *Phys. Rev. B*, **83** (8), 085402-1–085402-7.

105 Spitsyn, B.V., Bouilov, L.L., and Derjaguin, B.V. (1981) Vapor growth of diamond on diamond and other surfaces. *J. Cryst. Growth*, **52**, 219–226.

106 Ravi, K.V.E. and Dismukes, J.P.E. (1995) Diamond materials. Fourth International symposium: 187th Meeting: selected papers. The Electrochemical Society, Pennington, New Jersey, 1993.

107 Thoms, B.D., *et al.* (1994) Production and characterization of smooth, hydrogen-terminated diamond C(100). *Appl. Phys. Lett.*, **65** (23), 2957–2959.

108 Stallcup, R.E. and Perez, J.M. (2001) Scanning tunneling microscopy studies of temperature-dependent etching of diamond (100) by atomic hydrogen. *Phys. Rev. Lett.*, **86** (15), 3368–3371.

109 Lee, C.L., *et al.* (2008) Etching and micro-optics fabrication in diamond using chlorine-based inductively-coupled plasma. *Diamond Relat. Mater.*, **17** (7–10), 1292–1296.

4
Refractive and Diffractive Diamond Optics
Fredrik Nikolajeff and Mikael Karlsson

4.1
Introduction

The engineering of materials for optical purposes is by no means a recent discovery, with early mirrors of polished copper and bronze having survived from ancient Egypt and a planar convex glass lens being recovered in Pompeii. Although glass has for many years been – and to a large extent still is – the dominant optical material for the visible regime, other materials such as polymers and ceramics have now come into play. Indeed, during the past century an extension of the usable electromagnetic spectrum into the ultraviolet (UV) and infrared (IR) wavelengths has led to a veritable explosion in the number of industrial and consumer devices that take advantage of optical phenomena. This, in turn, has necessitated the palette of optical materials to be expanded to include oxides and nitrides, semiconductors, and even rare earth-doped materials. The importance of materials within the science of optics cannot be emphasized enough – a fact which is also evidenced by the increasing number of journals dedicated to the broad area where optics and the materials sciences overlap.

In this chapter, topics of refractive and diffractive optics will first be discussed, including classical optical components such as windows, lenses, prisms, filters and coatings, as well as modern inventions such as subwavelength gratings, diffractive Fresnel lenses, and fan-out elements. The notation will be introduced of *linear optics* for this class of optical elements (nonlinear effects are discussed elsewhere; see in particular Chapters 8 and 9). The advantages that diamond offers as a linear optical material include the following: optical transmission over a large wavelength range; the ability to withstand high optical powers; and superior mechanical and thermal properties. Unfortunately, slow progress in the production of synthetic diamond of good optical quality has for a long time prohibited the advance of diamond optics. It is natural, therefore, that some of the first suggested experiments with diamond optics were conducted in extreme environments requiring very high-power lasers, and also in space, where conventional materials had proven previously to be insufficient. As a consequence, with a steady decrease in the production costs of free-standing diamond substrates, optical engineers have more recently been allowed to exploit diamond for a wider variety of applications, including their use as internal reflection elements in evanescent vibrational spectroscopy. The high refractive index of diamond also makes it suitable for the confinement and collection of light, and this has proven valuable for applications such as high-density data storage systems and medical in-vivo imaging tools.

In the following sections, information will be provided regarding the use of diamond windows, the development of techniques for the micro- and nanostructuring of diamond, and how the utilization of such methods can lead to the realization of high-performance refractive and diffractive diamond optics. Other important engineering aspects, such as coatings and surface finishing, are also outlined, and an overview is provided of the practical applications of diamond optics in science, industry, and medicine. Finally, suggestions are made as to how the next generation of diamond linear optics might benefit humankind in particular.

4.2
Windows and Domes

Diamond has many unique characteristics, including the transmission of light from UV regions to the far IR, providing a high thermal conductivity, and demonstrating extreme durability and resistance against a wide variety of chemicals. Despite these properties making diamond extremely interesting for use in a wide range of optical applications, it was not until the late 1970s that a feasible means was determined to produce synthetic diamond windows of optical grade.

One of the earliest modern examples of diamond optics was provided in 1978, when a small circular window made from a natural diamond stone was produced for the NASA Pioneer probe destined to Venus [1, 2]. The window was 18.2 mm in diameter and had a thickness of 2.8 mm, and its purpose was to act as an IR-transmissive and mechanically protective shield in front of an IR radiometer. Diamond was the material of choice since it was considered to be the only material that could endure the corrosive, high-temperature, high-pressure environment, and still transmit IR light. The cost—estimated at more than US$ 250 000 in 1990 [3]—was of course significant; in fact, simply finding the correct diamond stone for the purpose was a time-consuming task.

The emergence of the chemical vapor deposition (CVD) synthesis of diamond with low impurities during the 1980s signaled the start of significant research activity to manufacture synthetic diamond windows. As diamond produced by CVD ("CVD diamond") is potentially much less expensive than natural diamond, the introduction of synthetic diamond opened up many possibilities for diamond optics in industrial segments other than exotic applications such as the above-described space mission. Although, for the Pioneer project the manufacturing cost of the diamond window was very high, it was still considered reasonable in comparison to the total cost of the entire mission. However, this type of application must be considered very rare.

By the 1990s, however, several research groups had been able to tune the CVD process in order to fabricate large-area freestanding polycrystalline diamond substrates of optical quality. Single-crystal diamond in its purest natural form (Type IIa) has a high optical transmission that extends from 220 nm to the far IR, except for an absorption region of moderate intensity in the mid-IR range due to multi-phonon bands. In many respects, optical grade polycrystalline diamond has characteristics that are comparable with Type IIa natural diamond; for example, the thermomechanical properties are similar between the two. Optical transmission in the IR range is also identical, but at shorter wavelengths single-crystal diamond begins to be more favorable for use. This benefit is due mainly to the loss of light caused by scattering in the polycrystalline diamond and absorption in nondiamond carbon located in the grain boundaries. In other words, at shorter wavelengths single-crystal diamond is often preferable, although for parts of the visible spectrum and further into the IR regime diffraction-limited optics can be fabricated from polycrystalline diamond of optical grade.

During the past ten years, polycrystalline diamond windows have found several applications in the IR range between 8 and 12 μm; an example is the exit windows of CO_2 lasers operating at a wavelength of 10.6 μm. This type of high-power industrial laser is widely used for machining, cutting, and welding. By introducing diamond as the exit window, the output power of the operating laser has been allowed to increase to several kilowatts, which in turn leads to increased production levels (e.g., faster laser machining or cutting). At the time of writing, diamond polycrystalline windows are available commercially with diameters up to 150 mm, and single-crystalline windows up to 10 mm.

Based on diamond's ability to transmit radiation across the wavelength spectrum, diamond windows have found use also for the shorter and longer wavelengths. In ongoing research for energy production by fusion processes, diamond windows are used for the transmission of very intense (ca. 1 MW) microwave radiation [4], while single-crystal diamond windows have been evaluated as high-heat-flux X-ray monochromators [5].

One of the first diamond optical components to be manufactured with a more complex geometry was the dome. In targeting the IR region, polycrystalline diamond domes are intended for use in high-speed flight systems [6, 7]. The shape of the dome is determined by a reasonable compromise between aerodynamic and optical considerations [8, 9]. The surface of such a diamond dome typically must be polished to an average roughness of less than 20 nm. Of special interest here are missile applications (notably heat-seeking missiles), where diamond domes exhibit an extraordinary resistance against impacts from rain and sand, thermal shock, and pressure changes. An IR missile dome made from polycrystalline diamond, demonstrated around 2002, is shown in Figure 4.1.

The lack of viable methods to structure diamond on the microscale proved to be another reason why most of the early demonstrations of diamond for optical purposes came in the form of simpler devices, such as windows. However, with the powerful microelectronics industry as a driving force for continuously improved lithographic patterning, the situation has since changed; hence, in the following

Figure 4.1 An infrared missile dome fabricated from highly polished polycrystalline diamond, in a joint project between the UK Ministry of Defence and Element Six Ltd (first demonstrated about 2002). From Ref. [7].

subsections details will be provided of the fabrication and applications of more sophisticated diamond optical components.

4.3
Refractive Devices

4.3.1
Lenses

Given their universal use in the world of optics, it is natural that lenses have been among the first components to be considered for diamond engineering. Various properties of lenses – both in the form of single macrolenses and microoptical counterparts, as well as being combined to form arrays – can be measured using standard techniques. Surface topologies and wavefront aberrations are common examples of important measures when characterizing lenses, whereas the specified *f*-number and numerical aperture guide the user to choose a certain lens for any given application. As such, lenses also represent information-rich optical structures from which the fidelity of the manufacturing process can be examined.

As with many other times in the history of optics, the suggestion of utilizing intrinsic material properties – like those offered by diamond – came way ahead of the engineering developments. In his book *Treatise on New Philosophical Instruments*, published in 1813, Sir David Brewster claimed that "... *We cannot, therefore, expect any essential improvement in the single microscope, unless from the discovery of some transparent substance, which, like the diamond, combines a high refractive power, with a low power of dispersion."* This prompted the London-based optician Andrew Pritchard to try to realize a jewel microscope by mechanically grinding natural diamond into a plano-convex refractive lens. Despite Pritchard actually succeeding in his attempts, the practical use of the fabricated lens element was mainly hindered by the associated cost and flaws in the diamond stone itself. Soon thereafter interest faded because of the invention of achromatic lenses, which enabled the creation of compound microscopes with unprecedented resolution. Nevertheless, Pritchard's achievement is probably the first published work on diamond lenses [10]. A curiosity here is that Pritchard, after completing the first diamond lens in 1824 and satisfactorily inspecting it, simply lost the piece: *"When thus far advanced, fate decreed that I should lose the stone."* As a consequence, he set out to create a new stone – also successfully – which is now on display at the Science Museum in London [11].

It was not until the recent development of CVD techniques, which can be used to deposit high-quality diamond onto flat or structured wafers, that interest for diamond optics regained interest. Much like the traditional way of grinding and polishing a mold to create an optical surface, Woerner *et al.* in 2001 used machined silicon substrates to demonstrate refractive CVD diamond lenses [12]. The steps of fabrication are shown diagrammatically in Figure 4.2. The spherical impressions on the 2 inch (5 cm) silicon wafer had diameters between 2 and 5 mm, and the maximum depth for an individual impression was 380 μm. By using methane and hydrogen, CVD diamond of high phase purity was next deposited onto the wafer, such that the impressions were filled. The growth side of the diamond layer was then polished, the silicon substrate removed, and laser cutting used to separate the lenses. An inspection of the curved diamond surfaces showed a deviation of less than 1% from a perfect spherical profile for all lenses, with rms surface roughness below 5 nm on both the flat and curved sides. Several CVD diamond lenses, with focal lengths ranging from 3.2 to 5.2 mm, are shown in Figure 4.3. Diamond lenses produced in this way are highly transparent and of good optical quality. Woerner *et al.* suggested the use of such lenses in fiberoptic laser surgery and as tips for length gauges under highly abrasive conditions [12].

These investigations took advantage of the major technological breakthrough offered by synthetic diamond deposition, which obviated the need to mechanically grind naturally grown diamond stones. Unfortunately, the high growth temperature involved (typically ca. 800 °C) limits which type of template substrate can be used; moreover, as the template is destroyed when separating it from the diamond layer a new one must be formed for each production. A further restriction is that the diamond must be of high optical quality from the very first layer deposited onto the template, as subsequent polishing of the curved diamond surface is extremely difficult to achieve.

Another strategy for producing diamond lenses, as presented by Karlsson and Nikolajeff in 2003, employed commercially grown and polished flat diamond substrates [13]. This method involved the structuring of a polymer layer coated onto a diamond substrate, followed by plasma etching to transfer the polymer

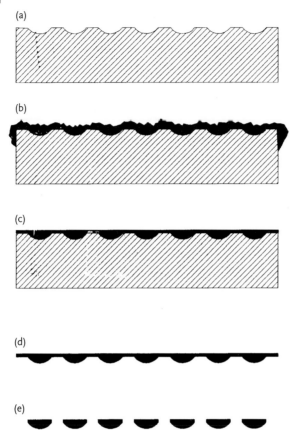

Figure 4.2 Processing steps for creating CVD diamond lenses. (a) Silicon substrate machined to act as a mold; (b) Deposition of diamond by CVD process; (c) Polishing of the diamond growth side; (d) Freestanding diamond film after separation from the Si mold; (e) Separate diamond lenses after laser cutting. From Ref. [12].

structure into the underlying diamond. Both of these steps may benefit from accurate and well-proven techniques normally used in the microelectronics industry, and several of the disadvantages of the method described by Woerner can be circumvented. However, the need for a structured polymer layer should lead to the process developed by Karlsson and Nikolajeff being better suited to the fabrication of microlenses having sag heights up to about 200 μm, rather than millimeter-thick diamond lenses. In the particular case of creating an array of microlenses, the method of lithographically patterning a polymer layer is a very fitting choice. As Popovic showed in 1988, a cylindrical pillar of photoresist on a suitable substrate may melt on heating and will – provided that the base of the pillar is preserved – form an almost perfect spherical cap due to surface tension [14]. The steps for generating an array of spherical resist microlenses are shown schematically in

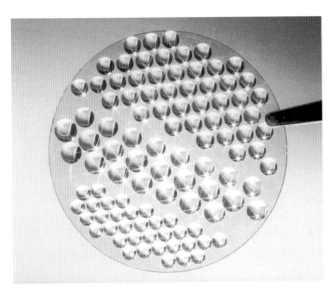

Figure 4.3 A 2-inch (5 cm) CVD diamond wafer with refractive lenses. From Ref. [12].

Figure 4.4a, while a scanning electron microscopy (SEM) image of a single fabricated microlens is shown in Figure 4.4b.

A spherical lens can be defined by its *f*-number, which is determined by the lens aperture and radius of curvature. From simple geometric relations, a link can be found between the radius of curvature R, the lens sag height s, and the lens half-aperture a:

$$R = s/2 + a^2/2s \quad (4.1)$$

Using volume conservation, the resist thickness T required to fabricate the cylindrical pillar can next be calculated:

$$T = s/2 + s^3/6a^2 \quad (4.2)$$

In this way, the process can be started by asking an optical designer for a specification of conventional lens parameters (focal length, lens aperture); these can then easily be translated into data (resist thickness, mask coordinates) that are useful to the lithographic engineer, thus enabling realization of the lens array. Finally, the engineer may use a gas chamber having preset parameters for diamond etching in order to rapidly (less than hours) transfer the lithographically defined surfaces into diamond substrates. One crucial parameter that must be well controlled in the latter step is the etch selectivity, which is defined as the ratio between the etch rates in the two materials. As a polymer is typically etched faster than diamond, this can result in an etched curvature which is different from that of the original, although for shallow lenses this effect will be of minor importance. The above relationships can also be used to determine the focal length of a fabricated

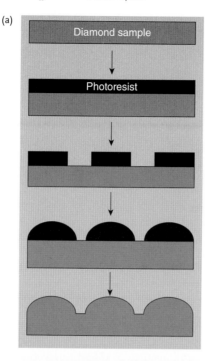

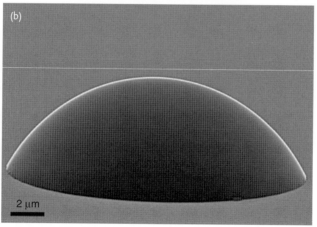

Figure 4.4 (a) Generation of spherical microlenses by melting small cylinders of photoresist (the last step involves diamond etching to transfer the optical structure from the molten resist into the substrate); (b) SEM image of a melted resist microlens.

spherical diamond microlens; for example, the paraxial rays of a plano-convex lens, the focal length is given by:

$$f = (a^2 + s^2)/2s(n-1) \qquad (4.3)$$

Hence, by measuring the aperture of a microlens and its sag height, using for example an atomic force microscope or a stylus profilometer, the focal length can be directly calculated using Equation 4.3.

In their report of 2003, Karlsson and Nikolajeff used commercial CVD polycrystalline diamond substrates that were 300 μm thick, had a diameter of 5 mm, and had polished surfaces with an rms roughness <15 nm. The fabricated microlenses, with apertures of 90 μm and sag heights of a few micrometers, were evaluated by using different interferometric techniques. The deviation from a perfect spherical surface was measured at about 7 nm over more than 90% of the lens area, and the phase error was less than 31 nm ($\lambda/20$). Notably, no noticeable increase was observed in the surface roughness when comparing untreated substrates with processed diamond surfaces.

Single-crystal diamond, although still currently more expensive than polycrystalline diamond by several orders of magnitude, exhibits outstanding material properties; consequently, single-crystal diamond microlenses would be expected to demonstrate better results. Indeed, in 2004 and 2005, when Gu et al. repeated the technique of reflown photoresist followed by plasma etching, they showed that the surface roughness of etched microlenses could be reduced to only a few nanometers when using single-crystal diamond rather than polycrystalline [15, 16]. The high purity of single-crystal diamond also allows a high UV transmission down to approximately 230 nm, an absence of IR absorption in the 7–10 μm band, and an exceptionally high thermal conductivity. On this basis, single-crystal diamond microlenses can provide both focusing and heat-dissipation services when integrated with light emitters.

In a further investigation conducted by Lee et al., microcylindrical and microring lenses were prepared using CVD polycrystalline diamond; such lenses again benefitted from the combination of standard photolithography, resist reflow, and plasma etching [17]. A technique developed by the same group, based on laser scanning reflection/transmission confocal microscopy, was used to characterize the optical properties of the fabricated diamond lenses.

A different approach for realizing concave, rather than convex, refractive diamond microlenses was demonstrated by Lee et al. in 2006 [18]. In this case, a silicon substrate was first structured to contain an array of spherical microlenses, after which the silicon was employed as a mold to emboss the pattern into a resist layer that was deposited onto a single-crystal diamond, followed finally by plasma etching. In this way, negative diamond microlenses could also be achieved (see Figure 4.5). In fact, a means of changing the polarity of refractive micro-optics had been reported earlier for the construction of GaP micromirrors, but with the intermediate step of first transferring the original pattern into a nickel mold before embossing [19]. By utilizing the electroplating processes adjusted for the CD/DVD industry (so-called family development of nickel shims), it is possible to change

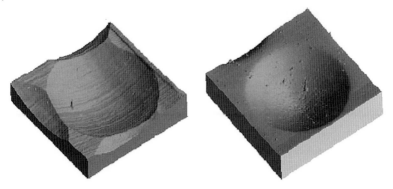

Figure 4.5 Atomic force microscopy images of concave microlenses in HPHT diamond. From Ref. [18].

the polarity from convex to concave (or vice versa) of almost any originated refractive surface without losing fidelity, and then to emboss the desired topography into a resist.

All of the microlenses described so far typically had radii of curvature much larger than the apertures of the lenses or, in other words, high f-numbers. Fabricating lenses with millimeter-sized apertures and low f-numbers by the technique of reflowed resist provides a challenge, as very thick resist layers must be used. If such lenses are required, however, it might be preferable to utilize a process of CVD polycrystalline diamond deposition onto preshaped substrates (as shown in Figure 4.2). For certain applications, such as solid immersion lenses, a single-crystal diamond lens with a low f-number may be required; however, due to constraints in the deposition of single-crystal diamond, the process of applying a preshaped surface from another material cannot be used. An industrial process for polishing single-crystal diamond, based on a combination of laser and mechanical processing, has recently been applied to create millimeter-sized diamond refractive lenses with very low f-numbers [20].

4.3.2
Prisms

Today, both polycrystalline and single-crystal diamond prisms are also produced, typically employing either CVD or high-pressure, high-temperature (HPHT) methods, followed by laser cutting or mechanical grinding. These prisms are useful for certain IR-based spectroscopic measurements (see Section 4.8 for further details).

4.3.3
Other Devices

Refractive optical devices may, of course, take shapes other than those of spherical lenses or prisms. Anamorphic lenses and axicons are examples of more unusual –

but nevertheless useful – components in applications such as cinema projection systems and atomic traps, respectively. Currently, freeform optical surfaces are fabricated mainly using computer numerical control (CNC) machines or single-point diamond turning tools; indeed, a variety of freeform optical devices have been produced in this way, using materials such as germanium, aluminum, copper, glass, and fused silica. Unfortunately, however, these techniques tend to be inadequate for machining freeform shapes from diamond substrates, due simply to the mechanical hardness of diamond. A more likely approach would be to use laser structuring, with excimer laser ablation having proved useful for patterning shallow diffractive structures in diamond (see Section 4.4 and Chapter 12). The deep three-dimensional (3-D) micromachining of bulk diamond rather than thin substrates is, of course, a highly challenging task, while the details of an interesting experiment using two-photon UV laser etching of diamond have recently been reported by Mildren *et al.* [21]. Clearly, further developments and improvements in the laser-assisted freeform shaping of diamond should be expected in the near future.

4.4
Diffractive Components

Diffractive optics is an important subfield of optical science, in which diffraction is used to create the desired optical function. Diffractive optical elements (DOEs) make use of the wave nature of light by harnessing the diffracted light from a fabricated optical structure. Typically, a DOE is wavelength-sensitive, and the invention of the laser in 1960s heralded the start of extensive investigations into these types of optical structure. As a consequence, a broad range of information is currently available, covering the theory, design, and manufacture of DOEs [22–24].

In this section, a review of diamond DOEs is presented, together with a short introduction to diffractive optics (the reader might be less familiar with this topic than more traditional refractive optics). The DOEs described here are all optical components that modulate the phase of the incident light to produce a desired wavefront or light pattern. Another approach would be to modulate the amplitude of the light, typically by blocking light in certain areas, though this may lead to low diffraction efficiencies. Depending on the design freedom of phase-modulating DOEs, they are able to manipulate light in ways that would be impossible by traditional optics. An additional advantage is that DOEs have a surface relief with a thickness of the same order as the illuminating wavelength, which in turn means that they can be made very thin and compact. A downside, however, is that the surface reliefs often require contours with complex geometries and small features, which means that their fabrication can be very demanding.

A *blazed grating* is a conceptually simple but powerful DOE, which has an abrupt phase swing of 2π that is typical of most DOEs. A blazed grating can redirect an incoming wavefront to a specific angle; moreover, its dispersive power is greater than that of a prism, which has made it useful for example in spectral analyzers.

For a transmissive blazed grating made from a material with refractive index n, the optimal depth h for diffraction of monochromatic light with wavelength λ into the m:th order (assuming the ambient medium to be air) is given by:

$$h = m\lambda/(n-1) \qquad (4.4)$$

This expression can easily be derived by combining Snell's law and the grating equation:

$$m\lambda = p \sin \Theta \qquad (4.5)$$

where p is the grating period and Θ the angle of diffraction for the m:th order. If there is a deviation from the optimal depth h or the design wavelength λ, then the wavelets diffracted from each grating facet will not be perfectly matched in the m:th order, and this will reduce the amount of light diffracted into the desired order; light will also be diffracted into other orders. In an analog or continuous-relief DOE, the phase of the DOE surface relief is continuously varying between 0 and 2π, whereas a multilevel DOE has a fixed number of values between 0 and 2π. The desired phase shift for a pure binary structure is π, and hence the optimal relief height becomes half the depth of a true continuous structure (i.e., $h/2$). Typically, binary and multilevel DOEs suffer from a lower diffraction efficiency than continuous DOEs. Plugging in relevant numbers for a continuous diamond blazed grating to function as a beam steering device (1st diffraction order) for a CO_2 laser, gives $h = 7.68\,\mu m$.

The possibility of designing a diffractive equivalent structure from the refractive lens is shown in Figure 4.6. By dividing the lens curvature into thicknesses that correspond to a phase shift of 2π, and then joining the curvature segments from each slice, a diffractive Fresnel lens (named after the French physicist Augustin Jean Fresnel) is generated. As the thickness corresponding to a 2π phase delay only holds true for a particular wavelength of the illuminating light, a diffractive Fresnel lens is normally used for monochromatic laser light. In fact, the method of dividing a curvature into thinner segments can be used to transform any type

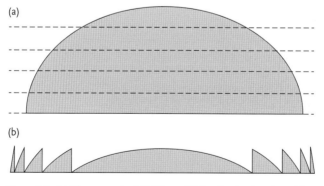

Figure 4.6 (a) The concept of dividing a thick refractive lens into thinner segments; (b) Realization of the much thinner counterpart, a Fresnel lens.

of refractive surface into a diffractive structure (as an exercise, the reader is encouraged to deduce the shape of a blazed grating from a prism). The diffractive Fresnel lens has much the same optical function as a classical refractive lens, and is widely used for focusing or collimating laser light. The main reasons for using a diffractive Fresnel lens rather than a refractive lens are that the former can be made more compact (allowing easier system integration), thinner (less bulk material used and therefore lower optical absorption), and lightweight (especially important in space optics).

The fabrication of a diamond diffractive Fresnel lens was described by Kononenko et al., using an excimer laser [25, 26]. The excimer laser process, which employs short, high-intensity laser pulses in the UV domain, results in an ablation of diamond caused by local heating and graphitization, but without thermal damage to the surrounding areas. The Fresnel lens was designed for focusing the laser beam from a CO_2 laser operating at 10.6 μm. Several focal distances between 25 and 100 mm were demonstrated, showing diffraction efficiencies that were 1–2% below the theoretical values. These diffractive lenses were multi-level lenses (up to eight levels), and one such diffractive lens is shown in Figure 4.7. One strength of UV laser ablation is that it is a direct fabrication method, and does not involve the use of many lithographic steps in a cleanroom environment. However, in contrast to the rapid single-step wafer illumination offered by photolithography, the excimer laser beam has to move pixel by pixel over a diamond substrate, which results in rather long processing times.

In another approach to creating a diffractive Fresnel lens, Björkman et al. [27] deposited diamond onto a fused silica template, in similar fashion to the process shown in Figure 4.2. The diamond diffractive lens produced in this way had a true analog relief, but no details of optical testing were presented, most likely due to

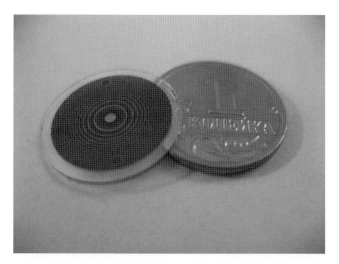

Figure 4.7 A diffractive Fresnel lens made by UV laser ablation, before annealing in ambient air. From Ref. [26].

the fact that the diamond was grown via hot filament chemical vapor deposition (HFCVD). The latter growth technique is typically more suited to mechanical or tribological applications of diamond, as the resulting optical absorption is often too high.

Shortly thereafter, another fabrication method which employed rapid inductively coupled plasma (ICP) etching was described by Karlsson et al. [28]. In this case, two types of diamond DOE – a Fresnel lens and a blazed grating – were first prepared as surface reliefs in a thin resist layer, whereby the surface reliefs were defined with electron beam lithography (EBL); this was followed by transfer to the underlying diamond substrates using an oxygen- and argon-based plasma chemistry. Both elements were of the analog type, and the blazed grating with a period of 45 µm showed a diffraction efficiency of just a few percentage points below the theoretical value.

Some years later, Fu and Bryan reported the fabrication of a diffractive diamond Fresnel lens using focused ion beam (FIB) milling [29]. The lens was prepared in a HFCVD diamond film, and the experiment was carried out under vacuum with combined SEM and FIB instrumentation, with a small amount of water vapor being used to enhance the etch rate of the diamond. Although the FIB milling technique proved to be very time-consuming, and thus more suitable for prototype or mold fabrication, one clear advantage was an ability to fabricate structures down to 10 nm in size, directly in diamond.

Another important diffractive element is the so-called "moth-eye" structure or subwavelength grating. From the grating equation it can be seen that, assuming the surrounding medium to be air and with illumination perpendicular to the surface normal, for values of the grating period p which are shorter than the illuminating wavelength λ, light cannot be diffracted into any higher orders except for the zeroth order. However, if the higher diffraction orders are absent and no deviation of the light occurs, what is then the purpose of such a device? In some sense, it might be said that light will not be able to resolve the fine pattern of the grating as the optical resolving power is limited by the light wavelength; instead, light will "see" the grating region as a homogeneous layer with an artificial refractive index. This artificial refractive index will have a value between the refractive indices of the grating material and air, the value being determined by the geometry of the grating (grating shape, grating depth, and fill factor). The exact calculation is nontrivial, and hence a rigorous coupled wave analysis typically must be used [30]. So, the answer to the question posed is that with the ability to generate a matching value of a layer's refractive index by varying the physical parameters of the subwavelength grating, the possibility will arise (among other points) to reduce unwanted surface reflections (Fresnel losses). Indeed, periodic subwavelength structures already exist in Nature, being present on the cornea of a certain night-flying moth with the purpose of diminishing reflections from the moth's eye and thus avoiding its discovery by predators [31, 32].

Subwavelength gratings therefore present an alternative means of providing antireflection (AR) treatments for optical components. The normal process employed for AR treatment is to coat the optical surfaces with either a quarter-wave film (monochromatic illumination) or a multilayer stack coating (for broadband light), matching the difference in the refractive indices of an optical surface and its surrounding medium. Subwavelength gratings allow the AR treatment of optical components without adding any new material, and this is often vitally important for diamond optics. Any addition of thin film coatings may otherwise reduce the benefits provided by diamond's extraordinary properties.

The first demonstration of subwavelength AR structures in diamond was presented by Harker and DeNatale in 1992 [33]. In this case, two-dimensional (2-D) tapered microstructures were fabricated in germanium and silicon substrates by photolithography, followed by etching; thin free-standing polycrystalline diamond films were then replicated from these textured surfaces by CVD deposition. When such a structured diamond film was adhered to a germanium substrate, measurements at $\lambda=10\,\mu m$ showed a decrease in Fresnel loss, from 18% to 7%. Later, Ralchenko *et al.* demonstrated 2-D diamond pyramid structures (grating period $=4.5\,\mu m$ for a design wavelength of about $11\,\mu m$), also by using the method of CVD deposition onto a silicon template [34]. The free-standing diamond film was in this case $80\,\mu m$ thick, and measurements showed an increase in the optical transmission from 71% to 78%. The maximum transmission through a diamond substrate with a perfect AR treatment on one side is 83%. The nonoptimal transmission value measured was attributed to problems with forming the correct depth of the negative pyramidal shapes in the silicon mold. It should be noted that it is impossible to fabricate subwavelength AR structures on both sides of a diamond film using the method of deposition onto a template; hence, this technique is restricted to one-side only AR treatment.

Diamond subwavelength gratings fabricated by selective laser ablation of 300- to $400\,\mu m$-thick polycrystalline CVD diamond films were reported by Kononenko *et al.* [35]. In these investigations, both a KrF excimer laser ($\lambda=248\,nm$) and a Nd:YAP laser ($\lambda=1078\,nm$) were used to produce one-dimensional (1-D) sawtooth-like structures and 2-D pits. When measured with a CO_2-laser, the optical transmission could be raised to 80% after structuring both sides of a diamond substrate. Similar structures were also fabricated with an ArF excimer laser ($\lambda=193\,nm$) [36]. At this short wavelength, there is no need to deposit a thin absorptive gold layer on the diamond surface to initiate the ablation process, as the photon energy exceeds the band gap of diamond.

Karlsson and Nikolajeff demonstrated the manufacture of a subwavelength 2-D square pattern in diamond using ICP etching [13]. For this, the pattern was first fabricated in resist using EBL, which served as a mask for the Al layer on top of a diamond substrate. The Al mask was opened by chlorine-based ICP plasma etching, and the diamond grating finally realized with oxygen/argon etching. The grating had a period of $4\,\mu m$ and a depth of $1.8\,\mu m$, while the AR-structured area was $2.5\times2.5\,mm^2$ on both sides of a $300\,\mu m$-thick polycrystalline diamond substrate. Spectrophotometric measurements at wavelengths around $10\,\mu m$ showed an increase in transmission from 71% for blank diamond to 97% for a

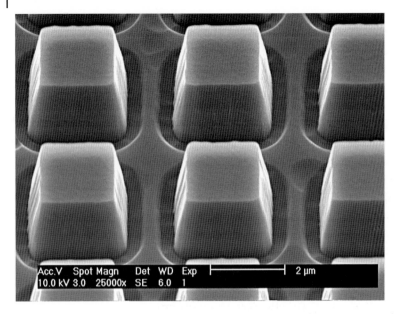

Figure 4.8 SEM image of a diamond subwavelength grating, designed to reduce optical reflections at a wavelength of 10.6 μm.

double-sided, AR-treated diamond substrate. An SEM image of the subwavelength structure made in diamond is shown in Figure 4.8.

The three types of diffractive components described here, namely the blazed grating, the Fresnel lens, and the subwavelength grating, all represent practical examples of diamond diffractive optics. With the power at hand to harness diffraction phenomena, and based on a "happy marriage" between accurate optical theories and modern fabrication methods, the reader is encouraged to further explore the wealth of diffractive optics that offer functions not always achievable using classical refractive optics [24]. A *diffractive fan-out element* is one such example (see Figure 4.9a). The phase relief required to generate this type of light pattern can today easily be calculated on a computer using specially developed optimization algorithms [37]. The diamond diffractive fan-out element shown in Figure 4.9b was presented by Karlsson and Nikolajeff [38]. The designed DOE was intended to produce 16 spots with equal intensities in the far field (image plane). Due to the symmetry of the ring pattern, the phase relief could be designed as a binary structure yielding either 0 or π phaseshift, which allows relaxed fabrication demands. Subsequently, two diamond fan-out elements were manufactured: one for illumination with 633 nm wavelength, and one for 10.6 μm wavelength. UV-photolithography followed by diamond etching with an ICP reactor was employed for the DOE fabrication, whereby the measured diffraction efficiency at 633 nm was 64.5% (by comparison, the calculated value was 65.7%). The second DOE was evaluated with a CO_2 laser operating in continuous wave mode; a subsequent

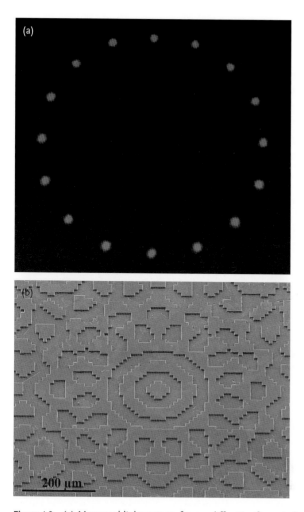

Figure 4.9 (a) Measured light pattern from a diffractive fan-out element; (b) SEM image of a binary diamond fan-out element.

visual inspection of laser-machined bulk polymer substrates provided satisfying results.

As a final example demonstrating the value of diffractive optics, reference should be made to the beam-shaping elements in diamond created by Kononenko *et al.* [39, 40], who designed two DOEs: one for shaping the Gaussian beam profile outcoupled from a CO_2 laser into a homogeneous-intensity focal rectangle; and the other for shaping the Gaussian beam into a focal square contour. The diamond structures were prepared using UV laser patterning, and despite the eight-level diamond DOEs having feature sizes greater than 40 μm, the evaluation with a CO_2 laser showed a good optical performance.

4.5
Polishing

In order to obtain a curved surface from a flat substrate (e.g., the spherical curvature of a lens), or to make a surface more smooth so as to minimize scattering and reflection losses, polishing is normally the method of choice in optical engineering. Hence, it would not come as a surprise that polishing is an important aspect also for diamond optics. As a more extensive review of this topic is provided in Chapters 3 and 12, only a brief introduction will be made at this point.

The traditional technique used for glass optics involves a polishing powder (e.g., silicon carbide or diamond), a rotating wheel, and a glass workpiece. Due to the random motion of the wheel relative to the workpiece, the surface of the workpiece will eventually gain a spherical form. Another more recent method, which allows shapes other than spherical also to be formed, is that of single-point diamond turning, where material is removed using a very sharp and hard turning tool (most often a diamond tip). Of these two techniques, only the powder polishing method

Table 4.1 Various diamond polishing techniques. From Ref. [42].

Technique	Bulk processing temperature (°C)	Polishing mechanism	Shape limitations	Size limitations	Special requirements
Mechanical lapping	Room	Abrasive wear	Only planar surfaces	No limit	None
Thermomechanical	800–900	Graphitization; diffusion	Only planar surfaces	Plate size	Need environmental control
CAMPP	≥350	Oxidation	Only planar surfaces	Plate size	None
Laser	Room	Etching, evaporation	Nonplanar surfaces possible	No limit	Need scanning of samples
Ion beam	Room	Sputtering, etching	Nonplanar surfaces possible	Beam size	Need high vacuum
RIE	700	Sputtering	Nonplanar surfaces possible	Plasma size	Need high vacuum

CAMPP, chemical-assisted mechanical polishing and planarization; RIE, reactive ion etching.

would be applicable to diamond substrates. The reason that diamond turning falls short is due to the inevitable wear of the diamond tip itself during its continuous contact with a diamond substrate. On the other hand, the crystal nature of diamond (as opposed to glass, which is an isotropic amorphous material) makes it more difficult to polish diamond in certain directions, and as a result also the powder method faces difficulties when used for polishing diamond substrates. Recent industrial progress in combining laser cutting and mechanical polishing has enabled single-crystal diamond lenses to be produced, and this approach holds much promise also for achieving shapes other than spherical [41].

As CVD diamond, regardless of whether it is grown as a film or onto a preshaped substrate, is normally too rough to be used directly for optical purposes, the diamond surface must be smoothed in some way. In the past, many polishing concepts have been examined to reduce the surface roughness of diamond films; an excellent review of this topic is available [42] in which an overview is provided of the various techniques used for polishing CVD diamond (see Table 4.1).

Set-up	Equipment cost	Large-area processing cost	Processing time (per cm^2)	Reported roughness (nm)	Parametric freedom	Surface change (contamination)
Rigid and geometry-sensitive	Low	Low	Few days	210	Small	Yes
Rigid and geometry-sensitive	Low	Low	Tens of minutes	5.5	Restricted	Yes
Simple	Low	Low	A few hours	49	Restricted	Yes
Simple	High	Medium	A few seconds	500	Large	Yes
Complex	High	High	Tens of hours	5	Restricted	Yes
Rigid	High	Medium	Tens of minutes	71	Large	Yes

Acceptable values of roughness for an optical surface depend, of course, on the application and the wavelengths being used. Measurements of the scattering, transmission, and reflection properties from unpolished versus polished diamond surfaces have shown that current polishing techniques are indeed adequate in achieving a sufficient optical finish of diamond substrates for most applications.

Finally it should be noted that, as much of the current progress related to diamond polishing has been made in the industrial sector, the developed knowledge may be proprietary and out of the public domain [43].

4.6
Micromachining

The extreme properties of diamond are not always beneficial, especially with regards to its machining. Due to the hardness and chemical inertness of diamond, it has always been nontrivial to manufacture any type of geometry in this material. The removal of diamond material by manual polishing is still considered an art requiring skilled craftsmen, and consequently the fabrication of diamond structures on a micro- or nanoscale in a controlled fashion was long considered impossible. During the past few decades, however, tremendous developments have been made in the microtexturing of diamond, with several methods (albeit with different pros and cons) having been introduced to the diamond community. Although one of the simplest processes for microtexturing a material is wet chemical etching, there is still no feasible wet etchant for diamond in existence. Some reports have been made on the etching of diamond with molten rare earth metals (Ce or La) [44], but this is a high-temperature process (>900 °C) which, until now, has not been used for the micro- or nanostructuring of diamond. Rather, the first successful techniques used to microstructure diamond were plasma etching, ion beam milling, and laser-based processing. Some early studies in the field of diamond texturing were described by Rothschild *et al.*, who used an excimer laser operating at a wavelength of 193 nm to etch micrometer-sized structures in diamond [45]. A more detailed discussion of the laser-assisted texturing of diamond is available in Chapter 12.

An outline of the early studies of diamond structuring is provided in the following subsections, together with a detailed description of state-of-the-art plasma etching for creating micro- and nanostructures in diamond. Here, some fabrication aspects will be addressed that may be useful for research groups/engineers about to commence investigations in the field of microstructured diamond.

One of the first demonstrations of microstructured natural diamond (Type IIa) was performed by Efremow *et al.* in 1985 [46], who used ion beam-assisted etching that involved a beam of Xe^+-ions and a reactive gas flux of NO_2 to achieve diamond etch rates of up to $200\,nm\,min^{-1}$. At about the same time, diamond removal by sputtering with an argon and oxygen beam was also reported [47]. In 1989, Sandhu *et al.* first described the reactive ion etching (RIE) of diamond with O_2 and H_2, achieving etch rates of about $35–40\,nm\,min^{-1}$ [48]. During the 1990s, the first reports were made of the RIE of polycrystalline diamond films using oxygen plasma, but the etched diamond surfaces were found to be very rough due to the formation of columnar structures [49]. Although, in an attempt to overcome this problem SF_6 or CF_4 were added [50, 51], the major drawback of this approach proved to be the reduced selectivity when Al was used as the mask material. Despite many reports being made between 2000 and 2005 which described pattern transfer into diamond with RIE [52–57], there proved to be some severe limitations associated with this approach. The most notable points were that the etch rates were low (normally in the range of 10–60 nm min^{-1}), and that the diamond surfaces produced tended to be quite rough, especially when considered for use in optical applications.

Subsequently, improvements in both etch rates and surface smoothness became possible as a result of more recently developed high-density plasma etching systems, which included ICP systems and electron cyclotron reactor (ECRs). The ECR oxygen etching of diamond was demonstrated by Pearton et al., with etch rates of over 300 nm min^{-1}, and the etched surface morphology being similar to that before etching [58]. The ICP etching of diamond was demonstrated by Karlsson et al. some years later, and showed a similar etch performance as with ECR [28]. Practical disadvantages with the ECR technique, such as challenges in automatic tuning of the microwave energy and wafer damage outside the resonance conditions, have led to ICP becoming the preferred technology by current major suppliers of plasma equipment for high-rate/low-damage etch applications.

One main advantage with ICP etching over RIE systems is that it offers a user-independent control of the plasma density and the substrate bias; this is in contrast to the RIE system, where the bias is related directly to the plasma density. The bias regulates how strong the ion bombardment of the substrate surface will be, while the plasma density corresponds to the chemical etching of the substrate. As a result of such independent control of the bias and plasma density, an ICP system will allow the etch selectivity between the mask material and diamond to be varied over a larger range, which is especially important when fabricating micro-optical structures with both shallow and deep features. With an ICP system, it is also possible to achieve higher etch rates in diamond, as the combination of chemical and physical processes can be optimized. Moreover – and perhaps the most important feature – the ICP etching of diamond has proven to generate smooth etched diamond surfaces. Indeed, if the ICP process is tuned correctly, the etched diamond surface can be equally smooth as that before etching [59], and in some cases even smoother [60]. Several reports have described a smooth pattern transfer by ICP plasma etching of diamond, both for single-crystal and polycrystalline substrates [61–64]. The etching of single-crystal diamond means that only pure diamond must be etched, whereas polycrystalline diamond also contains grain boundaries with sp^2-bonded carbon. Consequently, the etching of polycrystalline diamond will require the etching of two slightly different materials at similar rates, or the surface roughness will be increased during etching [65]. Similar etch "recipes" can normally be used for both single-crystal and polycrystalline diamond.

When selecting the correct plasma parameters for etching diamond, several points must be considered:

- The chemical stage of the etching process requires a gas that will react with carbon. Oxygen radicals are highly reactive species that will react with the carbon atoms on the diamond surface and form volatile etch products such as CO and CO_2; hence, oxygen is a suitable choice for this task.
- The physical stage of the etching process is important to achieve sufficiently high etch rates, and relies on ion bombardment. Inert gases represent a good choice for this stage, and argon is normally used.
- The applied bias enhances the diamond etching due to the impinging Ar ions, but without any bias the diamond etch rate will be reduced dramatically to values well below 10 nm min^{-1} [66]. In general, a bias of over 100 V must be applied to achieve a good diamond etching process, as the use of bias values below 100 V can lead to problems with redeposition of the etch mask. Grass-like diamond surfaces and the formation of thin pillars is a sign of redeposition. Redeposition can in fact be problematic in all process regions, especially when etching large open areas. A normal bias in diamond ICP etching is around 200 V, and this results in typical etch rates of 200 nm min^{-1}, although etch rates of up to 650 nm min^{-1} have been reported [67].
- The process pressures used for diamond etching are normally in the range of 2 to 20 mTorr. A lower pressure means that more of the ion bombardment will strike the diamond surface at normal incidence due to less probability of ion collisions, which in turn leads to a more directional etching. Highly anisotropic features can thus be formed when using low process pressures.

As the plasma etching of diamond depends heavily on physical processes (the energy of the ion bombardment), a so-called "trenching effect" can often be seen on a masked and etched diamond surface. *Trenching* is a phenomenon that occurs when the ions which hit etched walls (glancing incidence) yield an increased flux of ion bombardment at the bottom of an etched structure [68, 69]. As the etch rate is strongly related to the density of ion bombardment, this area will be etched at a faster rate. Although the ion–surface interaction (physical process) is a requisite for the plasma etching of diamond, the etch rate would be decreased drastically in the absence of oxygen radicals. The Ar ion bombardment greatly enhances the reaction between oxygen radicals and the carbon atoms on the diamond surface. In other words, the "trick" in diamond plasma etching is to identify a suitable combination of the physical and chemical processes.

Three very important parameters in the ICP etching of diamond are the ICP coil power, bias, and process pressure:

- An increase in bias will lead to a higher energy of the Ar ions, which in turn leads to an increase of the ion bombardment effect on the carbon atoms, such that the etch rate increases as a consequence. At a certain bias level, however, the increase in etch rate will begin to slow, an effect that can be attributed to a slower increase in sputter yield at these high voltages. Another saturation effect occurs when the reactive oxygen flux arriving at the diamond surface is insufficient, and can limit the chemical etching mechanism.

- One way of further increasing the etch rate would be to increase the ICP coil power, as this will lead to a higher density of both Ar ions and oxygen radicals in the plasma. An increment of Ar ion density will lead to a larger number of Ar ions hitting the diamond surface, and this increased flux of Ar ions – together with the increased amount of oxygen radicals – will lead to a higher etch rate.
- At pressures of between 1 and 10 mTorr, an increase in pressure will generally lead to an increment in the etch rate. Increasing the process pressure leads to a longer residence time of the oxygen radicals and an increase in the reactive species. However, at higher pressures (>10 mTorr) a further increase will instead lead to a decreased etch rate [18, 64]. The enhanced ion–electron recombination at higher pressures will quench Ar ion production [70] and, as described above, a reduced ion bombardment will lead to a reduced etch rate.

An example of a standard etching recipe is as follows: ICP coil power 900 W, platen power 220 W (DC bias −340 V), process pressure 5 mTorr, oxygen flow 40 sccm (standard cubic centimeter), and argon flow 20 sccm. The diamond etch rate is, of course, dependent on the particular design of the ICP plasma etch system used, but with these process parameters it should be in the range of 150–200 nm min^{-1} for polycrystalline diamond of optical grade.

Another point to consider when plasma etching diamond is the choice of mask material. Etch selectivity is defined as the ratio between the diamond etch rate and the mask material etch rate. Often, it is desirable to have a high etch rate in diamond and, simultaneously, a slow etch rate in the mask material (i.e., a high selectivity). The most commonly used mask materials are Al, SiO_2, and different types of photoresist. The selectivity can vary widely, depending on the etch process, but normal values are as follows: Al about 50, SiO_2 about 20, and for resist less than 0.3. As a rule of thumb, Al is typically a good choice as a mask material for achieving binary structures and resist for analog curvatures.

As the diamond substrate will be heated during etching, it is very important to ensure a correct cooling of the diamond substrate. Otherwise, there is a high probability that the mask material – especially when using a photoresist – will be burned and subsequently destroy the etch performance. The use of vacuum grease to attach the diamond substrate to a suitable carrier (or directly to the chuck of the etch equipment) will ensure sufficient cooling.

When etching polycrystalline diamond it may be worth considering which side should be processed, as the two sides normally differ with respect to their grain sizes. The nucleation side (where the diamond deposition was started) has small grain sizes and a high density of grain boundaries, whereas the other side has large diamond grains and a low density of grain boundaries. One report suggested that it may be slightly favorable to use the side with large diamond grains due to the minor difference in etch rates between etching grain boundaries and the diamond grains [65].

The micromachining of diamond often means working with small substrates (typically pieces with sizes down to 5 × 5 mm, either square or circular), and this places special demands on the lithographic patterning of diamond substrates. When applying a resist to a small substrate, an "edge bead" will be formed such that the thickness of the resist is much greater at the edge. In fact, the width of the edge bead may be up to several millimeters and make standard contact photolithography difficult to use, as an air gap may form between the photomask and the resist in the middle of the substrate and cause the pattern to be distorted due to diffraction. An alternative, more feasible, replication technique may be that of nanoimprint lithography (NIL) [71, 72].

In NIL, the desired pattern is first created in a standard material (e.g., silicon) and then transferred into a mold (normally nickel). The diamond substrate is next coated with a thermally curable NIL resist (possibly after the deposition of a mask material), placed in contact with the mold, and a pressure is then applied at elevated temperature. Typical NIL process parameters would be a pressure of 10–20 bar and a temperature of 100–180 °C. In this way, the original pattern can be truthfully copied into the NIL resist layer on top of the diamond substrate. When using this replication method it is possible to use the same mold for many imprints, which justifies the use of costly nanopatterning techniques such as EBL to produce a mold with very fine details. Previously, high-aspect ratio nanometer structures in polycrystalline diamond have been demonstrated using NIL [65].

A number of nanowires (200 nm diameter) fabricated in single-crystal diamond using direct-write EBL and RIE, and representing examples of state-of-the-art micromachined structures, are shown in Figure 4.10 [73].

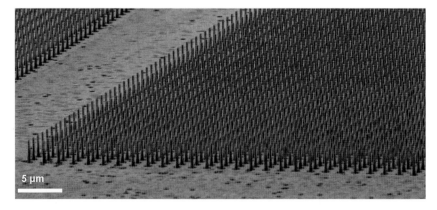

Figure 4.10 High-aspect ratio nanopillars made by electron-beam lithography and reactive ion etching in a single-crystal diamond. From Ref. [73].

4.7
Coatings

Coatings and diamond optics can be treated in two ways: (i) by applying single- or multilayer coatings onto diamond substrates; or (ii) by applying thin diamond coatings onto substrates of another material. In the first case, the purpose of the coating is typically to reduce surface reflections from a diamond–air interface, as Fresnel losses can otherwise dramatically lower the overall optical performance because of the high refractive index of diamond. In the latter case, a diamond coating can function as a protective layer, for instance to improve the resistance against detrimental impact by water droplets or sand particles.

Astronomical and remote sensing instrumentation is one area of particular interest for diamond optics, where the undisturbed high transparency from 8 μm into the far-IR (>50 μm), together with excellent thermal, mechanical and chemical resistance properties, make diamond a very promising material candidate. One challenge here, however, is the difficulty to obtain a good adhesion of the antireflective coatings or multilayer filters on diamond substrates. Mollart *et al.* have highlighted out the importance of ensuring a very clean diamond surface before applying a coating, and have suggested recipes for the removal of unwanted nondiamond contamination or hydrocarbons [74]. Mollart and coworkers, and also other groups, have shown that yttria coatings can be sputtered adherently to cleaned diamond surfaces, enabling functional quarter-wave antireflection coatings at wavelengths around 10 μm.

Thorium fluoride has proven to be a very useful antireflective coating for diamond windows at wavelengths between 8 and 12 μm, although concerns regarding the toxicity of thorium fluoride (it is mildly radioactive) are likely to limit its role in the future. Hawkins has investigated multilayer coatings using silicon oxide as an adhesion layer in attempts to use diamond as a substrate for mid-IR Q-band filters (wavelengths between 16 and 28 μm); however, the requirement for cryogenic cooling renders this approach currently impractical for such applications [75].

Germanium and zinc sulfide are commonly used materials for IR windows and domes, with good optical transmission also in the 3–5 μm band where diamond experiences phonon absorption. When using these materials in airborne systems, however, cracking and erosion can pose serious problems; hence, it has been suggested that diamond coatings are used for window protection, whereby thin diamond layers would result in an acceptably small absorption at 3–5 μm. Unfortunately, the thermal expansion mismatch between diamond and other IR materials can lead to large stresses and delamination of the coating. Miller and coworkers have developed techniques for a good adherence of diamond coatings on different IR window materials, using interlayer deposition and substrate patterning [76].

4.8
Applications

One of the first commercial sectors to show interest in diamond optics was the high-power laser industry. The key parameters, among others, for a high-power laser window are low levels of absorption and scatter. Typical breakdown effects when using too high laser powers can be attributed to fracture resulting from a build-up of thermal stresses or thermal lensing. The normally preferred material for outcoupling windows of CO_2 lasers is zinc selenide, though both theoretical modeling and experimental studies have shown that CVD diamond may be superior. Pickles *et al.* have, for instance, shown that CVD diamond windows can survive CO_2 power densities in excess of $160\,kW\,mm^{-2}$ [77]. As a comparison, the usable levels for industrial high-power lasers at the time of these studies was below $0.05\,kW\,mm^{-2}$. It should be noted that the absorption coefficient of CVD diamond is significantly higher (typically between 0.03–$0.1\,cm^{-1}$) than that of zinc selenide ($0.000\,57\,cm^{-1}$), and as a consequence much heat is generated in a CVD diamond window. It is important that this heat energy is removed very effectively from the diamond window, and this can be achieved by accessing the disc edge, utilizing the excellent thermal conductivity of diamond. Today, a steady decrease in the production costs of CVD diamond of optical quality is leading commercial vendors to offer diamond as replacement windows for high-power CO_2 lasers, polycrystalline diamond of good optical quality has so far proven sufficient for the above-mentioned applications, such that single-crystal diamond have, until now, not been required. A commercially available diamond outcoupling window is shown in Figure 4.11.

Figure 4.11 A diamond outcoupling window for CO_2 lasers. Courtesy of Element Six Ltd, UK. For a color version of this figure, please see the color plate at the end of this book.

Physics at the extreme also benefits from diamond optics. In order to realize hard X-ray free-electron lasers, which can offer unprecedented coherence and brightness, it is necessary to identify a material with a sufficiently high X-ray reflectivity. It has been predicted on a theoretical basis, and also very recently verified experimentally, that diamond crystals grown by HPHT methods can indeed exhibit the necessary X-ray reflectance under Bragg conditions [78]. Today, existing synchrotrons—such as the European Synchrotron Radiation Facility in Grenoble, France—utilize diamond monochromators that act as very precise filters for the passing of X-rays, while the Large Hadron Collider at CERN in Switzerland is using fast and radiation-tolerant diamond sensors to monitor beam conditions in the hunt for the Higgs boson.

One practical and useful biochemical analysis technique that takes advantage of refractive diamond optics is Fourier transform infrared attenuated total reflectance (FTIR-ATR) spectroscopy. This technique employs broadband mid-IR light (with wavelengths typically in the range of 2–12 µm) coupled into an IR-transparent crystal. If the crystal is made from a material having a refractive index higher than that of the surrounding media, and the incoupled beam is under the condition of total internal reflection, then there will leak out an evanescent wave that penetrates only a few microns above the diamond surface. This evanescent wave can be used to interact with a sample—either in liquid or powder form—that is applied in close contact with the surface. In regions of the IR spectrum where the sample absorbs energy, the evanescent wave will be either attenuated or altered. By employing mid-IR light, the transmitted signal will carry information about the specific fundamental vibrational properties of molecules. Diamond crystals are preferred over other IR-transparent materials (e.g., germanium or zinc selenide), as diamond is scratch-proof and resistant against chemicals, and is also considered to be biocompatible. Currently, FTIR-ATR spectroscopy is used routinely in the pharmaceutical and food industries for quality control purposes. Vibrational spectroscopy is also useful for protein studies [79], with recent investigations demonstrating schemes for the selective binding of proteins on diamond surfaces [80] perhaps leading to new medical devices, such as sensitive proteomic assays or diagnostic tools.

Sophisticated diamond micro-optical components have also found use in astrophysics. The imaging and characterizing of faint sources in the close vicinity of bright astronomical objects is very demanding; for example, extrasolar planets are typically 10^4- to 10^7-fold fainter than their host star in the thermal IR range (wavelengths between 3.5 and 20 µm); consequently, very high rejection ratios are needed. In order to image and detect such exoplanets, a coronagraph can be used; an example is the Lyot coronagraph, which modulates the amplitude by using an opaque spot centered on the optical axis. Unfortunately, such a component will limit the optical throughput, and a more efficient solution was introduced by Mawet et al. in terms of the annular groove phase mask (AGPM) coronagraph [81]. The AGPM is a subwavelength component that consists of a circular grating which induces an artificial birefringence that modifies the polarization of the radiation in a circular symmetric way. A phase ramp is generated between the s- and p-polarization, which can lead to an almost total starlight rejection. Recent initial

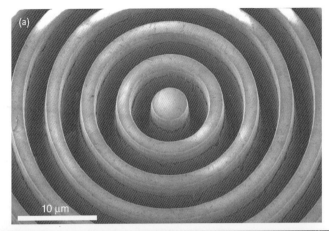

Figure 4.12 (a) SEM image of a diamond annular groove phase mask (AGPM) component; (b) The Very Large Telescope in Chile, as operated by the European Southern Observatory. Courtesy of the European Southern Observatory. For a color version of this figure, please see the color plate at the end of this book.

attempts to fabricate diamond AGPMs have shown much promise [82]. The main reasons for using diamond are its high refractive index (which allows shallower structures than low-index materials) and a high transmittance in the far-IR region. Currently, one AGPM has been developed for the N-band (wavelength around 12 μm) and one for the L-band (wavelength around 4 μm). The desired grating periods for these AGPMs are in the range of 1 to 5 μm, with etch depths ranging from 4 to 15 μm. As these are very demanding structures to fabricate, a combination of electron-beam lithography, NIL and ICP etching was developed in collaboration with a company focusing on the optical applications of diamond [83]. An SEM image of a diamond AGPM structure is shown in Figure 4.12a; this is intended for use at the Very Large Telescope in Chile (Figure 4.12b).

Yet another application is that of next-generation optical disk systems, for which high-density, near-field photon detection has been demonstrated using diamond solid-immersion lenses with a high numerical aperture [84]. The increased light collection efficiency offered by diamond lenses has also found use in medical applications such as endoscopes. Within these diagnostic tools, which are used to examine the interior of the human body, diamond lenses can improve the image quality and allow finer details to be visualized.

So, the journey from the development of linear diamond optics by Sir David Brewster in 1813, and the first concepts of diamond lenses, has concluded with these final examples. Today, synthetic diamond of sufficiently high optical quality can be prepared routinely and shaped into desired optical structures. Clearly, Brewster would have been delighted to have developed many more ingenious ideas for diamond optics, but that must be left to the optical engineers of today, and of the future.

4.9
Conclusions and Outlook

In this chapter, progress in linear diamond optics has been reviewed, from the early stage of windows and lenses made from natural diamond stones to current state-of-the-art components that include both refractive and diffractive structures micromachined in synthetic diamond, and of good optical quality. The fabrication techniques utilized to realize shallow, deep, convex, concave, binary, and analog surfaces have been discussed, with particular focus having been placed on the plasma etching of diamond. A brief introduction to polishing and coating technologies was provided, together with examples of real and practical applications for linear diamond optics. Continued improvements in diamond microfabrication technologies should lead to linear diamond optics becoming more common in future optical systems. In fact, several of the micromachining methods described in this chapter can already be considered quite mature, and new methods for the free-form shaping of bulk diamond are under way.

Although much progress has been made in synthetic diamond production, one hurdle that remains is the cost of the diamond material itself. Diamond optics has, until recently, been used mainly for applications where production cost is not a priority, but rather the performance of the component (e.g., space applications and energy fusion experiments). During the past years, however, diamond optics has begun to be more commercially accepted in markets such as high-power lasers (diamond outcoupling windows for CO_2-lasers) and IR spectroscopy (diamond ATR crystals). With a prediction of further decreased production costs for synthetic diamond, it can be foreseen that diamond optics will soon begin to enter certain high-volume markets. One such area might be medical devices, where the biocompatibility of diamond makes the material very attractive for both in-vivo and in-vitro solutions. In order to enclose fluidic channels made in diamond substrates, it would then be highly desirable to determine a means of bonding diamond to itself or to substrates of other materials.

Finally, if the deposition temperature of diamond can be decreased further, then a wide range of new applications in optics may begin to be exploited, allowing an integration between diamond and many other materials or systems.

References

1 Ditchburn, R.W. (1982) Diamond as an optical material for space optics. *Opt. Acta*, **29** (4), 355–359.

2 Bienstock, B.J. (2004) Pioneer Venus and Galileo entry probe heritage, in A. Wilson (ed.), *1st International Workshop on Radiation of High Temperature Gases in Atmospheric Entry, October 8–10, 2003, Lisbon, Portugal*, ESA Special Publications.

3 Harlow, G.E. (ed.) (1998) *The Nature of Diamonds*, Cambridge University Press, Cambridge.

4 Thumm, M. (2011) Progress on gyrotrons for ITER and future thermonuclear fusion reactors. *IEEE Trans. Plasma Sci.*, **39** (4), 971–979.

5 Freund, A.K. (1995) Diamond single crystals: the ultimate monochromator material for high-power X-ray beams. *Opt. Eng.*, **34** (2), 432–440.

6 Savage, J.A., Wort, C.J.H., Pickles, C.S.J., Sussmann, R.S., Sweeney, C.G., McClymont, M.R., Brandon, J.R., Dodge, C.N., and Beale, A.C. (1997) Properties of free-standing CVD diamond optical components, in R.W. Tustison (ed.), *Proceedings, 5th International Conference on Window and Dome Technologies and Applications Location, April 21–22, 1997, Orlando, USA*, Society of Photo-Optical Instrumentation Engineers (SPIE), Bellingham.

7 Mollart, T.P. (2003) The development of CVD infrared optics from planar windows to missile domes, in R.W. Tustison (ed.), *Proceedings, 8th Annual International Conference on Electro-Optic Windows, Domes, and Materials, April 22–23, 2003, Orlando, USA*, Society of Photo-Optical Instrumentation Engineers (SPIE), Bellingham.

8 Sparrold, S.W., Mills, J.P., Knapp, D.J., Ellis, K.S., Mitchell, T.A., and Manhart, P.K. (2000) Conformal dome correction with counterrotating phase plates. *Opt. Eng.*, **39** (7), 1822–1829.

9 Knapp, D.J. (2002) Fundamentals of conformal dome design, in P.K. Manhart and J.M. Sasian (eds), *Proceedings, International Optical Design Conference, June 3–5, 2002, Tucson, USA*, Society of Photo-Optical Instrumentation Engineers (SPIE), Bellingham.

10 Pritchard, A. (1832) *The Microscopic Cabinet of Select Animated Objects, with a Description of the Jewel and Doublet Microscope*, Whittaker, Treacher and Arnot, London.

11 Nuttall, R. and Frank, A. (1972) Jewel lenses – a historical curiosity. *New Sci.*, **53** (778), 92–93.

12 Woerner, E., Wild, C., Mueller-Sebert, W., and Koidl, P. (2001) CVD-diamond optical lenses. *Diamond Relat. Mater.*, **10** (3–7), 557–560.

13 Karlsson, M. and Nikolajeff, F. (2003) Diamond micro-optics: microlenses and antireflection structured surfaces for the infrared spectral region. *Opt. Express*, **11** (5), 502–507.

14 Popovic, Z.D., Sprague, R.A., and Connell, N. (1988) Technique for monolithic fabrication of microlens arrays. *Appl. Opt.*, **27** (7), 1281–1284.

15 Gu, E., Choi, H.W., Liu, C., Griffin, C., Girkin, J.M., Watson, I.M., Dawson, M.D., McConnell, G., and Gurney, A.M. (2004) Reflection/transmission confocal microscopy characterization of single-crystal diamond microlens arrays. *Appl. Phys. Lett.*, **84** (15), 2754–2756.

16 Choi, H.W., Gu, E., Liu, C., Griffin, C., Girkin, J.M., Watson, I.M., and Dawson, M.D. (2005) Fabrication of natural diamond microlenses by plasma etching. *J. Vac. Sci. Technol. B*, **23** (1), 130–132.

17 Lee, C.L., Gu, E., and Dawson, M.D. (2007) Micro-cylindrical and micro-ring lenses in CVD diamond. *Diamond Relat. Mater.*, **16**, 944–948.

18 Lee, C.L., Choi, H.W., Gu, E., Dawson, M.D., and Murphy, H. (2006) Fabrication and characterization of diamond micro-optics. *Diamond Relat. Mater.*, **15**, 725–728.

19 Nikolajeff, F., Ballen, T., Leger, J.R., Gopinath, A., Lee, T.C., and Williams, R. (1999) Spatial-mode control of vertical-cavity lasers with micromirrors fabricated and replicated in semiconductor materials. *Appl. Opt.*, **38**, 3030–3038.

20 Siyushev, P., Kaiser, F., Jacques, V., Gerhardt, I., Bischof, S., Fedder, H., Dodson, J., Markham, M., Twitchen, D., Jelezko, F., and Wrachtrup, J. (2010) Monolithic diamond optics for single photon detection. *Appl. Phys. Lett.*, **97** (24), 241902.

21 Mildren, R.P., Downes, J.E., Brown, J.D., Johnston, B.F., Granados, E., Spence, D.J., Lehmann, A., Weston, L., and Bramble, A. (2011) Characteristics of 2-photon ultraviolet laser etching of diamond. *Opt. Mater. Express*, **1**, 576–585.

22 Hutley, M.C. (1982) *Diffraction Gratings*, Academic Press, London.

23 Goodman, J.W. (1996) *Introduction to Fourier Optics*, McGraw-Hill, New York.

24 Turunen, J. and Wyrowski, F. (1997) *Diffractive Optics for Industrial and Commercial Applications*, Akademie Verlag GmbH, Berlin.

25 Kononenko, V.V., Konov, V.I., Pimenov, S.M., Prokhorov, A.M., Pavelev, V.S., and Soifer, V.A. (1999) Diamond diffraction optics for CO_2 lasers. *Quantum Electron.*, **29** (1), 9–10.

26 Kononenko, V.V., Konov, V.I., Pimenov, S.M., Prokhorov, A.M., Pavelyev, V.S., and Soifer, V.A. (2000) CVD diamond transmissive diffractive optics for CO_2 lasers. *New Diam. Front. Carbon Technol.*, **10** (2), 97–107.

27 Björkman, H., Rangsten, P., and Hjort, K. (1999) Diamond microstructures for optical micro electromechanical systems. *Sens. Actuators A*, **78** (1), 41–47.

28 Karlsson, M., Hjort, K., and Nikolajeff, F. (2001) Transfer of continuous-relief diffractive structures into diamond by use of inductively coupled plasma dry etching. *Opt. Lett.*, **26** (22), 1752–1754.

29 Fu, Y.Q. and Bryan, N.K.A. (2003) Investigation of diffractive optical element fabricated on diamond film by use of focused ion beam direct milling. *Opt. Eng.*, **42** (8), 2214–2217.

30 Raguin, D.H. and Morris, G.M. (1993) Antireflection structured surfaces for the infrared spectral region. *Appl. Opt.*, **32** (7), 1154–1167.

31 Bernhard, C.G. (1967) Structural and functional adaptation in a visual system. *Endeavour*, **26** (98), 79–84.

32 Clapham, P.B. and Hutley, M.C. (1973) Reduction of lens reflection by moth eye principle. *Nature*, **244** (5414), 281–282.

33 Harker, A.B. and DeNatale, J.F. (1992) Diamond gradient index moth-eye antireflection surfaces for LWIR windows, in P. Klocek (ed.), *Proceedings, International Conference on Infrared Windows and Domes, July 22–23, 1992, San Diego, USA*, Society of Photo-Optical Instrumentation Engineers (SPIE), Bellingham.

34 Ralchenko, V.G., Khomich, A.V., Baranov, A.V., Vlasov, I.I., and Konov, V.I. (1999) Fabrication of CVD diamond optics with antireflective surface structures. *Phys. Status Solidi A*, **174** (1), 171–176.

35 Kononenko, T.V., Kononenko, V.V., Konov, V.I., Pimenov, S.M., Garnov, S.V., Tishchenko, A.V., Prokhorov, A.M., and Khomich, A.V. (1999) Formation of antireflective surface structures on diamond films by laser patterning. *Appl. Phys. A*, **68** (1), 99–102.

36 Gloor, S., Romano, V., Luthy, W., Weber, H.P., Kononenko, V.V., Pimenov, S.M., Konov, V.I., and Khomich, A.V. (2000) Antireflection structures written by excimer laser on CVD diamond. *Appl. Phys. A*, **70** (5), 547–550.

37 Wyrowski, F. and Bryngdahl, O. (1988) Iterative Fourier-transform algorithm applied to computer holography. *J. Opt. Soc. Am. A*, **5** (7), 1058–1065.

38 Karlsson, M. and Nikolajeff, F. (2003) Fabrication and evaluation of a diamond diffractive fan-out element for high power lasers. *Opt. Express*, **11** (3), 191–198.

39 Kononenko, V.V., Konov, V.I., Pimenov, S.M., Prokhorov, A.M., Pavelyev, V.S., Soifer, V.A., Ludge, B., and Duparre, M. (2002) Laser shaping of diamond for IR diffractive optical elements, in I. Miyamoto, Y.F. Lu, K. Sugioka, and J.J. Dubowski (eds), *Proceedings, 2nd International Symposium on Laser Precision Microfabrication, May 16–18, 2001, Singapore, Singapore*, Society of Photo-Optical Instrumentation Engineers (SPIE), Bellingham.

40 Pavelyev, V.S., Soifer, V.A., Golovashkin, D.L., Kononenko, V.V., Konov, V.I., Pimenov, S.M., Duparre, M., and Luedge,

B. (2004) Diamond diffractive optical elements for infrared laser beam control, in F. Wyrowski (ed.), *Conference on Photon Management, April 27–28, 2004, Strasbourg, France*, Society of Photo-Optical Instrumentation Engineers (SPIE), Bellingham.

41 Nelissen, W., Godfried, H., Houwman, E., Kriele, P., Lamers, J., Pels, G., Oerle, B., Spaay, P., Mcclymont, M., Wort, C., and Scarsbrook, G. (2006) Single crystal diamond elements having convex surfaces and methods of its fabrication. WO Patent 2007/007126A1, filed July 7, 2006 and issued January 18, 2007.

42 Malshe, A.P., Park, B.S., Brown, W.D., and Naseem, H.A. (1999) A review of techniques for polishing and planarizing chemically vapor-deposited diamond films and substrates. *Diamond Relat. Mater.*, **8**, 1198–1213.

43 Mollart, T.P., Wort, C.J.H., Pickles, C.S.J., McClymont, M.R., Perkins, N., and Lewis, K.L. (2001) CVD diamond optical components, multi-spectral properties and performance at elevate temperatures, in R.W. Tustison (ed.), *Conference on Window and Dome Technologies and Materials VII, April 16–17, 2001, Orlando, USA*, Society of Photo-Optical Instrumentation Engineers (SPIE), Bellingham.

44 Jin, S., Graebner, J.E., McCormack, M., Tiefel, T.H., Katz, A., and Dautremont-Smith, W.C. (1993) Shaping of diamond films by etching with molten rare-earth-metals. *Nature*, **362** (6423), 822–824.

45 Rothschild, M., Arnone, C., and Ehrlich, D.J. (1986) Excimer-laser etching of diamond and hard carbon films by direct writing and optical projection. *J. Vac. Sci. Technol. B*, **4** (1), 310–314.

46 Efremow, N.N., Geis, M.W., Flanders, D.C., Lincoln, G.A., and Economou, N.P. (1985) Ion-beam-assisted etching of diamond. *J. Vac. Sci. Technol. B*, **3** (1), 416–418.

47 Whetten, T.J., Armstead, A.A., Grzybowski, T.A., and Ruoff, A.L. (1984) Etching of diamond with argon and oxygen ion beams. *J. Vac. Sci. Technol. A*, **2** (2), 477–480.

48 Sandhu, G.S. and Chu, W.K. (1989) Reactive ion etching of diamond. *Appl. Phys. Lett.*, **55** (5), 437–438.

49 Dorsch, O., Werner, M., Obermeier, E., Harper, R.E., Johnston, C., and Buckleygolder, I.M. (1992) Etching of polycrystalline diamond and amorphous carbon films by RIE. *Diamond Relat. Mater.*, **1** (2–4), 277–280.

50 Dorsch, O., Werner, M., and Obermeier, E. (1995) Dry etching of undoped and boron doped polycrystalline diamond films. *Diamond Relat. Mater.*, **4** (4), 456–459.

51 Shiomi, H. (1997) Reactive ion etching of diamond in O-2 and CF4 plasma, and fabrication of porous diamond for field emitter cathodes. *Jpn. J. Appl. Phys.*, **36** (12B), 7745–7748.

52 Masuda, H., Watanabe, M., Yasui, K., Tryk, D., Rao, T., and Fujishima, A. (2000) Fabrication of a nanostructured diamond honeycomb film. *Adv. Mater.*, **12** (6), 444–447.

53 Leech, P.W., Reeves, G.K., and Holland, A. (2001) Reactive ion etching of diamond in CF_4, O_2, O_2 and Ar-based mixtures. *J. Mater. Sci.*, **36** (14), 3453–3459.

54 Ando, Y., Nishibayashi, Y., Kobashi, K., Hirao, T., and Oura, K. (2002) Smooth and high-rate reactive ion etching of diamond. *Diamond Relat. Mater.*, **11** (3-6), 824–827.

55 Nohammer, B., Hoszowska, J., Freund, A.K., and David, C. (2003) Diamond planar refractive lenses for third- and fourth-generation X-ray sources. *J. Synchrotron Radiat.*, **10**, 168–171.

56 Ando, Y., Nishibayashi, Y., and Sawabe, A. (2004) "Nano-rods" of single crystalline diamond. *Diamond Relat. Mater.*, **13** (4-8), 633–637.

57 Ding, G.F., Mao, H.P., Cai, Y.L., Zhang, Y.H., Yao, X., and Zhao, X.L. (2005) Micromachining of CVD diamond by RIE for MEMS applications. *Diamond Relat. Mater.*, **14** (9), 1543–1548.

58 Pearton, S.J., Katz, A., Ren, F., and Lothian, J.R. (1992) ECR plasma etching of chemically vapor deposited diamond thin films. *Electron. Lett.*, **28** (9), 822–824.

59 Enlund, J., Isberg, J., Karlsson, M., Nikolajeff, F., Olsson, J., and Twitchen,

D.J. (2005) Anisotropic dry etching of boron doped single crystal CVD diamond. *Carbon*, **43** (9), 1839–1842.

60 Lee, C.L., Gu, E., Dawson, M.D., Friel, I., and Scarsbrook, G.A. (2008) Etching and micro-optics fabrication in diamond using chlorine-based inductively-coupled plasma. *Diamond Relat. Mater.*, **17** (7–10), 1292–1296.

61 Yoshikawa, H., Shikata, S., Fujimori, N., Sato, N., and Ikehata, T. (2006) Smooth surface dry etching of diamond by very high frequency inductively coupled plasma. *New Diamond Front. Carbon Technol.*, **16** (2), 97–106.

62 Yamada, T., Yoshikawa, H., Uetsuka, H., Kumaragurubaran, S., Tokuda, N., and Shikata, S. (2007) Cycle of two-step etching process using ICP for diamond MEMS applications. *Diamond Relat. Mater.*, **16** (4–7), 996–999.

63 Isakovic, A.F., Stein, A., Warren, J.B., Narayanan, S., Sprung, M., Sandy, A.R., and Evans-Lutterodt, K. (2009) Diamond kinoform hard X-ray refractive lenses: design, nanofabrication and testing. *J. Synchrotron Radiat.*, **16**, 8–13.

64 Moldovan, N., Divan, R., Zeng, H.J., and Carlisle, J.A. (2009) Nanofabrication of sharp diamond tips by e-beam lithography and inductively coupled plasma reactive ion etching. *J. Vac. Sci. Technol. B*, **27** (6), 3125–3131.

65 Karlsson, M., Vartianen, I., Kuittinen, M., and Nikolajeff, F. (2010) Fabrication of sub-micron high aspect ratio diamond structures with nanoimprint lithography. *Microelectron. Eng.*, **87** (11), 2077–2080.

66 Grot, S.A., Ditizio, R.A., Gildenblat, G.S., Badzian, A.R., and Fonash, S.J. (1992) Oxygen based electron cyclotron resonance etching of semiconducting homoepitaxial diamond films. *Appl. Phys. Lett.*, **61** (19), 2326–2328.

67 Hwang, D.S., Saito, T., and Fujimori, N. (2004) New etching process for device fabrication using diamond. *Diamond Relat. Mater.*, **13** (11–12), 2207–2210.

68 Vossen, J.L. and Kern, W. (eds) (1991) *Thin Film Processes II*, Academic Press, Inc., San Diego.

69 Hoekstra, R.J., Kushner, M.J., Sukharev, V., and Schoenborn, P. (1998) Microtrenching resulting from specular reflection during chlorine etching of silicon. *J. Vac. Sci. Technol. B*, **16** (4), 2102–2104.

70 Jiang, L.D., Plank, N.O.V., Blauw, M.A., Cheung, R., and van der Drift, E. (2004) Dry etching of SiC in inductively coupled Cl_2/Ar plasma. *J. Phys. D*, **37** (13), 1809–1814.

71 Chou, S.Y., Krauss, P.R., and Renstrom, P.J. (1996) Imprint lithography with 25-nanometer resolution. *Science*, **272** (5258), 85–87.

72 Guo, L.J. (2007) Nanoimprint lithography: methods and material requirements. *Adv. Mater.*, **19** (4), 495–513.

73 Babinec, T.M., Hausmann, B.J.M., Khan, M., Zhang, Y., Maze, J.R., Hemmer, P.R., and Loncar, M. (2010) A diamond nanowire single-photon source. *Nat. Nanotechol.*, **5**, 195–199.

74 Mollart, T.P., Lewis, K.L., Wort, C.J.H., and Pickles, C.S.J. (2001) Coatings technology for CVD diamond optics, in R.W. Tustison (ed.), *Conference on Window and Dome Technologies and Materials VII, April 16–17, 2001, Orlando, USA*, Society of Photo-Optical Instrumentation Engineers (SPIE), Bellingham.

75 Hawkins, G., Sherwood, R., and Djotni, K. (2008) Mid-infrared filters for astronomical and remote sensing instrumentation, in N. Kaiser, M. Lequime, and A. Macleod (eds), *Conference on Advances in Optical Thin Films III, September 2–3, 2008, Glasgow, UK*, Society of Photo-Optical Instrumentation Engineers (SPIE), Bellingham.

76 Miller, A.J., Reece, D.M., Hudson, M.D., Brierley, C.J., and Savage, J.A. (1997) Diamond coatings for IR window applications. *Diamond Relat. Mater.*, **6**, 386–389.

77 Pickles, C.S.J., Madgwick, T.D., Sussmann, R.S., and Hall, C.E. (2000) Optical applications of CVD diamond. *Ind. Diamond Rev.*, **60** (587), 293–299.

78 Shvyd'ko, Y.V., Stoupin, S., Cunsolo, A., Said, A.H., and Huang, X. (2010) High-reflectivity high-resolution X-ray crystal optics with diamonds. *Nat. Phys.*, **6**, 196–199.

79 Andersson, P.O., Lundquist, M., Tegler, L., Borjegren, S., Baltzer, L., and

Österlund, L. (2007) A novel ATR-FTIR approach for characterisation and identification of ex situ immobilised species. *ChemPhysChem*, **8** (5), 712–722.

80 Fromell, K., Forsberg, P., Karlsson, M., Larsson, K., Nikolajeff, F., and Baltzer, L. (2012) Designed protein binders in combination with nano crystalline diamond for use in high-sensitivity biosensors. *Anal. Bioanal. Chem.*, **404**, 1643–1651.

81 Mawet, D., Riaud, P., Absil, O., and Surdej, J. (2005) Annular groove phase mask coronagraph. *Astrophys. J.*, **633** (2), 1191–1200.

82 Delacroix, C., Forsberg, P., Karlsson, M., Mawet, D., Absil, O., Hanot, C., Surdej, J., and Habraken, S. (2012) Design, manufacturing, and performance analysis of mid-infrared achromatic half-wave plates with diamond subwavelength gratings. *Appl. Opt.*, **51** (24), 5897–5902.

83 Adamantis (2004) Available at: http://www.adamantis.com (accessed 10 September 2012).

84 Shinoda, M., Saito, K., Kondo, T., Nakaoki, A., Furuki, M., Takeda, M., Yamatomo, M., Schaich, T., Van Oerle, B., Godfried, H., Kriele, P., Houwman, E., Nelissen, W., Pels, G., and Spaaij, P. (2006) High-density near-field readout using diamond solid immersion lens. *Jpn. J. Appl. Phys.*, **45**, 1311–1313.

5
Nitrogen-Vacancy Color Centers in Diamond: Properties, Synthesis, and Applications

Carlo Bradac, Torsten Gaebel, and James R. Rabeau

5.1
Introduction

Diamond is a semiconductor with a wide band-gap of 5.5 eV corresponding to the deep ultraviolet (UV) wavelength of 225 nm. Although it should be perfectly transparent and colorless in the visible-light range (wavelength 390–750 nm), diamond normally displays a variety of different colorations that include yellow, brown, blue, green, black, pink, orange, purple, and red [1]. The different colors originate from diamond defects having characteristic optical absorptions. Such defects can be either inclusions of foreign atoms or, in cases such as brown diamonds, plastic deformations of the diamond crystal lattice [2].

Defects in diamond are intriguing. The observed color change is only a manifestation of a myriad of complex material changes caused by the impurities, but they can affect mechanical properties or change electrical conductivity, as occurs for example in diamonds doped with boron (p-type) or phosphorus (n-type) atoms [3, 4]. Besides the bulk effects, many of these diamond color centers display unique properties of their own that might be exploited for a wide variety of specific applications.

To date, hundreds of defects have been identified and characterized in diamond [5]. In this chapter, attention is focused on one defect in particular, namely the nitrogen-vacancy (NV) center. This center, which has captured the interest of science and technology research groups over the past few years, possesses unique optical and spin properties [6, 7] that have led to it being a promising candidate for several technologies ranging from quantum science [8], through biology [9] and magnetic sensing [10–12].

In this chapter, the NV center and its properties are analyzed in detail, and the main techniques developed to synthesize diamond and enhance the concentration of color centers are summarized. Details are then presented of the novel technologies that have been identified exploiting the NV center. This is followed by a discussion of the feasibility of such technologies relative to advances in the state of the art of diamond material processing, with regards to some observed

Optical Engineering of Diamond, First Edition. Edited by Richard P. Mildren and James R. Rabeau.
© 2013 Wiley-VCH Verlag GmbH & Co. KGaA. Published 2013 by Wiley-VCH Verlag GmbH & Co. KGaA.

"anomalies" of the defect when hosted in ultrasmall (size <10 nm) nanodiamonds, or in the proximity of the diamond surface.

5.2
Defects in Diamond

Diamond is host to a wide variety of crystallographic defects which disrupt the tetrahedral periodicity of the carbon atoms in the lattice [5]. In the recent past, considerable efforts have been made to identify and characterize such defects, often with the intent of exploiting their properties in cutting-edge technologies.

In diamond, defects are generally classified as either intrinsic or extrinsic, according to their nature and origin.

5.2.1
Intrinsic and Extrinsic Defects

Intrinsic defects (also known as lattice irregularities) occur when carbon atoms are displaced from their nominal position in the periodic crystalline structure of diamond (Figure 5.1a–c). These can be: (i) interstitial carbon atoms or complexes where two or more carbon atoms share a single lattice site [13, 14]; (ii) vacancies or multivacancy complexes consisting of carbon atoms missing from

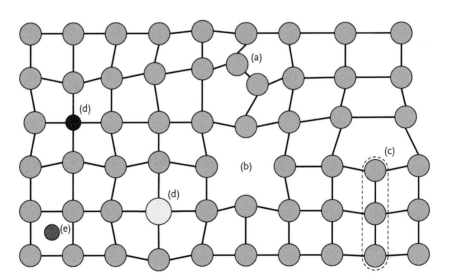

Figure 5.1 Two-dimensional schematic of some point defects (a, b, d, e) and linear defects (c) in diamond. (a) Self-interstitial carbon atom; (b) Vacancy; (c) Edge dislocation, with extra net plane; (d) Foreign substitutional atoms, for example, N, P, B; (e) Foreign interstitial atom, for example, Li.

crystallographic sites [2, 15–20]; (iii) dislocations with broken bonds between atoms of the same index (*shuffle set*) or between layers of atoms of different indexes (*glide set*) [21]; (iv) platelets of regular arrays of carbon interstitials [22–24]; and (v) voidites made of nanometer-sized carbon clusters [25, 26]. Intrinsic defects can be found in natural diamond; they can also be artificially produced or enhanced through irradiation with high-energy particles.

Extrinsic defects mostly occur due to the presence of substitutional and interstitial impurities of foreign elements embedded in the carbon matrix (Figure 5.1d, e). They are present in diamond as isolated atoms or organized in small atomic clusters, and they can be found naturally dispersed in diamond or artificially introduced in the structure, for example, by ion-implantation or by addition during the diamond growth processes. More than 150 vibrational, and more than 500 electronic optical centers, have been detected in diamond within the spectral range from 20 μm to 0.17 μm. Many impurities of foreign elements are known to form optically active defects in diamond, including H, He, Li, B, N, O, Ne, P, Si, As, Ti Cr, Ni, Co, Zn, Zr, Ag, W, Xe, and Tl [5, 27, 28].

5.2.2
Nitrogen-Related Defects in Diamond

Among the hundreds of types of luminescent extrinsic impurities that can form in diamond, the nitrogen-related impurities call for special attention. Nitrogen is in fact the most common impurity in diamond, and is responsible for a large number of impurity-related centers.

Nitrogen as a diamond impurity was first identified in 1959 by W. Kaiser and W.L. Bond of Bell Telephone Laboratories [29], who showed experimentally that the nitrogen content was responsible for the infrared (IR) absorption of diamond around 8 μm, and for the UV absorption below 400 nm. Kaiser and Bond also observed that different concentrations of nitrogen could influence some diamond properties, such as photoconductivity, thermal conductivity, and birefringence.

Nitrogen can form various complexes in diamond. It can exist either as single substitutional impurities, called C-centers, or in aggregated structures, called A- and B-centers (Figure 5.2). Nitrogen complexes can also coexist in combination with other defects such as platelets and voidites. Apart from voidites, which can be observed using transmission electron microscopy (TEM) [25], all the other aggregates have characteristic optical spectroscopy signatures.

Diamonds containing mainly C-center defects are classified as Type Ib (N concentration up to 500 ppm, 0.05%), while diamonds containing mainly A- and B-centers are classified respectively as Type IaA and Type IaB (N concentration up to 3000 ppm, 0.3%). When A and B aggregates occur in similar concentrations, diamonds are generally identified as Type IaAB. Type IIa diamonds are almost devoid of impurities (N concentration <20 ppm, 0.002%), while Type IIb diamonds are characterized by the presence of boron impurities (B concentration ~0.25 ppm, 0.000025% in natural diamonds, and up to 270 ppm, 0.027%, in doped synthetic diamonds) (Figure 5.3).

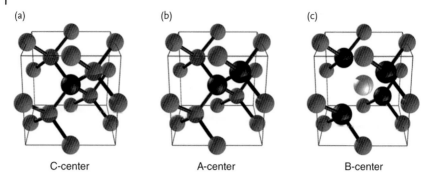

Figure 5.2 Structure of nitrogen-related defects in diamond. (a) C-center; (b) A-center; (c) B-center. Nitrogen atoms are shown in black, carbon atoms in gray, and vacancies in white with no bonds (black sticks) indicated.

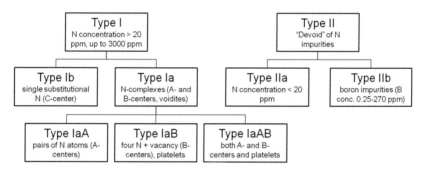

Figure 5.3 Schematic representation of the different types of diamond.

5.2.2.1 C-Center

The C-center defect consists of a single substitutional nitrogen atom, either neutral or positively charged, in the diamond lattice (Figure 5.2a). The nitrogen atom forming the C-center alters the properties of diamond local to the center.

- The neutral nitrogen defect does not possess the diamond tetrahedral (C_{3v}) symmetry. It undergoes a trigonal distortion with the nitrogen and one of the carbon atoms moving away from each other along the [111] direction. The estimated length of the nitrogen–carbon bond for the C-center is about 1.92 Å, which is longer than the nearest-neighbor distance of 1.54 Å between carbon atoms in diamond [30].

- The center has an identifiable IR local mode of vibration at 1344 cm^{-1} [31].

- Because the nitrogen atom has five available electrons (one more than the carbon atom it replaces), it acts as a deep donor, forming a donor energy level within the diamond band gap, 1.7 eV below the conduction band edge [30].

- The C-center shows an electron paramagnetic resonance (EPR) signal called P1 [32]. This signal identifies the C-center to be an electron paramagnetic system with electron spin angular momentum $S = 1/2$ in the ground state [33].
- The C-center is responsible for the yellow/brown coloration of diamonds [1].

5.2.2.2 A-Center

The A-center defect consists of a nearest-neighbor pair of nitrogen atoms substituting for the carbon atoms (Figure 5.2b).

- The nitrogen–nitrogen bond length is estimated to be 2.14 Å [34].
- The A-center gives vibrational resonances at 1282, 1203, 1093, and 480 cm^{-1} [33].
- The A-center does not cause discoloration on its own [35].

5.2.2.3 B-Center

The B-center defect consists of four nitrogen atoms surrounding a vacancy (Figure 5.2c).

- The nitrogen–carbon bond length in this structure is estimated to be 1.49 Å [34].
- The defect, sometimes indicated as VN_4, gives vibrational bands at 1332, 1171, 1093, 1003, and 780 cm^{-1} [34].
- The B-center defect does not cause discoloration on its own [35].

5.2.2.4 Platelets

Platelets are planar defects in the <100> lattice planes, and can coexist with nitrogen complexes [23, 36, 37].

- Platelets generally possess a length scale of the order of 10 to 100 nm, but some can extend to the range 1–10 μm [36].
- They show a characteristic absorption peak at 328 cm^{-1} and an IR absorption band with a maximum between 1358 cm^{-1} and 1373 cm^{-1}, labeled B' [23, 34].
- Platelets do not affect the diamond color [35].

5.2.2.5 Other Nitrogen-Vacancy Complexes

Other complexes composed of nitrogen and vacancies can be found in diamond; these are shown in Figure 5.4.

- VN_2 or H_3 center, which has absorption with a zero phonon line (that is, pure electronic transition with no phonons involved) at 2.463 eV, and belongs to point group C_{2v} [5, 27, 38].
- V_2N_4 or H_4 center, which has absorption with a zero phonon line at 2.498 eV, and belongs to point group C_{1h} [5, 27, 34, 38–40].
- VN_3 or N_3 center, which has absorption with a zero phonon line at 2.985 eV, and belongs to point group C_{3v} [5, 23, 27].

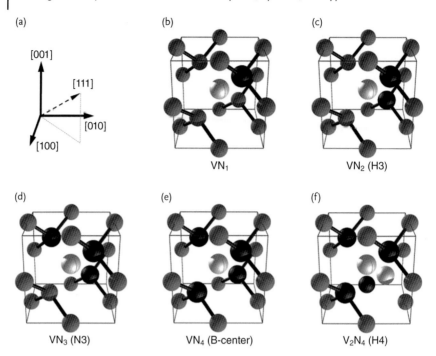

Figure 5.4 Schematic structure of nitrogen-vacancy complexes in diamond. (a) Diamond unitary cell crystallographic axes; (b) VN_1 defect; (c) VN_2 (H3) defect; (d) VN_3 (N3) defect; (e) VN_4 (B-center) defect; (f) V_2N_4 (H4) defect. Nitrogen atoms are shown in black, carbon atoms in gray, and vacancies in white with no bonds (black sticks) indicated.

- VN_1 center, which has absorption with a zero phonon line at 1.945 eV, and belongs to point group C_{3v} [5, 27, 38].

The most extensively studied peaks in the absorption spectrum of diamond due to defects, aggregates and impurity atoms are summarized in Table 5.1. In Table 5.2, which relates to the vacancy-related and nitrogen-related defects, the peaks are normally designated by letters.

5.2.3
Nitrogen-Vacancy (NV) Center

Among all the structures that nitrogen can form in diamond, probably the most extensively studied is the VN_1 structure (Figure 5.4b). Generally addressed as the nitrogen-vacancy center or simply NV, this diamond defect has been the subject of intense study in the recent past. The NV center possesses unique room-temperature optical and spin properties which make it a fascinating object for practical applications in many fields of research, including quantum information

Table 5.1 Nomenclature for defect-related systems in diamond.

Designation	Significance	Range	Reference
N	Natural diamond	N1 (1.50 eV) to N9 (5.26 eV)	[27]
GR	General radiation (induced in diamonds by irradiation)	GR1 (1.67 eV) to GR8 (3.00 eV)[a]	[27]
R	Radiation (masked in Type Ia diamond by the secondary absorption edge)	R9 (3.04 eV) to R11 (3.99 eV)[b]	[27]
TR	Type II radiation (not observed in Type I diamond)	TR12 (2.64 eV) to TR17 (2.83 eV)	[27]
H	Heat treatment (preceded by irradiation)	H1 (0.18 eV) to H18 (3.56 eV)	[27]

a) GR1 to GR8 is the most extensively studied system: these peaks are caused by neutral vacancies.
b) R10 is more frequently called ND1.

Table 5.2 Vacancy-related and nitrogen-related defects in diamond.

Designation	Type of defect	Zero phonon line (77 K)	References
GR1	Neutral vacancy (V^0)	741.1 nm/1.673 eV	[5, 27, 38]
ND1 (R10)	Negative vacancy (V^-)	393.6 nm/3.150 eV	[5, 27, 38]
NV	Vacancy plus single nitrogen (VN_1)	637.3 nm/1.945 eV	[5, 27, 38]
H3	Vacancy and two nitrogens (VN_2)	503.2 nm/2.463 eV	[5, 27, 38]
H4	Two vacancies and four nitrogens (V_2N_4)	496.2 nm/2.498 eV	[5, 27, 38]
N3	Vacancy and three nitrogens (VN_3)	415.3 nm/2.985 eV	[5, 23, 27]

technology, nanomedicine and biotechnology, and high-resolution magnetic sensing (cf. Sections 5.4.1–5.4.3).

5.2.3.1 Structure

The NV center is an extrinsic diamond crystallographic defect incorporated in the tetrahedral diamond structure of four carbon atoms where two adjacent sites are altered: one carbon atom is replaced by a nitrogen atom, and the other is replaced by a vacant space (vacancy). The defect has a trigonal symmetry of point group C_{3v}, where the C_3 principal axis is along the NV pair in the crystallographic [111] direction (Figure 5.5a).

Being in Group V of the Periodic Table, nitrogen has five valence electrons. In the NV complex, three of the nitrogen valence electrons are shared with the three nearest-neighbor carbon atoms, while the last two occupy the dangling bond in

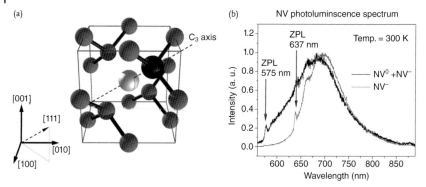

Figure 5.5 NV center crystallographic structure and emission spectrum. (a) The NV center has a trigonal symmetry of point group C$_{3v}$, where the C$_3$ principal axis of symmetry is along the NV pair in the crystallographic [111] direction. The nitrogen atom is shown in black, carbon atoms are in gray, and the vacancy in white with no bonds (black sticks) indicated; (b) Fluorescence emission spectra at room temperature of an NV$^-$ center (gray) and of NV^0 + NV$^-$ (black). The zero phonon lines (ZPL) at 575 nm (2156 eV) for the NV^0 and at 637 nm (1.945 eV) for the NV$^-$ are highlighted (arrows); they are accompanied by vibronic side bands at lower energies (longer wavelengths).

the direction of the vacancy. Looking at the vacancy, there are five unsatisfied active bonding electrons that belong to the dangling bonds of the atoms adjacent to the vacancy itself – two from the nitrogen and three from the surrounding carbons. Each of the three carbon atoms has nominally four sp^3 symmetry bonds: one dangling bond near the vacancy, and three back bonds pointing to the other carbons of the lattice. Similarly, the nitrogen atom has one dangling bond pointing to the vacancy and three back bonds pointing to other carbons of the lattice [41]. For the carbon, the sp^3 hybridization is not complete; the orbital is 90% p-character and 10% s-character, against the general 75% p-character and 25% s-character of the fully hybridized tetrahedral sp^3 orbital. For nitrogen, the ratio of p to s orbitals is almost the same ($p/s = 1.2$), while for carbon the p orbital dominates ($p/s = 9.8$). It has been estimated that, in the NV complex, approximately 72% of the charge density is at the three nearest-neighbor carbon atoms surrounding the vacancy, only about 0.2% is at the nitrogen, and approximately 28% is spread over the lattice [41, 42]. This means that, in a naive view, the electrons spend most of their time in the dangling bonds on the three carbon atoms neighboring the vacancy.

5.2.3.2 Charge State

The NV center exists in diamond in two charge states. The charge state of the defect depends mainly on its local lattice environment; the presence of close surrounding impurities acting as electron acceptors or donors causes the NV to be either in its neutral state NV^0 or in its negatively charged state NV^- [43, 44]. In the case of NV^-, it is assumed that the neutral NV^0 defect, with five electrons occupying the dangling bonds in the neighborhood of the vacancy complex, acquires an additional sixth electron from elsewhere in the lattice, for instance from another nitrogen atom acting as a donor [43]:

$$2N^0 + V^0 \rightarrow NV^- + N^+ \tag{5.1}$$

The ground state of the neutrally charged NV^0 has an electron spin angular momentum $S = 1/2$ (singlet), while EPR measurements show that the negatively charged NV^- has a paramagnetic ground state with electron spin angular momentum $S = 1$ (triplet) [32]. This is compatible with having respectively an uneven (five for the NV^0) and even (six for the NV^-) number of electrons surrounding the vacancy in the NV complex.

The charge state of the NV is a local effect depending on the presence of nearby electron donors or acceptors. This means that NV^0 and NV^- can coexist in the same diamond, depending on the local environment surrounding each center [44]. It has also been observed, in emission, that the same NV defect can undergo photochromic switching between the two charge states. Two mechanisms can induce this photoconversion: (i) directly, via ionization of a negatively charged NV^-; or (ii) indirectly, via ionization of the substitutional nitrogen atoms surrounding the NV center in the crystal and acting as donors [45, 46]. The two charge states of the NV can be clearly identified spectrally due to their different and characteristic optical transition [5].

The NV^- has a strong optical transition with a zero phonon line (ZPL) – that is, the pure electronic transition – at 637 nm (1.945 eV) (Figure 5.5b). Uniaxial strength measurements have shown that the ZPL is associated with an electric dipole transition between isolated states of spatial symmetry $^3A_2$ (ground state) and ^3E (excited state) at a site of trigonal symmetry [47, 48]. The transition is also accompanied by a vibronic side band to high energy in absorption, and to low energy in emission. Absorption and emission do not show the same vibronic band; in absorption, the first feature of the vibronic sideband gives a double peak, whereas in emission it gives only a single peak. This spectral difference has been explained by tunneling of the nitrogen atom, enabling it to interchange positions with the vacancy in the NV center [47]. Recently, this idea has been disputed, with ab initio calculations showing that the nitrogen binds strongly with the three neighboring carbon atoms, but that the tunneling of the nitrogen would require too much energy (>4 eV) to make this mechanism likely. An alternative proposal considers the Jahn–Teller effect for the excited state of the NV center [49]. The Jahn–Teller effect is a geometric distortion which occurs in nonlinear molecules and takes the form of an elongation or shortening of certain chemical bonds, since the removal of orbital and electronic degeneracies lowers the total energy of the complex, for instance by reducing the electrostatic repulsion among electrons [50]. In the case of the NV center, the excited state would undergo a Jahn–Teller splitting: two local modes a_1

and e would couple in the vibration band of absorption giving rise to the observed double peak, whereas only the a_1 local mode would couple in the vibration band of emission with a corresponding single spectral peak [51].

The NV^0 shows a ZPL at 575 nm (2.156 eV) corresponding to a transition between the ground state of spatial symmetry ^2E and the excited state of symmetry $^2A_1$ (Figure 5.5b) [52, 53].

5.2.3.3 Energy Level Scheme

Considerable efforts have been made to shed light on one intriguing aspect of the NV center that remains an intense object of debate, namely its electronic structure. At this point, attention is focused on the negatively charged state of the center (NV^-), as this has emerged as being much more interesting in terms of spin, and magnetic and optical properties.

Detailed descriptions of the NV^- energy level scheme, based on both experimental and theoretical considerations, have recently been provided [54–56], and a simplified version is considered here (Figure 5.6). Despite being a simplification, this scheme is sufficiently accurate and is therefore often used to describe the dynamic of the system and its properties from a working point of view.

The NV^- center is treated as a three-level system, with a ground triplet state ^3A, a triplet excited state ^3E, and an intermediate singlet state ^1A [57–67]. The separa-

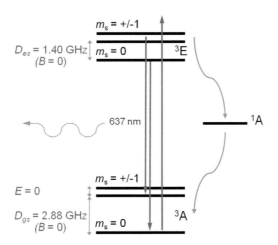

Figure 5.6 Simplified electronic-level scheme of the NV^- center. The ground ^3A and the excited ^3E states are triplets, and their separation in energy is 1.945 eV (ZPL at 637 nm). They both show a zero-field splitting between spin sublevels of electronic spin quantum number $m_s = 0$ and $m_s = \pm 1$; the ground-state zero-field splitting is $D_{gs} = 2.88$ GHz. The sublevels $m_s = +1$ and $m_s = -1$ are degenerate in energy for symmetry reasons, but a local change in the symmetry can lift the degeneracy, making them distinguishable. The intermediate energy level ^1A is a singlet. The straight arrows indicate the optical excitation and the subsequent radiative decay, while the curved arrows indicate the nonradiative decay via the metastable singlet state ^1A.

tion in energy between the ground and the excited triplet states ^3A and ^3E is 1.945 eV (ZPL at 637 nm). Both states are manifold in energy, showing a zero-field splitting between the spin sublevels of the electronic spin quantum number $m_s = 0$ and $m_s = \pm 1$. For the ground state, the zero-field splitting has been measured as $D_{gs} = 2.88\,\text{GHz}$, which means that this transition is accessible via microwave excitation. Recently, the fine structure of the NV⁻ excited state has also been investigated; pulsed optically detected magnetic resonance measurements have revealed the occurrence of spin transitions in the emitting ^3E excited state [68]. The sublevels $m_s = +1$ and $m_s = -1$ are degenerate in energy for reasons of symmetry. A local change in the symmetry due to lattice irregularities, strain or external fields can lift the degeneracy, causing the sublevels to be distinguishable.

5.2.4
Optical and Spin Properties of NV⁻ Centers in Diamond

The NV⁻ center is a single photon emitter in the visible range [69, 70], is photostable and immune to photo-bleaching [71], and can undergo spin-sensitive optical transitions under ambient conditions [54, 72]. It also shows a long spin coherence time [73].

The NV⁻ properties central to the interest around new applications may now be summarized.

5.2.4.1 Polarization

The first interesting property of the NV⁻ center is its optical polarizability, which is crucial if the NV⁻ center is to be used both as a spin quantum bit (qubit) and as a probe for high-resolution magnetometry.

Polarization preferentially populates a certain state. In the case of the NV⁻, optical excitation preferentially populates the $m_s = 0$ ground spin state of the center. Optical pumping causes a non-Boltzmann steady-state spin alignment of the NV⁻ in the ground state consisting of a reorientation of the ground state spins. The direct implication of this is that the initial state of the system can be easily and reliably "set."

The polarization mechanism is explained in Figure 5.7. Initially, the NV⁻ is unpolarized (Figure 5.7a) and the ground spin sublevels $m_s = 0$ and $m_s = \pm 1$ are equally populated. In reality, however, the occupation probability of the two sublevels obeys the Boltzmann steady-state distribution; in this example it is assumed for simplicity (without losing generality) that the two sublevels are equally populated (the concept can be easily extended to the case with the actual Boltzmann steady-state distribution for the system). By pumping the NV⁻ center with a true laser (e.g., 514 nm or 532 nm), the population is optically excited from the ground spin sublevels $m_s = 0$ and $m_s = \pm 1$ to the corresponding optically excited spin levels (Figure 5.7b). The optical transition is spin-conserving ($\Delta m_s = 0$), which means that the quantum spin number m_s is conserved. From the optically excited electron state the system can decay radiatively, either directly towards the ground state with emission of photons ($\lambda = 637\,\text{nm}$), or nonradiatively, towards the ground state

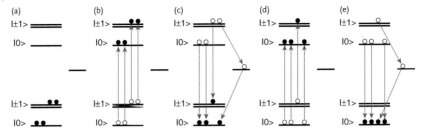

Figure 5.7 Schematic representation of the NV⁻ polarization mechanism. (a) Initially, the NV⁻ population is evenly distributed between the ground spin sublevels $m_s = 0$ and $m_s = \pm 1$ (indicated, respectively, as |0⟩ and |±1⟩); (b) A green-wavelength laser excitation drives the optical transition (which is spin-conserving, i.e., $\Delta m_s = 0$) towards the excited state; (c) The system relaxes, mainly radiatively (again $\Delta m_s = 0$), by emitting photons at 637 nm; however, some part of the population from the excited spin sublevel $m_s = \pm 1$ relaxes nonradiatively, via the intermediate singlet level, towards the ground sublevel $m_s = 0$ (in this case the relaxation mechanism is not spin-conserving); (d) In the following cycle, the green laser excitation drives again the optical transition (which is spin-conserving, $\Delta m_s = 0$) towards the excited state; (e) As per step (c), part of the population from the excited spin sublevel $m_s = \pm 1$ relaxes nonradiatively towards the ground sublevel $m_s = 0$. After a few optical cycles, all of the population ends up in the ground spin sublevel $m_s = 0$, and the NV spin state is therefore initialized. Note the clear difference between the unpolarized system in (a) and the polarized system in (e). (Note: This schematic representation is only indicative, and applies rigorously to an ensemble of NV⁻ centers. An alternative interpretation is to consider the population of the states as that of a single NV⁻ center, but at different instants of time, shown simultaneously.)

through the singlet intermediate spin level (Figure 5.7c). The radiative optical transition is again spin-conserving ($\Delta m_s = 0$). The non-radiative decay is allowed by symmetry, and is not necessarily spin-conserving. Although it can involve both the excited sublevels with spin numbers $m_s = 0$ and $m_s = \pm 1$, it emerges – from symmetry considerations (i.e., the involved levels/orbitals must share the same symmetry) – that the most probable nonradiative path is the one from the excited state with $m_s = \pm 1$, through the intermediate singlet level, and eventually to the ground state with $m_s = 0$ (Figure 5.7c). After a few optical cycles, the NV⁻ finally becomes polarized in the ground-state sublevel with the quantum number $m_s = 0$ (Figure 5.7d, e). More precisely, such an excitation and decay cycle causes a non-Boltzmann steady-state spin alignment of the NV⁻ center in the ground state, consisting of a reorientation of the ground-state spins. It is important to note the clear difference in the population distribution between the unpolarized system in the schematic representation of Figure 5.7a and the polarized system of Figure 5.7e after a few optical polarizing cycles. For the sake of completeness, the NV⁻ spin polarization is actually not complete (~70–80%), due to a slight mixing of the spin sublevels [74].

5.2 Defects in Diamond

The key point to remember about the polarization of the NV⁻ center is that, starting from a random spin population density, the NV⁻ can be initialized (polarized) into a specific spin state. This is necessary to prepare the system for further manipulations.

5.2.4.2 Control

Considering an NV⁻ center initialized (polarized) in its ground state with $m_s = 0$, manipulation of the NV⁻ spin state consists of driving the ground-state spin sublevel transition from $m_s = 0$ to $m_s = \pm 1$ (Figure 5.8).

The controlled ground-state spin transition is achieved by applying a resonant microwave field (~2.88 GHz at zero magnetic field) with the transition to access

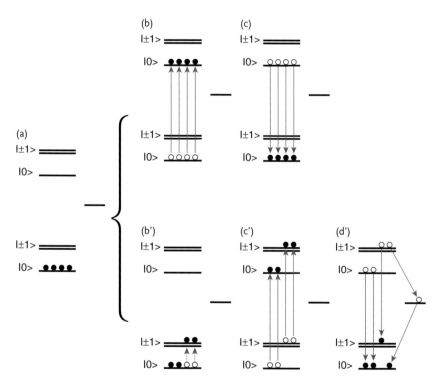

Figure 5.8 Schematic representation of the NV⁻ spin manipulation and read-out mechanisms. (a) Initially, the NV⁻ center is polarized by optical excitation; (b, c) With no microwave field applied, all the population cycle radiatively emit photons (ZPL at 637 nm) with a maximum in the corresponding detected fluorescence; (b') A resonant (2.88 GHz) microwave field (dashed lines) is used to raise part of the population from the ground spin sublevel $m_s = 0$ to the ground- state spin sublevel $m_s = \pm 1$ (indicated, respectively, as |0⟩ and |±1⟩); (c', d') This fraction of raised population can relax to the ground state nonradiatively through the intermediate singlet level: the nonradiative relaxation mechanism causes a detectable drop (~30%) in the fluorescence of the NV⁻ center. Note the difference between (c), where all the population cycles radiatively, and (d'), where only a fraction of it cycles radiatively.

(Figure 5.8b′). Averaging in time, the electron spin nutates between the $m_s = 0$ and $m_s = \pm 1$ ground spin states due to the coherent interaction of the spin with the microwave field. The spin nutation can be described as [72, 75]:

$$r_3(t) = r_3(0)(1/\varpi^2)(\Delta\omega^2 + \omega_1^2 \cos \varpi t) \tag{5.2}$$

where $r_3(t)$ is the difference between the probability to find the system in the $m_s = 0$ and $m_s = \pm 1$ sublevel, $\Delta\omega$ is the detuning between the microwave frequency and the actual transition frequency, ω_1 is the microwave Rabi frequency, and $\varpi = \sqrt{\Delta\omega^2 + \omega_1^2}$.

5.2.4.3 Measurement

Considering an NV⁻ center that has been polarized in its ground-state spin sublevel $m_s = 0$ (Figure 5.8a), the application of a resonant microwave field causes an increase in the population of the $m_s = \pm 1$ spin sublevels (Figure 5.8b′) which leads to an overall decrease in the fluorescence because of the nonradiative decay via the intermediate metastable state (Figure 5.8c′ and d′). The fluorescence intensity carries the information about the spin state of the NV⁻ center [optically detected electron spin resonance (ESR) spectrum] which can be therefore read-out optically. In Figure 5.8, the difference should be noted between (c), where there is no microwave field applied and all the population cycle radiatively, and (d′), where only a fraction of it cycles radiatively.

5.2.4.4 Relaxation Time

The control and measurement of the NV⁻ center spin-state requires the spin relaxation time to be taken into account. This refers to the time over which the spin stays in a certain state before it starts randomly to lose its phase, or is involved in a spin-flip event due to coupling to the environment (i.e., interaction with another electron or a nuclear spin). More specifically, when placed in a coherent superposition of spin-up and spin-down states, the spin of an isolated NV⁻ center undergoes (in time) both energy and phase relaxation processes. These relaxation processes occur on different and characteristic time scales, generally indicated by T_1, T_2, and T_2^*.

T_1 is known as the spin-lattice or longitudinal (spin-flip) relaxation time, and involves the interaction of a single spin with the surrounding lattice field. The lattice field is the complex overall magnetic field caused by the nuclei in the crystal lattice structure undergoing vibrational and rotational thermal motion around their equilibrium positions. The spin-lattice relaxation mechanism refers to the process in which a spin, brought to a higher energy spin-state for instance by a radiofrequency (RF) pulse, returns the energy it obtained back to the surrounding lattice (i.e., the lattice excitation modes carry away energy quanta on the scale of the spin Larmor frequency), thereby restoring its equilibrium state (spin-flip).

T_1 is affected by the mobility of the lattice. In general, a higher lattice mobility, together with associated higher vibrational and rotational frequencies of the nuclei, are more likely to induce the transition (relaxation) from high to low spin states. On the other hand, the relaxation probability decreases for extremely high lattice mobility, as the extremely high vibrational and rotational frequencies become too off-resonance to induce the transition between the spin-states. In other words, only lattice vibrations with frequency matching the Larmor frequency of the spin can induce a spin-flip with a rate [73, 76]:

$$\frac{1}{T_1} = \left(\frac{54\pi v^2}{\hbar \rho v_0^5}\right)\left(\frac{\mu_B^2}{r_0^3}\right)^2 kT \quad (5.3)$$

where ρ is the mass density, v is the oscillator frequency, v_0 is the velocity of sound, r_0 is the lattice separation of neighbors, μ_B is the Bohr magneton, k is Boltzmann's constant, and T is the temperature. Due to the strong covalent bonds between the low-mass carbon atoms, which results in a low lattice mobility, diamond shows relatively long phonon-limited spin-lattice relaxation times $T_1 \sim 10^4$ s (at $T = 300$ K). The electron spin-flip process can also be induced by inelastic scattering of a higher frequency lattice phonon. In this case (Raman process), the relaxation rate is [73, 76]:

$$\frac{1}{T_1} = \left(\frac{163\pi^2 6!}{\rho^2 v_0^{10}}\right)\left(\frac{\mu_B^2}{r_0^3}\right)^2 \left(\frac{kT}{h}\right) \quad (5.4)$$

and the corresponding relaxation time is $T_1 \sim 10^2$ s (at $T = 300$ K). Another spin-lattice relaxation mechanism involves the interaction with optical phonons (Orbach process [77]). In diamond, this process is not efficient as optical phonons are not excited under ambient conditions. The spin-lattice relaxation time T_1 can be measured by using pulsed EPR spectroscopy and performing echo-detected inversion recovery or saturation recovery sequence experiments [78].

T_2 is known as spin-spin or transverse dephasing relaxation time. It is also commonly referred to as the "coherence time," although the two terms are not strictly interchangeable [79]. It involves the dipolar interaction of a single spin with the "spin bath" formed by the surrounding spins. The bath spins undergo spin-flip events and produce a fluctuating local field so that the dipolar coupling induces energy conserving spin flip-flop processes in the NV⁻ single spin; this results in a loss of the NV⁻ spin memory phase. In diamond, the spin bath can vary from sample to sample. In general, the spin bath is dominated by the electron spins of residual nitrogen atoms, these constitute the most significant source of dephasing for the NV⁻ center and shorten its T_2 [72, 80]. Temperature also plays a key role, in that the lower the temperature the lower will be the spin bath flip rate and the longer the NV⁻ coherence time T_2. The spin-spin relaxation time T_2 of the NV⁻ center also increases when the concentration of nitrogen impurities in the sample is reduced; for very low nitrogen concentration (e.g., 5×10^{-2} ppb) the main source of electron spin decoherence becomes the nuclear spin bath constituted by ^{13}C atoms, and the corresponding spin-spin relaxation time T_2 can be raised to ~0.5–2 ms [73, 78, 81]. The spin-spin relaxation time T_2 can be measured by using pulsed EPR spectroscopy and performing Hahn echo sequence experiments [82]. Nutation experiments have shown that the initialization, manipulation, and reading out of an NV⁻ center spin state is feasible. In this case, the NV⁻ center is initialized with a green laser, which is then switched off and a microwave pulse of certain duration is applied; the intensity of the fluorescence is then monitored. A transient nutation (periodic modulation) in the fluorescence intensity is observed according to the duration of the microwave pulse applied. Upon increasing the microwave pulse length, more of the population is inverted (from the $m_s = 0$ ground state to the $m_s = \pm 1$ ground state), until the full population inversion is reached (for a pulse, indicated as a π-pulse, of specific duration) and therefore the minimum fluorescence is observed [7, 72]. Usually, in a nutation experiment the harmonic oscillation pattern of the fluorescence appears to be modulated by an exponentially decaying function: the decay envelope represents the decoherence or dephasing time T_2^* of the NV⁻ center. A long coherence time is a technologically desirable characteristic for a quantum spin system, since it means that once the spin state has initially been set, there is a window of time sufficiently long to actually manipulate and read-out the state before the information carried is lost due to interaction with the environment. The NV⁻ has a long coherence time T_2, of up to 1.8 ms [73]; this is a few orders of magnitude larger than the duration of a π-pulse (which is as fast as a few nanoseconds, depending on power and resonant frequency of the microwave field applied) required to induce the NV⁻ spin-state inversion, and which sets the time basis for performing single quantum operations. In comparison, the spin coherence times T_2 of, for instance, semiconductor quantum dots are much shorter, in the range of ~1 μs [83], while those of single molecules are of the order of ~1–10 ns [84, 85].

For the coherence time T_2, a distinction is generally made between the spin-spin relaxation times T_2 and T_2^*. T_2 refers rigorously to the homogeneous relaxation time for a single quantum system, and T_2^* to the inhomogeneous decay time averaged over an ensemble of quantum systems. The inhomogeneous coherence time T_2^* is also used for single quantum systems in place of T_2 to account for variation of the system over time, under the hypothesis of ergodicity (i.e., the average of the variations of T_2 for a single system in time coincides with the average of the variations over an ensemble of systems in space).

In general, the following relationships hold between the relaxation times T_1, T_2, and T_2^*:

$$\begin{cases} T_2 \leq 2T_1 \\ T_2^* \leq T_2 \end{cases} \tag{5.5}$$

The bounds in Equation (5.5) show that the ultimate limit for the coherence time is given by the spin-lattice relaxation process (T_1), which is expected to occur on a seconds timescale, and that local field inhomogeneities result in a T_2^* relaxation time shorter than T_2.

To summarize, the properties of the negative charge state (NV⁻) of the NV center presented above are the reason why such a diamond defect has attracted so much interest in the recent past. The center's unique energy-level structure and associated capability of preparing, manipulating and reading out its electron spin state, together with the relatively long spin coherence time, are in fact characteristics which are quite rarely found combined in the same quantum system.

5.3
Synthesis of Diamond

Because of its exceptional properties, it is not surprising that over the past two centuries many scientists and businessmen have pursued the dream of synthesizing artificial diamonds.

In 1772, Antoine Lavoisier conducted an experiment to show that diamond and charcoal were chemically alike, by burning about 150 mg of diamond in a sealed container and determining that the gas produced was simply carbon dioxide (CO_2). Lavoisier then showed that the very same amount of carbon dioxide was produced when 150 mg of amorphous carbon in the form of charcoal was burned, but was reluctant to publicize such a striking, and at first counterintuitive, result. Some 25 years later, in 1797, the English chemist S. Tennant performed further combustion experiments and concluded that diamond *". . . consists entirely of charcoal, differing from the usual state of that substance only by its crystallized form"* [86].

The discovery that diamond is made purely of carbon fueled the quest for its artificial synthesis. Over the following 150 years, at least 30 claims of success were advanced, the most famous probably being those of J.B. Hannay in 1879 [87] and of F.F.H. Moissan in 1893 [88]. Although none of these early attempts was successful, they were clearly the forerunners of what has today become reality, namely the synthesis of man-made diamonds. Currently, at least three main processes are known and widely used for the synthesis of artificial diamonds: high-pressure, high-temperature (HPHT); chemical vapor deposition (CVD); and detonation.

The details of these methods (as described in the following subsections) are presented at this point in recognition that the ability to synthesize diamond has led to changes in its use for technological purposes. Natural diamonds occur as several different types, with properties that differ significantly from type to type. Until recently, such variability may have prevented diamond from being regarded as a suitable material for technological applications. Today, however, the synthesis of diamond with dramatically improved qualities has led to a major boost in the realization of diamond-based technologies. Nanocrystalline diamond is available both for commercial and research applications. Its price can vary considerably according to the quality of the diamond itself. Detonation and HPHT diamond can cost tens of US$ per gram, but single crystalline CFD diamond of electronic quality can go up to thousands US$ per gram.

5.3.1
High-Pressure, High-Temperature (HPHT) Synthesis

The synthesis of artificial diamond via HPHT processes was historically the first to be developed and to be realized on a practical basis [89]. The technique involves subjecting a carbon-containing material to a high temperature (>1500 °C) and high pressure (>5 GPa), so as to induce the phase transition of graphite to diamond.

The method is inspired by—and actually mimics—the geological processes involved in the formation of natural diamond. The required high pressure and high temperature are achieved by using press systems of different types [90], including belt systems [89], cubic systems [91], and the bars systems [92].

All HPHT press technologies employ a core reaction cell which contains a carbon source and some seed crystals to facilitate the growth of diamonds. During the process a metal solvent or *flux* is used as a solvent to transport the material of the carbon source to the seeds; typically, fluxes may be iron (Fe), nickel (Ni), or a mixture of the two. The use of a flux allows diamond to be synthesized at much lower temperatures and pressures than would be necessary for a direct conversion of graphite into diamond. The exact composition of the flux may have a major influence on the properties of the synthetic diamond. Although only a few minutes are required to convert graphite into powder-sized diamond, it takes about three days under well-controlled growth conditions to obtain a rough gem-quality stone.

The main advantages of the HPHT method for growing diamonds are the relatively low cost and high throughput.

5.3.2
Chemical Vapor Deposition (CVD) Synthesis

An alternative method by which artificial diamonds can be synthesized is that of CVD. In this process, the diamonds are literally grown on a substrate from a hydrocarbon gas mixture, inside a chamber where the temperature and pressure can be tightly controlled.

The first successful documented attempt at growing diamond at low pressure was made by W.G. Eversole at the Union Carbide Corporation (USA) in 1952, which led to Eversole being the very first "diamond-maker" in history [93]. Unfortunately, however, the growth rates of the CVD process were initially very slow and the method was eclipsed by the contemporary success of the HPHT method.

The basic principle of CVD diamond growth is fairly straightforward. A substrate—generally single-crystal diamond, silicon, quartz, sapphire, or some type of metal—is chosen and placed inside a growth chamber where the pressure and temperature are controlled. A mixture of hydrocarbon gas, typically methane and hydrogen, is then injected into the chamber and heated using either a microwave beam, a hot filament, an arc discharge, a welding torch, a laser, an electron beam, or other means. At temperatures above 800 °C the mixture is ionized into chemically active radicals such that a plasma forms; the radicals then react and begin to form diamond crystals around the substrate defects, which behave as trigger seeds for the diamond growth process. Typically, the substrate is seeded with nanodiamond powder. Initially, small diamond crystals can nucleate randomly but, as the growth proceeds, the nuclei become larger and eventually coalesce to form a continuous polycrystalline film. During the process the methane serves as the carbon source, while the hydrogen preferentially etches.

5.3.3
Detonation

Another approach to the synthesis of diamonds is that of detonation. Diamond crystals, nominally a few nanometers in size, have been shown to originate from the detonation of carbon-containing explosives in a metal chamber, and are commonly referred to as detonation nanodiamonds (DNDs).

The detection of diamond in a preserved shock-compressed graphite sample was first observed in the USA by P.S. DeCarli and J.C. Jamieson in 1961 [94]. Two years later, in July 1963, the synthesis of detonation diamond nanocrystals was actually realized by K.V. Volkov, V.V. Danilenkov, and V.I. Elin, working at the USSR All Union Research Institute of Technical Physics (VNIITF), where nuclear weapons had previously been developed [95]. The VNIITF team found that a popular military explosive made from 40% trinitrotoluene (TNT) and 60% hexogen (RDX) is oxygen-deficient and hence, upon detonation in an inert medium such as argon or water, would produce up to 12% soot by incomplete combustion, and that this surprisingly contained up to 75% diamond carbon. In the USA, detonation nanodiamond synthesis was not discovered until 1988 [96], and only recently has the striking interest in newborn nanotechnologies led to a renewed attention of detonation synthesis. Since 2003, a few factories specializing in detonated nanodiamond production have begun to emerge in different regions, including Russia, Ukraine, Belarus, and China [97].

Detonation diamond crystals originate from the detonation of TNT and RDX inside a metal chamber. The correct conditions of high pressure (~20–30 GPa) and high temperature (~3000–4000 K) for the thermodynamic stability of diamond are achieved at the shock front of the explosion, and under such conditions the carbon atoms contained in the explosives themselves assemble to form diamonds. The chamber must be rapidly cooled after the detonation to suppress the tendency for carbon to form graphite. As the time during which the pressure and temperature are suitable for carbon to form diamond is limited (fractions of microseconds), detonation nanodiamonds are consequently very small (ca. 5 nm) and tend to have an extremely narrow distribution in size. Generally, commercial detonation nanodiamond powders contain up to 98% diamond crystals in the range of ~2 to 10 nm, and with very small level of sp^2 impurity.

Detonation nanodiamonds are complex objects, and possess a layered-structure where at least three zones can be distinguished: (i) an inner (~4–6 nm) sp^3-hybridized carbon atoms core; (ii) an intermediate (~0.4–1 nm) inhomogeneous sp^2-hybridized carbon sheath; and (iii) an outer layer consisting of carbon and various functional groups (up to 10% of the total particle mass), mainly oxygen-containing groups such as hydroxyl, carboxyl, ketone, and lactone [98, 99].

Although its discovery was contemporary to that of HPHT and CVD methods, the synthesis of artificial diamonds by detonation has taken a decisive upturn during the past few years. Notably, although the greatly reduced size of detonation diamonds means that they are less attractive than HPHT or CVD diamonds for generic industrial purposes or as gems, they are regarded as highly promising for a variety of applications in the emerging field of nanotechnology.

5.3.4
Enhancement of Color Center Concentration

The synthesis of artificial diamonds represents only one aspect of tailoring the material for industrial and technological applications, or even just for the gem-trade market. As defects and impurities can significantly alter the properties of diamond, or be directly exploited for their optical and spin properties, in parallel with studies of different diamond-synthesis techniques a variety of methods have been developed to enhance and control the concentration of impurities and color centers within the material.

Different approaches have been perfected to ensure that color centers are either formed or included in diamond. Each of these aims for the highest possible control in terms of concentration, position, and isolation of the color centers. In fact, the quantity and quality of luminescent centers in diamond must be accurately controlled to make certain technological applications feasible. A list of some defect-enhancing techniques is provided in the following subsections, together with a brief analysis of methods that are specific to the NV center (though some can be applied to other types of diamond defect).

5.3.4.1 Irradiation

One of the most common approaches to induce the formation of NV centers in diamond is by *irradiation*. In this case, a focused beam of ions or electrons is used to irradiate the target diamond material. The incident particles penetrate the diamond lattice and create vacancies in the crystalline structure, after which the diamond undergoes a subsequent annealing process at temperatures in the region of 850 °C. The annealing enhances the migration of the formed vacancies, which become trapped by nitrogen atoms already present in the diamond lattice; as a consequence, stable NV centers are formed [43]. This process functions mainly for two reasons:

- The diffusion of vacancies in diamonds begins at about 650 °C, whereas the diffusion of nitrogen atoms begins at about 1500 °C. This facilitates a migration of the vacancy towards fixed nitrogen atoms in their lattice positions and enhances the NV trapping mechanism [100].

- The formation of NV pairs is guaranteed by the fact that they are energetically favored over having nitrogen atoms and vacant carbons separately and randomly dispersed in the diamond lattice.

The irradiation technique is often used on nitrogen-rich Type I diamond, which naturally contains nitrogen concentrations over 20 ppm.

The number of NV centers created by irradiation depends on the energy and nature of the focused particles striking the diamond. As a rule of thumb, higher energies and smaller incident ions will result in a deeper penetration with the subsequent enhanced probability of creating vacancies. On the other hand, the small penetration depths of the incident particles causes a significant fraction of vacancies to diffuse to the surface during annealing, where they annihilate without forming stable NV centers [101].

5.3.4.2 Ion Implantation

NV defects can be directly created in nitrogen-free Type IIa diamonds by *ion implantation*. In this case, the target diamond material is irradiated with N^+ ions and then annealed to enhance the migration of vacancies towards the implanted nitrogen atoms. Again, the annealing is performed at ~850 °C to mobilize the vacancies while keeping the nitrogen atoms fixed. This is desirable as it leads to stationary NV centers being located at positions where the N^+ ions have been initially implanted. Although an ability to localize the NV center represents one of the major strengths of this technique [102, 103], this advantage is unfortunately diminished by two effects: (i) the stochastic nature of the ion beam, in obeying a Poissonian statistic, causes some parts of the sample to contain more than one NV center, while others contain none; and (ii) the straggle of the ions – that is, the random spreading out of the incident ions when they penetrate into the sample – causes a reduction in the control of the ion beam spot size. The resolution can be increased by lowering the kinetic energy of the ions, but in so doing the lateral displacement is limited and the spot dimension is reduced; nevertheless, the lower ion energy causes a reduction in the efficiency of NV center formation during the subsequent annealing process [104]. The yield of NV creation as a function of the ion energy can be less than 1% for energies <5 keV, and up to 45% for energies >18 MeV [105].

5.3.4.3 Incorporation during CVD Growth

When artificial diamonds are synthesized by CVD, the inclusion of NV centers in the diamond lattice occurs spontaneously during crystal growth. This is due to the fact that nitrogen is abundant in the atmosphere (molecular nitrogen constitutes 78% of the air), and that NV centers are energetically stable in diamond. As a result, the incorporation of nitrogen and the subsequent formation of NV centers in CVD diamond growth are inevitable. However, the correct choice of gas mixture in the CVD chamber, and also of the process parameters, can be employed to exert a certain degree of control over the concentration of NV centers incorporated [106].

5.4 Applications of Color Centers in Diamond

After having analyzed the characteristics and properties of the NV^-, and describing the methods used to synthesize artificial diamond containing color centers, an overview is now provided of some applications in which the unique room-temperature optical and spin properties of the NV^- are exploited.

Diamond color NV^- centers appear to offer interesting possibilities in at least three main research areas, namely quantum information technology, biomedicine, and high-resolution magnetometry.

5.4.1
Quantum Information Technology

Currently, the NV⁻ center is being investigated as a candidate qubit for solid-state quantum technologies, as it may provide a feasible hardware system for the generation of addressable and easy-to-manipulate quantum states, even at room temperature [107]. In fact, as discussed previously (see Sections 5.2.3 and 5.2.4), the NV⁻ center is a single-photon emitter which shows a high oscillator strength for the dipole transition between the ground state and the first optically excited state. Moreover, its spin state can be optically initialized, coherently manipulated, and read out with optical and microwave excitation; in addition, it shows a desirable long spin coherence time at room temperature.

Some of the achievements exploiting the NV⁻ center for quantum information technologies are described below. Simple logic gates have been demonstrated at room temperature using NV centers [72, 108], while coherent coupling between electronic spins of the NV center and the nuclear spin of proximal ^{13}C atom have been examined experimentally to prove that a selective addressing and manipulation of single isolated nuclear spins is possible [109]. The coupling of the NV center to a neighbor single nitrogen defect [110], as well as to another NV center [111], in diamond at room temperature has also been demonstrated, and may open the way towards a viable solid-state quantum device operating under ambient conditions. NV centers have also been explored in quantum optics and in quantum cryptography as single-photon sources to realize quantum cryptographic protocols [112, 113].

Although NV centers in diamond appear to show great potential as building blocks for quantum devices, the latter are still far from actual implementation. For example, although the NV center can produce a relatively high number of single-photon pulses, its wavelength and bandwidth are not ideal for the eventual requirements of optical quantum information processing. In the case of the NV center, the overall linewidth is of the order of 200 nm, while the ZPL at 637 nm itself has a width of several nanometers at room temperature [71, 106]. In this context, several studies have been conducted aiming to enhance both the emission rate and photon collection efficiency, as well as to manipulate the spectral shape of the NV center. These include coupling the radiating NV to a localized mode of a high-Q resonator such as a microsphere [114, 115], a microdisc [116] or a photonic crystal cavity [117–120] or, alternatively, by using a solid immersion lens to collect the NV emission [121, 122]. Different schemes have been also proposed to exploit, for example, a coupling of the NV center with localized surface plasmons to achieve an enhancement of the center emission rate [123, 124]. Another technological problem here would be in positioning several single NV centers with sufficient control to guarantee an interaction of their spins with adjacent NV spins. Although regular arrays of single NV centers have been created, the accuracy achieved has not been sufficient to realize actual applications [103, 104].

5.4.2
Life Sciences

In the life sciences, investigations towards exploiting color centers in diamond for biological and medical purposes have become much more extensive during the past few years.

Fluorescent nanodiamonds possess several attractive properties; for example, they are biocompatible and non-cytotoxic, their surface can be functionalized and, when excited by a laser, defect centers within the nanodiamonds emit photons that are capable of penetrating tissues [125–129]. Unfortunately, however, these characteristics are difficult to find in combination (e.g., in biological probes), and current studies are being targeted towards the use of fluorescent nanodiamonds as probes for biolabeling, and also as drug-delivery vehicles [125, 130–132].

5.4.3
High-Resolution Magnetometry

Another intriguing area in which single NV centers are the subject of active interest is that of high-resolution magnetometry and single-spin sensing techniques.

The use of NV centers to detect single spins and weak magnetic fields is a recent advance, the key concept being that external fields acting on a single NV center can measurably alter its fluorescence. The ability to detect induced changes in NV fluorescence can also be used to probe and measure the surrounding fields, including even the weak magnetic fields generated by single electron or nuclear spins (at least in theory). Other promising applications have been identified in the physical sciences for the read-out of single spin-based magnetic memories [133], and in the life sciences for high-resolution imaging techniques [10, 11].

5.5
Feasibility of NV Center-Based Nanotechnologies

Although the potential for feasible NV center-based applications is real, as the materials sciences continue to advance, new challenges will be posed. In particular, some applications which exploit NV center in diamond have specific material requirements. For example, diamond must be of ultra-high purity for quantum technologies where the spin coherence is critical, whereas for high-resolution magnetometry and biomedical imaging the small size of the diamond nanocrystals and the optical stability of the hosted NV centers are crucial. Recently, for instance, major efforts have been expended pursuing the synthesis of ultrasmall (<10 nm) nanodiamonds that are still capable of hosting color centers. It has in fact been realized that for many of the identified applications, it is essential for the NV^- to be hosted in the smallest possible nanodiamonds and/or be as close as possible to the diamond surface, and this is true in general. The questions of whether: (i) the NV center is considered for biomarking applications (where the reduced size of the nanodiamonds would advantageous for noninvasive interactions with tissue and cells); (ii) it is considered as a probe for sensing single spins (where a reduced distance between these and the probing NV would result in an increased resolution); or (iii) it is considered as a single photon source in quantum technologies involving the coupling of the NV with photonic waveguides or microcavities (where the center proximity to the cavity would enhance the coupling efficiency), remain unclear.

Today, the fabrication of ultrasmall nanodiamonds hosting active NV centers is a practical challenge that is being met. Nevertheless, as the ability to control diamond and color centers is improved, new questions are being raised regarding the behavior of the NV center in such small nanoparticles. Consequently, some of the "anomalies" displayed by NV centers, either when they are hosted specifically in ultrasmall nanodiamonds or when they are in close proximity to the diamond surface, are outlined in the following subsections. An understanding of these anomalies represents the next pivotal step towards the implementation of NV center-based nanotechnologies.

5.5.1
Fabricating Ultrasmall (<10 nm) Fluorescent Nanodiamonds

During the past few years, a variety of techniques have been optimized to fabricate nanodiamonds that were as small as possible (only a few nanometers in size) but still able to host active color NV centers. These techniques were similar to those described above (see Sections 5.3.1–5.3.4), but focused more on the reduced size of the nanodiamonds and the optical stability of the hosted centers. Ultimately, such a task proved to be non-trivial, with theoretical calculations showing that, in a diamond nanocluster with nitrogen inclusions, a minimization of the crystal energy favors the nitrogen to be located at the surface rather than in the core of the particle [134–136]. Consequently, in very small nanodiamonds the NV centers tend to be unstable – either because the vacancy of the NV complex is likely to diffuse and annihilate at the near surface, or because the NV center undergoes a structural distortion which could render it optically inactive. This seemed to justify the limited observations of fluorescent NV centers in CVD [106] and HPHT [136] diamond grains less than 40 nm in size. Shortly before the latest discoveries in the field [137–139], several studies – both theoretical [134] and experimental [140] – had indeed manifested serious concerns regarding the possibility that nanodiamonds of <10 nm could have barely contained any optically active NV centers.

The thermodynamic stability of the NV center in nanodiamonds as small as 5 nm has ultimately been demonstrated experimentally [137–139, 141]. Some of the most recent studies conducted in the field are summarized as follows:

- Luminescent nanodiamonds (≤10 nm) have been synthesized via a HPHT method, in combination with additional treatments [137, 141, 142].

- Several studies have been conducted on the synthesis of CVD nanodiamonds, despite it having been argued that the incorporation of nitrogen in small CVD nanocrystals is inefficient [106, 143, 144].

- Diamond synthesis by detonation has recently attracted renewed interest for the production of ultradispersed and ultrasmall (ca. 4–5 nm) nanodiamonds [138, 139, 145–148].

- Dispersed nanodiamonds of average about 2–6 nm diameter have been successfully synthesized using carbon black as a carbon source, using long-pulse-width laser irradiation in water at room temperature and normal pressure [149].

- Diamond nanocrystals of about 25–50 nm diameter have been produced using a microwave plasma torch technique with methane (CH_4) and argon (Ar) or molecular nitrogen (N_2) as catalysts [150].

5.5.2
Some NV Luminescence "Anomalies" at the Nanoscale

The fabrication of fluorescent ultrasmall nanodiamonds shows that diamond materials sciences is finally – and very quickly – being recognized as a mature approach to testing the requirements of some NV center-based technologies. Although, the thermodynamic stability of the NV center has been established in 5 nm-sized nanodiamonds [137–139, 141], the most recent investigations have revealed that the center optical stability might in fact not be preserved. Indeed, within ultrasmall nanodiamonds, or in close proximity to the diamond surface, the physics of the NV centers appears to obey specific rules that do not necessarily hold for larger diamond crystals, or in bulk. Some recent studies addressing this issue are outlined in the following subsections.

It has been shown that, in high-purity diamond, the preferred charge state of NV centers created within close proximity to the surface (within 200 nm) is the neutral (NV^0) charge state, rather than the negative state (NV^-) [45, 151]. It is however the NV^- charge state of the center that possesses the attractive optical and spin properties for the majority of the identified NV center-based applications. Hence, an understanding of, and an ultimate ability to control the mechanism of the center charge state conversion, becomes imperative within the perspective of engineering the material for applications that rely on having NV^- in close proximity to the diamond surface. Recently, K.-M.C. Fu and colleagues investigated this issue [152] by showing that the annealing of a single-crystal CVD diamond surface in oxygen (O_2) at 465 °C would induce the NV^0-to-NV^- charge conversion, but that the conversion would not take place if the surface of the same sample was masked with a 125 nm-thick layer of silicon dioxide (SiO_2) that prevented the sample from being exposed to oxygen. Hence, it was proposed that oxygen annealing at elevated temperatures would result in both surface termination with oxygen-containing functional groups and the selective removal of sp^2 carbon from the diamond surface [153]. This would have a counter-effect on the electronic depletion which favors the neutral NV^0 charge state, and which is generally observed in high-purity implanted samples in relation to graphitic damage and/or to hydrogen termination of the surface.

Subsequently, Rondin et al. further investigated the process leading to the NV charge state conversion in nanodiamonds [154]. In this case, it was shown that the proportion of negatively (NV^-) against neutrally (NV^0) charged centers decreased as the size of the nanodiamond was reduced, thus confirming that surface-related effects are involved in the conversion between NV^- and NV^0. Rondin and colleagues suggested that graphitic-related defects formed on the diamond surface would act as possible electron traps and so reduce the availability of electrons necessary for the stability of the NV^- as opposed to the NV^0 [155]. As reported by Fu et al., however, Rondin et al. were able, by removing graphitic-related defects through thermal oxidation, to enhance the proportion NV^-/NV^0. This was observed especially in the smallest nanodiamonds, where the centers are more likely to be located in proximity to the surface and thus be exposed to surface-related effects.

Very recently, these results have been supported by the findings of M.V. Hauf and collaborators, who investigated the effect of the surface termination of diamond on the charge state of the NV center [156]. In an exquisite experiment, Hauf and colleagues demonstrated a contrast in the number of NV luminescent centers between areas of the same sample that were oxygen-terminated compared to areas that were hydrogen-terminated. As opposed to the oxygen-terminated regions, their hydrogen-terminated counterparts did not show fluorescence from either NV^- or NV^0. However, the fluorescence could be recovered by exposing the hydrogen-terminated regions to oxygen plasma. Moreover, the fact that there was almost no measurable fluorescence in the hydrogen-terminated regions suggested that the NV centers were neither negatively nor neutrally charged, but rather were in an unknown and nonfluorescent charge state. Hauf and colleagues also observed that the deeper the defect was within the diamond, the weaker was the surface termination-discharging effect.

In a similar fashion, the crucial role played by the proximity of the NV center to the diamond surface was confirmed by C. Bradac et al., who reported a fluorescence intermittency or "blinking" of NV^- centers in 5 nm-sized nanodiamonds [139]. In a blinking NV, the center switches between bright "on" intervals and dark "off" intervals on a millisecond to second time scale (the lifetime of the NV excited state is in the nanosecond range). Unfortunately, the mechanism of such blinking is not (yet) fully understood.

5.6
Conclusions and Outlook

During recent years, NV color centers in diamond have attracted considerable attention, with several potential NV center-based technologies having been identified to exploit the unique optical and spin properties of this defect. As noted in this chapter, these possible applications span an impressive interdisciplinary range of research, including fields of quantum information technology, high-resolution magnetometry, biotechnology, and nanomedicine.

To date, the majority of NV-related studies have shown optimistic results with regards to the feasibility of the applications, some of which have specific material requirements. Whilst such requirements have always been taken very carefully into account, they often could have not been tested directly due to limitations of an ability to process the diamond material in an appropriate fashion. For example, it is only very recently that ultrasmall (<10 nm) nanodiamonds containing NV centers have been practically synthesized, and previously the behavior of defects in such small nanoparticles could only be speculated. Clearly, diamond materials sciences is finally – and very quickly – maturing to the point where at least some of these requirements can be verified.

The most recent studies of NV centers appear to have led to the conclusion that factors such as the *size* of the nanodiamonds, the *surface* proximity of the NV centers, and the characteristic of the *substrate* and those of the *surrounding media*, are all key parameters which influence the properties of the hosted NV centers. Within ultrasmall nanodiamonds, or in close proximity to the surface, the physics of NV centers obeys specific rules that do not necessarily hold for larger diamond crystals, or in bulk. Although "encaged" in small nanodiamonds, the interaction of the NV centers with the surrounding environment is possible, and the NV is not an "imperturbable" entity. Whilst this may upset certain present-day convictions, and even force a reassessment of the feasibility of certain proposed NV center-based technologies, it might at the same time lead to a novel range of possibilities which, potentially, could lead to an exploitation of this newly acquired control variable.

References

1 Harlow, G.E. (1998) *The Nature of Diamonds*, Cambridge University Press.
2 Hounsome, L.S., et al. (2006) Origin of brown coloration in diamond. *Phys. Rev. B*, **73** (12), 125203.
3 Achard, J., et al. (2011) Thick boron doped diamond single crystals for high power electronics. *Diamond Relat. Mater.*, **20** (2), 145–152.
4 Koizumi, S., et al. (1997) Growth and characterization of phosphorous doped {111} homoepitaxial diamond thin films. *Appl. Phys. Lett.*, **71** (8), 1065–1067.
5 Zaitsev, A.M. (2001) *Optical Properties of Diamond: A Data Handbook*, Springer-Verlag.
6 Jelezko, F. and Wrachtrup, J. (2006) Single defect centres in diamond: a review. *Phys. Status Solidi A*, **203** (13), 3207–3225.
7 Jelezko, F. and Wrachtrup, J. (2004) Read-out of single spins by optical spectroscopy. *J. Phys. Condens. Matter*, **16** (30), R1089–R1104.
8 Milburn, G.J. (2010) Quantum measurement and control of single spins in diamond. *Science*, **330** (6008), 1188–1189.
9 Schrand, A.M., Hens, S.A.C., and Shenderova, O.A. (2009) Nanodiamond particles: properties and perspectives for bioapplications. *Crit. Rev. Solid State Mater. Sci.*, **34** (1–2), 18–74.
10 Balasubramanian, G., et al. (2008) Nanoscale imaging magnetometry with diamond spins under ambient

conditions. *Nature*, **455** (7213), 648–651.

11 Maze, J.R., *et al.* (2008) Nanoscale magnetic sensing with an individual electronic spin in diamond. *Nature*, **455** (7213), 644–647.

12 Chen, Y.-Y., *et al.* (2011) Measuring Forster resonance energy transfer between fluorescent nanodiamonds and near-infrared dyes by acceptor photobleaching. *Diamond Relat. Mater.*, **20** (5–6), 803–807.

13 Hunt, D.C., *et al.* (2000) Identification of the neutral carbon <100>-split interstitial in diamond. *Phys. Rev. B*, **61** (6), 3863.

14 Smith, H.E., *et al.* (2004) Structure of the self-interstitial in diamond. *Phys. Rev. B*, **69** (4), 045203.

15 Iakoubovskii, K., Dannefaer, S., and Stesmans, A. (2005) Evidence for vacancy-interstitial pairs in Ib-type diamond. *Phys. Rev. B*, **71** (23), 233201.

16 Kiflawi, I., Collins, A.T., Iakoubovskii, K., and Fisher, D. (2007) Electron irradiation and the formation of vacancy–interstitial pairs in diamond. *J. Phys. Condens. Matter*, **19** (4), 046216.

17 Iakoubovskii, K., *et al.* (2003) Annealing of vacancies and interstitials in diamond. *Phys. B Condens. Matter*, **340–342**, 67–75.

18 Dannefaer, S. and Iakoubovskii, K. (2008) Defects in electron irradiated boron-doped diamonds investigated by positron annihilation and optical absorption. *J. Phys. Condens. Matter*, **20** (23), 235225.

19 Twitchen, D.J., *et al.* (1999) Electron-paramagnetic-resonance measurements on the divacancy defect center R4/W6 in diamond. *Phys. Rev. B*, **59** (20), 12900.

20 Iakoubovskii, K. and Stesmans, A. (2002) Dominant paramagnetic centers in ^{17}O-implanted diamond. *Phys. Rev. B*, **66** (4), 045406.

21 Hanley, P.L., Kiflawi, I., and Lang, A.R. (1977) On topographically identifiable sources of cathodoluminescence in natural diamonds. *Philos. Trans. R. Soc. Lond. Ser. A Math. Phys. Sci.*, **284** (1324), 329–368.

22 Kiflawi, I., *et al.* (1998) "Natural" and "man-made" platelets in type-Ia diamonds. *Philos. Mag. B*, **78** (3), 299–314.

23 Goss, J.P., *et al.* (2003) Extended defects in diamond: the interstitial platelet. *Phys. Rev. B*, **67** (16), 165208.

24 Iakoubovskii, K. and Adriaenssens, G.J. (2000) Characterization of platelet-related infrared luminescence in diamond. *Philos. Mag. Lett.*, **80** (6), 441–444.

25 Chen, J.H., *et al.* (1998) Voidites in polycrystalline natural diamond. *Philos. Mag. Lett.*, **77** (3), 135–140.

26 Kiflawi, I. and Bruley, J. (2000) The nitrogen aggregation sequence and the formation of voidites in diamond. *Diamond Relat. Mater.*, **9** (1), 87–93.

27 Walker, J. (1979) Optical absorption and luminescence in diamond. *Rep. Prog. Phys.*, **42** (10), 1605.

28 Collins, A.T. (2003) The detection of colour-enhanced and synthetic gem diamonds by optical spectroscopy. *Diamond Relat. Mater.*, **12** (10–11), 1976–1983.

29 Kaiser, W. and Bond, W.L. (1959) Nitrogen, a major impurity in common type I diamond. *Phys. Rev.*, **115** (4), 857.

30 Kajihara, S.A., *et al.* (1991) Nitrogen and potential n-type dopants in diamond. *Phys. Rev. Lett.*, **66** (15), 2010.

31 Titus, E., *et al.* (2006) Nitrogen and hydrogen related infrared absorption in CVD diamond films. *Thin Solid Films*, **515** (1), 201–206.

32 Loubser, J.H.N. and van Wyk, J.A. (1978) Electron spin resonance in the study of diamond. *Rep. Prog. Phys.*, **41** (8), 1201–1248.

33 Davies, G. (1994) *Properties and Growth of Diamond Data Review*, IET, London, p. 427.

34 Jones, R., Briddon, P.R., and Oberg, S. (1992) First-principles theory of nitrogen aggregates in diamond. *Philos. Mag. Lett.*, **66** (2), 67–74.

35 Read, P.G. (2005) *Gemmology*, Butterworth-Heinemann.

36 Lang, A.R. (1964) A proposed structure for nitrogen impurity platelets in diamond. *Proc. Phys. Soc.*, **84** (6), 871.

37 Woods, G.S. (1986) Platelets and the infrared absorption of type Ia diamonds. *Proc. R. Soc. Lond. Ser. A Math. Phys. Sci.*, **407** (1832), 219–238.

38 Davies, G., et al. (1992) Vacancy-related centers in diamond. *Phys. Rev. B*, **46** (20), 13157.

39 Mainwood, A. (1994) Nitrogen and nitrogen-vacancy complexes and their formation in diamond. *Phys. Rev. B Condens. Matter*, **49** (12), 7934–7940.

40 Jones, R., et al. (1994) Theory of nitrogen aggregates in diamond: the H3 and H4 defects. *Mater. Sci. Forum*, **143–147**, 45–49.

41 Gali, A., Fyta, M., and Kaxiras, E. (2008) Ab initio supercell calculations on nitrogen-vacancy center in diamond: electronic structure and hyperfine tensors. *Phys. Rev. B*, **77** (15), 155206.

42 He, X.-F., Manson, N.B., and Fisk, P.T.H. (1993) Paramagnetic resonance of photoexcited N-V defects in diamond. II. Hyperfine interaction with the {14}N nucleus. *Phys. Rev. B*, **47** (14), 8816.

43 Mita, Y. (1996) Change of absorption spectra in type-Ib diamond with heavy neutron irradiation. *Phys. Rev. B Condens. Matter*, **53** (17), 11360–11364.

44 Collins, A.T. (2002) The Fermi level in diamond. *J. Phys. Condens. Matter*, **14** (14), 3743.

45 Gaebel, T., et al. (2006) Photochromism in single nitrogen-vacancy defect in diamond. *Appl. Phys. B*, **82** (2), 243–246.

46 Manson, N.B. and Harrison, J.P. (2005) Photo-ionization of the nitrogen-vacancy center in diamond. *Diamond Relat. Mater.*, **14** (10), 1705–1710.

47 Davies, G. and Hamer, M.F. (1976) Optical studies of the 1.945 eV vibronic band in diamond. *Proc. R. Soc. Lond. Ser. A Math. Phys. Sci. (1934–1990)*, **348** (1653), 285–298.

48 Redman, D.A., et al. (1991) Spin dynamics and electronic states of N-V centers in diamond by EPR and four-wave-mixing spectroscopy. *Phys. Rev. Lett.*, **67** (24), 3420.

49 Fu, K.-M.C., et al. (2009) Observation of the dynamic Jahn–Teller effect in the excited states of nitrogen-vacancy centers in diamond. *Phys. Rev. Lett.*, **103** (25), 256404.

50 Jahn, H.A. and Teller, E. (1937) Stability of polyatomic molecules in degenerate electronic states. I. Orbital degeneracy. *Proc. R. Soc. Lond. Ser. A Math. Phys. Sci.*, **161** (905), 220–235.

51 Gali, A., et al. (2011) An ab initio study of local vibration modes of the nitrogen-vacancy center in diamond. *New J. Phys.*, **13** (2), 025016.

52 Davies, G. (1979) Dynamic Jahn–Teller distortions at trigonal optical centres in diamond. *J. Phys. C Solid State Phys.*, **12** (13), 2551.

53 Felton, S., et al. (2008) Electron paramagnetic resonance studies of the neutral nitrogen vacancy in diamond. *Phys. Rev. B*, **77** (8), 081201.

54 Manson, N.B., Harrison, J.P., and Sellars, M.J. (2006) Nitrogen-vacancy center in diamond: model of the electronic structure and associated dynamics. *Phys. Rev. B Condens. Matter Mater. Phys.*, **74** (10), 104303–104311.

55 Manson, N.B. and McMurtrie, R.L. (2007) Issues concerning the nitrogen-vacancy center in diamond. *J. Lumin.*, **127** (1), 98–103.

56 Tamarat, P., Manson, N.B., Harrison, J.P., McMurtrie, R.L., Nizovtsev, A., Santori, C., Deausoleil, R.G., Neumann, P., Gaebel, T., Jelezko, F., Hemmer, P., and Wrachtrup, J. (2008) Spin-flip and spin-conserving optical transitions of the nitrogen-vacancy centre in diamond. *New J. Phys.*, **10** (4), 045004.

57 Redman, D., Brown, S., and Rand, S.C. (1992) Origin of persistent hole burning of N-V centers in diamond. *J. Opt. Soc. Am. B*, **9** (5), 768–774.

58 Harley, R.T., Henderson, M.J., and Macfarlane, R.M. (1984) Persistent spectral hole burning of colour centres in diamond. *J. Phys. C Solid State Phys.*, **17** (8), L233.

59 Reddy, N.R.S., Manson, N.B., and Krausz, E.R. (1987) Two-laser spectral hole burning in a colour centre in diamond. *J. Lumin.*, **38** (1–6), 46–47.

60 Holliday, K., et al. (1989) Optical hole-bleaching by level anti-crossing and cross relaxation in the N-V centre in

diamond. *J. Phys. Condens. Matter*, **1** (39), 7093.

61 van Oort, E., Stroomer, P., and Glasbeek, M. (1990) Low-field optically detected magnetic resonance of a coupled triplet-doublet defect pair in diamond. *Phys. Rev. B*, **42** (13), 8605.

62 van Oort, E., et al. (1991) Microwave-induced line-narrowing of the N-V defect absorption in diamond. *J. Lumin.*, **48–49** (2), 803–806.

63 Hiromitsu, I., Westra, J., and Glasbeek, M. (1992) Cross-relaxation dynamics of the N-V center in diamond as studied via optically detected microwave recovery transients. *Phys. Rev. B*, **46** (17), 10600.

64 He, X.-F., Manson, N.B., and Fisk, P.T.H. (1993) Paramagnetic resonance of photoexcited N-V defects in diamond. I. Level anticrossing in the 3A ground state. *Phys. Rev. B*, **47** (14), 8809.

65 Oort, E.V., et al. (1988) Optically detected spin coherence of the diamond N-V centre in its triplet ground state. *J. Phys. C Solid State Phys.*, **21** (23), 4385.

66 Holliday, K., et al. (1990) Raman heterodyne detection of electron paramagnetic resonance. *Opt. Lett.*, **15** (17), 983–985.

67 Manson, N.B., He, X.-F., and Fisk, P.T.H. (1990) Raman heterodyne detected electron-nuclear-double-resonance measurements of the nitrogen-vacancy center in diamond. *Opt. Lett.*, **15** (19), 1094–1096.

68 Neumann, P., et al. (2009) Excited-state spectroscopy of single NV defects in diamond using optically detected magnetic resonance. *New J. Phys.*, **11** (1), 013017.

69 Brouri, R., et al. (2000) Photon antibunching in the fluorescence of individual color centers in diamond. *Opt. Lett.*, **25** (17), 1294–1296.

70 Beveratos, A., et al. (2001) Nonclassical radiation from diamond nanocrystals. *Phys. Rev. A*, **64** (6), 061802.

71 Kurtsiefer, C., et al. (2000) Stable solid-state source of single photons. *Phys. Rev. Lett.*, **85** (2), 290–293.

72 Jelezko, F., et al. (2004) Observation of coherent oscillations in a single electron spin. *Phys. Rev. Lett.*, **92** (7), 076401.

73 Balasubramanian, G., et al. (2009) Ultralong spin coherence time in isotopically engineered diamond. *Nat. Mater.*, **8** (5), 383–387.

74 Harrison, J., Sellars, M.J., and Manson, N.B. (2004) Optical spin polarisation of the N-V centre in diamond. *J. Lumin.*, **107** (1–4), 245–248.

75 Breiland, W.G., Brenner, H.C., and Harris, C.B. (1975) Coherence in multilevel systems. I. Coherence in excited states and its application to optically detected magnetic resonance in phosphorescent triplet states. *J. Chem. Phys.*, **62** (9), 3458–3475.

76 Pake, G.E. (1962) *Paramagnetic Resonance, An Introductory Monograph*, W. A. Benjamin, New York.

77 Orbach, R. (1961) Spin-lattice relaxation in rare-earth salts. *Proc. R. Soc. Lond. Ser. A Math. Phys. Sci.*, **264** (1319), 458–484.

78 Takahashi, S., et al. (2008) Quenching spin decoherence in diamond through spin bath polarization. *Phys. Rev. Lett.*, **101** (4), 047601.

79 Zurek, W.H. (2003) Decoherence, einselection, and the quantum origins of the classical. *Rev. Mod. Phys.*, **75** (3), 715–775.

80 Hanson, R., Gywat, O., and Awschalom, D.D. (2006) Room-temperature manipulation and decoherence of a single spin in diamond. *Phys. Rev. B*, **74** (16), 161203.

81 Maze, J.R., Taylor, J.M., and Lukin, M.D. (2008) Electron spin decoherence of single nitrogen-vacancy defects in diamond. *Phys. Rev. B*, **78** (9), 094303.

82 Schweiger, A., and Jeschke, G. (2001) *Principles of Pulse Electron Paramagnetic Resonance*, Oxford University Press.

83 Petta, J.R., et al. (2005) Coherent manipulation of coupled electron spins in semiconductor quantum dots. *Science*, **309** (5744), 2180–2184.

84 Hill, S., et al. (2003) Quantum coherence in an exchange-coupled dimer of single-molecule magnets. *Science*, **302** (5647), 1015–1018.

85 Loubens, G.D., et al. (2008) High frequency EPR on dilute solutions of the single molecule magnet Ni_4. *J. Appl. Phys.*, **103** (7), 07–910.

86 Tennant, S. (1797) On the nature of the diamond. By Smithson Tennant, Esq. F. R. S. *Philos. Trans. R. Soc. Lond.*, **87**, 123–127.
87 Hannay, J.B. (1879) On the artificial formation of the diamond. *Proc. R. Soc. Lond.*, **30** (200–205), 450–461.
88 Moissan, H. (1894) Nouvelles experiences sur la reproduction du diamant. *C. R. Acad. Sci., Paris*, **118**, 320–326.
89 Bundy, F.P., et al. (1955) Man-made diamonds. *Nature*, **176** (4471), 51–55.
90 Ito, E. and Gerald, S. (2007) Theory and practice – multianvil cells and high-pressure experimental methods, in *Treatise on Geophysics* (ed. S. Gerald), Elsevier, Amsterdam, pp. 197–230.
91 Hall, H.T. (1958) Some high-pressure, high-temperature apparatus design considerations: equipment for use at 100 000 atmospheres and 3000 °C. *Rev. Sci. Instrum.*, **29** (4), 267–275.
92 Pal'yanov, N., et al. (2002) Fluid-bearing alkaline carbonate melts as the medium for the formation of diamonds in the Earth's mantle: an experimental study. *Lithos*, **60** (3–4), 145–159.
93 Eversole, W.G. (1962) Synthesis of diamond. US Patents Nos 3030187 and 3030188.
94 DeCarli, P.S. and Jamieson, J.C. (1961) Formation of diamond by explosive shock. *Science*, **134** (3472), 92.
95 Danilenko, V.V., Trefilov, V.I., and Danilenko, V.N. (1991) USSR Patent No. SU 181, 329 AZ.
96 Greiner, N.R., et al. (1988) Diamonds in detonation soot. *Nature*, **333** (6172), 440–442.
97 Danilenko, V.V. (2004) On the history of the discovery of nanodiamond synthesis. *Phys. Solid State*, **46**, 595–599.
98 Dolmatov, V. (2006) Development of a rational technology for synthesis of high-quality detonation nanodiamonds. *Russ. J. Appl. Chem.*, **79** (12), 1913–1918.
99 Ho, D. (ed.) (2009) Nanodiamonds: applications in biology and nanoscale medicine, in *Technology and Engineering* (ed. D. Ho), Springer, pp. 79–116.
100 Meijer, J., et al. (2006) Concept of deterministic single ion doping with sub-nm spatial resolution. *Appl. Phys. A Mater. Sci. Proc.*, **A83** (2), 321–327.
101 Martin, J., et al. (1999) Generation and detection of fluorescent color centers in diamond with submicron resolution. *Appl. Phys. Lett.*, **75** (20), 3096–3098.
102 Kalish, R., et al. (1997) Nitrogen doping of diamond by ion implantation. *Diamond Relat. Mater.*, **6** (2–4), 516–520.
103 Meijer, J., et al. (2005) Generation of single color centers by focused nitrogen implantation. *Appl. Phys. Lett.*, **87** (26), 1–3.
104 Rabeau, J.R., et al. (2006) Implantation of labelled single nitrogen vacancy centers in diamond using ^{15}N. *Appl. Phys. Lett.*, **88** (2), 023113–023113.
105 Pezzagna, S., et al. (2010) Creation efficiency of nitrogen-vacancy centres in diamond. *New J. Phys.*, **12** (6), 065017.
106 Rabeau, J.R., et al. (2007) Single nitrogen vacancy centers in chemical vapor deposited diamond nanocrystals. *Nano Lett.*, **7** (11), 3433–3437.
107 DiVincenzo, D.P. (1995) Quantum computation. *Science*, **270** (5234), 255–261.
108 Jelezko, F., et al. (2004) Observation of coherent oscillation of a single nuclear spin and realization of a two-qubit conditional quantum gate. *Phys. Rev. Lett.*, **93** (13), 130501.
109 Childress, L., et al. (2006) Coherent dynamics of coupled electron and nuclear spin qubits in diamond. *Science*, **314** (5797), 281–285.
110 Hanson, R., et al. (2006) Polarization and readout of coupled single spins in diamond. *Phys. Rev. Lett.*, **97** (8), 087601.
111 Neumann, P., et al. (2010) Quantum register based on coupled electron spins in a room-temperature solid. *Nat. Phys.*, **6** (4), 249–253.
112 Wrachtrup, J. and Jelezko, F. (2006) Processing quantum information in diamond. *J. Phys. Condens. Matter*, **18** (21), S807.
113 Gisin, N., et al. (2002) Quantum cryptography. *Rev. Mod. Phys.*, **74** (1), 145.
114 Park, Y.-S., Cook, A.K., and Wang, H. (2006) Cavity QED with diamond

115 Schietinger, S. and Benson, O. (2009) Coupling single NV-centres to high-Q whispering gallery modes of a preselected frequency-matched microresonator. *J. Phys. B At. Mol. Opt. Phys.*, **42** (11), 114001.

116 Santori, C., et al. (2010) Nanophotonics for quantum optics using nitrogen-vacancy centers in diamond. *Nanotechnology*, **21** (27), 274008.

117 Stewart, L.A., et al. (2009) Single photon emission from diamond nanocrystals in an opal photonic crystal. *Opt. Express*, **17** (20), 18044–18053.

118 Wolters, J., et al. (2010) Enhancement of the zero phonon line emission from a single nitrogen vacancy center in a nanodiamond via coupling to a photonic crystal cavity. *Appl. Phys. Lett.*, **97** (14), 141108.

119 Van der Sar, T., et al. (2011) Deterministic nanoassembly of a coupled quantum emitter-photonic crystal cavity system. *Appl. Phys. Lett.*, **98** (19), 193103.

120 Englund, D., et al. (2010) Deterministic coupling of a single nitrogen vacancy center to a photonic crystal cavity. *Nano Lett.*, **10** (10), 3922–3926.

121 Schroder, T., et al. (2011) Ultrabright and efficient single-photon generation based on nitrogen-vacancy centres in nanodiamonds on a solid immersion lens. *New J. Phys.*, **13** (5), 055017.

122 Castelletto, S., et al. (2011) Diamond-based structures to collect and guide light. *New J. Phys.*, **13** (2), 025020.

123 Schietinger, S., et al. (2009) Plasmon-enhanced single photon emission from a nanoassembled metal-diamond hybrid structure at room temperature. *Nano Lett.*, **9** (4), 1694–1698.

124 Kolesov, R., et al. (2009) Wave-particle duality of single surface plasmon polaritons. *Nat. Phys.*, **5** (7), 470–474.

125 Chao, J.-I., et al. (2007) Nanometer-sized diamond particle as a probe for biolabeling. *Biophys. J.*, **93** (6), 2199–2208.

126 Fu, C.-C., et al. (2007) Characterization and application of single fluorescent nanodiamonds as cellular biomarkers. *Proc. Natl Acad. Sci. USA*, **104** (3), 727–732.

127 Huang, L.C.L. and Chang, H.-C. (2004) Adsorption and immobilization of cytochrome c on nanodiamonds. *Langmuir*, **20** (14), 5879–5884.

128 Schrand, A.M., et al. (2007) Differential biocompatibility of carbon nanotubes and nanodiamonds. *Diamond Relat. Mater.*, **16** (12), 2118–2123.

129 Schrand, A.M., et al. (2006) Are diamond nanoparticles cytotoxic? *J. Phys. Chem. B*, **111** (1), 2–7.

130 Treussart, F., et al. (2006) Photoluminescence of single colour defects in 50 nm diamond nanocrystals. *Phys. B Condens. Matter*, **376-377**, 926–929.

131 Barnard, A.S. (2009) Diamond standard in diagnostics: nanodiamond biolabels make their mark. *Analyst*, **134** (9), 1751–1764.

132 Chang, Y.-R., et al. (2008) Mass production and dynamic imaging of fluorescent nanodiamonds. *Nat. Nano*, **3** (5), 284–288.

133 Dutt, M.V.G., et al. (2007) Quantum register based on individual electronic and nuclear spin qubits in diamond. *Science*, **316** (5829), 1312–1316.

134 Barnard, A.S. and Sternberg, M. (2005) Substitutional nitrogen in nanodiamond and bucky-diamond particles. *J. Phys. Chem. B*, **109** (36), 17107–17112.

135 Turner, S., et al. (2009) Determination of size, morphology, and nitrogen impurity location in treated detonation nanodiamond by transmission electron microscopy. *Adv. Funct. Mater.*, **19** (13), 2116–2124.

136 Bradac, C., et al. (2009) Prediction and measurement of the size-dependent stability of fluorescence in diamond over the entire nanoscale. *Nano Lett.*, **9** (10), 3555–3564.

137 Tisler, J., et al. (2009) Fluorescence and spin properties of defects in single digit nanodiamonds. *ACS Nano*, **3** (7), 1959–1965.

138 Smith, B.R., et al. (2009) Five-nanometer diamond with luminescent nitrogen-vacancy defect centers. *Small*, **5** (14), 1649–1653.

139 Bradac, C., et al. (2010) Observation and control of blinking nitrogen-vacancy centres in discrete nanodiamonds. *Nat. Nano*, **5** (5), 345–349.

140 Vlasov, I.I., et al. (2010) Nitrogen and luminescent nitrogen-vacancy defects in detonation nanodiamond. *Small*, **6** (5), 687–694.

141 Boudou, J.-P., et al. (2009) High yield fabrication of fluorescent nanodiamonds. *Nanotechnology*, **20** (23), 235602.

142 Gaebel, T., et al. (2011) Size-reduction of nanodiamonds via air oxidation. *Diam. Relat. Mater.*, **21** (0), 28–32.

143 Stacey, A., et al. (2009) Controlled synthesis of high quality micro/nano-diamonds by microwave plasma chemical vapor deposition. *Diamond Relat. Mater.*, **18** (1), 51–55.

144 Kennedy, T.A., et al. (2003) AIP. Long coherence times at 300 K for nitrogen-vacancy center spins in diamond grown by chemical vapor deposition. *Appl. Phys. Lett.*, **83**, 4190–4192.

145 Krüger, A., et al. (2005) Unusually tight aggregation in detonation nanodiamond: identification and disintegration. *Carbon*, **43** (8), 1722–1730.

146 Osawa, E. (2008) Monodisperse single nanodiamond particulates. *Pure Appl. Chem.*, **80** (7), 1365–1379.

147 Osswald, S., et al. (2008) Increase of nanodiamond crystal size by selective oxidation. *Diamond Relat. Mater.*, **17** (7–10), 1122–1126.

148 Iakoubovskii, K., et al. (2008) High-resolution electron microscopy of detonation nanodiamond. *Nanotechnology*, **19** (15), 155705.

149 Hu, S., et al. (2009) Synthesis and luminescence of nanodiamonds from carbon black. *Mater. Sci. Eng. B*, **157** (1–3), 11–14.

150 Ting, C.-C., Young, T.-F., and Jwo, C.-S. (2007) Fabrication of diamond nanopowder using microwave plasma torch technique. *Int. J. Adv. Manuf. Technol.*, **34** (3), 316–322.

151 Santori, C., et al. (2009) Vertical distribution of nitrogen-vacancy centers in diamond formed by ion implantation and annealing. *Phys. Rev. B*, **79** (12), 125313.

152 Fu, K.M.C., et al. (2010) Conversion of neutral nitrogen-vacancy centers to negatively charged nitrogen-vacancy centers through selective oxidation. *Appl. Phys. Lett.*, **96** (12), 121907.

153 Osswald, S., et al. (2006) Control of sp2/sp3 carbon ratio and surface chemistry of nanodiamond powders by selective oxidation in air. *J. Am. Chem. Soc.*, **128** (35), 11635–11642.

154 Rondin, L., et al. (2010) Surface-induced charge state conversion of nitrogen-vacancy defects in nanodiamonds. *Phys. Rev. B*, **82** (11), 115449.

155 Ristein, J. (2000) Electronic properties of diamond surfaces – blessing or curse for devices? *Diamond Relat. Mater.*, **9** (3–6), 1129–1137.

156 Hauf, M.V., et al. (2011) Chemical control of the charge state of nitrogen-vacancy centers in diamond. *Phys. Rev. B*, **83** (8), 081304.

6
n-Type Diamond Growth and Homoepitaxial Diamond Junction Devices
Satoshi Koizumi and Toshiharu Makino

6.1
n-Type Diamond Growth and Semiconducting Characteristics

6.1.1
Background

The control of n- and p-type doping is a key technology for diamond-based optoelectrical devices. The p-type doping of diamond is a well-established technology that is performed by doping boron atoms in the lattice; p-type diamond also exists in Nature, known as blue diamond. In contrast, n-type diamond does not exist in Nature and remains difficult to produce, even after many years of research. Doping is difficult for many wide bandgap materials, the reasons being that the energy level of the conduction band minimum is close to vacuum level, or that the energy level of the valence band maximum is very deep. In the case of diamond, n-type doping is thought to be difficult because the conduction band minimum is at a higher energy level than that of vacuum – in other words, the negative electron affinity nature of diamond. In the same way that the successful p-type doping of gallium nitride led to blue light-emitting diodes and power device applications, it is important to establish the technology of conduction-type control.

During the early 1980s, the successful chemical vapor deposition (CVD) of diamond was reported, and this triggered much research into the use of CVD to grow n-type diamond for new industrial applications. Among the elements in Group V of the Periodic Table, nitrogen is known to result in a deep donor level due to local distortion of the lattice near nitrogen atoms at substitutional sites [1]. Phosphorus (P) is the most attractive candidate for doping, as it is predicted theoretically to give a shallower donor level at 0.2–0.4 eV below the conduction minimum [2–4]. Between the 1980s and early 1990s, the P doping of polycrystalline diamond films underwent intensive investigation, but clear evidence of n-type conductivity was not obtained, most likely because of a low crystalline perfection and the existence of grain boundaries [5]. As the presence of a large amount of hydrogen was thought to suppress the ionization of P atoms [6], one of these authors focused their attention on achieving perfectly crystalline P-doped

Optical Engineering of Diamond, First Edition. Edited by Richard P. Mildren and James R. Rabeau.
© 2013 Wiley-VCH Verlag GmbH & Co. KGaA. Published 2013 by Wiley-VCH Verlag GmbH & Co. KGaA.

homoepitaxial diamond thin films, and succeeded in growing n-type diamond during the mid-1990s [7]. In these studies, the growth of higher crystalline perfection P-doped films was obtained on the {111}-oriented diamond substrates which are normally unfavored to grow high-quality B-doped or intrinsic diamond layers. In 2005, a Japanese group succeeded in growing {100} P-doped n-type diamond films [8], and today n-type diamond can be grown on both {111}- and {100}-oriented diamond substrates. In contrast, shallower n-type doping studies have been reported, such as sulfur doping and boron–deuterium (B-D) complex formation in diamond [9, 10]. Unfortunately, however, these results were not reproducible and could not be used for n-type diamond growth [11]. For studies of the B-D complex, a possible experimental error in Hall measurements was reported [12]. Other studies of co-doping (i.e., multi-element doping to form specific complexes in the crystalline lattice) using theoretical modeling have been reported, but very few experiments have been conducted in this area [13–16].

6.1.2
Chemical Vapor Deposition of n-Type Diamond Thin Films

For the advanced control of purity in doping experiments, microwave plasma-assisted CVD has been used for diamond growth. In the initial experiments performed during the 1990s, the old-style quartz tube-type CVD system was used with a background vacuum of 10^{-7} Torr. In this case, methane (CH_4) and hydrogen (H_2) were used as the source gas for diamond growth, and PH_3 was used as a doping source gas. Each of these gases had a purity of 6N, and P-doped diamond particles were grown on molybdenum substrates and observed using scanning electron microscopy (SEM). The SEM images of a typical P-doped diamond particle, grown under conditions of methane concentration (CH_4/H_2) 0.3%, gas pressure 80 Torr, phosphine concentration (PH_3/CH_4) 8000 ppm, and substrate temperature 980 °C, are shown in Figure 6.1. On the particle surface, it is possible to identify {111} and {100} facets forming most likely a cubo-octahedral shape, as shown in Figure 6.1a. It is clear that the {111} facets formed are very flat, in contrast to the rough and complicated surface morphology of {100} facets. As shown in Figure 6.1b, numerous hillocks that seem to be surrounded by {111} surfaces have been formed on {100}. This implies that the step flow growth has not proceeded favorably, and that only the defective or contaminated portions may serve as stable nucleation sites for the three-dimensional (3-D) growth of diamond. On {111} facets, diamond tends to grow in two-dimensional (2-D) fashion, as shown in Figure 6.1c. A series of growth experiments performed under various growth conditions of CH_4/H_2 of 0.15–2% and substrate temperatures of 800–1000 °C revealed that only the {111} surface is formed flat, and that a low methane concentration and a high substrate temperature are required for growth. When the methane concentration exceeds 0.3%, an unfavorable nucleation takes place on {111} and the grown crystals become highly defective. Although it is possible to grow flat {100} facets by using a high methane concentration (2%) and a lower substrate temperature (800 °C), the doping efficiency of P is about one-hundredth

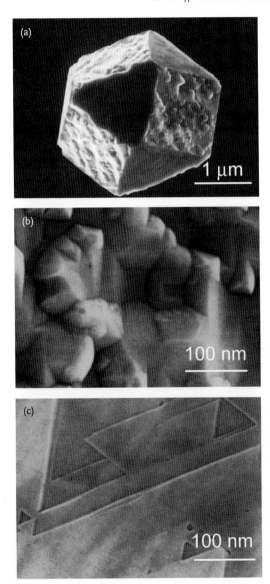

Figure 6.1 SEM images of a typical P-doped diamond particle. (a) Overall image of the particle; (b) {100} surface of the particle; (c) {111} surface of the particle.

that for {111}. However, the epitaxial diamond films were highly resistive and the growth conditions for n-type diamond on {100} could not be optimized; moreover, these films also demonstrated the co-incorporation of hydrogen along with P atoms. Thus, it was concluded that {111} is the most favorable surface orientation to form high-quality P-doped layers [7].

Table 6.1 Growth conditions of {111} P-doped diamond thin films.

Condition	Value
Source gases	PH_3, CH_4, H_2
CH_4/H_2	0.05%
PH_3/CH_4	10–5000 ppm
Total gas pressure	100 Torr
Total gas flow rate	1000 sccm
Substrate temperature	850–950 °C
Substrate	Ib diamond {111} ($2 \times 2 \times 0.5\,mm^3$)
Growth rate	400–500 nm h^{-1}

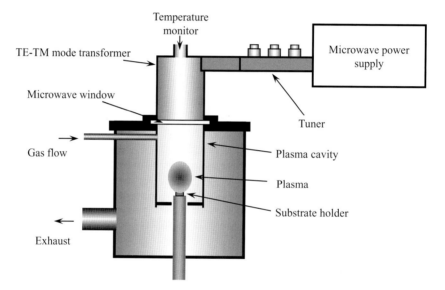

Figure 6.2 Schematic drawing of the microwave plasma CVD system used for high-purity n-type diamond growth.

Further growth experiments have been performed using {111} single-crystal diamond substrates to obtain high-perfection crystalline films. Experiments to optimize {111} P-doped diamond growth have revealed that the conditions listed in Table 6.1 achieve a reproducible growth of highly crystallized P-doped films [17]. For these experiments, metal-chamber-type microwave plasma-assisted CVD systems were used to achieve a much greater control of purity. A schematic diagram of the growth system is shown in Figure 6.2. The propagation mode of the microwave can be seen to change from rectangular mode to circular mode, through a cylindrical wave-transformer. The microwave is then introduced through a quartz glass window on the top of the chamber, in which a plasma cavity of a size tuned for 2.45 GHz microwave is attached beneath the window. The source gases are then introduced from the top of the cavity, while the substrate is placed on the molybdenum holder that penetrates into the cavity through a hole at the center of the reflector disk. Plasma is then formed stably on the substrate. The H_2, CH_4 and PH_3 are supplied independently as source gases, with the flow rate being controlled precisely by using mass flow controllers. For this, premixed phosphine/hydrogen (PH_3/H_2, 100 ppm) is used, with a final purity of 6N, while hydrogen gas with a purity of 9N is supplied through a Pd diffuser, and the purity of the methane is 6N. The growth chamber is evacuated using a turbo-molecular pump with an oil mechanical pump to a base pressure of less than 1×10^{-9} Torr. The chamber is baked out before each run of the growth experiment in order to reduce the water partial pressure, and Type Ib high-pressure high-temperature (HPHT) synthetic {111} single-crystal diamonds with a size of $2 \times 2 \times 0.5$ mm^3 are used as substrates. The substrates are chemically oxidized using $HNO_3:H_2SO_4$ (1:1, 220 °C, 30 min) to remove any organic contamination, followed by boiling and rinsing in ultra-pure water. This process is important as it initializes the condition of the substrate surface uniformly into an oxygen-terminated state. After installing the substrate into the chamber and evacuating, the growth experiments are performed as follows. The plasma is first ignited with a hydrogen flow at 10–20 Torr; the pressure is then increased to 100 Torr while keeping the plasma stable, such that after only a few minutes the substrate temperature rises to 850 °C. Methane is then introduced and the substrate temperature is allowed to rise further to 900 °C, and then to stabilize within a few minutes. At this stage, the microwave power is normally about 600 W. Phosphine is introduced and the P-doped diamond growth is started. The temperature of the substrate is monitored through a top window, using an optical pyrometer. The accuracy of the absolute value of the temperature is very poor because of optical emission from the plasma. The growth rate of P-doped diamond films was 400–500 nm h^{-1} at a methane concentration of 0.05%, and was proportional to the methane concentration within a range of 0.025–0.15% CH_4/H_2. The phosphine concentration in the gas phase had no major influence on the growth rate.

In order to check the performance of P-doping control and the doping profile in the grown layers, secondary ion mass spectroscopy (SIMS) measurements were performed in a deposition experiment for a sample that had been grown with several different doping levels which were changed sequentially. The SIMS depth profiles of impurities in a P-doped film grown consecutively at 10, 100, and 200 ppm of PH_3/CH_4 are shown in Figure 6.3. A good controllability of the P concentration was confirmed at doping levels as low as 1×10^{16} cm^{-3}, and the P concentration was estimated at approximately 1×10^{16} cm^{-3} for 10 ppm, 7×10^{16} cm^{-3} for 100 ppm, and 1×10^{17} cm^{-3} for 200 ppm. The boron, nitrogen, and hydrogen concentrations were below the detection limits, and the phosphorus doping efficiency (P atoms in the layer versus P atoms in the gas phase) was about 0.3%.

As with the atomic displacement of P atoms in diamond lattice, a combined technique of particle induced X-ray emission (PIXE) and Rutherford backscattering

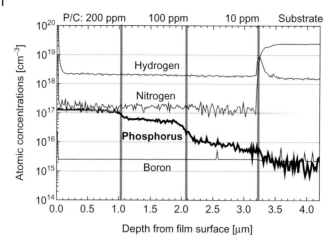

Figure 6.3 Secondary ion mass spectroscopy depth profiles of impurities in a P-doped film.

(RBS) analysis has been reported [18]. The results clearly showed that over 90% of the P atoms were located exactly at lattice substitutional sites, and there was no obvious displacement from the sites. Electron spin resonance (ESR) measurements also confirmed the existence of a local distortion of the lattice due to P incorporation [19].

6.1.3
Electrical Properties of n-Type Diamond [20]

Hall-effect measurements are necessary to clarify the conductivity nature of semiconductors. For the Hall measurement of n-type diamond thin films, metallic contacts are needed; however, it is difficult to form ohmic contacts for n-type diamond and hence precise studies have not been conducted to date. Especially for samples with a low doping level, the nature of the contact significantly influences the electrical measurements. To solve this problem, heavily P-doped diamond layers were grown selectively on a lightly P-doped film to reduce the contact resistance practically by enhancing tunneling probabilities. The detailed process of selective doping has been described previously [21]. In brief, before forming metal contacts, the surface of the sample was chemically treated to remove the surface conductive layer by strong oxidation using HNO_3/H_2SO_4 (1/1 v/v, 220 °C, 30 min). Ohmic electrodes composed of Ti capped with Au were formed on heavily P-doped diamond layers by electron-beam vacuum deposition. Hall-effect measurements using the van der Pauw method were carried out in the temperature range between 300 and 873 K at an alternating current (AC) magnetic field of 0.6 T with a frequency of 0.1 Hz.

P-doped films with phosphorus concentrations over 1×10^{16} cm^{-3} showed negative Hall coefficients very stably and reproducibly in the temperature range from

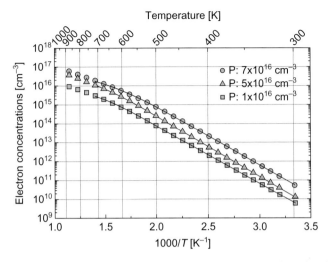

Figure 6.4 Temperature dependence of carrier concentration for P-doped diamond films. Solid curves indicate the fitting results using Equation (6.1).

300 to 873 K, which indicates an n-type conduction of the films. In the case of P-doped films with a P concentration of $1 \times 10^{16}\,\text{cm}^{-3}$ or less, although n-type conductivity of the films was observed at higher temperatures, negative Hall coefficients around room temperature have not been convincingly detected due to the high resistivity of the films. The temperature dependence of the carrier concentrations of the P-doped diamond films at various P concentrations is shown in Figure 6.4. The carrier concentration is exactly proportional to the P concentration in the film in the present doping range. The carrier concentration and resistivity of a P-doped film with a P concentration of $7 \times 10^{16}\,\text{cm}^{-3}$ were $5 \times 10^{10}\,\text{cm}^{-3}$ and $2 \times 10^{5}\,\Omega\cdot\text{cm}$ at room temperature, and $6 \times 10^{16}\,\text{cm}^{-3}$ and $2\,\Omega\cdot\text{cm}$ at 873 K, respectively. For a non-degenerate semiconductor, the activation energy and the compensation ratio can be calculated using the following equation:

$$\frac{n(n+N_A)}{N_D-N_A-n} = \frac{N_C}{2}\exp\left(-\frac{E_D}{kT}\right), \tag{6.1}$$

where n is the carrier concentration, N_D and N_A are the donor and acceptor concentrations, N_C is the effective density of states in the conduction band, g is the degeneracy factor for the donor, E_D is the activation energy of the donor, k is the Boltzmann constant, and T is the temperature. N_C is calculated by:

$$N_C = 2M_C\left(\frac{2\pi m^*_{\text{dos}}k_B T}{h^2}\right), \tag{6.2}$$

where h is Planck's constant, m^*_{dos} is the effective mass of the density of state, M_C is the number of conduction band minima in the energy state diagram (multi-valley structure) for which the value of 6 is used. The effective mass is calculated as:

$$m^*_{dos} = \left(m^*_l m^*_t m^*_t\right)^{1/3}, \quad (6.3)$$

where m^*_l is longitudinal effective mass and m^*_t is transverse effective mass. Values of $m^*_l = 1.81 m_0$ and $m^*_t = 0.306 m_0$ (m_0 is the mass of a free electron) were applied, which were deduced from the effective mass approximation corrected with excited states analysis from infrared absorption spectrometry experiments of n-type diamonds [22]. The temperature dependencies of carrier concentrations shown here closely agreed with the equations. For P-doped films with a P concentration of $7 \times 10^{16}\,\mathrm{cm}^{-3}$, values of $E_D = 0.57\,\mathrm{eV}$, $N_D = 6.8 \times 10^{16}\,\mathrm{cm}^{-3}$, and $N_A = 8.8 \times 10^{15}\,\mathrm{cm}^{-3}$ were obtained, and the compensation ratio $N_A/N_D = 0.13$. In a film with a P concentration of $1 \times 10^{16}\,\mathrm{cm}^{-3}$, N_A was $6.4 \times 10^{15}\,\mathrm{cm}^{-3}$. When the P concentration was decreased below $1 \times 10^{16}\,\mathrm{cm}^{-3}$, the P-doped films became fully compensated and had a high resistance. These results suggested that, in the present study, the concentration of compensating defects that exist naturally in the {111} CVD film is of the order of $10^{15}\,\mathrm{cm}^{-3}$.

The Hall mobility of the P-doped films as a function of temperature is shown in Figure 6.5. The Hall mobility at room temperature increases from 410 to $660\,\mathrm{cm^2\,V\text{-}s^{-1}}$ with decreasing P concentration from 5×10^{17} to $7 \times 10^{16}\,\mathrm{cm}^{-3}$ (this value is the highest ever reported for n-type diamond films). At P concentrations below $7 \times 10^{16}\,\mathrm{cm}^{-3}$, however, no improvement in Hall mobility has been achieved, a fact which may be ascribed to the existence of a large amount of compensating acceptors. The Hall mobility for the P-doped film with a P concentration of $7 \times 10^{16}\,\mathrm{cm}^{-3}$ decreases with increasing temperature as $T^{-1.4}$ up to 450 K, indicating

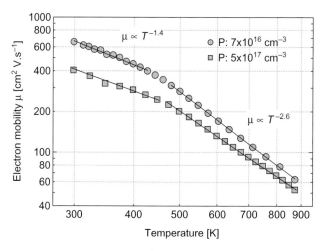

Figure 6.5 Temperature dependence of Hall mobility for P-doped diamond films.

that acoustic phonon scattering dominates. At temperatures above 450 K, the Hall mobility is proportional to $T^{-2.6}$. It is assumed that the Hall mobility at higher temperatures is dominated by various scattering mechanisms, such as intervalley scattering and acoustic phonon scattering, and is similar to that detected in drift mobility experiments with high-quality natural diamond [23]. The feature of electron mobility in n-type diamond was precisely investigated by taking Hall measurements for about 100 samples with different doping concentrations [24, 25]. From a detailed investigation of the electron scattering behavior of n-type diamond, the ultimate electron mobility was predicted to be 1000 cm^2 V-s^{-1} at room temperature [24].

{100} P-doped diamond grown using microwave plasma CVD clearly shows an n-type conductivity, as confirmed by Hall measurements [8]. The activation energy of donors was 0.57 eV, which was the same as for {111}, and the highest mobility of {100} P-doped diamond was reported as 780 cm^2 V-s^{-1} at room temperature, which was even better than the results for {111} [26]. However, the compensation ratio was up to 50–80%, and exceptionally large [27]. The electrical activity of P atoms was low at 20%, and in contrast to the value of >75% reported for {111} P-doped diamond [28]. The surface morphology of a {100} P-doped film shows a strong step bunching feature, which disturbs the formation of junction devices for optoelectrical applications.

6.2
Electrical Properties of Diamond pn Junctions

Diamond pn junctions can be formed by growing boron (B)-doped diamond and P-doped diamond in a stacking form [29]. As described previously, {111} gives a higher crystalline perfection for P doping, and so {111} single crystalline diamond substrates are preferred for forming diamond pn junctions. Both, B- and P-doped diamond thin films have been grown independently using individual microwave plasma-enhanced CVD systems. Metal-chamber-type CVD systems with a background pressure of 1×10^{-8} Torr were used to avoid the incorporation of unexpected impurities such as silicon and nitrogen (see Section 6.1.2 for a detailed description). The mechanically polished {111} surface of B-doped single-crystal Type Ib diamond (IbB) containing boron atoms at a concentration exceeding 100 ppm and showing high conductivity was used as a substrate to form a vertical pn diode in the present case. The size of the substrate was 2×2 mm^2, with 0.5 mm thickness. The pn structure is formed by the following process: (i) cleaning of the substrate surface; (ii) growth of the B-doped layer; (iii) cleaning of the sample; (iv) growth of the P-doped layer; (v) cleaning of the sample; and (vi) the formation of ohmic electrodes for both layers. For steps (i), (iii), and (v), a chemical oxidation process was employed, using an acid mixture of HNO$_3$/H$_2$SO$_4$ (1/1 v/v, 220 °C, 30 min). This resulted in the removal of any nondiamond adsorbents from the sample surface and termination of the surface with oxygen instead of hydrogen. Each B-doped and P-doped layer was grown using diborane (B_2H_6) and PH_3, respectively, as impurity source gases. The growth conditions were almost the same as for {111} P-doped diamond growth, as described above. Finally, the ohmic electrodes were formed for both the p-type and n-type layers by the electron-beam evaporation of a Ti film (thickness 50 nm) with an Au film (thickness 100 nm) for protection. The films were deposited at 400 °C in order to obtain better ohmic interfaces. The annealing process had no effect on the electrical properties of the p-type and n-type layers; in addition, at this annealing temperature no significant inter-diffusion of impurities would be expected at the pn junction interface. The Au/Ti-deposited contacts were made through the mask, with 150 μm diameter on the treated surfaces. Finally, the pn sample was processed by reactive ion etching (RIE). The aluminum vacuum-deposited film was formed on the sample to cover the electrode area as the RIE mask. RIE was performed using a conventional capacitance-coupled sputtering (etching) system operating at 13.56 MHz with a power of 50 W. Molecular oxygen was used as the reactant gas, at a pressure of 2–5 Pa, and the etching rate was approximately 20 nm min^{-1}. An aluminum film of 250 nm thickness functioned well as a protection mask in the etching process, with an etching depth of up to 2 μm.

The current–voltage (I–V) characteristics of the pn junction show a clear rectifying property (Figure 6.6a). The I–V data have been acquired for the voltage applied to the electrode of the p-type layer. Breakdown was not observed for a reverse voltage of over 100 V, and the turn-on voltage was about 5 V. The rectification ratio was over 10^{10} when a forward (n-type negative) and reverse (n-type positive) voltage of ±10 V was applied (Figure 6.6b). The ideality factor (n) was about 3.5, as deduced from the clear exponential regime shown in Figure 6.6b, and improved as the operation temperature increased; in fact, at a temperature of 600 °C the ideality factor was 1.9. The temperature dependence of the I–V characteristics suggests that the inferior feature of current transport is due to a tunneling of electrons through deep level defects that existing close to the pn interface [30]. There is also a strong deviation of the I–V profile from the exponential relationship above a current of 10^{-6} A, and an applied voltage of 5–6 V which is due to the high series resistance of the sample (mainly of the P-doped layer) of about 10^5 Ω. The resistance related to the B-doped layer and the substrate was negligibly small compared to that of the P-doped layer.

Capacitance–voltage measurements showed a clear voltage-dependent increase in capacitance in the reverse voltage region, thus confirming the existence of a depletion region and its narrowing as the voltage increased toward the forward voltage. The $1/C^2$–V plots showed a linear relationship in most cases, indicating that an abrupt junction interface was formed. The built-in potential was about 4.7 V, as estimated from the profiles. By utilizing an electron beam-induced current (EBIC) analysis, a clear current response was observed at the pn junction, confirming the existence of a depletion region at the pn junction [31]. These results confirmed that the diode characteristics observed in the present sample had originated from the pn junction interface.

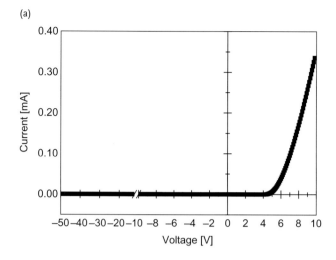

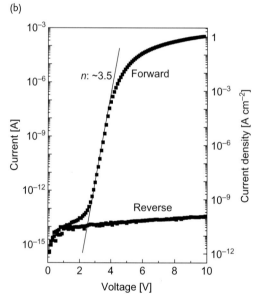

Figure 6.6 Current–voltage characteristics of diamond pn junction. (a) Linear scale plots; (b) Semi-logarithmic plots.

6.3
Diamond Deep-UV LEDs

Currently, ultraviolet (UV) light sources with a wavelength shorter than 350 nm are in high demand for a variety of applications that include sterilization, water and air purification, high-resolution photolithography, and biomedical research.

Although mercury lamps are currently used as general UV light sources, UV light-emitting diodes (LEDs) offer various advantages such as compact size, nontoxic material, and long life [32].

Diamond, as a well-known, wide-bandgap (5.5 eV), indirect-transition semiconductor, emits deep-ultraviolet (DUV) light of 235 nm by free-exciton radiative recombination accompanied by a transverse optical (TO) phonon at room temperature [33–35]. In diamond, the binding energy of the excitons is large (80 meV) and the Bohr radius small (1.57 nm), due to its smaller dielectric constant compared to other wide-bandgap semiconductors. Thus, diamond has the potential to attain a very dense exciton state on the order of 10^{18} cm^{-3}, even at room temperature [36, 37], and can emit intense DUV light even despite being an indirect-transition semiconductor. Clearly, diamond has emerged as a strong candidate for a DUV light source.

Diamond, like silicon, is a single-element semiconductor, and can be grown as higher-quality single crystals when compared to compound semiconductors. Diamond also possesses excellent physical and electrical characteristics, notably a high thermal conductivity and excellent radiation hardness [38]. Taken together, these advantageous characteristics of diamond lead to its being highly suitable as a base material for DUV-LEDs.

In order to develop LEDs, several fundamental technological issues must first be resolved, including intrinsic diamond growth, p- and n-type doping, contact fabrication, and etching technology. In particular, it is difficult to optimize n-type doping and its associated technologies in diamond synthesis. At this point, some details are provided of investigations into diamond DUV-LEDs that consist of diamond pn junction diodes and pin junction diodes, in which the intrinsic layer is sandwiched between the p- and n-type layers. These are based on the successful n-type diamond growth achieved via P-doping, as described in Sections 6.1 and 6.2.

6.3.1
Diamond pn Junction LEDs

6.3.1.1 {111} Diamond pn Junction LEDs

The first report of a diamond DUV LED was presented by Koizumi *et al.* [29] in 2001, who succeeded in fabricating homoepitaxial diamond pn junction LEDs with B-doped p-type and P-doped n-type layers on a HPHT-synthesized {111}-oriented Type IbB diamond substrate. Details of the fabrication processes have been described in Section 6.2. A schematic diagram of the LED structure is shown in Figure 6.7, where the pn junction shows diode characteristics with a rectification ratio of over 10^3 at ±15 V. The current-injected light emission spectra of the {111} pn junction LEDs at room temperature are shown in Figure 6.8. For forward currents exceeding 5 mA, a DUV light with a sharp emission peak located at 235 nm (5.27 eV) is observed; this emission peak is attributed to free-exciton radiative recombination coupled with a TO phonon (FE$_{TO}$). The small peak (shoulder) located at around 243 nm (5.10 eV) is attributed to the phonon replica of FE$_{TO}$ peak

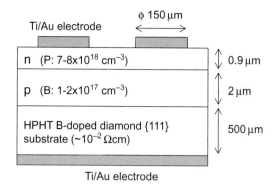

Figure 6.7 Schematic structure of {111} diamond pn junction LEDs.

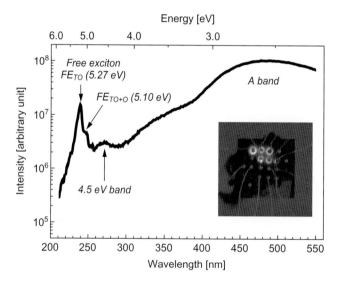

Figure 6.8 Current-injected light-emission spectra of the {111} pn junction LEDs at room temperature. The inset shows a typical optical image of the LEDs, with light emission.

(FE_{TO+TO}). The inset of Figure 6.8 shows a typical optical image of the LED with light emission, whereby a bluish-white light emitted from a deep level was observed. The spectrum in Figure 6.8 suggests that DUV light due to FE_{TO} should be emitted simultaneously.

As deep-level emissions, broadband emission in the UV region with peak energy located at 4.5 eV and broadband emission in the visible region were observed. The origin of the 4.5 eV band emission is not clear, but it is often observed for B- and P-doped CVD diamond thin films by cathodoluminescence (CL) and photoluminescence (PL) analysis [39, 40]. The band in the visible region is attributed to

A-band emission, which is often observed – using CL and PL – emitting from CVD diamond thin films.

These data represent the first report of light emission from diamond LEDs, while the successful observation of an excitonic emission by current injection to diamond pn junctions at room temperature is also a major milestone.

6.3.1.2 {100} Diamond pn Junction LEDs

At present, {100} diamond substrates are larger and easier to mechanically polish than {111} substrates. However, in order to bring diamond LEDs closer to markets and applications, not only {111} but also {100} diamond LEDs must first be developed.

Horiuchi et al. [41] developed diamond pn junction LEDs on {100} surfaces by using sulfur (S)-doped thin films as the n-type layer [9]. These LEDs showed a rectification ratio of 10^3 at ±10 V at room temperature, and a relatively high-performance excitonic emission was obtained. The best performance values reported for this LED structure to date have included a 0.027% external quantum efficiency and 7 µW output power at 10 mA [42]. However, Hall-effect measurements failed to show any clear evidence of n-type conduction for the S-doped CVD diamond layer [11, 43]; clearly, further investigations of n-type diamonds are required to improve this LED.

In contrast, in 2005 Kato et al. created {100} P-doped diamond films that showed a clear n-type conductivity, by optimizing the conditions for CVD growth [8]. This proved to be a breakthrough towards eliminating the restrictions on crystal orientation for n-type doping, and for fabricating LEDs with a wide range of applications.

Based on this technique, the successful fabrication of {100} diamond LEDs was achieved in 2006 [44, 45]. The structure of the {100} pn junction LEDs is shown schematically in Figure 6.9, where the B- and P-doped layers were grown inde-

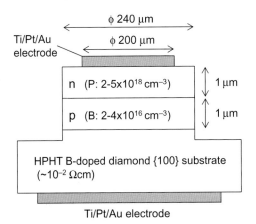

Figure 6.9 Schematic structure of {100} diamond pn junction LEDs.

pendently by CVD using a mixture of CH_4 and H_2 gas on HPHT-synthesized {100} IbB single-crystal diamonds with a resistivity of approximately $10^{-2}\,\Omega\cdot$cm. As impurity source gases, PH_3 diluted with H_2 was used for P-doping, while B-doping was achieved with CH_4 and H_2 plasma by an after-effect following heavy B-doping deposition by B_2H_6. After the growth, mesa structures were fabricated by conventional photolithography and inductively coupled plasma (ICP) etching processes, in order to decrease the leakage current. The pn junction layers were then maintained in a mixture of H_2SO_4 and HNO_3 at 200 °C for 60 min in order to remove any surface contamination and to terminate the surface with oxygen. Finally, Ti (30 nm)/Pt (30 nm)/Au (100 nm) electrodes were formed on the n-type layer and the reverse side of the Type Ib {100} B-doped substrate, and annealed at 420 °C for 30 min in Ar ambient.

A typical I–V characteristic of the {100} pn junction LEDs is shown in Figure 6.10, where a clear rectification characteristic with a rectification ratio of about 10^6 at ±30 V was observed. The gradual increase in current at the forward bias voltage of >6 V originates from the high series resistance of the pn junction LEDs, especially the n-type layer with resistivity of the order of $1\,M\Omega\cdot$cm [8]. The contact resistance is almost equivalent to that of the device impedance.

The current-injected light emission spectra of the {100} pn junction LEDs under currents of 23 and 35 mA and forward bias voltages of 43 and 47 V, respectively, at room temperature, is shown in Figure 6.11. The free-exciton emission (FE_{TO}) can be observed clearly, which indicates the good crystalline quality of the p- as well as n-type layers on the {100} substrate. The peak wavelengths of broad deep-level emissions are almost the same as those of the {111} pn junction LEDs shown in Figure 6.8. The relative integrated intensity of FE_{TO} to deep-level emissions was as small as ~0.01 at the forward current of 35 mA. This value was also the same order of magnitude as that of the {111} pn junction LEDs.

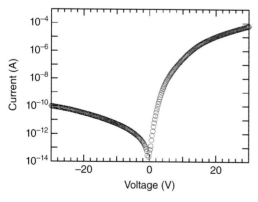

Figure 6.10 Typical current–voltage characteristic of {100} pn junction LEDs at room temperature.

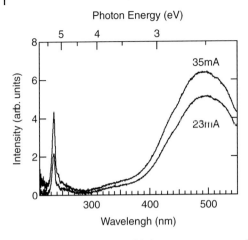

Figure 6.11 Current-injected light-emission spectra of {100} pn junction LEDs at room temperature.

6.3.2
Diamond pin Junction LEDs

In order to increase DUV light emission due to free-exciton recombination, and simultaneously to suppress the deep-level emission, {100} [44] and {111} [46, 47] homoepitaxial diamond pin junction diodes were fabricated by sandwiching the intrinsic layer, which has a low impurity concentration and low density of deep levels, between p- and n-type layers, and their electrical and light-emitting characteristics were characterized.

6.3.2.1 {100} Diamond pin Junction LEDs

The processes of fabricating the {100} pin junction LEDs are almost the same as those of {100} pn junction LEDs [44]. Here, HPHT-synthesized Type Ib {100} single-crystal diamond was used as a substrate, and a heavily B-doped p^+-type layer was deposited as an under-electrode layer, using an impurity source gas of B_2H_6 diluted with H_2. In order to grow the non-doped intrinsic layer, O_2 was added to the feeding gas to reduce the slight contamination of B and other impurity atoms [48–50]. The schematic structure of the {100} pn junction LEDs and the SIMS depth profiles of the pin junction are shown in Figure 6.12a and b, respectively. The SIMS data revealed that an intrinsic layer with a thickness of about 0.1 μm was clearly formed between the p- and n-type layers, without any intermixture area of B and P atoms. The impurity concentrations of B and P atoms in the intrinsic layer were below the detection limit.

For the current density–voltage (J–V) measurements, good diode characteristics with a rectification ratio of about 10^{10} at ±25 V were observed (see Figure 6.16).

The current-injected light-emission spectra of the {100} pin junction LEDs under currents of 9, 20, 34, and 79 mA and bias voltages of 22, 24, 27, and 30 V,

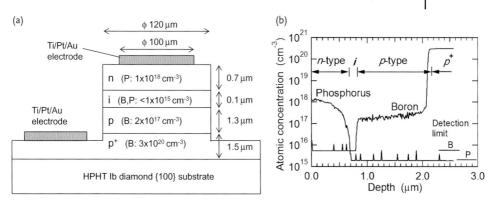

Figure 6.12 (a) Schematic structure of {100} diamond pin junction LEDs; (b) SIMS depth profiles of {100} diamond pin junction LEDs.

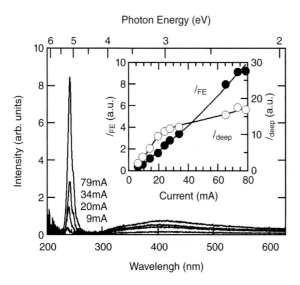

Figure 6.13 Current-injected light-emission spectra of {100} pin junction LEDs at room temperature. The inset shows the integrated intensity of free-exciton emission (I_{FE}) and deep-level emission (I_{deep}) as a function of the current.

respectively, at room temperature, are shown in Figure 6.13. At currents exceeding 9 mA, emissions of FE_{TO} at 235 nm and FE_{TO+TO} at 243 nm were observed. The significant result of this demonstration was that the excitonic emission intensity was considerably higher than the deep-level emissions; in other words, the deep-level emissions were significantly suppressed. This result was supported by an analysis of the SIMS data (Figure 6.12b) and the J–V characteristic (see Figure 6.16).

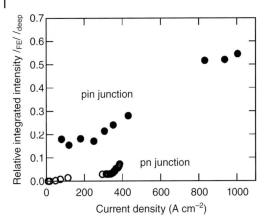

Figure 6.14 Relative integrated intensity of the excitonic emissions to the deep-level emissions (I_{FE}/I_{deep}) for {100} pin junction LEDs as a function of the forward current density. I_{FE}/I_{deep} for pn junction LEDs is also shown for comparison.

A very interesting trend was found (as shown in the inset of Figure 6.13) that the integrated intensity of free-exciton emission (I_{FE}) increases superlinearly, whereas the slope of integrated intensity of deep-level emission (I_{deep}) decreases with increasing current. The same trend was also observed for the pn junction LEDs. The relative integrated intensity of the excitonic emissions to the deep-level emissions (I_{FE}/I_{deep}) for the {100} pn and pin junction LEDs as a function of the forward current density is shown in Figure 6.14. Here, it was found that I_{FE}/I_{deep} increased nonlinearly from 0.17 to 0.55 as the current density was increased from 76 to 1019 A cm^{-2} for the {100} pin junction LEDs. In contrast, for the pn junction LEDs the I_{FE}/I_{deep} ratio was increased from 0.002 to 0.08 as the forward current density was increased from 48 to 398 A cm^{-2} (see Figure 6.11). Thus, the value of I_{FE}/I_{deep} for the pin junction LEDs was improved about 20-fold compared to that of the pn junction LEDs. Thus, the DUV light emission characteristics of homoepitaxial diamond LEDs could be significantly improved by sandwiching the high-quality intrinsic layer between p- and n-type layers.

6.3.2.2 {111} Diamond pin Junction LEDs

As a next step, the absolute intensity of excitonic emission must be increased; that is, the injected current must be increased. For the {100} diamond pin junction LEDs, the series resistance remained high, and especially the high resistivity (of the order of 10^6–10^7 $\Omega \cdot$cm) and high specific contact resistance (the order of 10^5 $\Omega \cdot$cm^2) of the n-type layer [51]. Unfortunately, these findings posed a problem when the injected current was increased, as they indicated that the junction temperature would also be increased [52]. Consequently, in order to increase the excitonic emission intensity it is necessary first to reduce the series resistance of the pin junction, and especially that of the n-type layer.

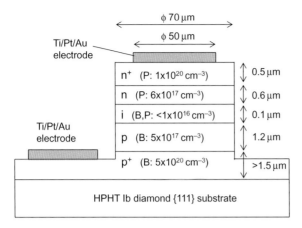

Figure 6.15 Schematic structure of {111} diamond pin junction LEDs.

In 2008, Kato et al. achieved a lower resistivity (of the order of $10^2 \, \Omega \cdot cm$) and a lower specific contact resistance (of the order of $10^{-3} \, \Omega \cdot cm^2$) of {111} n-type diamond than those of {100} n-type diamond at room temperature by increasing the P-doping concentration up to the order of $10^{20} \, cm^{-3}$ [53], under which conditions hopping conduction also occurred. The resistivity of {111} n-type diamond was also lower by one order of magnitude compared to that of {100} n-type diamond under the lightly P-doped condition of the order of $10^{17} \, cm^{-3}$, because the compensation ratio of {111} n-type diamond was smaller than that of {100} diamond [54]. As a result, studies were conducted of not only {100} but also {111} pin junction diodes [46].

The schematic structure of the {111} pin junction LEDs is shown in Figure 6.15, and a typical J–V characteristic of the {111} pin junction at room temperature in Figure 6.16. For comparison, the J–V characteristic of the {100} pin junction [44] is also shown in Figure 6.16, where the forward current density of the {111} junction can be seen to increase sharply compared to that of the {100} junction. These results indicated that the series resistance of the {111} junction would be clearly improved by the low resistivity and low specific contact resistance of the n- and n^+-layers.

The current-injected light emission spectra of the {111} pin junction LEDs under currents of 20, 40, and 60 mA (current densities of 1020, 2040, and 3060 A cm^{-2}) for bias voltages of 16.2, 18.0, and 19.0 V, respectively, at room temperature, are shown in Figure 6.17. Here, a considerably higher excitonic emission intensity than the deep-level emission intensity on not only {100} but also {111} surfaces, was achieved by improving the device structure from the pn junction to the pin junction.

The integrated intensities of free-exciton emission (I_{FE}) and deep-level emission (I_{deep}) of the {111} pin junction LEDs as a function of the forward current are shown in Figure 6.18; for comparison, the I_{FE} of the {100} pin junction LEDs

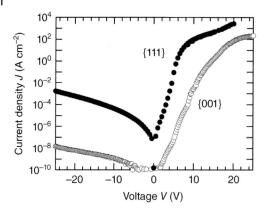

Figure 6.16 Typical current density–voltage (J–V) characteristic of {111} pin junction LEDs at room temperature. The J–V characteristic of {100} pin junction LEDs is shown for comparison.

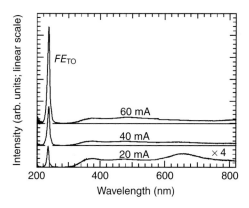

Figure 6.17 Current-injected light-emission spectra of {111} pin junction LEDs at room temperature.

is also shown. Below a current of about 10 mA (region I), the I_{FE} of the {111} pin junction LEDs was increased linearly, and continued to increase superlinearly over 10 mA (region II). The I_{FE} of the {100} pin junction LEDs was the same order of magnitude as that of the {111} pin junction LEDs at the current region of 10–20 mA, and decreased at currents exceeding 30 mA due to an increase in junction temperature caused by the high series resistance. On the other hand, for the {111} pin junction LEDs, the I_{FE} increased superlinearly at least up to 60 mA. A one order of magnitude higher absolute I_{FE} than that of the {100} pin junction LEDs was achieved. Thus, in order to increase the absolute excitonic emission intensity, it is important to suppress the junction temperature by decreasing the series resistance, and especially the resistivity and specific contact resistance of the n-type layer.

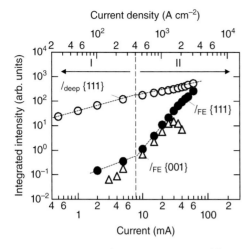

Figure 6.18 Integrated emission intensities of free exciton (I_{FE}) and deep-level (I_{deep}) for {111} pin junction LEDs as a function of the forward current. I_{FE} for {100} pin junction LEDs is also shown for comparison.

As shown in Figure 6.18, the I_{deep} of the {111} junction LEDs was increased sublinearly in region II, whereas I_{FE} was increased superlinearly. Thus, with the low series resistance of the junction and the superlinear increase of I_{FE}, a strong excitonic emission was achieved while maintaining a high relative integrated intensity I_{FE}/I_{deep} of 0.5. This was comparable with the previously reported value for the {100} high-quality diamond pin junction, at the forward current of 60 mA for the {111} pin junction LEDs.

6.3.2.3 {111} Diamond pin Junction LED with Thick i-Layer

For diamond pin junction LEDs, the exciton lifetime within the i-layer is relatively long because of the indirect-transition semiconductor, which means that the injected carriers and/or excitons in the i-layer have a wide distribution region. In that case, a thicker i-layer of the same order as the wide distribution region of injected carriers would be expected to produce a diamond LED with a high emission efficiency. Thus, {111} diamond pin junction LEDs were fabricated with a thick i-layer that exceeded several micrometers, and their characteristics studied [47]. The structure of the improved {111} pin junction LEDs with an i-layer thickness of 14 μm is shown schematically in Figure 6.19. Here, the p-type substrate was a HPHT-synthesized Type Ib {111} B-doped single-crystal diamond, while the intrinsic layer and heavily P-doped n⁺-type layer were grown independently using microwave plasma-assisted CVD. For the intrinsic layer, O_2 was added to the feed gas in order to reduce the contamination of B and other impurity atoms [48–50]. The resultant LED structures were simple, and no mesa structures were fabricated.

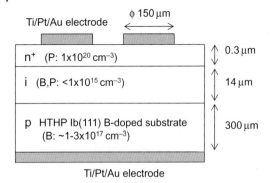

Figure 6.19 Schematic structure of the improved {111} diamond pin junction LED.

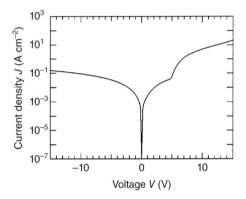

Figure 6.20 Typical current density–voltage characteristic of improved {111} pin junction LEDs at room temperature.

The typical J–V characteristics for the improved {111} pin junction LEDs at room temperature are shown in Figure 6.20. In the reverse bias region, a relatively high leakage current due to surface leakage was observed, but this could be suppressed by fabricating mesa structures. In the forward bias region, the evaluated differential resistance dV/dJ at 10 V was $0.6\,\Omega\cdot cm^2$, which was much lower than the bulk resistance of the i-layer ($>10^8\,\Omega\cdot cm^2$). This result indicated that conductivity modulation would be induced by the injection of excess carriers into the i-layer, such that the effective resistance of the i-layer would become low. This would be valuable for the operation of diamond LEDs, mainly because of the low series resistance in the forward bias region, even in the presence of a thick i-layer.

The current-injected light emission spectra of the improved {111} pin junction LEDs under pulsed currents of 160, 200, and 300 mA (pulse duration 10 ms, duty 10%) at room temperature, are shown in Figure 6.21a. The strong DUV emission due to FE_{TO} (235 nm) and its TO phonon replicas (240–250 nm) and a significant

6.3 Diamond Deep-UV LEDs

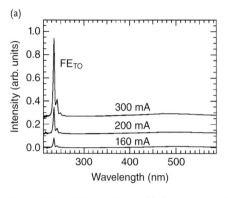

Figure 6.21 (a) Current-injected light-emission spectra of improved {111} pin junction LEDs under pulsed-current operation (pulse duration 10 ms, duty 10%) at room temperature; (b) Light-emission image for improved {111} pin junction LEDs under a pulsed current of 150 mA.

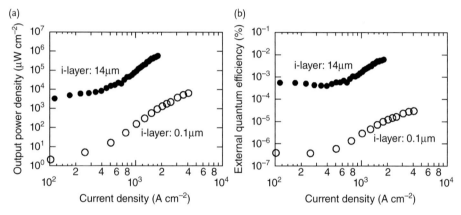

Figure 6.22 (a) Output power density and (b) external quantum efficiency of excitonic emission as a function of current density for improved {111} pin junction LEDs with i-layer thickness of 14 μm under pulsed-current operation (pulse duration 10 ms, duty 10%). The data for the LED with i-layer thickness of 0.1 μm are also shown for comparison.

suppression of deep-level emission were observed under high-current conditions. A light-emission image for the diamond LED at a pulsed current of 150 mA is shown in Figure 6.21b, where a very bright bluish-white light emitting from a deep level was observed. According to the spectra in Figure 6.21a, this intense DUV light should be emitted simultaneously.

The output power density and external quantum efficiency (EQE) of FE_{TO} as a function of current density for the improved LED with an i-layer thickness of 14 μm under pulsed-current operation (pulse duration 10 ms, duty 10%) are shown in Figure 6.22a and b, respectively. The data for the LED with an i-layer thickness

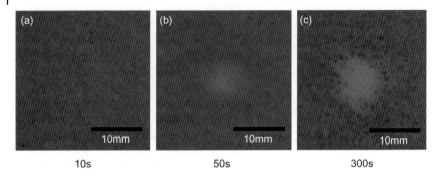

Figure 6.23 Inhibition of E. coli growth after irradiation with DUV light for (a) 10 s, (b) 50 s, and (c) 300 s from improved {111} pin junction LEDs, followed by incubation at 36 °C for 24 h.

of 0.1 μm are also shown for comparison [50]. The output power density and the EQE for the improved LED were increased more than 500-fold compared to those for the previously reported LED. Taken together, these results indicated that the ratio of the number of carriers distributed within the i-layer to the number of injected carriers was increased with increasing i-layer thickness. Moreover, the output power density and the EQE for the improved LED were increased superlinearly, even at a high current density of over 1000 A cm^{-2}. This robust characteristic was considered due to the superior material properties of diamond – that is, the highest thermal conductivity among bulk materials. For the improved LED, the output power and the EQE reached 0.1 mW and 0.006%, respectively.

In an actual application of DUV light sources, sterilization of the bacterium *Escherichia coli* was investigated by irradiating the organisms with an excitonic emission from diamond LEDs. The irradiation of *E. coli* for only 50 s with a DUV light source caused a significant inhibition of growth, with an area of ~10 mm diameter, when the bacteria were incubated at 36 °C for 24 h (Figure 6.23). Thus, diamond LEDs may prove useful as a germicidal light source.

The fabricated diamond LED has a simple structure without mesa structures (see Figure 6.19), and most of the excitonic emission from the i-layer is either blocked by the n$^+$-type electrode or reflected at the surface of the n$^+$-type layer, mainly because of the small total reflection angle (~24°) due to the high refractive index. Consequently, the light-extracting efficiency for diamond LEDs is still low (less than a few %). The next issue would be to increase the light-extracting efficiency, and to optimize the i-layer thickness. By improving electrode shapes and/or fabricating mesa structures, diamond DUV LEDs with higher EQE should easily be achieved.

6.3.3
Light-Emission Mechanisms

As described in Section 6.3.2, a superlinearity of I_{FE} and a sublinearity of I_{deep} were observed, depending on the current at room temperature, but these characteristics cannot be explained by a simple Shockley–Read–Hall-type recombination process [55] of electron-hole pairs via deep levels in indirect-transition semiconductors. At this point, a discussion is included as to why nonlinear effects were observed for the pin junctions.

As a simple case, two states should be considered: the exciton state and the deep-level state. The actual lifetime of excitons in diamond τ_{ex} is given by:

$$\tau_{ex} = \left(\tau_r^{-1} + \tau_{deep}^{-1}\right)^{-1} \tag{6.4}$$

where τ_r is the radiative recombination time of excitons and τ_{deep} is the trapping time of electrons or holes consisting of excitons to deep levels. I_{FE} and I_{deep} are given by

$$I_{FE} \propto \frac{G\tau_{ex}}{\tau_r} = G\left(\frac{1}{1 + \tau_r/\tau_{deep}}\right) \tag{6.5}$$

$$I_{deep} \propto \frac{G\tau_{ex}}{\tau_{deep}} = G\left(\frac{1}{1 + \tau_{deep}/\tau_r}\right) \tag{6.6}$$

where G is the generation rate of excitons, which is almost proportional to the injection current. In general, for an indirect-transition semiconductor τ_r is much longer than τ_{deep} due to the requirement for momentum conservation in optical transitions. Indeed, τ_r of diamond has been reported to be of the order of microseconds [56] and is insensitive to G, while τ_{deep} is of the order of nanoseconds under low G conditions. This means that most excitons should be trapped to deep levels under low G conditions; that is, τ_{ex} of most excitons is given by τ_{deep} from Equation (6.4). In order for the nonlinear effects to occur, τ_{deep} needs to become longer with increasing G, as can be seen from Equations (6.5) and (6.6). Based on this consideration, the kinetics of the system was analyzed, taking into account exciton trapping and recombination processes via deep levels.

In the case of charged deep levels, bound carrier states are formed at those levels by the Coulomb potential of electric charge reduced by the screening effect of the permittivity. In a crystal with a low relative dielectric constant, the screening effect of the Coulomb potential is weak and the energy level of the bound states deep. In the case of diamond, Gheeraert et al. reported that the binding energies of second excited (2P) states for both donor and acceptor centers are about 80 meV by infrared absorption and photothermal ionization spectroscopy [57]. This indicates that the bound states at charged deep levels in diamond are stable even at high temperature, and that the trapping and recombination processes of electron-hole pairs via the bound states should be considered when discussing exciton trapping to deep levels.

Based on this analysis, the following mechanisms of the nonlinear emission characteristics could be clarified. Generally, in Equations (6.4)–(6.6), τ_{deep} is inversely proportional to the unoccupied deep level's density $N_{unoccupied}$; that is, $\tau_{deep} = (C_{ex}N_{unoccupied})^{-1}$, where C_{ex} is the capture rate of excitons to deep levels. After the transition of excitons to deep levels, the occupied deep levels are charged and form the bound carrier states at those levels as described above, while the trapped carriers recombine with counter-charged free carriers, not directly but rather via the bound states. Under high-excitation conditions, free-exciton density exceeds the deep-level density; however, because the recombination of trapped carriers via bound states is constant and relatively slow, almost all deep levels are trapped by a part of the high-density excitons; that is, $N_{unoccupied}$ becomes low and τ_{deep} becomes longer. Then, as shown in Equations (6.5) and (6.6), a superlinearity of I_{FE} and a sublinearity of I_{deep} occur, depending on the current.

This kinetic analysis also allowed a successful explanation of the nonlinear effect, a phenomenon which is important for high-efficiency DUV-LED actions. In addition, strong excitonic emissions for diamond LEDs can be achieved, even though diamond is an indirect-transition semiconductor. It should be noted at this point that higher-density excitons than the deep-level density would be required in order for almost all of the deep levels to be trapped by carriers. Diamond is one of the few materials to realize this condition, because of its superior excitonic nature (high density and high stability of excitons, even at high temperature).

6.4
Recent Progress of Diamond Junction Devices

Diamond has excellent physical and electrical characteristics, such as a wide bandgap, a high breakdown electric field, and a high thermal conductivity. As noted above, diamond DUV-LEDs have been developed by utilizing not only these excellent characteristics but also the unique characteristic of high-density excitonic states. But diamond also has other unique characteristics, such as a heavily doped layer with a high crystalline quality, a negative electron affinity (NEA), and a long spin relaxation time at room temperature. It is these unique characteristics that lead to diamond junction devices with interesting properties.

An example of this is the development of novel diamond diodes – namely Schottky-pn diodes (SPNDs) – by utilizing the unique characteristic of heavily doped layer with high crystalline quality [58, 59]. The structure of the diamond SPND is shown schematically in Figure 6.24. The SPND is a combination of an n-type Schottky barrier diode (SBD) with a pn junction diode (PND) in tandem.

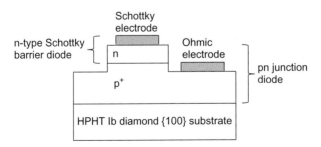

Figure 6.24 Schematic structure of the diamond SPND.

The device was designed such that the n-type layer would always be fully depleted under both forward and reverse bias conditions, and this represents the key structure of the SPND. From an analysis based on Poisson's equations, the SPND has been shown to possess the following five characteristics:

1) The SPND is a unipolar (hole) device.

2) The switching (turn-off) characteristic is fast (on the order of nanoseconds).

3) The built-in voltage V_{bi} at forward bias is ~1 eV, which is lower than that of PND, at 4.5 V.

4) The specific on-resistance $R_{on}S$ mainly depends on the B concentration in the p^+-type layer. To decrease $R_{on}S$, a higher B concentration can be designed in the p^+-type layer.

5) The breakdown voltage V_{break} mainly depends on the thickness of the n-type layer. To increase V_{break}, a thicker n-type layer can be designed while maintaining the fully depleted condition by decreasing the P concentration. By combining characteristics (4) and (5), V_{break} and $R_{on}S$ can be designed independently.

For low-loss power switching devices, low $R_{on}S$, low V_{bi}, high V_{break}, and fast switching characteristics are required simultaneously. The conventional SBD has a trade-off characteristic between $R_{on}S$ and V_{block}, however, whereby the conventional diamond PND has a relatively high $R_{on}S$, high $V_{built-in}$, and a slow switching speed because of the high resistivity of the n-type layer, wide bandgap, and slow recovery speed of stored charges, respectively. In contrast, diamond SPND can overcome these weaknesses of conventional devices, as shown in characteristics (1) to (5). Thus, diamond SPND is suitable for power switching devices. As noted above, the unique characteristics of diamond can be utilized in this device; that is, a good crystalline quality can be preserved at a high B concentration of $>10^{20}$ cm^{-3}. In other words, it is possible to fabricate high junction–quality homoepitaxial p^+n junctions with a heavily B-doped p^+-type layer in order to achieve a low $R_{on}S$. This is a major advantage for fabricating SPND by diamond. Indeed, a diamond SPND was fabricated with a p^+-type layer and a B concentration of 5×10^{20} cm^{-3}, and an n-type layer with a thickness of 160 nm and a P concentration of $<2 \times 10^{17}$ cm^{-3}. This SPND showed a low V_{bi} of ~1.2 V, an extremely low $R_{on}S$ of 0.03 mΩ·cm^2, a fast switching (turn-off) speed of <10 ns, and V_{break} of 55 V, simultaneously, as shown in Figure 6.25. In the future, it will become necessary to increase V_{break} while maintaining these good characteristics.

As other diamond junction devices utilizing the unique characteristics of diamond, electron emitters were recently reported by Koizumi et al. [60, 61] and Takeuchi et al. [62, 63]. These electron emitters are not the conventional field-emission type, but rather novel emitters that combine diamond pn and/or pin junctions with the characteristics of NEA. In this case, electrons were successfully extracted from the diamond pn and/or pin junctions at room temperature at a voltage exceeding $V_{built-in}$ (~4.5 V), which was much lower than that of the

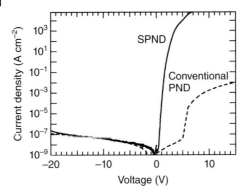

Figure 6.25 Typical current density–voltage characteristic for diamond Schottky-pn diodes at room temperature. The data for the conventional diamond pn junction diodes are also shown for comparison.

conventional field-emission type junctions. In future, this novel device will surely be applied to a variety of fields including electron emitters and switching devices operating in vacuum at extremely high voltages.

As noted above, diamond junction devices have great potential not only for LEDs but also for power devices and electron emitters, by utilizing the unique properties of diamond. Moreover, by combining diamond LEDs with these unique devices and/or the unique characteristics of diamond (e.g., long spin relaxation time), novel electronic and/or optoelectronic devices will surely be fabricated in the future.

6.5
Summary

Today, in the field of diamond research, it remains very difficult to grow highly conductive n-type diamond. At present, P-doped diamond is the best candidate, and n-type conductivity over a wide range of doping concentrations on {111} and {100} crystalline orientations has been obtained successfully. Further investigations have led to the precise recognition of opto-electronic properties and have revealed the high-quality nature of n-type diamond layers.

Having achieved n-type doping in diamond, high-quality homoepitaxial diamond pn and pin junction LEDs were fabricated, and their diode and light-emitting characteristics investigated.

In the case of diamond pn junction LEDs, sharp DUV light emissions were clearly observed at 235 nm and attributed to free-exciton radiative recombination coupled with a TO phonon. The spectra were also associated with strong, deep-level emissions. For diamond pin junction LEDs with an i-layer thickness of

0.1 μm, strong free-exciton emissions and significantly suppressed deep-level emissions were achieved. The relative integrated intensities of the free-exciton emission to the deep-level emission were about 20-fold higher than those of the pn junctions. Moreover, a higher absolute excitonic emission intensity (by one order of magnitude) was achieved by suppressing the increase in junction temperature by reducing the series resistance of the pin junction LEDs, especially the resistivity and specific contact resistance of the n-type layer. As a next step, the i-layer thickness was increased from 0.1 to 14 μm in order to increase the excitonic emission efficiency by raising the ratio of the number of carriers distributed within the i-layer to the number of injected carriers. As a result, excitonic emission intensity was enhanced more than 500-fold, and the maximum output power and external quantum efficiency of the excitonic emission were 0.1 mW and 0.006%, respectively. The ability to sterilize bacterial contaminations (e.g., of *E. coli*) was also demonstrated as a possible application for DUV light sources.

Among studies of diamond DUV-LEDs, it was found that the free-exciton and the deep-level emission intensity were increased superlinearly and sublinearly, respectively, depending on the current. These light-emitting characteristics can be explained by an analysis based on the trapping and recombination processes of electron-hole pairs via bound states at charged deep levels. Owing to these nonlinearities, strong excitonic emissions were attained for diamond LEDs, even though diamond is an indirect-transition semiconductor. Thus, these experimental findings have provided interesting new insights into the light-emitting properties and mechanisms of diamond pn and pin junction LEDs.

In future, improvements such as an increased DUV light extraction efficiency will enable the practical use of diamond deep-UV LEDs. Likewise, other unique diamond pn and pin junction devices, such as Schottky-pn diodes and electron emitters, have recently been developed by utilizing the unique characteristics of diamond such as heavily doped layer with a high crystalline quality, and a negative electron affinity. The combination of diamond LEDs with these unique devices and/or the unique characteristics of diamond will permit the fabrication of novel electronic and optoelectronic devices.

6.6
Acknowledgments

These studies were conducted with the valuable advice and encouragement of members of the Wide Band Gap Materials Group at NIMS and the Energy Enabling Technology Group at AIST. In addition, the authors would like to thank Mr H. Ozaki (Canon Corp.), Dr S. Mita and Dr M. Katagiri (Toshiba Corp.) for their great contribution to the study of n-type diamond while they were graduate school students, and Dr H. Kato of the Energy Technology Research Institute at AIST for the growth and characterization of n-type diamond. For the investigations of sterilization, the authors thank Mr K. Yoshino, Mr Uchida, Mr Sakai, and Dr Kinoshita of the Advanced Technology Center at Iwasaki Electric Co., Ltd, and Dr

S. Tanimoto for valuable discussions concerning diamond junction devices. These studies were partly supported by the Japan Science and Technology Agency (JST).

References

1 Farrer, R.G. (1969) *Solid State Commun.*, **7**, 685.
2 Kajihara, S.A., Antonelli, A., Bernhole, J., and Car, R. (1991) *Phys. Rev. Lett.*, **66**, 2010.
3 Wang, L.G. and Zunger, A. (2002) *Phys. Rev. B*, **66**, 161202.
4 Goss, J.P., Briddon, P.R., Jones, R., and Sque, S. (2004) *Diamond Relat. Mater.*, **13**, 684.
5 Okano, K., Kiyota, H., Iwasaki, T., Nakamura, Y., Akiba, Y., Kurosu, T., Iida, M., and Nakamura, T. (1990) *Appl. Phys. A*, **51**, 344.
6 Kamo, M., Yurimoto, H., Ando, T., and Sato, Y. (1991) SIMS analysis of epitaxially grown diamond. Proceedings of the 2nd International Conference on New Diamond Science and Technology, (eds R. Messier, J.T. Glass, J.E. Butler, and R. Roy), Materials Research Society Symposium Proceedings, Pittsburgh, PA, p. 637.
7 Koizumi, S., Kamo, M., Sato, Y., Ozaki, H., and Inuzuka, T. (1997) *Appl. Phys. Lett.*, **71**, 1065.
8 Kato, H., Yamasaki, S., and Okushi, H. (2005) *Appl. Phys. Lett.*, **86**, 222111.
9 Sakaguchi, I., Gamo, M.N., Kikuchi, Y., Yasu, E., Haneda, H., Suzuki, T., and Ando, T. (1999) *Phys. Rev. B*, **60**, R2139.
10 Teukam, Z., Chevallier, J., Saguy, C., Kalish, R., Ballutaud, D., Barbe, M., Jomard, F., Tromson-Carli, A., Cytermann, C., Butler, J.E., Bernard, M., Baron, C., and Deneuville, A. (2003) *Nat. Mater.*, **2**, 482.
11 Kalish, R., Reznik, A., Uzan-Saguy, C., and Cytermann, C. (2000) *Appl. Phys. Lett.*, **76**, 757.
12 Kumar, A., Pernot, J., Omnès, F., Muret, P., Traoré, A., Magaud, L., Deneuville, A., Habka, N., Barjon, J., Jomard, F., Pinault, M.A., Chevallier, J., Mer-Calfati, C., Arnault, J.C., and Bergonzo, P. (2011) *J. Appl. Phys.*, **110**, 033718.
13 Eaton, S.C., Anderson, A.B., Angus, J.C., Evstefeeva, Y.E., and Pleskov, Y.V. (2003) *Diamond Relat. Mater.*, **12**, 1627.
14 Miyazaki, T. and Okushi, H. (2002) *Diamond Relat. Mater.*, **11**, 323.
15 Lowther, J.E. (2003) *Phys. Rev. B*, **67**, 115206.
16 Yu, B.D., Miyamoto, Y., and Sugino, O. (2000) *Appl. Phys. Lett.*, **76**, 976.
17 Koizumi, S., Teraji, T., and Kanda, H. (2000) *Diamond Relat. Mater.*, **9**, 935.
18 Hasegawa, M., Teraji, T., and Koizumi, S. (2001) *Appl. Phys. Lett.*, **79**, 3068.
19 Katagiri, M., Isoya, J., Koizumi, S., and Kanda, H. (2006) *Phys. Status Solidi A*, **203**, 3367.
20 Katagiri, M., Isoya, J., Koizumi, S., and Kanda, H. (2004) *Appl. Phys. Lett.*, **85**, 6365.
21 Teraji, T., Katagiri, M., Koizumi, S., Ito, T., and Kanda, H. (2003) *Jpn. J. Appl. Phys.*, **42**, L882.
22 Gheeraert, E., Casanova, N., Koizumi, S., Teraji, T., and Kanda, H. (2001) *Diamond Relat. Mater.*, **10**, 444.
23 Nava, F., Canali, C., Jacoboni, C., Reggiani, L., and Kozlov, S.F. (1980) *Solid State Commun.*, **33**, 475.
24 Pernot, J., Tavares, C., Gheeraert, E., Bustarret, E., Katagiri, M., and Koizumi, S. (2006) *Appl. Phys. Lett.*, **89**, 122111.
25 Pernot, J. and Koizumi, S. (2008) *Appl. Phys. Lett.*, **93**, 052105.
26 Kato, H., Makino, T., Okushi, H., and Yamasaki, S. (2007) N-type diamond growth by phosphorus doping. MRS Proceedings, 26–30 November 2007, Boston, MA, p. 1039-P05-01.
27 Kato, H., Yamasaki, S., and Okushi, H. (2007) *Diamond Relat. Mater.*, **16**, 796.
28 Suzuki, M., Koizumi, S., Katagiri, M., Ono, T., Sakuma, N., Yoshida, H., Sakai, T., and Uchikoga, S. (2006) *Phys. Status Solidi A*, **203**, 3128.
29 Koizumi, S., Watanabe, K., Hasegawa, M., and Kanda, H. (2001) *Science*, **292**, 1899.

30 Garino, Y., Teraji, T., Lazea, A., and Koizumi, S. (2012) *Diamond Relat. Mater.*, **21**, 33–36.
31 Sekiguchi, T. and Koizumi, S. (2002) *Appl. Phys. Lett.*, **81** (11), 1987–1989.
32 Schubert, E.F. and Kim, J.K. (2005) *Science*, **308**, 1274–1278.
33 Dean, P.J., Lightowlers, E.C., and Wight, D.R. (1965) *Phys. Rev.*, **140**, A352–A368.
34 Kawarada, H., Matsuyama, H., Yokota, Y., Sogi, T., Yamaguchi, A., and Hiraki, A. (1993) *Phys. Rev. B*, **47**, 3633–3637.
35 Watanabe, H., Hayashi, K., Takeuchi, D., Yamanaka, S., Okushi, H., and Kajimura, K. (1998) *Appl. Phys. Lett.*, **73**, 981–983.
36 Nagai, M., Shimano, R., Horiuchi, K., and Gonokami, M.K. (2003) *Phys. Status Solidi B*, **238**, 509–512.
37 Okushi, H., Watanabe, H., and Kanno, S. (2005) *Phys. Status Solidi A*, **202**, 2051–2058.
38 (a) Berman, R. (1979) Part 1 Solid State 1. Thermal properties, in *The Properties of Diamond* (ed. J.E. Field), Academic Press, London; (b) Clark, C.D., Mitchell, E.W.J., and Parsons, B.J. (1979) Part 1 Solid State 2. Colour centers and optical properties, in *The Properties of Diamond* (ed. J.E. Field), Academic Press, London.
39 Lawson, S.C., Kanda, H., Kiyota, H., Tsutsumi, T., and Kawarada, H. (1995) *J. Appl. Phys.*, **77**, 1729–1734.
40 Sternschulte, H., Thonke, K., Sauer, R., and Koizumi, S. (1999) *Phys. Rev. B*, **59**, 12924–12927.
41 Horiuchi, K., Kawamura, A., Ide, T., Ishikura, T., Nakamura, K., and Yamashita, S. (2001) *Jpn. J. Appl. Phys.*, **40**, L275–L278.
42 Horiuchi, K., Kawamura, A., Okazima, Y., and Ide, T. (2004) *New Diamond*, **20**, 6–11. [in Japanese].
43 Garrido, J.A., Nebel, C.E., Stutzmann, M., Gheeraert, E., Casanova, N., and Bustarret, E. (2002) *Phys. Rev. B*, **65**, 165409.
44 Makino, T., Tokuda, N., Kato, H., Ogura, M., Watanabe, H., Ri, S.-G., Yamasaki, S., and Okushi, H. (2006) *Jpn. J. Appl. Phys.*, **45**, L1042–L1044.
45 Makino, T., Kato, H., Ogura, M., Watanabe, H., Ri, S.-G., Yamasaki, S., and Okushi, H. (2005) *Jpn. J. Appl. Phys.*, **44**, L1190–L1192.
46 Makino, T., Ri, S.-G., Tokuda, N., Kato, H., Yamasaki, S., and Okushi, H. (2009) *Diamond Relat. Mater.*, **18**, 764–767.
47 Makino, T., Yoshino, K., Sakai, N., Uchida, K., Koizumi, S., Kato, H., Takeuchi, D., Ogura, M., Oyama, K., Matsumoto, T., Okushi, H., and Yamasaki, S. (2011) *Appl. Phys. Lett.*, **99**, 061110.
48 Kawarada, H., Tsutsumi, T., Hirayama, H., and Yamaguchi, A. (1994) *Appl. Phys. Lett.*, **64**, 451–453.
49 Tachibana, T., Yokota, Y., Hayashi, K., and Kobashi, K. (1999) *J. Electrochem. Soc.*, **146**, 1996–1999.
50 Kadri, M., Araujo, D., Wade, M., Deneuville, A., and Bustarret, E. (2005) *Diamond Relat. Mater.*, **14**, 566–569.
51 Kato, H., Makino, T., Yamasaki, S., and Okushi, H. (2007) *J. Phys. D Appl. Phys.*, **40**, 6189–6200.
52 Makino, T., Tokuda, N., Kato, H., Kanno, S., Yamasaki, S., and Okushi, H. (2008) *Phys. Status Solidi A*, **205**, 2200–2206.
53 Kato, H., Umezawa, H., Tokuda, N., Takeuchi, D., Okushi, H., and Yamasaki, S. (2008) *Appl. Phys. Lett.*, **93**, 202103.
54 Kato, H., Takeuchi, D., Tokuda, N., Umezawa, H., Yamasaki, S., and Okushi, H. (2008) *Phys. Status Solidi A*, **205**, 2195–2199.
55 Sze, S.M. and Ng, K.K. (2007) *Physics of Semiconductor Devices*, 3rd edn, John Wiley & Sons, Inc., Hoboken, NJ.
56 Fujii, A., Takiyama, K., Maki, R., and Fujita, T. (2001) *J. Lumin.*, **94–95**, 355–357.
57 Gheeraert, E., Koizumi, S., Teraji, T., Kanda, H., and Nesladek, M. (1999) *Phys. Status Solidi A*, **174**, 39–51.
58 Makino, T., Tanimoto, S., Hayashi, Y., Kato, H., Tokuda, N., Ogura, M., Takeuchi, D., Oyama, K., Ohashi, H., Okushi, H., and Yamasaki, S. (2009) *Appl. Phys. Lett.*, **94**, 262101.
59 Makino, T., Kato, H., Tokuda, N., Ogura, M., Takeuchi, D., Oyama, K., Tanimoto, S., Okushi, H., and Yamasaki, S. (2010) *Phys. Status Solidi A*, **207**, 2105–2109.
60 Koizumi, S., Ono, T., and Sakai, T. (2006) Extended abstract of the 20th Diamond

Symposium, New Diamond Forum, Tokyo, pp. 262–263. [in Japanese].

61 Koizumi, S. and Kono, S. (2009) Proceedings, 16th International Display Workshop (IDW '09), The Institute of Image Information and Television Engineers and The Society for Information Display, Tokyo, pp. 1479–1480.

62 Takeuchi, D., Makino, T., Ri, S.-G., Tokuda, N., Kato, H., Ogura, M., Okushi, H., and Yamasaki, S. (2008) *Appl. Phys. Express*, **1**, 015004.

63 Takeuchi, D., Makino, T., Kato, H., Ogura, M., Tokuda, N., Oyama, K., Matsumoto, T., Hirabayashi, I., Okushi, H., and Yamasaki, S. (2010) *Appl. Phys. Express*, **3**, 041301.

7
Surface Doping of Diamond and Induced Optical Effects
Vladimira Petrakova, Miroslav Ledvina, and Milos Nesladek

7.1
Introduction

Today, diamond doping—that is, the fabrication of p- and n-type diamond—is a well-established technology. Diamond can be doped by using either boron or phosphorus to produce p- and n-type semiconducting diamond, for both {100} and {111} and even 110 surfaces [1–3], as discussed previously in Chapter 6. Doping over a wide concentration range, from about $10^{-15}\,\text{cm}^{-3}$ (or lower) to above $3 \times 10^{20}\,\text{cm}^{-3}$, that corresponds to a semiconductor–semimetallic transition, can be achieved and used to fabricate p(i)n junctions and other attractive semiconducting devices, as are frequently employed for high-power–high-temperature operations [4–6]. Yet, at the same time doping can provide very interesting possibilities (as discussed later in the chapter) as a powerful tool to control the optical activity of defect centers in ultrapure diamond crystals. Such control can be achieved by engineering the Fermi level position; that is, by changing the defect occupation and/or by manipulating with the quasi Fermi levels in the junction region of semiconducting devices. An example of such a device is shown in Figure 7.1, where a deep-laying defect center as a single-photon, single-spin nitrogen-vacancy (NV) defect can be switched from its negative charged state (NV⁻) to a neutral state (NV⁰) simply by movement of the quasi Fermi levels [7]. Moreover, it can be assumed that, in the injection mode, such devices would allow the control of NV-occupation by electron injection, thus permitting changes in the occupancy and optical activity of defects. Recently, the first of these devices for driving an electric field has been realized by a research group in Japan [8]. An additional function of the donor or acceptor is to provide a free charge as well as nuclear spin baths which may have a strong influence on, for example, the T_2 time.

Fortunately, the principles that are valid for engineering the Fermi level position by doping can be applied equally to band bending at diamond surfaces. For example, the electronic occupation of impurities that are shallow-implanted and form defects with vacancies (such as SiV or NV centers) can be influenced by the electric field. Within this scenario, by changing the quasi Fermi level position, it should be possible to achieve the same effect as for pn junctions (see Figure 7.1).

Optical Engineering of Diamond, First Edition. Edited by Richard P. Mildren and James R. Rabeau.
© 2013 Wiley-VCH Verlag GmbH & Co. KGaA. Published 2013 by Wiley-VCH Verlag GmbH & Co. KGaA.

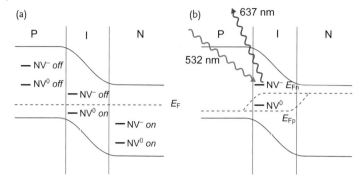

Figure 7.1 Schematics of driving luminescence by the electric injection. E_F, E_{Fn} and E_{Fp} are the Fermi level, and quasi Fermi levels for electrons and holes respectively; NV^0 and NV^- are the neutral and negatively charged NV centers.

For defects located at an approximate depth of 10 nm from the surface, the surface band bending (i.e., the electric field profile at the surface influenced by the surface charges) [7] can affect the occupation of shallow-implanted centers in the surface band bending zone. In this way – and in similar fashion to doping – it is possible to perform manipulations with the color centers occupied, and in fact this technique – known as "surface transfer doping" [9–11] – was used recently to control the occupancy of NV centers in diamond [7, 12]. This, in turn, raises the possibility of driving the luminescence of any NV centers that lie at an approximate depth of 10 nm from the surface [7].

7.2
NV Centers in Diamond

The NV defect is a naturally occurring impurity in diamond, and is responsible for the pink coloration of diamond crystals when present in high concentration (see Figure 7.2). The NV center consists of a nearest-neighbor pair of a substitutional nitrogen atom and a lattice vacancy, and can exist in two charged states: neutral (NV^0) and negative (NV^-), each with different optical properties. The overall structure of the center is an axial C_3V symmetry, whereby a nitrogen atom at the center has three valence electrons bonded to the carbon atoms, while two unbonded valence electrons form a lone pair. Two of the three electrons from the vacancy form an electron pair with a nitrogen lone pair, which results in one remaining unpaired electron. In order to form the NV^- center, the additional electron is required. The NV center has a wide absorption spectrum in the visible range that allows the use of various lasers (from blue to yellow) to excite the emission of the center. NV^0 and NV^- differ in their absorption and emission spectra. The NV^-

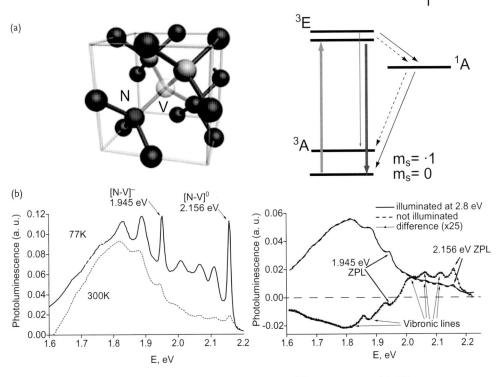

Figure 7.2 (a) Left: Schematic structure; Right: energetic levels of the NV center (right) [13]; (b) Left: Room-temperature PL spectra with and without 2.8 eV illumination; Right: Difference spectrum, illustrating light-induced charge transfer of electrons from NV⁻ to NV⁰ centers [14].

center emits bright red with a zero phonon line (ZPL) at 637 nm; this is followed by a broad side band luminescence with the highest intensity at about 700 nm. This contrasts with the orange luminescence from the NV^0 center, with a ZPL around 575 nm that is also followed by broad side bands. Recently, the application of confocal microscopy techniques has enabled the examination of single NV centers; this, in turn, has led to a series of new experiments involving single narrow optical transitions, coupled to nearby spins and other effects that are difficult or impossible to observe in ensemble studies (such as nonclassical photon correlations).

One interesting fact here is a possibility of a direct charge transfer of these two states, accompanied by changes of the ZPL from 575 nm to 636 nm – that is, from NV^-. Charge transfer has been shown previously to demonstrate manipulation with occupation of the NV center by a monochromatic light absorption of a defined energy; this in turn allowed quenching of the radiative transition, from the ^3E excited state to the ^3A ground-state triplet (see Figure 7.2b).

7.3
Theoretical Considerations of Surface Manipulation with Optical Defects by Transfer Doping

H-terminated diamond surfaces possess a high electric dipole moment which attracts negative polar ions, such as surface adsorbates [9, 15], and this leads to the creation of a hole accumulation layer at the surface for undoped diamond of high purity. As a consequence, there is an upwards surface band bending, which causes the generation of a two-dimensional hole gas (2-DHG) at the diamond surface and a pinning of the Fermi level at the valence band maximum (E_{VBM}) [11, 16]. This effect is referred to as "surface transfer doping" [10], and is used in many devices such as solution gate field-effect transistors (FETs) [17, 18]. A very similar effect t–that is, changes of chemical potential of variously terminated diamond surfaces–can also be employed to modify the surface band bending and consequently to influence the occupation of NV centers lying close to the surface that are present in the band-bending zone. This effect has been clearly documented for single-crystal, polycrystalline, and also nanocrystalline diamond [19, 20]. In order to determine exact band bending profiles, however, it is necessary to solve the Schrodinger–Poisson equation, calculating with N donors and charged NV centers that are present in the diamond and which are influencing the band-bending zone. Based on a solution of the Schrödinger equation using density functional theory (DFT) calculations, NV^0 and NV^- centers have different ground state energies of 1.2 eV and 2.0 eV, respectively (as calculated by Goss et al. [21, 22]). If the Fermi level (E_F) energy is to be lowered below the energy of the NV^- ground state, then this state will lose its electron and become unoccupied; in this way the center will become inactive and cannot then contribute to the luminescence (electrons cannot be excited from this state). However, by controlling the band bending at the diamond surface it is possible to control the NV (or SiV and other) charge states to a certain depth and, consequently, the photoluminescence (PL) spectra. Color shifts in diamond that are affected by surface termination have been reported [23] and were also observed by others [24], although no explanation of their origin was provided. In general, hydrogen termination followed by adsorbate attachment onto an undoped diamond surface will lead to an upwards band bending (i.e., the hole accumulation layer), whereas oxidation will cause the opposite effect, namely a downwards band bending. If any particular defects are located at a distance of approximately 10 nm from such a surface, then surface transfer doping [10, 11] could influence the defect occupancy and, consequently, its luminescence. In this way it might be possible to induce luminescence changes via the relative occupation weights of NV^0/NV^- centers, influenced by the surface charges. This situation is illustrated graphically in Figure 7.3.

7.3.1
Band Bending Model Calculations

The electronic transition (i.e., optical excitation) from NV^- or NV^0 ground states (^3A) to an excited state (^3E) occurs by the absorbance of a photon of energy,

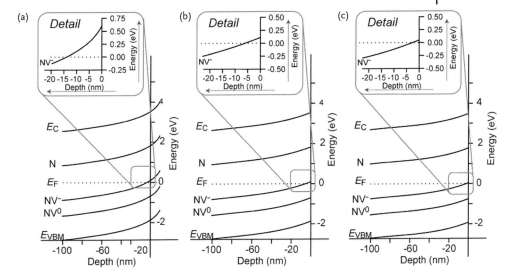

Figure 7.3 Schematic model of surface band bending for hydrogen-terminated diamond containing NV centers with (a) a low content of nitrogen and (b) a high content of nitrogen. Nitrogen in the diamond lattice acts as an electron donor; electrons will compensate the holes accumulated at the surface, effectively reducing the surface band bending and inhibiting the influence of hydrogen termination on photoluminescence. Additional nitrogen influences the position of the Fermi level E_F in the band gap; (c) Oxidized diamond surface with downward band bending, NV^0 and NV^- are occupied. x (nm) and y (nm) are regions at which the Fermi level crosses the NV^- and NV^0 energy—that is, regions where the ground levels of NV^- and NV^0 are not occupied and therefore luminescence cannot occur.

$h\nu = E_{3_E} - E_{3_A}$. This is only possible if the NV^0 or NV^- ground states are originally occupied by a single electron (neutral NV^0) or two electrons (negative NV^-), and this means that E_F must lay above the dark ground level of a particular NV^0 or NV^- for the center [14]. In order to determine exactly the position of E_F at the surface in N implanted diamond, it is essential to take into account the influence of NV^-, NV^0, N defects and the electric field at the surface induced by H-termination and surface adsorbates. All of these effects will contribute to the total charge balance and influence the surface band bending (see Figure 7.3). If the concentration of the donor-like ls (N or NV^-) is increased, then the holes from the accumulation layer at an H-terminated surface can be transferred to deep donors by converting N to N^+, and NV^- to NV^0; the upwards band bending is then consequently reduced, depending on the concentration of adsorbates.

The following model is calculated for N-shallow implanted diamond with energies 2–5 keV, followed by subsequent annealing to generate NV centers. The occupation of NV^0 and NV^- states is calculated from the Boltzmann's statistics (the concentration of NV centers is in the range of 10^{17} cm^{-3}), taking into account the presence of N and NV centers at various concentrations. The electric field profile on an H-terminated surface with surface adsorbates is calculated using the Poisson equation for a dielectric medium, and the energy of the $NV^{0/-}$ centers is determined from DFT calculations by solving the Schrödinger equation [21, 22]. These values for NV^0 and NV^- energy levels (1.2 eV and 2.0 eV, respectively) are relative to the valence band maximum. In this model, N acts as a deep electron donor with an energy level at 1.7 eV below the conduction band [25] (Figure 7.3). It is evident that, in case of doped diamond, ionized impurities such as B^- or P^+ can also be counted on.

In order to study, on a quantitative basis, the occupation of NV^- and NV^0 ground states, the mathematical modeling outlined below has been carried out using Boltzmann statistics and a solution of the Schrödinger equation for NV level energies. The boundary conditions are as follows:

The detail charge balance includes the surface acceptors at distance $x = 0$ with a surface density of $10^{13}\,cm^{-2}$ [26]. For an H-terminated surface on a nonimplanted single-crystal diamond (SCD) with a surface conductivity in the range of $10\,k\Omega$ (as measured in the present case), it can be assumed that the Fermi level is pinned at the surface at E_{VBM} [9].

Thus, it is possible to write for the depth (x)-dependent total space charge density $\rho(x)$:

$$\rho(x) = eN_V \exp\left(-\frac{(E_F - E_V)_x}{kT}\right), \qquad (7.1)$$

Where N_V is the temperature-dependent effective density of states at E_{VBM}, $N_V = 2.7 \times 10^{19}\,cm^{-3}$ at room temperature, e^e is an elementary charge, k is the Boltzmann constant, and T is thermodynamic temperature. E_F and E_V are the energetic levels of Fermi and valence band:

$$\frac{d^2(E_F - E_V)_x}{dx^2} = \frac{e}{\varepsilon_r \varepsilon_0} \rho(x), \qquad (7.2)$$

where $\varepsilon_r \varepsilon_0$ is the permittivity of material and $\rho(x)$ is total space charge density given by the Boltzmann distribution [Equation (7.1)].

The Poisson equation was solved numerically with boundary conditions

$$(E_F - E_V)_0 = kT \ln\left(\frac{N_V}{p_0}\right) \qquad (7.3)$$

$$p_0 = e(N_A - N_D(x)(1-\zeta) - N_D(x)\cdot\zeta\cdot g) \qquad (7.4)$$

where p_0 is the total unscreened positive charge at $x = 0$ from Equation (7.1) and N_A is the effective density of the surface acceptors. The terms $eN_D(x)(1-\zeta)$ and $eN_D(x)\cdot\zeta\cdot g$ represent the net original concentrations of N and NV^- centers respectively in the sample after annealing, while ζ is the conversion efficiency of N to NV centers for thermal annealing, which is different for SCD and for nanodiamond (ND) [27, 28], g is the relative occupation of NV^- and NV^0 centers in bulk diamond, and e is the unitary electron charge.

The concentration of nitrogen $N_D(x)$ is, in the case of implanted SCD, the density of donors (i.e., N and NV$^-$ centers), and is given by the TRIM simulation of the implantation nitrogen (see Figure 7.5). For the case of high-pressure, high-temperature (HPHT) NDs, it is assumed that the nitrogen and NV centers are homogeneously distributed in the sample thickness, and that the irradiation of 5.4 MeV protons is also homogeneous in the volume. Only a part of the implanted nitrogen is converted to NV centers during annealing [13, 28], and therefore the parameter ζ is introduced to the model. This parameter represents a bulk dependent conversion efficiency of N in the lattice to NV centers. It has been shown recently [13] that the parameter ζ depends strongly on the method of implantation, which may also be the reason for the observed difference in the NV PL intensity in both SCD and HPHT ND (as discussed later in the chapter). In these calculations, ζ was set to 0.1 for Type Ib HPHT NDs [28] and to 0.01 for Type IIa implanted and annealed SCDs [13]. The most recent results (see later in this chapter) have shown that an efficacy as high as 2–5% can be achieved for the generation of NV centers. Upon annealing and diffusion of vacancies, it can be estimated that the N in the bulk of a diamond will be converted proportionally to NV^0 and NV$^-$ centers; thus, a parameter g can be introduced which in standard stimulations is set to 0.5 (i.e., the concentration of NV$^-$ and NV^0 centers would be equally produced without any surface bending). The band bending was then calculated using these input parameters.

It is interesting to evaluate electric field distribution not only in flat samples but also in round nanoparticles that have important applications in quantum informatics or nanobiology. In order to calculate the surface band bending in the case of spherical symmetric ND particles, the above equations are transformed to spherical coordinates. The surface charge is assumed to be homogeneously distributed at the surface of the sphere:

$$\frac{d^2(E_F - E_V)_r}{dr^2} = \frac{e}{\varepsilon \varepsilon_0} 4\pi e N_V \exp\left(-\frac{(E_F - E_V)_r}{kT}\right), \tag{7.5}$$

where r is the radius of the sphere, and the boundary conditions are used as previously.

The results of the numerical simulation for the case of the H-termination of diamond containing N and NV centers are shown in Figure 7.4. Electrons are transferred from the valence band of diamond to compensate for a charge induced at the diamond surface by adsorbates, which contribute to an upwards surface band bending. The total electric field profile is then the balance between the surface adsorbates and the deep donors (N, NV$^-$). This calculation yields the value of about 2 eV band bending at $x = 0$ for undoped diamond (i.e., diamond without any NV/N defects), which agrees well with the measured value [17]. However, if additional donors (N, NV$^-$) are available near the surface, then the band bending will be reduced in proportion to the defect or dopant concentrations.

Band-bending calculations, for two model situations – namely, for HPHT NDs and for implanted SCDs – are compared in Figure 7.4. On comparing the modeled data in Figure 7.4 and experimental data in Figures 7.6 and 7.11, it is clear that

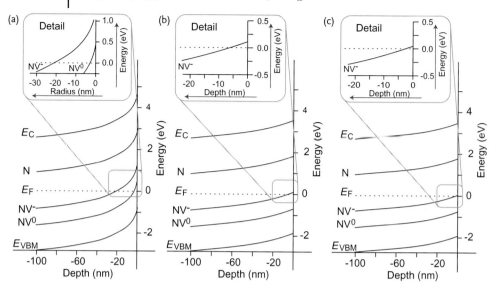

Figure 7.4 Simulation of electron statistics in hydrogen-terminated surfaces using a density of states (DOS) model. Band diagrams show energetic profiles of the conduction band minimum (E_C), valence band maximum (E_{VBM}), NV⁻, NV⁰ and N impurities relative to the Fermi level (at zero energy). (a) Type Ib HPHT diamond contained 200 ppm of nitrogen with a surface carrier density of $10^{13}\,cm^{-3}$, where 10% of nitrogen is converted to NV centers [27, 28]. In this case, the Poisson equation for ND particles is solved using spherical coordinates and diamond is approximated as a spherical particle; (b) Type IIa diamond plate implanted with nitrogen (6 keV, 10^{13} ions cm^{-2}), surface carrier density $10^{13}\,cm^{-3}$ with 1% yield of conversion to NV centers [13]; (c) Type IIa diamond plate implanted with nitrogen (1 keV, 10^{13} ions cm^{-2}), surface carrier density $10^{13}\,cm^{-3}$ with 0.1% yield of conversion to NV centers [13]. Cases (b) and (c) were solved using a one-dimensional model.

the observed effect of a PL shift is much weaker in SCD than in ND. The influence of H/O termination on PL in implanted SC chemical vapor deposition (CVD) diamond has been discussed previously [29], and also more recently [12, 30]. The enhancement of the PL shift in ND is in agreement, at least in theory, with a size-dependent effect in spherical particles, as outlined above.

7.3.2
Implication of Surface Functional Changes on Luminescent Properties of NV Centers Implanted to Different Depths

In order to demonstrate the possibility of luminescence driving by changing the surface charges and, consequently, the occupation of shallow NV centers, ultra-high-purity single-crystal CVD diamond has been used, implanted by N. In this case, four different sectors in a high-purity SCD were shallow implanted with nitrogen, using energies of 1 keV and 6 keV. The NV centers created were localized

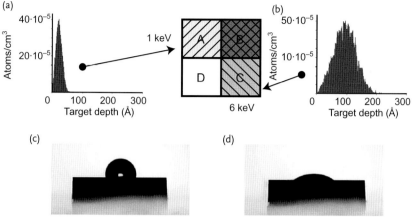

Figure 7.5 Schematic map and TRIM simulation for SCD diamond with four sectors implanted by N⁺ ions. (a, b) Sectors A and B were implanted with the same dose 10^{13} cm^{-2}, but different energies (1 and 6 keV, respectively). Sectors C and D were implanted twice at the same conditions (dose 10^{13} cm^{-2}, energies 1 and 6 keV, respectively). The y-axis scale shows implanted atoms in atoms cm^{-3} per one carbon atom; (c) Contact angle of hydrogenated and (d) oxidized SCD.

from around 3 nm to 10 nm of depth (Figure 7.5), and the surface of the SCD was later oxidized and hydrogenated.

The typical Raman and PL spectra of hydrogenated and oxidized SCDs for implantation energies of 1 and 6 keV are shown in Figure 7.6. Although the NV⁻/NV⁰ ratio changed with the surface termination, the observed changes were rather small; the PL emission from the NV⁻ ZPL was also clearly influenced by the implantation energy. For H-terminated SCD on sectors with 6 keV implantation energy (and a higher surface conductivity), the PL spectra showed a decreased emission from NV⁻ ZPL compared to 1 keV implantation (Figure 7.6a). Subsequently, a mathematical modeling was performed to explain the reduction in the effect when compared to ND.

In parallel, the possibility of influencing the occupation of NV centers chemically was reported by Hauf *et al.* [12] on NV centers implanted to the single crystal diamond, using various implantation energies and fluences (see Figure 7.6b). Recently, the same group described the control of a charge state of the single center by employing an electrolytic gate electrode [9].

7.4
Formation of Variously Charged NV Centers in Diamond

In seeking an ability to control NV centers not only in quantum computing but also for biological applications, the charged state of the NV centers, as well as

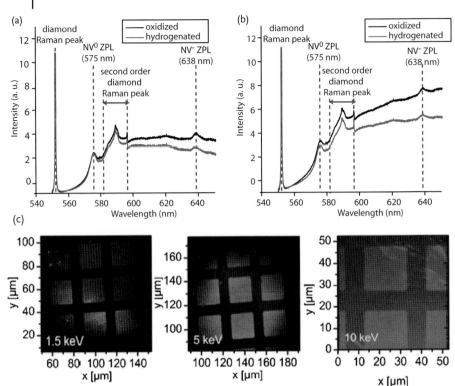

Figure 7.6 (a, b) 514 nm-pumped Raman and PL spectra of treated SCD, showing the influence of different surface termination on NV⁰ and NV⁻ related luminescence. NV⁰ ZPL (575 nm) peaks and NV⁻ ZPL (638 nm) peaks are visible for oxidized SCD. For SCD with hydrogenated surface, NV⁻-related luminescence is decreased with an increase in NV⁰-related luminescence, as shown in the difference spectra. All spectra were normalized to the diamond Raman peak (1332 cm^{-2}) and background-corrected. (a) 1 keV and (b) 6 keV N⁺ ion implantation energy. When normalized to second-order diamond Raman peak, the number of counts (i.e., intensity) of NV-related luminescence was significantly lower for 1 keV implantation energy compared to 6 keV implantation energy. This suggests a different efficiency of creation of NV centers, depending on the implantation energy for the same total dose. This effect is discussed in the text and evaluated by mathematical modeling; (c) Fluorescence of NV defects implanted with three different energies. The dark stripes mark areas where the diamond surface is hydrogenated, while bright areas indicate regions where the surface was oxygen-terminated [12]. For a color version of this figure, please see the color plate at the end of this book.

the number of active NV centers present, is of primary importance. Whereas, single NVs or their arrangement in interacting networks are required for quantum applications, very bright luminescent nanodiamonds are needed for biological applications. In this respect, there is a need to optimize the process of formation of NV centers in order to control not only their numbers but also their occupation. The fabrication methods employed to form NV centers in diamond include three

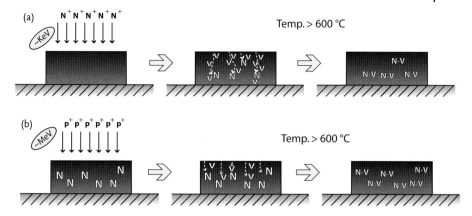

Figure 7.7 Fabrication of NV centers in diamond. (a) Implantation of nitrogen and annealing; (b) Irradiation of Type Ib diamond and annealing.

basic approaches: (i) the implantation of nitrogen and annealing in ultra-high-purity diamond; (ii) the irradiation of Type Ib diamond and annealing; and (iii) the incorporation of NV centers during the process of CVD growth (see Figure 7.7). In the case of N implantation (see Section 7.3.2), CVD growth is effective mainly for Si incorporation, as the incorporation of N (especially high concentrations of N) will normally lead to a deterioration in diamond quality. Consequently, at this point attention will be focused on the second of the above-listed methods.

The irradiation of Type Ib diamond (most commonly HPHT) is the most favored method used for the large-scale production of highly luminescent diamonds, by generating vacancies and recombining them with N. Currently, HPHT nanodiamonds are available commercially at low cost, and in sizes ranging from single-digit NDs to hundreds of nanometers. A further advantage is the natural content of P1 centers (isolated substitutional nitrogen in the diamond lattice), which allows problems such as the low penetration depth in the case of nitrogen implantation ($1\,\mu m\,MeV^{-1}$ [26]) to be overcome with the introduction of nitrogen, although this limitation may prevent large-scale production. High-energy irradiation leads to the creation of vacancies as the ion loses energy to the lattice. In general, electrons and protons produce predominantly isolated vacancies, while neutrons and heavy ions produce regions of multiple damage [31]. Additionally, electrons and neutrons exhibit a high penetration depth (on the order of millimeters), which is in contrast to the rather low penetration depth in the case of protons or other heavily charged particles (orders of micrometers for MeV energies). During annealing, single nitrogen is immobile and vacancies are mobile; however, when the vacancy is trapped to the nitrogen, a thermally stable NV center will be formed. Although the formation of defect centers in diamond by irradiation has been well described on SCDs, no systematic study has yet been conducted on nanodiamond particles. In order to obtain NDs with a high population of NV centers suitable for biolabeling applications, the influence of particle type, particle energy and annealing parameters was investigated.

7.4.1
Introduction of Vacancies to the Diamond Lattice: Different Irradiation Strategies

Accelerators – that is, isochronous cyclotron U-120M and microtron MT 25 – were used for the production of fluorescent nanodiamonds (fNDs). The irradiations were performed both on cyclotron beams (p^+, d^+, $^3He^{2+}$, $^4He^2$) with energies ranging from 5 MeV to 25 MeV, and on a microtron with electron beams of energy ranging from 7 MeV to 23 MeV. The particle density was chosen as approximately 10^{16} cm^{-2}.

Target holders were designed for the purpose of irradiation in the cyclotron, and the electron irradiation was performed using a glass vial as the target. A layer of ND was prepared on each target, while two different irradiation energies were chosen for each type of irradiation particle in the cyclotron (p^+, d^+, $^3He^{2+}$, $^4He^2$); energy deposit with (when the ion beam is stopped in the target) and without a Bragg peak (when the ion beam flies through the target). Two size ranges of NDs were used in the study, with size distributions of 35 and 130 nm. All particles were subsequently annealed at 700 °C for 2 h under an argon atmosphere and oxidized by wet oxidation [32]. The quality of the samples produced was analyzed by their characterization with luminescence and Raman spectroscopy.

The concentration of nitrogen in Type Ib HPHT ND was about 100–200 ppm. For the fluence of 10^{16} cm^{-2}, the maximal vacancy concentration of the atomic displacements was 10^{19} cm^{-3} (corresponding to 100 ppm) [26].

7.4.1.1 Creation of Vacancies: GR1 versus ND1
All irradiated samples used in the study showed clear PL originating from the GR1 center (Figure 7.8). The effect of the various irradiation conditions was studied after annealing, by a comparison of the NV luminescence emission normalized

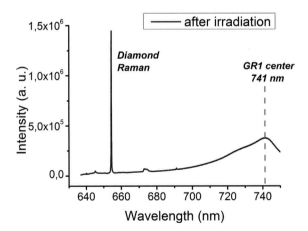

Figure 7.8 Photoluminescence spectra of proton irradiated nanodiamonds *before* annealing (130 nm particles), showing detectable emission from the GR1 center (measured at room temperature, excitation 636 nm).

to the Raman signal. In addition, amorphization of the samples was studied using Raman spectroscopy.

The presence and concentration of GR1 and ND1 centers in natural and synthetic SCDs, produced by radiation damage, has been studied using optical absorption spectroscopy and electron paramagnetic resonance (EPR). Although optical techniques – unlike photoluminescence spectroscopy – can be used to determine quantitatively the amount of defect centers, their use is limited in the case of ND particles (due to high reflection, scattering, low signal). The possible option for confirming the presence of a defect center would be photoluminescence; although this method cannot be used directly to evaluate the number of defects, it is possible to compare quantitatively the *number* of defect centers present when the photoluminescence is normalized to the Raman signal [31].

The irradiation of nitrogen-rich diamond causes the number of created negatively charged vacancies (ND1 center) to greatly exceed the number of neutral vacancies (GR1 center) [33, 34], because substitutional nitrogen acts as an electron donor to the vacancy. If the concentration of nitrogen is higher than a few parts per million, however, the number of ND1 centers will dominate over the GR1 centers [34], but when the concentration of nitrogen is 75 ppm, the GR1 absorption is undetectable [31]. GR1 centers have a characteristic absorption and emission ZPL around 741 nm, whereas ND1 has a characteristic absorption line at 3.15 eV and does not exhibits any PL; rather, it emits an electron.

7.4.1.2 Resulting NV PL of Variously Irradiated NDs

The influence of NV luminescence on irradiation energy has been studied by depositing the ND by impacting it with high-energy particles, and by varying the irradiation geometry, including the Bragg deposition.

In the case of 130 nm particles, the only positive influence of a Bragg peak deposit was in the case of deuteron irradiation, whereas in the case of 30 nm particles the influence of a Bragg peak deposit was negative in all cases. The best luminescent characteristics were obtained by proton beam irradiation with the energy deposit and without a Bragg peak. In this case, the creation efficiency was approximately 10-fold greater in the case of larger particles. When applying Raman spectroscopy, there was a systematic increase in the sp2 content (G band) in samples irradiated with a Bragg peak when compared to samples without s Bragg peak (Figure 7.9).

The evolution of an amorphous structure upon thermal annealing depends critically on the damage density created during irradiation. In regions where the vacancy density is above a critical threshold, the diamond crystalline structure will be permanently converted to an sp^2-bonded phase; however, when the damage density is below the critical threshold, thermal annealing has the effect of converting the amorphous structure back to a crystalline diamond phase, with the residual point defect formed in the crystal [35]. The critical thresholds in previous studies performed on SCDs were set to a fluency of $5 \times 10^{15}\,cm^{-2}$ for the case of helium irradiation [35], to $5 \times 10^{16}\,cm^{-2}$ for the case of proton irradiation [36], and to $2 \times 10^{12}\,cm^{-2}$ for the case of nitrogen implantation to Type IIa diamond [13].

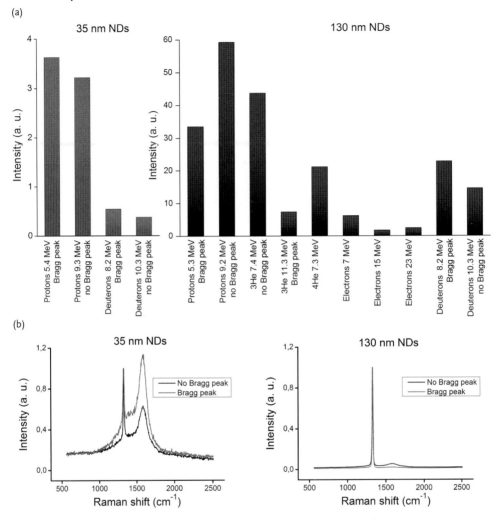

Figure 7.9 (a) Normalized luminescence intensity of NV⁻ center for various irradiation parameters measured on 35 nm ND (gray columns) and 130 nm ND (black columns); (b) Raman spectra of proton-irradiated ND show increased sp^2 content after irradiation with the Bragg peak (gray line) compared to irradiation without the Bragg peak. The damage is much lower in the case of 130 nm NDs.

The presence of a high sp^2 content in the samples, where the high stopping power at the Bragg peak created the maximum concentration of vacancies, showed that the critical threshold for amorphization was reached in these samples. In this respect, it could be predicted that irradiation with lower fluency than in the present case ($9 \times 10^{15}\,\text{cm}^{-2}$) would lead to a higher creation efficiency, especially in the case of smaller ND particles.

7.4.2
Formation of NV Centers: An Annealing Study

In biological studies, the use of very bright luminescent ND particles induces requests for the optimization of NV generation efficiency. The critical step when preparing a luminescent ND is the annealing procedure but, again, no systematic study has yet been performed on ND particles. The most commonly used annealing parameters of ND particles are in the range of 700–800 °C, with an annealing time of approximately 1–2 h [37–39].

The two-dimensional (2-D) matrix of annealing conditions, consisting of various combinations of temperatures and respective times (700–950 °C, 0.5–8 h), are shown in Figure 7.10 for two sizes of Type Ib ND particles of 35 and 130 nm diameter. The NDs were irradiated by protons (ca. 5 MeV), after which the samples were annealed in an argon atmosphere and subsequently oxidized in a mixture of HNO_3 and H_2SO_4 (85 °C, 3 days).

When comparing the data presented in Figure 7.10a and b, a clear difference can be observed in PL that results as a function of annealing conditions for the 35- and 130-nm particles. A longer annealing time will lead to a higher creation efficiency of NV centers in 130 nm NDs for temperatures below 850 °C, whereas the 35 nm particles failed to show such a trend over this temperature range. In fact, annealing for more than 1–2 h did not lead to any further increase in the intensity of the NV luminescence. Likewise, the trend of a moderate increase in NV PL with increasing temperature (up to 900 °C), as observed with 130 nm particles, was not apparent with 35 nm particles, although the latter did show a sharp and discrete maximum in PL, observed at 900 °C. When GR1 and ND1 absorption were monitored as a function of the annealing time, the 130 nm particles were seen to behave similarly to SCD. In this case, the losses in the GR1 and ND1 absorption and the increase in NV absorption continued to mount for up to about 20 h when annealing the irradiated Type Ib diamond sample [40].

It was supposed that the increase in NV PL intensity for longer annealing times in the case of 130 nm NDs was related to the diffusion and charge state of the vacancies. This assumption was based on the fact that, in Type Ib HPHT diamond with a high concentration of nitrogen, mainly ND1 (negatively charged vacancy) centers are formed, yet the capture of vacancies only occurs through the motion of GR1 (neutral vacancy). During annealing, the ND1 must first be converted to GR1, which is mobile [40]. For these reasons, longer annealing times are more suitable for the NV center formation in Type Ib diamond. In the case of 35 nm NDs, the reason for the decrease in NV PL intensity for longer annealing times and higher temperatures could be explained by a trapping of the vacancies to the surface that would be reasonably higher for smaller NDs.

A decrease in the intensity of NV PL after annealing above 900 °C can also be explained by a formation of the NVN center. This proposal is supported by the presence of an NVN (H3) center that was monitored in the small fraction (below 15%) of NDs annealed at temperatures of 900 °C and higher, which was in agreement with studies performed on SCD [31]. On comparing the ratio of NV^-/NV^0

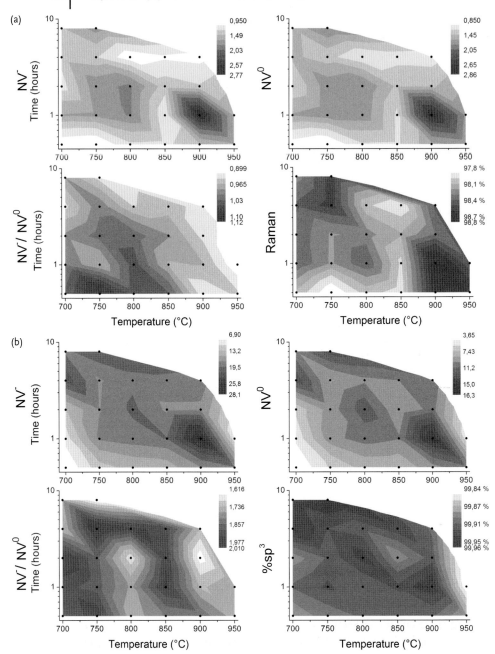

Figure 7.10 Normalized relative luminescence intensity of 35 nm ND (a, upper) and 130 nm ND (b, lower) particles as a function of annealing time and temperature. The black dots represent the matrix of samples, the darker color represents brighter samples. The lower right-hand corner shows the % of sp^3 counted from the Raman spectra.

luminescence in Figure 7.10, the NV⁻ were formed preferentially at lower temperatures and shorter times. This effect was more pronounced with 35 nm NDs, but a similar general trend was also observed for 130 nm NDs.

The NV-induced PL of 35 nm particles is, on average, approximately 10-fold lower than the NV-induced PL of 130 nm particles. When normalized to the Raman signal, however, this corresponds to the difference in the volume fraction between 35 and 130 nm particles. It should be noted that, besides the number of NV centers created, other effects may be involved in the total luminescent yield, such as the quenching effect of the surface, leading to the "dark state" of NV centers; this is of particular importance for small particles. However, as various studies [13, 28] have also employed PL as an evaluation method, the results presented here, which discuss only optically active NV centers, are equally comparable. In the present case, the efficacy of generating NV centers was typically about 2% for the conditions used, while the highest efficiency achieved in these studies was 7%.

Nonetheless, questions continue to arise regarding the presence of a discrete maximum of NV PL observed in samples annealed at 900 °C for 1 h (termed 900/1). In general, the NV PL intensity of these samples was two- and threefold higher for the 130-nm and 35-nm NDs, respectively, in comparison to an average sample. In order to confirm the repeatability of this result, and to rule out any experimental error, the NV PL was monitored on 20 additionally prepared samples (five times each on 900/1 and 700/1 samples for both 130- and 35-nm NDs), and similar results were obtained in all cases.

For very small NDs, the surface can influence the formation efficiency of NV centers if it is charged [13]. This effect is difficult to study because, during the irradiation process, the surface charges can be fully compensated by incoming ion fluxes, or the occupancy of negatively charged vacancies can be changed by them emitting an electron and becoming neutral.

7.5
Transfer Doping Effects: Luminescent Properties of NV Centers in Variously Terminated Nanodiamonds

At this point, a description is provided of how surface chemistry effects can cause the ND bulk to become luminescence-sensitive to on-going chemical processes being conducted at the ND surface. The proposed method is based on the control of electronic chemical potential at the ND surfaces, influencing the surface band bending (as calculated in Section 7.3.1). In that way it is possible to change the occupation of the luminescent NV centers that exist in either neutral (NV^0) or negative charge states (NV^-) [34, 41] with a different PL–ZPL emission wavelength – that is, 575 nm for NV^0 (ZPL) and 637 nm for NV^- (ZPL).

This mechanism is demonstrated on ND with a hydrogenated and oxidized surface that exhibits important differences in the surface chemical potential. The model and ND size effect can be validated by comparing the results with a defect-free, CVD-grown Type IIa SCD with electronic-grade surface polish (<0.1 nm) which was implanted to nanometer depth with nitrogen and subsequently annealed to convert N centers to NV centers by the trapping of vacancies. The electric field penetration-related and corresponding size effects of ND particles on luminescence is modeled mathematically. Thus, it can be shown clearly why the size effect of nanoparticles is important with comparison to bulk single-crystal CVD diamond [12, 29], as this allows a significant increase in the color shift magnitude.

7.5.1
Quenching of NV-Luminescence on ND Particles

The above-described effect of NV luminescence color shift is especially interesting for applications to biology. Particles of 20–50 nm in size, composed of HPHT-synthesized Type Ib diamond are highly suitable for biological imaging applications, due to the possible production of large amounts of stable NV centers (see Section 7.3.5). These centers provide a sufficient luminescence contrast that is comparable to that of other biomarkers, while retaining advantages related to their nanodimensions for cellular tracking (see also Chapter 13). In order to investigate the effect of PL shifts, HPHT ND particles containing approximately 100 ppm of N and which had a peak size distribution of between ~40 nm and 100 nm, were irradiated with protons using an energy of 5.4 MeV to produce vacancies, and subsequently annealed to produce NV centers. Raman spectra recorded at each step of the process confirmed a high quality of ND, similar to SCD, but showed no visible nondiamond sp^2 bonds that could have a negative influence on the surface charge interactions – that is, to reduce the sensitivity of the surface termination to the surface chemical potential, as is the case for detonation ND [42]. These measurements were recorded in a biological buffer solution (pH 7). The experimental data obtained for 40 nm ND for three different situations – oxidized surface, hydrogenated surface, and diamond after subsequent oxidation performed by annealing at 400 °C in air – are shown in Figure 7.11a. According to this model, the NV^- luminescence with ZPL at 638 nm was fully quenched after hydrogen termination, while the NV^0 luminescence remained visible. The NV^- luminescence could be restored again by annealing in air (leading to oxidized surfaces). The PL spectra then recorded at room temperature of H-terminated ND, upon heating in air at gradually elevated temperatures, are shown in Figure 7.11b. A reverse process – that is, a backwards transition from NV^0 luminescence to an NV^--dominated luminescence could be generated on heating (Figure 7.11b). The first change occurred at about 200 °C, a temperature which agreed well with the experimentally observed desorption of adsorbates [43], and confirmed by loss of the 2-DHG surface conductivity (i.e., band bending). At 400 °C, permanent changes in the surface termination occurred, leading to a loss of surface hydrogen due to oxidation, leaving NV^- to dominate the PL spectra.

In addition, a large influence of ND size on the PL intensity of both NV^0 and NV^- defect centers was observed. The size effect for hydrogenated ND is shown

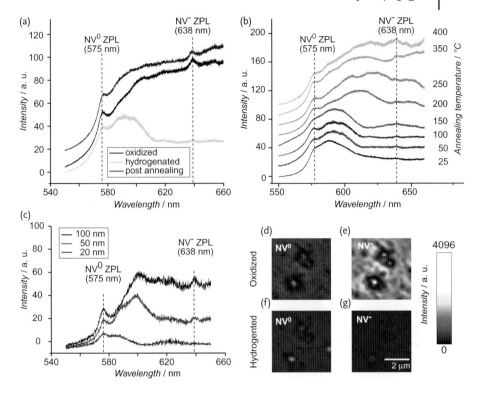

Figure 7.11 (a) Changes in NV⁻ and NV⁰ luminescence induced by various terminations of 40 nm-sized ND particles. Spectra of oxidized, hydrogenated, and annealed surfaces at 400 °C (leading to restoration of original surface termination). The hydrogenation of larger particles (~40 nm) resulted in a luminescence shift towards NV⁰ luminescence (hydrogenated 1), for the smaller particles (<20 nm), the luminescence quenched completely (hydrogenated 2); (b) Luminescence changes of hydrogenated ND (40 nm) upon annealing in air at different temperatures. Samples were heated to the target temperature (using a ramp of 25 °C per min), maintained at the set temperature for 30 min, and then cooled to room temperature. All spectra were recorded at room temperature and baseline-corrected; (c) PL spectra of hydrogenated ND of various sizes (20–100 nm) measured in physiological solution (pH 7), showing the ND size effect on the PL changes in hydrogenated/oxidized diamond and the possibility to monitor PL changes in liquid, which is essential for biosensor usage. The NV⁻ to NV⁰ ratio could be tuned with the size of ND. All spectra were normalized to diamond Raman peak, and the water Raman background was subtracted; (d–g) Confocal microscopy images of single ND particles upon various treatments. For oxidized NDs, both NV⁰ (d) and NV⁻ (e) luminescence is clearly visible in contrast to hydrogenated ND, where NV⁰ luminescence (f) is dominant and NV⁻ luminescence (g) is barely visible. The confocal PL images correspond to luminescence collected from a 570 nm to 610 nm spectral range for NV⁰, and from a 630 nm to 750 nm range for NV⁻. For a color version of this figure, please see the color plate at the end of this book.

in Figure 7.11c, where NDs of a defined size were engineered by an extraction process which employed centrifugation. All NDs used, and containing both NV^- and NV^0 centers, were derived from the same irradiation badge, and the PL intensity was normalized to the Raman line signal for each batch of ND with a defined size. In the case of very small particles, even the NV^0 luminescence could be strongly reduced, whereas for larger particles the NV^-/NV^0 ratio was affected by the ND size, in accordance with the model described in Section 7.3.1.

Biomedical applications of NDs require specific particle sizes, depending on their nature. For example, the endosomatic penetration of fNDs through the cellular membrane is relatively efficient using medium-sized NDs (35–130 nm), but in order for NDs to penetrate the cell nuclei, ultra-small fNDs are required. Clearly, size effects – including luminescent brightness and the NV^0/NV^- luminescence ratio – are especially important for the biological application of very small particles.

7.5.2
Surface and Size Effects in the Variously Terminated Nanodiamond

Given the impact that hydrogen and oxygen termination has on the photoluminescence of NV centers close to the surface, it is to be expected that various possible functional groups on the ND surface – or molecules attached to them – might be useful for tuning the NV^-/NV^0 luminescence ratio. The changes in luminescence are related to modifications in the surface chemistry (and likely also to molecular attachment), such that the on/off switching of luminescence or switching between NV^- and NV^0 could become a practical nanodetection tool in biomedical applications.

7.5.2.1 Fluorinated NDs
The differences in the PL properties of hydrogen- and oxygen-terminated surfaces are the direct consequence of the electron affinity of hydrogenated and oxidized surfaces. A hydrogen-terminated surface has a highly negative electron affinity, and hence the upwards band bending is reached after adsorbate attachment. In contrast, oxidized diamond surfaces possess a positive electron affinity, leading to a downward band bending and stabilization of the luminescence characteristics. The variation in electron affinity is closely connected to the surface electric dipoles that are induced by the electronegativity of functional groups and surface bonds terminating the surface. Therefore, it might be expected that by introducing a highly electronegative element (e.g., fluorine) to the surface, the quenching effect of the surface on the luminescence could be reduced.

In addition to the possible impact of fluorination on luminescence properties, fluorinated nanodiamonds could be also used as a probe for nuclear magnetic resonance (NMR) imaging, and have a major impact on novel nucleic acid grafting strategies. Another application of fluorinated NDs is in the development of the highly hydrophobic NDs required for specific drug-delivery systems. Although specific plasma techniques exist for fluorination, these have mainly been applied to flat CVD diamond plates. Hence, in order to achieve a high surface coverage also on HPHT NDs, an original method was developed based either on the reaction of diamond with gaseous fluorine in the presence of hydrogen. Alternately, a modified reaction of diamond with gaseous fluorine in liquid hydrogen fluoride at elevated temperatures was used. The fluorination reaction scheme is shown in Scheme 7.1.

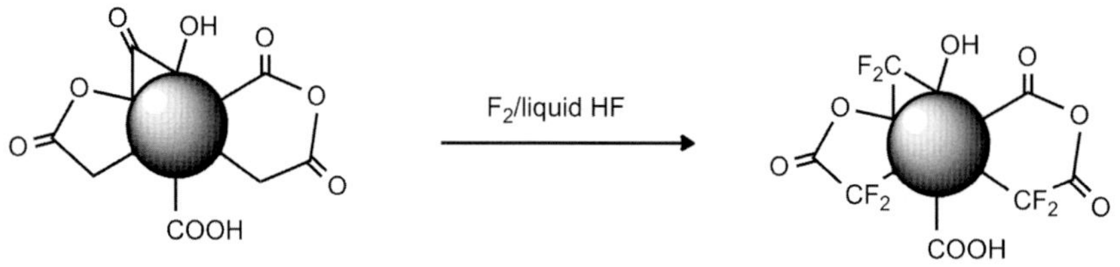

Scheme 7.1 Reaction scheme of fluorination.

The latter procedure resulted in high-surface coverage density, fluorinated HPHT nanodiamonds in which, according to an elementary analysis, the fluorine content was 1.7%, which corresponded to an atomic coverage above 37%. The presence of fluorine was further confirmed with Fourier transform infrared (FTIR) spectroscopy and ^{19}F NMR spectroscopy. In order to address the basic mechanism of the process of fluorination, and to examine the optical properties of the mixed fluorine–hydrogen- and fluorine–oxygen-terminated surfaces, the oxidized and hydrogen plasma-reduced nanodiamonds were used primarily as a source for fluorination. The final density of the fluorine substitution of oxidized ND was doubled (37%) in comparison to that for hydrogenated ND.

7.5.2.2 Size Dependence of NV Luminescence on Variously Terminated ND

The size dependence of NV luminescence (as introduced above by theoretical modeling) was further elaborated for variously terminated NDs. The ND fractions were separated by centrifugation from the original polydispersed ND solutions, creating ND groups of 8–10 nm, 10–20 nm, 20–50 nm, and >100 nm. Special care was taken to prevent the aggregation of NDs in aqueous solution, and this led to stable dispersions of NDs within all size ranges used in the experiment. A strong size dependence of PL properties (Figure 7.12) was identified; in the case of hydrogenated NDs the luminescence of H-terminated NDs was quenched completely for 8–10 nm NDs, the 15–20 nm NDs showed a clear NV^0-related luminescence with no NV^- luminescence, while for larger NDs the NV^-/NV^0 ratio could be precisely tuned. The PL of oxidized NDs was found to be less sensitive to the size of the NDs, but changes in NV^-/NV^0 ratio were also observed, with a favor towards NV^0 luminescence for smaller NDs. The NV^- luminescence remained dominant in all oxidized samples, and the PL of fluorinated NDs was found to be stable over all size ranges.

These results (see Figure 7.12) pointed towards the possibility of a precise tuning of the NV^-/NV^0 ratio, and also of the conversion of NV centers to the dark state. As noted above, the charge state of NV centers depends on the position of the

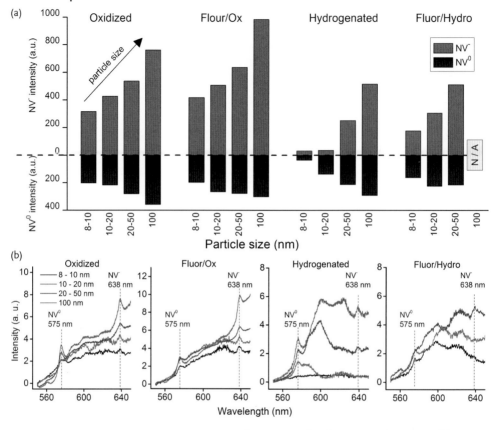

Figure 7.12 Luminescence intensity of variously terminated NDs as a function of size. Fluor/Ox represents NDs that were fluorinated after oxidization, leading to partially oxidized/partially fluorinated surface. Fluor/Hydro represents NDs with hydrogenated surfaces. (a) Peak intensity of the NV- and NV⁰ ZPL; (b) The corresponding photoluminescence spectra. For a color version of this figure, please see the color plate at the end of this book.

Fermi level with respect to the energy level of the defect that defines whether it either takes up or loses an electron. If the Fermi level is shifted above the ground state energy of the defect, an electron will be taken up by the defect, and this results in a charge switching-off the defect center. The results of modeling the NV⁻/NV⁰ ratio for the case of spherical ND particles as a function of the surface band bending are shown in Figure 7.13 (the ND sizes used in the experiment are marked in figure). For particle sizes of ~8 nm or less, a dark state can be created, as shown clearly in Figure 7.13.

A different situation occurs for fluorinated NDs, however, where it can be seen clearly that fluorinated NDs do not show any quenching effect on the NV⁰ or NV⁻ luminescence, even for small particles. This effect can be explained by the

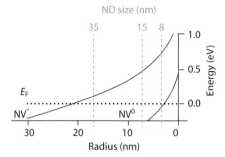

Figure 7.13 Energy levels of NV⁰ and NV-centers in hydrogenated spherical ND particles, showing the relative position to the Fermi level as a function of radius, modeled as described above. The solid lines represent the mean size distributions of NDs used in the experiment.

downward band bending – that is, fixing the E_F below the NV⁰ states. Also, when comparing fluorinated ND with oxidized ND, the ND that is partially covered with fluorine exhibits a higher NV⁻ luminescence. On average, the intensity of NV⁻ emission of fluorinated ND was 20% more intense than that of oxidized ND. The technique described in Scheme 7.1 allow the fluorination of both H-terminated or oxidized NDs. The fluorination of hydrogenated NDs led to a surface that was partially covered by fluorine and partially by hydrogen. In contrast, the maximum surface coverage by fluorine for originally H-terminated NDs was 23%, which allowed for a tuning of the surface electron affinity [41, 44] and could also be used to tune the NV⁻/NV⁰ PL ratio for small-sized NDs. Such electron affinity tuning provides the possibility of developing ND sensors that are sensitive to both positive and negative charges, and this aspect of the detection principle is addressed in the following section.

7.5.3
Fluorescent Nanodiamond Particles as a Sensor of Charged Molecules

The switching between charge state occupation of NV centers (i.e., NV⁰ or NV⁻) is related to the position of the Fermi level with respect to the energy of particular ground state, and can be enhanced by charged molecules or particles close to the diamond surface. An example of this is a charged polymer attached to a diamond surface where, if the charged molecule is strongly attracted to the surface, a charge transfer can occur depending on the chemical potential of the highest occupied molecular orbital (HOMO) and lowest unoccupied molecular orbital (LUMO) energetic levels of the charged molecule and the diamond surface electronegativity.

7.5.3.1 Interactions with Charged Polymers
The principles of charge-interaction of molecules in close proximity to the diamond can be verified by the detection of charged molecules (polymers) that are bound

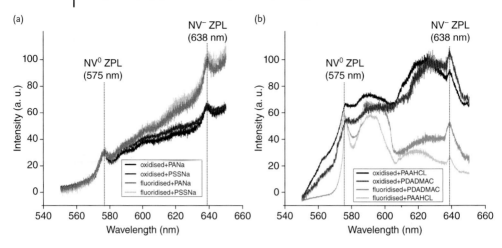

Figure 7.14 Photoluminsecence spectra showing the interaction of oxidized and fluorinated NDs with (a) negatively charged and (b) positively charged polymers. A clear decrease in NV⁻ luminescence was observed for F-fND after the attachment of positively charged polymers.

noncovalently to the ND surface. Four different types of polymer were selected for this experiment: polydiallyldimethyl ammonium chloride (PDADMAC) and polyallylamine hydrochloride (PAA.HCl) were chosen as positively charged molecules, while polyacrylic acid sodium salt (PANa) and polystyrene sulfonic acid sodium salt (PSSNa) were used as negatively charged molecules. As the polymer size did not exceed the size of ND, this favored the creation of stronger bonds. The colloidal solution of ND was mixed with polymers in 10-fold higher concentrations, which enabled a saturation of the surface with polymers.

The fluorinated NDs were found to be the most sensitive to the charged interactions, and showed a significant decrease in NV⁻ PL following the addition of positively charged polymers (Figure 7.14). The changes in the PL spectra of fluorinated and oxidized NDs, upon interaction with two different types of charged polymer, are shown in Figures 7.14 and 7.15. The luminescence of the NV-centers was clearly decreased upon interaction with positively charged molecules whereas, following the addition of negatively charged polymers, the luminescence was restored to the original level.

7.5.4
NDs for Biology: Monitoring of Transfection by ND Carriers

The possibility exists of using ND particles as active biological sensors, based on the luminescence changes induced by transfer doping. Typically, ND particles in the size range of 5 to 100 nm can serve as a new type of optical marker for cellular imaging [37–39, 45]. The luminescence emitted from NV centers is extremely

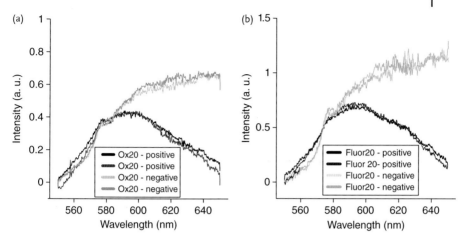

Figure 7.15 Photoluminsecence spectra showing interaction of (a) oxidized and (b) fluorinated NDs with negatively and positively charged polymers. A clear decrease in NV⁻ luminescence was observed for oxidized and fluorinated ND after the attachment of positively charged polymers. All spectra were normalized, and the background water and polymer spectra were subtracted.

stable without any photobleaching or photoblinking [14, 46, 47], and compared to the better known quantum dots, NDs bring about additional advantages such as a high biocompatibility [48, 49] and simple carbon-surface chemistry [50, 51]. This allows the grafting of biomolecules that are of interest in cellular targeting [52, 53] or biomolecular drug delivery [54–56]. Unfortunately, in the case of very small ND particles (5 nm) the blinking of NV centers was observed [57], which showed that the surface effects are important for the stabilization of NV luminescence properties (see Section 7.4.3).

Another highlight in the application of NDs to medicine is their use as drug-delivery systems. Indeed, recent studies conducted by Dean Ho and colleagues have shown that when doxorubicin was bound to detonation NDs, the toxicity of the complexes created was dramatically reduced compared to unmodified doxorubicin [58]. Recently, the same group also demonstrated the ability of NDs to deliver therapeutic nucleic acids, whereby polyethylenimine (PEI) -coated NDs showed an increased efficiency of siRNA transfection compared to plain PEI of the same molecular weight, while remaining biocompatible [59].

Based on the above-described data, the surface-charge induced driving of luminescence could allow NDs to be used as a tool for the optical tracking of drug-delivery events. The PEI used above [59] is a positively charged polymer; thus, the use of fluorinated NDs and the possible charge transfer from ionically bound polymers to NDs would allow the development of optical sensors for the subsequent monitoring of DNA attachment/release. This sensor may be based on the simple fact that DNA or RNA molecules are strongly negatively charged, and their use in combination with the PEI (positively charged) would lead to a reversible charge switching upon the attachment/release of oligonucleic acids.

The reaction pathway for the attachment of cationic PEI as a transfection agent is shown in Scheme 7.2, where the positively charged PEI is seen to condense the (negatively charged) DNA into a complex that can be taken up by the cell via endocytosis. Although, in general, PEI is increasingly cytotoxic in line with its molecular weight, the use of small NDs enables the transfection mechanism to be more effective whilst at the same time promoting the use of less-cytotoxic PEI of a lower molecular weight. The formation of PEI–ND complexes is based on the electrostatic attraction of positively charged PEI and the oxidized or fluorinated ND surface. Such polycationic NDs have been used successfully for the grafting of oligonucleotides (see Scheme 7.2).

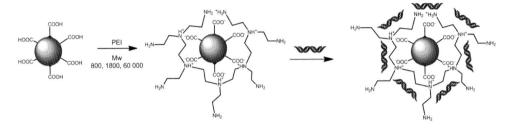

Scheme 7.2 Achieved noncovalent DNA grafting used for the design of fND–polymer biosensors operating contactless in cells.

The prepared PEI- and DNA-coated NDs were used with the expectation that an alternation of differently charged analytes would result in changes of the NV^0/NV^- PL ratio. These experiments were performed in a cell medium and, as expected, the presence of positively charged PEI led to a decrease or quenching of NV^- PL. The subsequent addition of negatively charged DNA compensated the positive charge, and this resulted in the enhancement or restoration of NV^- luminescence. A strong increase in luminescence after DNA attachment on PEI-coated ND compared to ND with PEI was apparent (as shown by the blue and red curves, respectively in Figure 7.16). Ultimately, this principle may be applied to the detection of DNA and RNA in cells.

The successful use of NDs as a detection tool for DNA was demonstrated for fNDs ranging from 8 to 100 nm in size. Ultimately, this proved to be a significant advantage over the Förster resonance energy transfer (FRET) technique, for which only very small ND particles (5–8 nm or smaller) are effective. With regards to detection by standard confocal microscopy (as frequently used in biological research), it is much easier to detect larger ND particles as they contain more NV centers. For FRET-based detection, the NV concentration must be enhanced, or special techniques such as single particle detection systems must be employed.

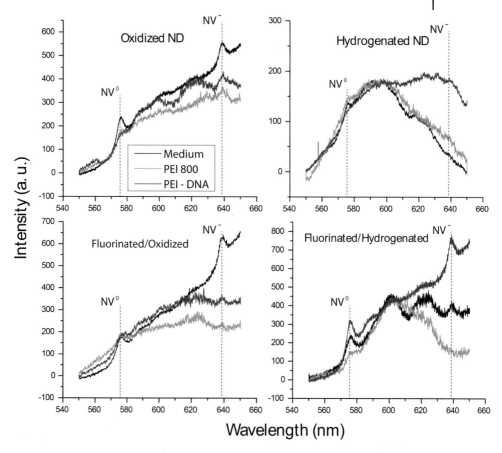

Figure 7.16 Photoluminsecence spectra of fluorescent NDs with different surface terminations. Large differences (e.g., the NV-fluorescence of the fluorinated/ hydrogenated nanodiamonds coated with PEI) could be used to monitor the interaction of the particles with strongly negatively charged biopolymers, such as DNA inside cells. For a color version of this figure, please see the color plate at the end of this book.

7.6
Conclusions

In this chapter, attention was focused on the luminescence control of optical defects in diamond, possibly by manipulating their occupation via influencing the position of the Fermi level relative to the defect transition level. Such optical engineering is important not only for quantum computing applications but also for applications in nano-biophotonics. Two types of optical defect control have been discussed: (i) control by bulk diamond doping; or (ii) by the use of "transfer doping." In the latter case, the surface charges can induce an electric field in the diamond subsurface, which in turn would lead to a specific field profile (i.e., band bending) at close range (~10 nm) to the diamond surface.

As a further point of discussion, the NV center in diamond is regarded as the "workhorse" of quantum computing physics, and also for nanodiamond applications in medicine. Also discussed were the charge transfer properties that occur at the NV center and lead to either neutral or negatively charged NV states (NV^0 and NV^-) that can be controllably engineered. The preparation of NV doped ND particles of 8 to 130 nm in size, as well as NV centers engineered in ultrapure N-implanted SCD, has also been demonstrated. In addition, the luminescent properties of shallow implanted diamonds to control surface termination, including hydrogen, oxygen and fluorine surface functional groups, have been demonstrated. Luminescence detected in variously terminated NDs shows distinct changes between NV^0 and NV^- occupation by shifts in the Fermi level, while transfer doping, which involves charges in close proximity to the diamond surface, can be used to control and modify the original NV^-/NV^0 ratio. The latter effect has been modeled mathematically for diamond nanoparticles and plates. In diamond nanoparticles, it has been established that PL enhancement can be observed due to nanometric size effects, as well as to an enhanced surface of the nanoparticle compared to flat, 2-D substrates. Such enhancement was modeled by solving the Poisson equation in 3-D coordinates and showing that, for ND particles, the surface-enhanced luminescence changes would be effective up to a particle size of ~40 nm. For very small particles (≤8 nm), both modeling and experimental data have shown that NV^0 luminescence can be significantly reduced due to size effects (i.e., when the particle size is comparable to the depletion zone at the diamond surface). In this case, the NV center enters a zone where the Fermi level is above the ground state of the original NV^0 center, and this dark state is expected to be positively charged. The controllable switching of luminescence brings about new possibilities for molecular imaging using NDs, that complement the presently used FRET technique. It has also been shown that this effect can be used as a nanosensor for monitoring the charged environment of ND particles in liquids, as clear luminescent shifts are observed when NDs interact with charged polymers. Finally, this effect has been used to monitor DNA release in transfection experiments performed on PEI-modified ND for nanomedical applications or real-time imaging in cells.

References

1 Nesladek, M. (2005) *Semicond Sci. Technol.*, **20** (2), R19–R27.
2 Koizumi, S., Watanabe, K., Hasegawa, M., et al. (2001) *Science*, **292** (5523), 1899–1901.
3 Haenen, K., Lazea, A., Barjon, J., et al. (2009) *J. Phys. C*, **21** (36), 364204.
4 Makino, T., Tanimoto, S., Kato, H., et al. (2009) *Phys. Status Solidi A*, **206** (9), 2086–2090.
5 Makino, T., Tanimoto, S., Hayashi, Y., et al. (2009) *Appl. Phys. Lett.*, **94** (26), 262101.

6 Kato, H., Makino, T., Ogura, M., et al. (2009) *Appl. Phys. Exp.*, **2** (5), 055502.
7 Petrakova, V., Taylor, A., Kratochvilova, I., Fendrych, F., Vacik, J., Kucka, J., Stursa, J., Cigler, P., Ledvina, M., Fiserova, A., Kneppo, P., and Nesladek, M. (2012) *Adv. Funct. Mater.*, **22** (4), 812–819.
8 Mizuochi, M., Makino, T., Kato, H., et al. (2012) *Nat. Photonics*, **6** (5), 299–303.
9 Maier, F., Riedel, M., Mantel, B., Ristein, J., and Ley, L. (2000) *Phys. Rev. Lett.*, **16** (85), 3472–3475.
10 Strobel, P., Riedel, M., Ristein, J., and Ley, L. (2004) *Nature*, **430** (6998), 439–441.
11 Nebel, C.E., Sauerer, C., Ertl, F., Stutzmann, M., Graeff, C.F.O., Bergonzo, P., Williams, O.A., and Jackman, R. (2001) *Appl. Phys. Lett.*, **79** (27), 4541–4543.
12 Hauf, M.V., Grotz, B., Naydenov, B., Dankerl, M., Pezzagna, S., Meijer, J., Jelezko, F., Wrachtrup, J., Stutzmann, M., Reinhard, F., and Garrido, J.A. (2011) *Phys. Rev. B*, **8**, 83.
13 Pazzagna, S., Naydenov, B., Jelezko, F., Wrachtrup, J., and Meijer, J. (2010) *New J. Phys.*, **12**, 065017.
14 Iakoubovskii, K., Adriaenssens, G.J., and Nesladek, M. (2000) *J. Phys. Condens. Matter*, **12**, 189–199.
15 Ristein, J., Riedel, M., and Ley, L. (2004) *J. Electrochem. Soc.*, **151** (10), 315–321.
16 Nebel, C.E. (2005) *New Diam. Front. Carbon Technol.*, **15** (5), 247–264.
17 Rezek, B., Shin, D., Watanabe, H., and Nebel, C.E. (2007) *Sens. Actuators B Chem.*, **122**, 596–599.
18 Garrido, J.A., Hardl, A., Kuch, S., Stutzmann, M., Williams, O.A., and Jackmann, R.B. (2005) *Appl. Phys. Lett.*, **86**, 073504.
19 Hartl, A., Schmich, E., Garrido, J.A., Hernando, J., Catharino, S.C.R., Walter, S., Feulner, P., Kromka, A., Steinmuller, D., and Stutzmann, M. (2004) *Nat. Mater.*, **3**, 736–742.
20 Rezek, B., Kratka, M., Kromka, A., and Kalbacova, M. (2010) *Biosens. Bioelectron.*, **26**, 1307–1312.
21 Goss, J.P., Briddon, P.R., Jones, R., and Sque, S. (2004) *Diamond Relat. Mater.*, **13**, 351–354.
22 Goss, J.P., Briddon, P.R., Jones, R., Sque, S., and Rayson, M.J. (2005) *Phys. Rev. B*, **72**, 035214.
23 Kratochvilova, I., Taylor, A., Kovalenko, A., Fendrych, F., Rezacova, V., Petrak, V., Zalis, S., Sebera, J., and Nesladek, M. (2010) *Mater. Res. Soc. Symp. Proc.*, **1203**, J03–J05.
24 Fu, K.M.C., Santori, C., Barclay, P.E., and Beausoleil, R.G. (2010) *Appl. Phys. Lett.*, **96**, 121907.
25 Ristein, J., Riedel, M., Stammler, M., Mantel, B.F., and Ley, L. (2002) *Diamond Relat. Mater.*, **11**, 359–364.
26 Zeigler, J. (2008) SRIM-2008. Available at: http://srim.org (accessed 13 August 2011).
27 Boudou, J.P., Curmi, P.A., Jelezko, F., Wrachtrup, J., Aubert, P., Sennour, M., Balasubramanian, G., Reuter, R., Thorel, A., and Gaffet, E. (2009) *Nanotechnology*, **20** (23), 5602–5613.
28 Santori, C., Barclay, P.E., Fu, K.C., and Beausoleil, R.G. (2009) *Phys. Rev. B*, **79** (12), 5313–5321.
29 Nesladek, M., Rezacova, V., Kovalenko, A., Kratochvilova, I., and Kocka, V. (2010) EMRS Spring Meeting, Strasbourg. Available at: http://www.emrs-strasbourg.com/files/USB%2010/symposium_d.pdf (accessed 5 May 2011).
30 Grotz, B., Hauf, M.V., Dankerl, M., Naydenov, B., Pezzagna, S., Meijer, J., Jelezko, F., Wrachtrup, J., Stutzmann, M., Reinhard, F., and Garrido, J.A. (2012) *Nat. Commun.*, **3**, 729.
31 Collins, A.T. (2007) *New Diam. Front. Carbon Technol.*, **17**, 47–61.
32 Cigler, P., Ledvina, M., Tvrdonova, M., Rezacova, V., Nesladek, M., Kratochvilova, I., Fendrych, F., Stursa, J., Kucka, J., and Ralis, J. (2010) Abstracts, 2nd International Conference, Conference proceedings, Nanocon 2010 Olomouc, Czech Republic, October 12–14, 2010.
33 Davies, G., Lawson, S.C., Collins, A.T., et al. (1992) *Phys. Rev. B*, **46** (20), 13157–13170.
34 Lawson, S.C., Fisher, D., Hunt, D.C., et al. (1998) *J. Phys. Condens. Matter*, **10** (27), 6171–6180.

35 Waldermann, F.C., Olivero, P., Nunn, J., et al. (2007) *Diamond Relat. Matter.*, **16** (11), 1887–1895.

36 Botsoa, J., Sauvage, T., Adam, M.P., et al. (2011) *Phys. Rev. B.*, **84** (12), 125209.

37 Fu, C.C., Lee, H.Y., Chen, K., Lim, T.S., Wu, H.Y., Lin, P.K., Wei, P.K., Tsao, P.H., Chang, H.C., and Fann, W. (2007) *Proc. Natl Acad. Sci. USA*, **104** (3), 727–732.

38 Chang, Y.R., Lee, H.Y., Chen, K., Chang, C.C., Tsai, D.S., Fu, C.C., Lim, T.S., Tzeng, Y.K., Fang, C.Y., Han, C.C., Chang, H.C., and Fann, W. (2008) *Nat. Nanotechnol.*, **3**, 284–288.

39 Liu, K.K., Cheng, C.L., Chang, C.C., and Chao, J.I. (2007) *Nanotechnology*, **18**, 325102.

40 Davies, G. and Collins, A.T. (1993) *Diamond Relat. Matter.*, **2** (2–4), 80–86.

41 Davies, G. and Hamer, M.F. (1976) *Proc. R. Soc. Lond. A.*, **348**, 285–298.

42 Williams, O.A., Hees, J., Dieker, C., Jager, W., Kirste, L., and Nebel, C.E. (2010) *ACS Nano*, **4** (8), 4824–4830.

43 Nesladek, M., Stals, L.M., Stesmans, A., Iakoubovskij, K., and Adriaenssens, G.J. (1998) *Appl. Phys. Lett.*, **72** (25), 3306–3308.

44 Tiwari, A.T., Goss, J.P., Briddon, P.R., et al. (2011) *Phys. Rev. B*, **84** (24), 245305.

45 Ho, D. (2009) *ACS Nano*, **3** (12), 3825–3829.

46 Tisler, J., Balasubramanian, G., Naydenov, B., Kolesov, R., Grotz, B., Reuter, R., Boudou, J.B., Curmi, P.A., Sennour, M., Thorel, A., Burscht, M., Aulenbacher, K., Erdmann, R., Hemmer, P.R., Jelezko, F., and Wrachtrup, J. (2009) *ACS Nano*, **3** (7), 1959–1965.

47 Gaebel, T., Domhan, M., Wittmann, C., Popa, I., Jelezko, F., Rabeau, J., Greentree, A., Prawer, S., Trajkov, E., Hemmer, P.R., and Wrachtrup, J. (2006) *Appl. Phys. B.*, **82**, 243–246.

48 Yu, S.J., Kang, M.W., Chang, H.C., Chen, K.M., and Yu, Y.C. (2005) *J. Am. Chem. Soc.*, **127**, 17604–17605.

49 Yuan, Y., Chen, Y., Liu, Y.H., Wang, H., and Liu, Y. (2009) *Diamond Relat. Mater.*, **18**, 95–100.

50 Liang, Y., Ozawa, M., and Krueger, A. (2009) *ACS Nano*, **3** (8), 2288–2296.

51 Zhang, X.Q., Chen, M., Lam, R., Xu, X., Osawa, E., and Ho, D. (2009) *ACS Nano*, **3** (9), 2609–2616.

52 Faklaris, O., Joshi, V., Irinopoulou, T., Tauc, P., Sennour, M., Girard, H., Gesset, C., Arnault, J.C., Thorel, A., Boudou, J.B., Curmi, P.A., and Treussart, F. (2009) *ACS Nano*, **3** (12), 3955–3962.

53 Huang, H., Pierstorff, E., Osawa, E., and Ho, D. (2008) *ACS Nano*, **2** (2), 203–212.

54 Martn, R., Lvaro, M., Herance, J.R., and Garca, H. (2010) *ACS Nano*, **4** (1), 65–74.

55 Chen, M., Pierstorff, E.D., Lam, R., Li, S.Y., Huang, H., Osawa, E., and Ho, D. (2009) *ACS Nano*, **3** (7), 2016–2022.

56 Liu, K.K., Zheng, W.W., Wang, C.C., Chiu, Y.C., Cheng, C.L., Lo, Y.S., Chen, C., and Chao, J.I. (2010) *Nanotechnology*, **21**, 315106.

57 Bradac, C., Gaebel, T., Naidoo, N., Sellars, M.J., Twamley, J., Brown, L.J., Barnard, A.S., Plakhotnik, T., Zvyagin, A.V., and Rabeau, J.R. (2010) *Nat. Nanotechnol.*, **5**, 345–349.

58 Chow, E.K., Zhang, X.-Q., Chen, M., Lam, R., Robinson, E., Huang, H., Schaffer, D., Osawa, E., Goga, A., and Ho, D. (2011) *Sci. Transl. Med.*, **3**, 73ra21.

59 Chen, M., Zhang, X.-Q., Man, H.B., Lam, R., Chow, E.K., and Ho, D. (2010) *J. Phys. Chem. Lett.*, **1** (21), 3167–3171.

8
Diamond Raman Laser Design and Performance

Richard P. Mildren, Alexander Sabella, Ondrej Kitzler, David J. Spence, and Aaron M. McKay

8.1
Introduction and Background

Diamond has already been demonstrated as a forerunner in Raman crystal performance despite only being taken up in earnest during the past five years. The research so far has largely concentrated on designs developed since the 1970s based on metal oxide crystals operating in the visible and near-infrared spectral regions. Currently, the highest efficiency and output power for several of the most prominent Raman laser configurations are obtained using diamond. There is now an intense interest in exploiting diamond's clear advantages in many categories of optical and physical properties, most notably transmission bandwidth and thermal conductivity, with the aim of creating devices with capabilities well beyond the limits of those readily achievable using other materials.

Raman lasers build on mature pump laser technologies to generate output with wavelengths not easily generated by other means. Using the phenomenon of stimulated Raman scattering (SRS), materials with a high Raman cross-section are used to step down the frequency of the pump laser beam by multiples of the characteristic Raman frequency of the material. In practice, this is achieved either by inserting the Raman material within the laser resonator (along with an appropriate modification of the optics), or by converting the pump laser output in an external Raman laser. Developments in this area have most often been driven by wavelength-specific applications that require wavelengths not provided by the Nd or Yb-doped workhorses of solid-state laser technology near 1 μm and their harmonics, or near 2 μm using Tm-doped laser materials. As summarized in Figure 8.1, the most-often targeted wavelength regions are between 1.2 and 1.5 μm, where there is demand from defense and aerospace applications (e.g., laser radar and remote sensing), and between 550 and 700 nm, where there are important applications in medicine (e.g., laser treatments of the retina and skin).

The state-of-the-art of crystal Raman laser design has developed from a large catalog of knowledge that dates back most notably to the pulsed laser designs reported as early as 1975 (refer to the timeline of Figure 8.2a). Between 1975

Optical Engineering of Diamond, First Edition. Edited by Richard P. Mildren and James R. Rabeau.
© 2013 Wiley-VCH Verlag GmbH & Co. KGaA. Published 2013 by Wiley-VCH Verlag GmbH & Co. KGaA.

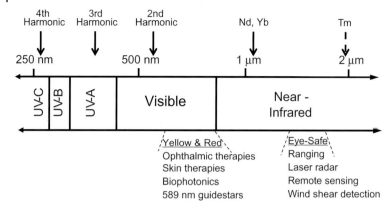

Figure 8.1 The major target wavelength regions for crystalline Raman laser materials are traditionally in the eye-safe region near 1.5 μm and in the yellow-to-red spectrum. The 2nd, 3rd and 4th harmonics of the fundamental 1064 nm laser line of Nd are also indicated.

and 1980, Ammann produced a series of articles [1–3] on the crystal lithium iodate which first introduced many of the major concepts that underpin device design today. By placing the Raman crystal inside the resonator of a Nd laser, Ammann demonstrated Stokes wavelength generation at 1.2 and 1.3 μm with output powers in excess of 1 W [1, 2]. In a further significant advance, Ammann demonstrated discretely tunable lasers in the visible range by utilizing lithium iodate's high $\chi^{(2)}$ nonlinearity to generate harmonic combinations of the cascaded Stokes orders oscillating in the cavity. Here, the harmonic output wavelength was selected according to the phase-matching conditions in the angle-tuned crystal (see Figure 8.2b), and visible average powers up to 0.7 W were achieved [3].

These pioneering studies were followed during the 1980s by demonstrations of alkali or alkali-earth nitrate Raman lasers ($NaNO_3$ and $Ba(NO_3)_2$) [4]–materials which have a high Raman gain coefficient very similar to diamond. With long lengths of these materials available (5–10 cm), high conversion efficiencies (with typically more than half of the pump photons being converted to Stokes photons) were readily achieved using a Raman resonator pumped by a separate laser (the so-called external cavity Raman laser). Although Raman lasers based on glass fibers doped with germanates and phosphates were first described at about this time, this development is mentioned here only in passing, as such devices have many properties distinct from their crystal counterparts.

During the past two decades, transition metal oxide materials – such as metal tungstates, vanadates, and molybdates – have proven to be excellent Raman materials providing a good compromise between gain and high damage resistance, thus alleviating some of the earlier problems encountered with material hydroscopicity and low laser damage thresholds. Subsequently, the field of crystalline Raman lasers has grown enormously as these crystals have become increasingly available,

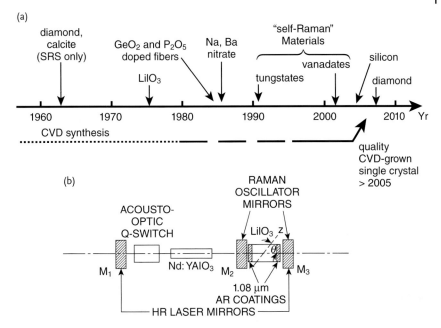

Figure 8.2 (a) Timeline for Raman laser material development and diamond synthesis by chemical vapor deposition (CVD). CVD growth of diamond was first demonstrated during the 1950s and progressed, notably after 1980, as methods to increase growth rates were developed. The timings shown indicate the approximate start of major research activity, although in several cases there were important earlier works; (b) The experimental layout from 1975 of the first intracavity Raman lasers based on lithium iodate. Output powers of approximately 1 W at kHz repetition rates were demonstrated for infrared [2] and visible wavelengths [3] using this arrangement. Part (b) reproduced with permission from Ref. [1]; © 1975, American Institute of Physics.

predominantly from suppliers in Eastern Europe, Russia, and China. Today, devices have been developed with pulsed energies and average powers up to approximately 100 mJ and 10 W, respectively [5–9].

Two of the most important recent developments have been the emergence of Raman crystals doped with active laser ions (e.g., Nd-doped transition metal oxides), and the improved fidelity of coatings to allow low peak power or continuous-wave (CW) operation. These two advances – when used in combination – have enabled the design of intracavity Raman lasers to be simplified and output powers to be raised [9]. Ammann's wavelength-selectable concepts have been applied and extended to enable recent demonstrations of visible lasers at average output powers of several watts in both pulsed [10] and CW [11] formats.

Although diamond was the first solid material in which SRS was observed (as early as 1963 [12]), serious efforts to develop it as a Raman laser material have only been made during the past five years [13–23], due mainly to the availability of large, high-quality synthetic single crystals. To date, synthesis by chemical vapor deposition (CVD) has been the favored source of laser-grade diamond, as it is able to yield low-loss crystals (i.e., low scatter, absorption and depolarization) more reproducibly than from natural sources or other synthesis techniques (see Chapter 2). This major achievement of CVD-enabled diamond growth follows a long history of attempts to synthesize diamond that date back to ancient times, and has emerged some 50 years after the first reports of homoepitaxial CVD growth and of isolated crystals using the high-pressure high-temperature (HPHT) technique. The quality attained through this method has opened up opportunities for new and improved optical devices across many fields in addition to Raman lasers.

The incorporation of diamond into well-established Raman laser designs has already revealed the material's suitability for addressing the aforementioned yellow and NIR applications (this will be discussed in the following subsections). The extremely robust properties and high Raman gain of diamond represent clear advantages, while access to slightly different wavelengths is provided due to its large Raman shift (1332 cm^{-1} compared to 750–1050 cm^{-1} for most other materials). More importantly, diamond portends extensions in performance that are substantially beyond the reach of other materials, enabling laser designers to tackle more challenging problems in optical source design. There are several pertinent optical, thermal and physical properties of diamond that are clearly outstanding, and, correspondingly, good prospects for diamond Raman laser technology to be highly disruptive.

The aim of this chapter is to review diamond laser design and to highlight some of the interesting directions for development. Following an introduction to the basic principles of Raman laser design and theory, a summary will be provided of diamond's properties pertinent to Raman laser design, and compared against other commonly used Raman laser crystals. The performance of diamond Raman lasers in conventional regimes of operation will then be reviewed. Finally, more recent achievements in extending Raman laser performance outside the usual remit of crystalline Raman lasers will be discussed, and the key challenges highlighted. The three main areas discussed, as depicted in Figure 8.3, relate to investigations into laser performance that utilize diamond's wide transmission range and excellent thermal properties to generate wavelengths longer than 1.5 μm, wavelengths shorter than the visible region, and at average powers above 10 W.

8.1.1
Crystalline Raman Laser Principles

Crystalline Raman lasers rely on the stimulated scattering of pump laser photons from optical phonons (or polaritons) excited in the lattice, creating optical gain at the beating Stokes difference frequency ω_s. (Note that in this chapter the special sum-frequency case of anti-Stokes scattering is not considered.) The Raman scattering cross-section is typically small, and intense pumps (typically $>10^6$ W cm^{-2}) are generally required to obtain efficient conversion. While the conservation of momentum dictates that the phonon wavevector equals the difference between the

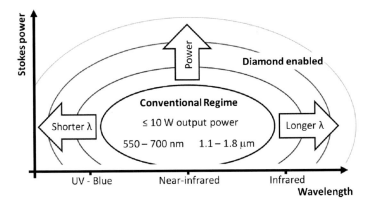

Figure 8.3 Major directions for diamond Raman laser development.

pump and the Stokes wavevectors, the wavevector magnitude is much smaller than the Brillouin zone (i.e., the phonon wavelength is much greater than the inter-atomic spacing), and there is generally always a phonon mode available with a matching wavevector at the Raman frequency. Thus, the scattering probability is independent of the pump and Stokes propagation directions, and SRS is considered to be automatically phase-matched. Optical phonons dephase in the material-characteristic time T_2, which is approximately 10^{-11} s in crystals, and ultimately decay to deposit heat in the crystal.

8.1.1.1 Basic SRS Theory

The theory of Raman lasers has undergone steady development since the early 1960s, and many different models have been applied to laser designs to predict parameters such as cavity losses and Raman gains from experimental data. The modeling approach required depends on the regime of SRS that applies. If the pump field or generated Stokes field has an amplitude or phase structure on timescales shorter than the dephasing time of the Raman transition, then the most general theory of transient Raman scattering must be considered, which requires modeling of the complex amplitudes of the pump, Stokes, and phonon fields. A general form of the complex envelopes of the optical and phonon fields, simplified from the wave equation using the slowly varying envelope approximations, and with several other generally well-satisfied assumptions [24, 25], can be written as:

$$\frac{\partial E_s}{\partial x} - \frac{1}{v_s}\frac{\partial E_s}{\partial t} = i\frac{g_s \omega_s}{n_s} E_F Q'^* \tag{8.1}$$

$$\frac{\partial E_F}{\partial x} - \frac{1}{v_F}\frac{\partial E_F}{\partial t} = i\frac{g_s \omega_F}{n_F} E_S Q' \tag{8.2}$$

$$\frac{\partial Q'}{\partial t} + \frac{1}{T_2} Q' = i\frac{c n_s n_F \varepsilon_0}{4\omega_s T_2} E_F E_S^* \tag{8.3}$$

in which $E(x,t)$ is the complex envelope of the field amplitude as a function of time and space, ω is the angular frequency, v is the group velocity, n is the refractive index, ε_0 is the permittivity of free space, and c is the speed of light. The F and S subscripts refer to the fundamental and Stokes fields respectively; s is the steady-state Raman gain coefficient, and $Q'(x,t)$ is proportional to the amplitude of the coherent phonon field. The coherently driven phonon field dephases with time T_2, or conversely takes a time T_2 to reach its maximum amplitude when driven by coherent pump and Stokes fields. It should be noted that power can flow either into or out of the Stokes field, depending on the relative phases of the fundamental and phonon fields. These basic equations can be generalized to include higher-order Stokes fields if needed.

Provided that the Stokes and pump fields vary in amplitude and phase slowly on the timescale of T_2, the assumption can be made that the phonon field is always in equilibrium with those fields; hence, Q' can be eliminated from Equations (8.1) and (8.2) to retrieve the standard steady-state relations for the field intensities:

$$\frac{\partial I_S}{\partial x} - \frac{1}{v_S}\frac{\partial I_S}{\partial t} = g_S I_F I_S \tag{8.4}$$

$$\frac{\partial I_F}{\partial x} - \frac{1}{v_F}\frac{\partial I_F}{\partial t} = -g_S \frac{\omega_F}{\omega_S} I_S I_F \tag{8.5}$$

for which the Stokes intensity sees an exponential gain, and there is no requirement for phase matching between the fundamental and Stokes fields. This steady-state approximation is accurate for pulses with duration or coherence length $\tau > 10T_2$.

These equations can be used to predict the behavior of a wide range of Raman interactions. For simple situations – for example, with a negligible depletion of the fundamental beam – the equations may be solved analytically, but for more complex situations they must be solved numerically. Finite difference methods can be used to convert the above time- and space-dependent partial differential equations into a set of coupled time-dependent ordinary differential equations for the values of the variables at each point on a spatial grid. Standard ordinary differential equation solvers can then be used to predict the time-behavior of a pulsed laser, or to determine the steady-state behavior of a CW system.

Compared to conventional optically pumped inversion lasers, certain key similarities and differences exist that are instructive:

1) The optical gain in the Raman medium is proportional to the product of the *pump intensity* and the material's characteristic *Raman gain coefficient*. The analogous factors for inversion lasers are the inversion density and the stimulated emission coefficient. Clearly, there is no energy storage in the Raman medium, and as a result there is no analog to Q-switching at the Stokes wavelength in the manner widely employed in inversion lasers.

2) As a consequence of the fact that SRS is automatically phase-matched, energy can be efficiently cascaded into multiple Stokes orders. For crystals, the Raman frequency is typically about 1000 cm^{-1}, which corresponds to approxi-

mately 30- to 40-nm steps in the visible range, and twice that in the NIR. This cascading property of SRS enables wavelength versatility by an appropriate selection of the output Stokes order. The generation of anti-Stokes wavelengths is also feasible (see e.g., Ref. [26]); however, in this case phase-matching is required to achieve gain, and high efficiencies are a challenge.

3) Gain relies on the coherent build-up of phonons in the Raman material, and is diminished for pumps of low temporal coherence (i.e., shorter than approximately T_2). This holds for crystals where dispersion is significant, but may be relaxed in gases and under special conditions. This is in contrast to inversion lasers where the maximum pump linewidth that can be efficiently used is determined by the absorption width of the inversion laser gain media. Pump coherence is an important consideration when using broadband pump lasers and ultrashort pump pulses.

4) The Raman scattering cross-section is typically small, and in general intense pumps are required (typically $>10^6\,\text{W}\,\text{cm}^{-2}$).

8.1.1.2 Raman Laser Design and Modeling

Designs take a variety of forms, with varying degrees of complexity. The simplest is the external cavity Raman laser, where the crystal is placed in a Fabry–Pérot resonator and pumped using a separate laser unit. Several examples of this configuration employing diamond have been reported for visible [13, 14] and infrared laser operation [17, 19, 21, 23]. Modeling of these types of systems [27, 28] is straightforward, with boundary conditions for the resonated Stokes fields determined by the cavity mirror reflectivities. Generally pumped by nanosecond or longer pulses, transient Raman theory is not required. Such a model can be used to identify the time behavior of a pulsed system by introducing a pulsed fundamental field; alternatively, the model can be run for a CW input fundamental field until a steady-state CW Stokes field behavior is determined.

Intracavity Raman lasers, in which the Raman crystal is placed inside the resonator of the pump, are typically more complicated in design due to the additional requirements on coatings to oscillate multiple wavelengths in the single resonator. When scaling to significant output powers, the interaction of thermal lenses in the Raman and fundamental gain media need to be managed to maintain beam quality and efficiency. However, an advantage of this form is that a lower threshold can be obtained due to the ability to resonate both the fundamental and Stokes fields within the Raman crystal. Researchers at the University of Strathclyde have used this configuration with much success to demonstrate a range of pulsed and CW devices with output powers up to approximately 5 W [15, 16, 20, 22, 29] (see Section 8.3.2). Modeling of these systems is more complex [30–33], as the laser population inversion must be modeled as well as its pumping mechanism. For low-gain lasers of this type, a standard technique can be used to distribute the laser gain, Raman gain, and losses throughout the cavity; cavity photon densities can then simply be modeled for the fundamental and Stokes fields, neglecting all spatial dependences and thus greatly simplifying the calculations [34].

The generation of pulses shorter than the transit time of the Raman crystal (i.e., less than approximately 10^{-10} s) requires mode-locking methods to be used, as is normally the case for conventional inversion lasers. Diamond Raman lasers with a pulse duration of the order of 20 ps and at megahertz repetition rates have been demonstrated using synchronous pumping of the diamond crystal placed in a resonator with a round trip time matching the interpulse period [18, 35] (see Sections 8.3.3 and 8.4.2). These lasers must be modeled using the full complex amplitude equations, because the picosecond or femtosecond pump pulses are short compared to T_2 [35–37]. Solving the equations in a reference frame that moves with the Stokes group velocity has been used to avoid numerical dispersion that would otherwise artificially lengthen the resonated Stokes pulse [35, 36].

8.2
Optical, Thermal, and Physical Properties of Diamond

Diamond consists of two interpenetrating face-centered cubic lattices of carbon atoms displaced by one quarter of the cube diagonal, and with symmetry properties in the International Symbol (Hermann–Mauguin) system class Fd$\bar{3}$m. The unit cell contains eight atoms in the cube of side dimension 0.36 nm, and is the most atomically dense of all solids. The symmetry and strong covalent bonding of the lattice constrains the optical phonon spectrum to frequencies greater than 1000 cm^{-1}. Indeed, the Debye temperature (2200 K), which marks the temperature for which the occupation number of optical phonons becomes significant, greatly exceeds that of most other insulating materials. Diamond's extremely low thermal resistance and low expansion coefficient are consequences of the high Debye temperature. The first-order Raman spectrum contains only the single triply-degenerate feature at 1332.3 cm^{-1}, corresponding to displacements between the two interpenetrating cubic lattices along the linking diagonal bond. The linewidth at room temperature is approximately 1.5 cm^{-1}, corresponding to $T_2 = 7.1$ ps (see Chapter 1). Symmetry thwarts the first-order lattice absorption that normally provides the long-wavelength transmission limit of most materials. As a result, diamond transmits at all wavelengths with energy less than the bandgap energy, with an exception in the 3- to 6-µm region, where moderate absorption (<10 cm^{-1}) prevails as a result of allowed higher-order multi-phonon processes. Diamond has a wide indirect bandgap of 5.47 eV providing transmission down to 225 nm. Moreover, being a cubic lattice the linear optical properties are isotropic.

Some of the optical and physical properties of importance to Raman laser design and performance are summarized in Table 8.1, along with representatives of the common Raman crystal classes of alkali-earth nitrates, alkali iodates, single-metal tungstates, double-metal tungstates, and metal vanadates. Silicon – the Group IV sibling of diamond – has received a great deal of attention due to its promise for on-chip optical processing, and its details are also included here. At first glance, the outstanding features of diamond are its high thermal conductivity, low thermal expansion coefficient, wide transmission range, and large Raman shift. Diamond's

8.2 Optical, Thermal, and Physical Properties of Diamond

Material	Diamond	Silicon	LiIO$_3$	Ba(NO$_3$)$_2$	KGd(WO$_4$)$_2$	BaWO$_4$	YVO$_4$
Crystal/optical class	Cubic/isotropic	Cubic/isotropic	Hexagonal/−uniaxial	Cubic/isotropic	Monoclinic/biaxial	Tetragonal/+uniaxial	Tetragonal/+uniaxial
Raman shift	1332.3	521	770, 822	1047.3	768, 901	924	890
Raman linewidth FWHM (cm^{-1})	1.5	1.24	5.0	0.4	7.8, 5.9	1.6	3.0
T_2 (ps)[a]	7	8.5	2.1	26	1.4, 1.7	6.6	3.5
Transmission range (μm)	>0.23[c]	>1.1[d]	0.31–4	0.35–1.8	0.34–5.5	0.26–5	0.4–5
dn/dT (10^{-6} K^{-1})	15	215	−95 −80	−20	−4.3–5.5[e]	<−9	3 (a) 8.5 (c)
Thermal conductivity (W mK^{-1})	2000	153	4	1.17	2.6 (a) 3.8 (b) 3.4 (c)	3	5.2
Thermal expansion (×10^{-6} K^{-1})	1.1	3	28 (a) 48 (c)	18.2	4.0 (a) 1.6 (b) 8.5 (c)[e]	4–6[a] 23[c]	4.43
Heat capacity (J g K^{-1})	0.52	0.75	0.58	0.6		0.31	
Refractive index n at 1064 nm	2.41	3.42	1.85 (o) 1.72 (e)	1.56	1.98 (p) 2.01 (m) 2.06 (g)	1.84	1.96 2.17
Density (g cm^{-3})	3.52	2.33	4.5	3.25	7.3	6.4	4.23
Raman gain coeff. @1064 nm	10–12	(20 @1550 nm)[b]	4.8	11	3.5	8.5	4.5
Growth method	CVD or HPHT	Czochralski or Float Zone	Aqueous	Aqueous	Flux or Czochralski	Czochralski	Czochralski
Useful references	Refer to Chapters 1 and 2	[38]	[39]	[40, 41]	[42]	[43–45]	[46]

a) The phonon-dephasing time T_2 and the linewidth $\Delta \nu$ are related through $T_2 = (\pi c \Delta \nu)^{-1}$.
b) 1064-nm radiation is absorbed at room temperature due indirect bandgap transitions.
c) There is moderate multi-phonon absorption in the range 2.5–6 μm.
d) There is moderate multi-phonon absorption in the range 6–25 μm.
e) There is a considerable range of reported values (see also Ref. [47]).

potential as an important Raman laser material depends upon many other variables, including its susceptibility for undesirable thermal effects, optical damage threshold, absorption, scatter, as well practical aspects such as manufacturability with robust optical coatings. Some of diamond's more important material properties for Raman laser design are discussed in the following subsections.

8.2.1
Raman Gain Coefficient

The high Raman gain coefficient in diamond is a fundamental advantage for Raman laser design, as it provides lower thresholds, relaxed constraints on the performance of optical coatings, and a greater freedom to use shorter crystal lengths. The steady-state gain coefficient g_s depends on the material deformation potential (usually characterized by the change in polarizability α with the oscillator displacement q) and the phonon dephasing time T_2 according to:

$$g_s = \frac{4\pi^2 \omega_S N T_2}{n_S n_F c^2 M \omega_R} \left(\frac{d\alpha}{dq}\right)^2 \tag{8.6}$$

where N is the density of scatterers, M is the reduced mass of the oscillating partners and ω_R is the angular frequency of the Raman shift. The g_s of diamond is high as a consequence of its closely packed lattice (high N) of relatively light scatterers (low M) and the three-fold degeneracy of the Raman mode (enhanced $d\alpha/dq$). Indeed, it is often considered to have the highest gain coefficient of all solids (however, as discussed below, this may not always be the case), and is sometimes used as a standard against which to benchmark other materials. The T_2 value for diamond is one of the longest at approximately 7 ps (refer Table 8.1) but still notably shorter than barium nitrate ($T_2 = 26$ ps), which enjoys an unusually low cross-talk between the phonon oscillator and the surrounding lattice.

A study of approximately 30 crystals [41] has ranked diamond as having the highest steady-state gain coefficient at the probe wavelength 488 nm. However, when comparing absolute coefficients it is important to consider that accurate measurements are invariably difficult to obtain. Techniques based on spontaneous scattering require careful photometric measurements and a good knowledge of the phonon dephasing time. Measurements based on the SRS threshold require special attention to surface losses and the spatio-temporal properties of the pump laser beam. A detailed comparison of gain measurements from the laser threshold is also often fraught by inconsistency in the definition used for the threshold, and in important experimental parameters such as crystal orientation, pump or Stokes beam linewidth, and the effects of backward Raman scattering. With these considerations, uncertainties in absolute values are often at least several tens of per cent.

Raman gain increases in proportion to the Stokes frequency, with added resonant contributions as the material bandgap is approached (leading to enhanced $d\alpha/dq$). According to the bandgap and nature of the electronic band structure, the wavelength region and magnitude of the enhancement for crystals can differ

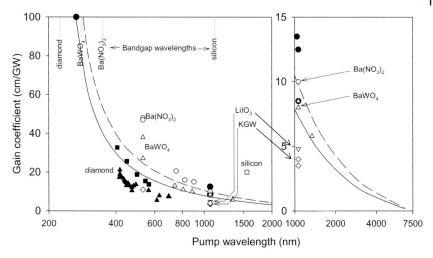

Figure 8.4 Wavelength dependence of selected gain coefficients for diamond (g_s<110>; solid symbols), barium nitrate [50], barium tungstate [51, 52], silicon [38] and potassium gadolinium tungstate (KGW) [42, 43]. The solid squares are values obtained from fits to the nonlinear dispersion of CARS around the Raman frequency [53]. The solid triangles show the coefficients determined from spontaneous scattering [54, 55] and using T_2 = 7 ps, except for the point at 694 nm which was determined in Ref. [12] using T_2 = 5.2 ps (2.04 cm^{-1}). The solid circles show coefficients obtained from measurements of SRS threshold [19, 35, 56] and our own pump-probe measured value of 8.5 cm/GW at 1064 nm. The solid line is the calculated g_s(<110>) using the $d\alpha/dq$ wavelength dependence of Ref. [55], as described in Chapter 1. The dashed line is g_s(<111>) and $1^1/_3$ times the solid line.

greatly. Indeed, very high near-bandgap values have been reported for silicon at 1064 nm (e.g., 190 cm GW^{-1} and at temperature 77 K [48, 49]). As a result, single wavelength gain measurements are not sufficient to characterize gain across the spectrum. A summary of reported diamond gain coefficients as a function of wavelength, alongside other prominent crystalline Raman laser materials, is shown in Figure 8.4 [50–56]. The diamond data (solid symbols) include values derived from spontaneous scattering (triangles) as well as the SRS threshold (circles). It should be noted that the data correspond to the gain coefficients for the pump polarization parallel to the <100> and <110> directions, denoted here as g_s(<100>) and g_s(<110>). Gain coefficients one-third higher can be obtained for g_s(<111>) [17]. As shown by the dashed line in Figure 8.4 (and as discussed in Section 8.2.1.1), this orientation provides the maximum gain coefficient for any orientation or polarization combination of Stokes and input beams, and has output polarization parallel to the pump field. Also included is a relation for the wavelength dependence for diamond, based on the semi-empirical relationship for $d\alpha/dq(\lambda)$ derived by Cardona et al. [55] (see also Section 1.6 of Chapter 1). Wavelength dependences for barium nitrate and barium tungstate have also been determined using a more empirical relationship, though this is not expected to be valid outside the measured range of 532–1064 nm [50, 51].

At the common pump wavelength of 1064 nm, reported values for g_s(<110>) for diamond range from approximately 8 to 13.5 cm GW^{-1} (see Figure 8.4), which are similar to (or perhaps marginally higher than) those of barium nitrate and tungstate, and approximately twice that for potassium gadolinium tungstate (KGW) and lithium iodate. The empirical wavelength dependence for Raman gain given by Cardona et al. [55], which appears to be generally well supported by measurements spanning the ultraviolet (UV) and infrared (IR) ranges, suggests that g_s(<111>) for diamond increases from approximately 10 cm GW^{-1} at 1064 nm to 20 cm GW^{-1} at 532 nm, 50 cm GW^{-1} at 355 nm, and 130 cm GW^{-1} at 266 nm. The smaller bandgaps for barium nitrate and barium tungstate appear to provide a significant gain enhancement for wavelengths less than 800 nm, and coefficients in the visible wavelength range are notably larger than those predicted for diamond. For example, at 532 nm the pump-probe gain measurements of Lisinetskii et al. [50] suggest that g_s for barium nitrate is 4.8 times higher than at 1064 nm. In contrast, the enhancement factor for diamond appears to be only two- to threefold over the same wavelength range. These results suggest that diamond's coefficient is lower than that of barium nitrate in the visible range (by about half); this is in contrast with the results of an earlier comparative crystal study performed by Basiev et al. [41], which suggested that the diamond coefficient was 1.59 times higher than that of barium nitrate at 488 nm. Clearly, there remains significant uncertainty in the absolute and relative gain coefficient measurements in the visible range. Apart from the values at 1064 nm, IR measurements are scarce. The formula provided in Ref. [55] suggests that the gain coefficient decreases to 5 cm GW^{-1} at 2 μm, and to approximately 2 cm GW^{-1} at 3.5 μm. Silicon appears to be the only material likely to have a higher gain coefficient at wavelengths shorter than approximately 2 μm.

8.2.1.1 Dependence on Polarization

The polarizability tensors, or equivalently Raman scattering tensors, describe the direction and magnitude of the induced polarization as a function of the incident polarization. The measured Raman scattering intensity is proportional to the sum of the contributions from each of the scattering matrices.

$$I_s \propto \left|e_s^T R_1 e_p\right|^2 + \left|e_s^T R_2 e_p\right|^2 + \left|e_s^T R_3 e_p\right|^2 \tag{8.7}$$

where e_s and e_p and e_s are vectors describing the input polarization and analyzer orientation, respectively. For the first-order diamond Raman modes, the tensors in the <100> coordinate system are represented by three scattering matrices:

$$R_1 = \begin{bmatrix} 0 & d & 0 \\ d & 0 & 0 \\ 0 & 0 & 0 \end{bmatrix} R_2 = \begin{bmatrix} 0 & 0 & 0 \\ 0 & 0 & d \\ 0 & d & 0 \end{bmatrix} R_3 = \begin{bmatrix} 0 & 0 & d \\ 0 & 0 & 0 \\ d & 0 & 0 \end{bmatrix} \tag{8.8}$$

where $d = d\alpha/dq$. (Refer to Section 1.6 in Chapter 1 for other useful crystal orientations.) The Raman gain coefficient follows the inversion of the above relationship, with e_p and e_s oriented with the pump and Stokes polarization directions of the laser [17]. The net effect of the three Raman scattering tensors can be formed into a Mueller matrix that allows a straightforward calculation of the polarization effects in Raman scattering [57]. The dependence of the Raman gain coefficient on the pump and Stokes polarization "Stokes" vectors P and S is obtained using the relationship $S = MP$, where

$$S = \begin{bmatrix} S_1 \\ S_2 \\ S_3 \\ S_4 \end{bmatrix} \quad P = \begin{bmatrix} P_1 \\ P_2 \\ P_3 \\ P_4 \end{bmatrix} \tag{8.9}$$

and the Mueller matrices M for the two commonly used propagation directions [57]

$$M_{<100>} = \begin{bmatrix} 1 & 0 & 0 & 0 \\ 0 & -1 & 0 & 0 \\ 0 & 0 & 1 & 0 \\ 0 & 0 & 0 & -1 \end{bmatrix} \quad M_{<110>} = \begin{bmatrix} 1.5 & 0.5 & 0 & 0 \\ 0.5 & -0.5 & 0 & 0 \\ 0 & 0 & 1 & 0 \\ 0 & 0 & 0 & -1 \end{bmatrix} \tag{8.10}$$

Several parameters can be derived from S, including the intensity of the polarized I_P and unpolarized I_U components of the scattered light:

$$I_P = \sqrt{S_2^2 + S_3^2 + S_4^2}; \quad I_U = S_1 - I_P. \tag{8.11}$$

Changes in the Raman laser threshold are proportional to the calculated changes in total intensity scattered into the laser polarization state. It is important to include the contribution from unpolarized light, half of which can map onto a given polarization. As shown below, it is found that $g_s \propto I_P + I_U/2$ with the angle of the output Stokes polarization given by

$$\theta_s = \frac{1}{2}\arg(S_2 + i \cdot S_3) \tag{8.12}$$

The maximum value $g_s(\theta_s)$ for a linearly polarized pump of any angle as a function of the pump direction is shown in Figure 8.5a. The maximum is obtained for pump directions parallel to <110> axes, or along directions on a straight line between the <110> axes. Compared to propagation directions along the <111> or <100> axes, the maximum gain is enhanced by a factor of $1^1/_3$.

The measured and calculated Raman laser threshold and polarization changes for a linearly polarized pump beam directed along a <110> axis are compared in Figure 8.5b. There are several points of interest here; the highest gain coefficient is obtained for pump polarizations aligned with a <111> axis of the crystal, which is the same direction as C–C bonds in the lattice and the induced vibrations. This result is also apparent in Figure 8.5a, where the directions of highest gain are normal to a <111> axis. In this case, the output Stokes polarization is parallel to the pump. For pump polarization aligned with the <100> and <110> axes, the gain coefficient is reduced and the Stokes output is no longer necessarily linearly

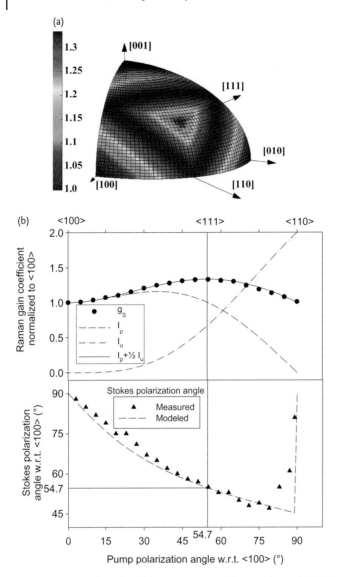

Figure 8.5 (a) Surface plot of the maximum Raman gain coefficient as a function of the propagation direction; (b) Measured polarization dependence of the gain coefficient g_S for propagation along <110> axes (upper) and the Stokes polarization (lower). The angle 54.7° corresponds to pump and Stokes alignment along the <111> direction. For a color version of this figure, please see the color plate at the end of this book.

polarized in the same direction as the input. For pump polarization aligned with <100>, the spontaneously scattered Stokes along the axis has polarization perpendicular to the pump, while for pump polarization aligned with <110> the spontaneously scattered Stokes is unpolarized. When operated as a Raman laser without polarization selective elements, the output polarization is found to fluctuate randomly from pulse to pulse as expected, since each laser pulse builds up from noise. Brewster-cut crystals for <110> polarization have been used [18, 35] to enable the Raman laser to stably run linearly polarized, without any loss in efficiency.

If the pump polarization state is generalized to include elliptically polarized cases, then several additional orientations with similarly high gain coefficients exist. Experimental measurements again have confirmed that the $g_s \propto I_P + I_U/2$ assumption still applies. As with the linearly polarized case, the highest Raman gain coefficients are achieved when the pump polarization matches the resultant Stokes polarization.

8.2.1.2 Dependence on Pump Linewidth

As SRS gain relies on the coherent build up of phonons, it is reasonable to expect that gain depends on the phase and frequency spread of the pump. This issue has been discussed in many experimental and theoretical reports, mostly in the context of gases, where it has been shown that gain is largely independent of pump linewidth up to limits set by dispersion (e.g., Refs [58, 59]). In crystals, however, the group velocity dispersion is much greater and the gain is expected to be much more sensitive to pump linewidth; in barium nitrate, for example, it was found that the gain is reduced for pump linewidths greater than the Raman linewidth [60].

The effect of pump laser linewidth on the Raman laser threshold for diamond was compared to barium nitrate, $0.4\,cm^{-1}$; and KGW $5.9\,cm^{-1}$ which have narrower and wider Raman linewidths than diamond ($1.5\,cm^{-1}$), respectively. The Raman crystals were placed in an external cavity resonator and the pump powers required to reach the Stokes thresholds were measured. The linewidth of the Nd:YAG pump laser was switched between approximately $1\,cm^{-1}$ and $0.1\,cm^{-1}$ (see Figure 8.6a), using an etalon inside the Q-switched resonator. It should be noted that the spectral lineshapes were not smooth, and there was a shot-to-shot variability which was averaged out by recording the average signal over 10^4 pulses. The spatial beam properties for each linewidth arrangement were verified to be essentially unchanged. For the $0.1\,cm^{-1}$ pump linewidth there was a notable increase in the pulse modulation depth, as expected, due to the reduced number of beating modes.

The measured change in threshold with narrowband pumping relative to the broader pump case is shown in Figure 8.6b. Diamond, with a Raman linewidth slightly wider than the broadband pump, experiences a minor reduction in threshold. Barium nitrate experiences a larger reduction in threshold when the pump linewidth is reduced to less than the Raman linewidth. For KGW, both pumping configurations are substantially narrower than the Raman linewidth, and a slight

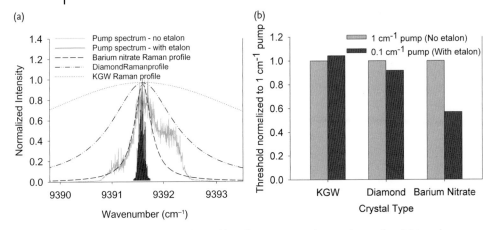

Figure 8.6 (a) Pump laser linewidths relative to Raman linewidths; (b) Comparison of external cavity Raman laser thresholds for KGW, diamond, and barium nitrate for a pump laser with 1 cm^{-1} and 0.1 cm^{-1} linewidth. The crystal lengths were 50 mm, 7 mm, and 50 mm, respectively.

increase in threshold is observed, providing an assurance that any changes in the temporal or spatial properties of the Nd:YAG laser upon insertion of the etalon are not responsible for the threshold improvements seen for diamond and barium nitrate. These results were consistent with the suggestion made elsewhere [60] that laser linewidths less than the Raman linewidth are needed to minimize the SRS threshold. In order to assess the suitability of various pump lasers, it would be of interest to fully investigate the dependence for a greater range of pump linewidths.

8.2.2
Thermal Properties

Optical phonons generated by Raman scattering decay rapidly to heat the sample, which in the example of a first Stokes diamond laser pumped at 1064 nm comprised 14% of the Stokes power. A parasitic absorption of the pump and Stokes beams in the sample may also contribute significantly to the heat load, especially for high-finesse resonators operating with low peak powers. From a Raman laser design perspective, diamond's exceptionally high thermal conductivity is one of its most outstanding and useful features, potentially enabling devices with much higher power handling than for metal oxide Raman crystals. Kemp et al. [16, 61] derived a thermal figure-of-merit for Raman materials to estimate lens strength based on the thermal conductivity and dn/dT. This analysis indicated a more than 20-fold advantage for diamond over the best of other Raman materials (excluding silicon).

Heating of the crystal affects the Raman response, the optical propagation, and the mechanical behavior of the crystal. The design parameters of most concern are induced lens strength, thermally induced stress birefringence (which affects the lens strength and beam depolarization), and stress fracture limits, as detailed below.

8.2.2.1 Thermal Lens Strength

In order to estimate the lens strength in diamond as a function of power, it is assumed that the power deposited in the crystal is uniform along the length of the crystal, with a spatial profile that is similar to the theory developed for fiber end-pumped lasers. In this case, the lens strength f^{-1} for a pump profile of radius ω_0 with power deposited P_{dep} is given by [62]

$$f^{-1} = \frac{P_{dep}}{2\pi\kappa\omega_0^2}\left(\frac{dn}{dT} + (n-1)(v+1)\alpha_T + n^3\alpha_T C_{r,\phi}\right) \quad (8.13)$$

where v is Poisson's ratio ($v = 0.069$ in diamond [63]), κ is the thermal conductivity and α_T is the thermal expansion coefficient. In order to obtain the photoelastic coefficients in cylindrical coordinates $C_{r,\phi}$, the result in Ref. [64] for Nd:YAG, which shares the lattice symmetry properties for the stress tensor, can be used. The radial and tangential components of the photoelastic stress tensor in the <111> direction for example are

$$C_r = \frac{(17v-7)p_{11} + (31v-17)p_{12} + 8(v+1)p_{44}}{48(v-1)}$$

$$C_\phi = \frac{(10v-6)p_{11} + 2(11v-5)p_{12}}{32(v-1)} \quad (8.14)$$

where the photoelastic tensor components for diamond are $p_{11} = -0.249$, $p_{12} = 0.043$, and $p_{44} = -0.172$ (see Chapter 1), giving $C_r = 0.015$ and $C_\phi = -0.032$.

Diamond has a moderate dn/dT value but a notably smaller α_T compared to other materials. The major contribution to the lens strength is due to dn/dT, whereas end-face curvature is an order of magnitude weaker, and photoelastic effects are two-orders of magnitude weaker for both radial and tangential polarizations. Using Equation (8.13), approximately 10 W of deposited power in a radius of 100 μm is required in order to induce a 1-diopter lens. The lens strength in diamond is typically two orders of magnitude lower than other laser and Raman materials for a given mode size.

8.2.2.2 Thermally Induced Stress Birefringence

Birefringence additionally caused by the stress induced from temperature gradients further depolarize the beams transmitted by the diamond crystal. This may affect the output beam properties, the gain in the Raman crystal (as gain is dependent on polarization), and also lead to additional losses for polarizers in the cavity. The birefringence is estimated using [64]

$$\Delta n_{r,\phi} = -n^3\alpha_T Q C_{r,\phi} r^2 / 2\kappa \quad (8.15)$$

where Q is the power deposited per unit volume and r is taken as the radius of the pumped region. The low expansion coefficient and high thermal conductivity in diamond combine to reduce the birefringence, for example, by approximately a factor of a thousand compared to Nd:YAG. More than 100 W of power deposited into a centimeter-long diamond crystal is required to induce birefringence ($\Delta n = \Delta n_r - \Delta n_\phi$) greater than the residual birefringence of the best materials currently available ($\Delta n \approx 10^{-7}$). Although the $C_{r,\phi}$ tensor components and Δn will vary for other propagation directions of interest, this result provides an estimate for the level at which thermally induced stress birefringence is likely to become important.

8.2.2.3 Thermal Stress Fracture

The combination of low thermal expansion coefficient and high thermal conductivity combine to yield an extremely high stress fracture limit for diamond. The deposited power limit for an end-pumped crystal is (see e.g., Ref. [65])

$$P_{\text{limit}} = 4\pi\sigma\kappa(1-v)/\alpha_T E\alpha \tag{8.16}$$

where α is the absorption coefficient, E is Young's modulus, and σ is the stress fracture limit. In order to estimate the limit, α is taken as $1\,\text{cm}^{-1}$, $E = 1100\,\text{GPa}$, and $\sigma = 4\,\text{GPa}$ [66] for diamond, giving $P_{\text{limit}} \approx 1\,\text{MW}$. This is approximately four orders of magnitude higher than for YVO$_4$ [65]. As discussed in Chapter 2, the stress fracture limit is markedly reduced for poor surface conditions, thus highlighting the importance for stressed facets to be well polished in high-power applications.

8.2.3
Laser Damage Threshold

As Raman lasers often involve placing the crystal in a beam waist where the resonator fields are highest, the laser damage threshold of the crystal often sets the upper limit to device output pulse energy and power. The short lengths of single-crystal diamond typically mean that the pump and Stokes beam Rayleigh ranges are longer than the crystal and that damage is initiated at the surfaces where the damage threshold is normally much lower than for the bulk. In many cases, as the crystal is anti-reflection coated, it is knowledge of the coated surfaces that is important. Unfortunately, coated damage thresholds are scarcely reported, but uncoated damage thresholds provide an indication of laser intensities that can be tolerated, and in most cases provide an upper bound to the coated damage threshold. In the case of pulsed laser damage, Klein [67] reviewed data from several earlier studies at 532 nm, 1064 nm, and 10.6 μm, considering in detail the effects of pulse duration, incident spot-size, and enhancements in the applied intensity that may be special to each experiment (such as facet reflections and self-focusing). The data over a wide range of pulse durations (t_p from 90 fs to 100 ns) and spot-sizes (of radius ω_0 2 to 100 μm) were remarkably consistent when normalizing the threshold against the Bettis–House–Guenther scaling law for the electric field for dielectric breakdown [68]:

$$E_{DB} \propto t_p^{1/4} \cdot \omega_0^{1/2} \tag{8.17}$$

Upon normalization, the incident irradiance for single-crystal surface damage for 1 ns pulses of radius 50 μm was 2 GW cm^{-2} at 532 nm, 8 GW cm^{-2} at 1064 nm, and 20 GW cm^{-2} at 10.6 μm, with an uncertainty on the order of 25%. Such thresholds are typical for other wide-band gap materials and hard optical crystals, and well above typical Raman laser thresholds (e.g., 0.05 GW cm^{-2}). However, it should be noted that in general for Raman lasers the surface facets are subjected to both the incident fundamental and generated Stokes fields, with the latter often consisting of high peak power spikes near threshold, resulting in damage at much lower incident pump intensities than expected.

Currenty, data are scarce for pulses longer than 100 ns and CW operation. Due to the interest in large area polycrystalline windows for use in high-power CW CO_2 lasers, the threshold has been investigated at 10.6 μm. For 1 mm thick windows and incident beam diameter of 0.6 mm, the CW damage threshold was 1.7 MW/cm^2 for steady-state irradiation of water-cooled polycrystalline windows, or up to 10 times that for 1-s exposures [69]. (The damage mode was thermal, rather than dielectric breakdown.)

In the present authors' CW studies, anti-reflection coated {110} facets have survived without damage under simultaneous exposure to an incident 1064 nm intensity up to 2 MW/cm^2 and 160 MW/cm^2 of circulating intracavity power at 1240 nm [70]. In pulsed format, damage free operation has been observed routinely for incident 10-ns pulse intensities of approximately 300 MW cm^{-2} and similar circulating intracavity first Stokes intensity. For a second Stokes laser (see Section 8.3.1 for operational details), damage was observed to the anti-reflection coatings when using an output coupler of relatively high reflectivity ($R = 60\%$ at the second Stokes) and with estimated cumulative (pump plus Stokes) intensities of approximately 1 GW cm^{-2}.

8.2.4
Design Implications for CVD-Grown Material

The chosen crystal orientation with respect to laser beam propagation direction depends on the directional gain dependence (as discussed above), and also on the cut orientation of the grown diamond sample. As discussed in Chapter 2, CVD-growth is usually fastest in the <100> direction, while the lowest residual birefringence is seen in directions perpendicular to the growth. In order to access high gain and lowest depolarization loss, it is therefore most convenient to place the laser axis along a <110> direction and the polarization parallel to the <111> [17]. Propagation along other axes is possible without making major compromises in gain (see Figure 8.5a); however, for designs requiring a long transit path through the material there may be large cost differences according to the anisotropic growth rates, and due to difficulties in polishing harder facets such as the {111}.

The amount of residual birefringence in the sample also influences design. The use of Brewster-facetted crystals may lead to large losses unless the retardance ($\Delta nl/\lambda$) upon transit through the crystal is much less than $\pi/4$ [13]. The use of anti-reflection-coated facets avoids this problem, but depolarization may still reduce the effective gain as well as affect the polarization purity of the output beam. This effect is exacerbated in high-finesse resonators, such as those used in CW lasers, in which photons undergo a large number of passes through the crystal before exiting the cavity.

8.3
Diamond Raman Laser Development

8.3.1
External Cavity Wavelength Conversion

The external cavity design refers here to the case where the Raman crystal is placed in a resonator and is pumped using a laser beam generated from an independent source. This concept enables increased flexibility of important output beam properties, including the wavelength, temporal properties, and spatial coherence. From the point of view of user applications, it is the wavelength shift that is often the main desired property, although increased output brightness or shortened output pulses are also beneficial and important attributes. The concept also allows wavelength selectivity among Stokes orders through choice of the output coupler's spectral response.

In order to describe the basic design of an efficient diamond Raman laser, consider the 1064 nm-pumped external cavity laser of Sabella *et al.* [17], as shown schematically in Figure 8.7. The laser used a diamond crystal that was 6.9 mm long and cut for propagation along a <110> crystal axis with the facets anti-reflection-coated at 1064 nm and 1240 nm. The resonator consisted of a flat input coupler that was highly transmitting at the input beam wavelength and highly reflecting at the Stokes wavelength. As the Stokes shift of diamond is larger than

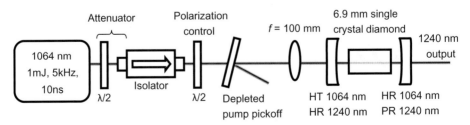

Figure 8.7 The experimental arrangement for an external cavity diamond Raman laser operating at the first Stokes wavelength of 1240 nm.

that of most other crystals, the design constraints on the dichroic input coupler coating are somewhat relaxed, which in turn leads to practical benefits such as reduced fabrication costs and an increased coating damage threshold. The output coupler was highly reflecting at the pump wavelength, 50% transmitting at the output Stokes wavelength of 1240 nm, and had a radius of curvature of approximately 1 m. The resonator mirrors were positioned close to the diamond crystal, resulting in an overall cavity length of approximately 1 cm. The input beam of $M^2 \approx 1.5$ was focused into the crystal to create a waist 115 μm in diameter and a confocal parameter longer than the crystal length. It should be noted that, when pumping using a laser operating at its fundamental (non-harmonic) wavelength, it is important to use an optical isolator between the pump and Raman lasers; otherwise, feedback from the Raman laser output coupler may lead to serious perturbations in the pump laser performance. For 10-ns pulses at a 5 kHz pulse repetition rate, the laser attained a threshold of approximately 0.5 W of pump power, and the output increased linearly (see Figure 8.8) with pump power up to 2.5 W at approximately five times the pump threshold. At high pump powers, the Stokes beam had an $M^2 \approx 1.5$ and contained a small amount (<5%) of the cascaded second Stokes at 1485 nm.

The quantum conversion efficiency, η_q, from the 1064 nm pump power incident on the crystal to the first Stokes output was 71% at the maximum pump power

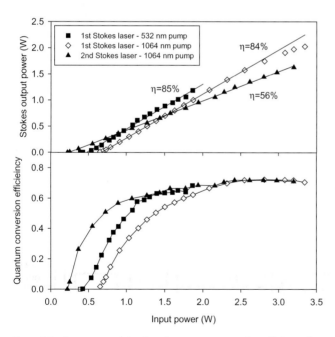

Figure 8.8 Output power (top) and quantum conversion efficiency (bottom) to a given Stokes wavelength for 5 kHz external cavity diamond Raman lasers configured for output at 573 nm [14], 1240 nm [17], and 1485 nm [21].

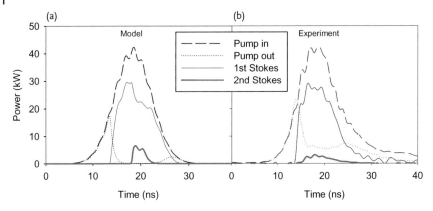

Figure 8.9 (a) Modeled and (b) experimental pump and Stokes traces for 0.44 mJ input pulse energy at 1064 nm.

(61% power conversion efficiency). On combining the contributions of both the first and a small amount of cascaded second Stokes, 78% of all pump photons were converted to Stokes (i.e., $\eta_q = 0.78$), which is believed to be the highest reported for any crystalline Raman laser. In comparison, the highest conversion efficiencies obtained using other materials, was $\eta_q = 0.69$ and $\eta_q = 0.675$ for barium tungstate and barium nitrate, respectively [71, 72]. Damage-free operation for pump powers more than fivefold above threshold was observed. Higher efficiencies may be obtained at higher pump powers, or by optimizing the temporal profile of the pump pulse, particularly on the leading edge. The pulse shapes for the pump, residual pump and output of Figure 8.9 reveal that a large fraction of the pump goes unconverted prior to the laser attaining threshold. There is also incomplete conversion during the pulse, which may be the result of a poor conversion of higher-order spatial modes. Notably, simple plane wave models (see Section 8.1.1.2) which use the pump pulse shape and mirror specifications as input reproduce many aspects of the behavior well (see Figure 8.9a).

Although 1240 nm may be of interest for some medical and remote sensing applications, the second Stokes wavelength at 1485 nm is perhaps of more interest due to the demand for lasers with reduced risks for eye damage ("eye-safe") while being able to be transmitted without major loss through the atmosphere. As described recently [21], efficient 1485 nm operation is achieved using essentially the same configuration as described elsewhere [17], except that the output coupler is exchanged to resonate the first and second Stokes, and with a suitably chosen output coupling at the second Stokes (as analyzed in detail using model calculations in Ref. [21]). Although the slope efficiency is lower for this system, as expected due to the twofold increase in loss by excited phonons, the quantum conversion efficiency ($\eta_q = 0.71$) was similar to the first-Stokes laser. The approach here for generating "eye-safe" laser wavelengths has some distinguishing advantages over technologies based on Er lasers and optical parametric oscillators;

notably, it is an add-on to mature Nd laser technology that is potentially more temperature stable in comparison to phase-matched parametric conversion alternatives. It is also likely to be more amenable for scaling up to higher average powers. One possible drawback, however, is that the converted wavelength using a Nd:YAG pump laser suffers increased atmospheric attenuation, so may be unsuitable for some applications that require long-distance transmission; in such a case the use of a longer wavelength Nd:YAP laser may be a better choice.

The efficiencies observed for the first and second Stokes 1064 nm-pumped diamond Raman lasers, and indeed for a similar 532 nm-pumped system operating at 573 nm [14], are each the highest for any Raman crystalline material. (The outright record for a Raman laser is $\eta_q = 0.91$, held by gas-filled hollow core microstructured fibers [73].) These high efficiencies in diamond Raman lasers are attributed to the greater design freedoms brought about by using a short, high-gain crystal with a large Raman shift, and prove the suitability of CVD-grown material for pulsed laser applications. There is excellent potential for further advances in performance range, including operation at higher output powers and increased wavelength versatility via the use of tunable pump lasers and intra-Raman-cavity harmonic generation.

8.3.2
Intracavity Diamond Raman Lasers

Raman conversion using low-power pumps can be efficiently achieved by placing the diamond crystal inside a pump laser with a resonator of high finesse at the pump wavelength. The design has been widely investigated for CW and high-pulse-repetition rate (>kHz) systems. In the push for output powers up to the multi-watt level, much attention has been paid to designs that allow stable operation in the presence of the dynamic lenses induced in the Raman and fundamental gain media (for a review, see Ref. [7]). "Self-Raman" lasers, in which the Raman and fundamental laser thermal lenses are essentially colocated, are simpler in design with a reduced number of intracavity elements that enables the demonstration of CW output powers of up to several watts (see e.g., Refs [11, 74]). Currently, the highest output powers of approximately 10 W have been obtained using separate Raman and gain crystals in high-repetition-rate, Q-switched lasers [5]. Diamond offers a means of increasing the stable operation range of these systems, by leveraging its high gain and reduced thermal effects, as well as providing a different range of output wavelengths by virtue of its larger Raman shift.

Intracavity diamond Raman lasers have been studied, largely in the CW regime, by a group at the University of Strathclyde [15, 16, 20, 22, 29]. In the early studies, output powers of 1.6 W at 1240 nm were reported, together with a slope efficiency with respect to the absorbed diode-laser pump power (at 808 nm) of 18%, using Nd:YVO$_4$ as the fundamental gain medium. In quasi-CW operation with 20% duty cycle, peak output powers of 2.8 W with a slope efficiency of 24% were observed, and a maximum absorbed diode-laser pump power to the Raman laser output power conversion efficiency of 13% [16, 20]. More recently, the Strathclyde group has reported CW powers of up to 5.1 W with excellent beam quality (M^2 <1.2) by utilizing the superior thermal properties of Nd:YLF as the fundamental gain medium [29]. The first Stokes output in this case was 1217 nm.

Relatively less effort has been expended in developing pulsed intracavity devices, however. The Strathclyde group described an arrangement using a diamond crystal inside an acousto-optically Q-switched Nd:YVO$_4$ laser that generated a first Stokes (1240 nm) average output power of approximately 0.4 W and a repetition rate of 6.3 kHz. Notable losses within the diamond (ca. 1% per single pass) were cited in this case as the main reason for the limited efficiency (4% from optical diode input power) [15]. However, improvements in diamond quality and resonator design are expected to enable higher efficiencies approaching the 10–20% values seen when using other materials.

One interesting development has been the use of vertical-external cavity surface-emitting lasers (VECSELs) as the fundamental gain medium. VECSELs have the advantage of being able to tailor the output wavelength range with more freedom than by using different active laser dopants, and the systems also offer tunability over a range of about 10 nm. Parrotta *et al.* have demonstrated a diode-pumped InGaAs VECSEL with intracavity diamond Raman shifting and a maximum output power of 1.3 W at the first Stokes wavelength of 1227 nm [22]. In this case, the optical conversion efficiency was 14.4% with respect to the absorbed diode laser pump power. Broad tuning of the Raman laser output between 1217 and 1244 nm was achieved by tuning the fundamental wavelength with an intracavity birefringent filter [22].

The characteristic design challenges for intracavity Raman lasers relate to the management of resonator stability for the multiple laser fields and thermal lenses, but without intensities exceeding the material and coating damage thresholds. In the case of diamond, the small Raman lens simplifies the design for higher output powers. The system is also amenable for generation of harmonic outputs with potential user selectability by including second-order nonlinear media within the resonator [3, 10, 11].

8.3.3
Synchronously Pumped Mode-Locked Lasers

The Raman conversion of picosecond and femtosecond pulses has a number of new considerations compared to the CW or nanosecond laser sources considered above. Although efficient conversion can be achieved without a resonator, single- and double-pass conversion have high thresholds and provide limited control over beam quality and cascading. The conversion of each pulse in a short external resonator is not appropriate, as the pulses are too short to allow an oscillating Stokes pulse to build up within the pump pulse duration. Intracavity schemes are possible where the Raman crystal is placed in the cavity of a mode-locked laser, but this has the disadvantage of potentially disrupting the pulse formation through nonlinear loss and additional dispersion presented to the pump laser.

A simple scheme to avoid these issues is to use a synchronously pumped external resonator, to generate a mode-locked train of Stokes pulses [18, 35, 75–78]. For

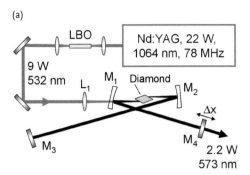
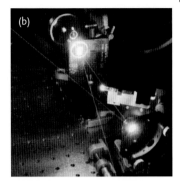

Figure 8.10 (a) Schematic diagram of the 573 nm synchronously pumped diamond Raman laser; (b) Photograph (courtesy of Eduardo Granados) of the center of the resonator, showing the diamond crystal and the intracavity yellow beam. For a color version of this figure, please see the color plate at the end of this book.

this, the external resonator must have its length matched to that of a mode-locked pump laser. Then, rather than Stokes-shifting each pump pulse independently, a circulating Stokes pulse is established in the resonator that, on each round trip, is amplified by successive pump pulses. This synchronous pumping can be viewed as a mode-locking mechanism for the Stokes laser, and provides a method for generating Stokes pulses that can be somewhat shorter than the pump pulses.

The use of synchronous pumping for Raman lasers has been investigated using diamond [35]. The set-up for a synchronously pumped diamond Raman laser pumped with 26-ps, 532-nm pulses frequency-doubled from a 78 MHz mode-locked Nd:YAG laser, is shown schematically in Figure 8.10a. The diamond external resonator was a z-fold cavity to achieve a small waist in the diamond crystal, while allowing an overall cavity length of approximately 1.9 m in order to match the cavity length with that of the pump laser.

With the cavity length adjusted for maximum conversion efficiency, the output at 573 nm was 2.2 W with a pulse duration of 21 ps, compared to the 532 nm pump power of 9 W with a pulse duration of 26 ps. This shortening of the pump pulses is typical for synchronously pumped lasers; indeed, by detuning the cavity length, pulses as short as 9 ps can be generated, albeit with a substantial loss of efficiency.

This approach to converting mode-locked trains of short pulses to a new wavelength is flexible, particularly using diamond (for which pump pulses from the deep UV can be converted; see Section 8.4.2). The technique can, in principle, be extended from the picosecond example described above down to femtosecond lasers. For 10 ps pulses there is no need to consider any dispersion compensation, though this would be a consideration for pulses of much less than 1 ps duration. As the pulses become shorter than the phonon dephasing time (7 ps for diamond), the transient regime of SRS is entered; this would in turn lead to a reduction in the effective Raman gain, and so a careful design would be imperative to achieve an efficient operation.

8.4
Extending the Capability of Raman Lasers Using Diamond

8.4.1
Long-Wavelength Generation

The transmission band of most Raman crystals typically extends from approximately 0.35 μm to about 4 μm. Although silicon transmits at wavelengths of up to 6 μm, the output wavelengths demonstrated to date have been limited to the range 1–2 μm and the efficiency has been poor due to parasitic nonlinear absorption. Diamond's wide transmission band and high IR gain coefficient provide an opportunity to generate laser wavelengths outside these usual bands. As shown in Figure 8.11, the IR transmission properties of diamond sit apart from those of other IR materials, such as low-index alkali and alkali-earth halides, and high-index semiconductor materials.

To the present authors' knowledge, the longest wavelength diamond Raman laser to date is the first Stokes 1.63 μm external resonator device pumped at 1.34 μm [23]. However, there is potential to develop sources at longer wavelengths closer towards the onset of strong absorption near 3.8 μm or at wavelengths longer than 6 μm. The latter includes the important molecular fingerprint region (6.5–20 μm) and long-wavelength IR transmission window (8–12 μm), where there is currently intense interest in developing practical laser sources for applications in civil security and defense, environmental mapping, and surgery. The major technologies addressing this wavelength region include quantum cascade lasers and wavelength conversion in $ZnGeP_2$ or, more recently, orientation-patterned GaAs.

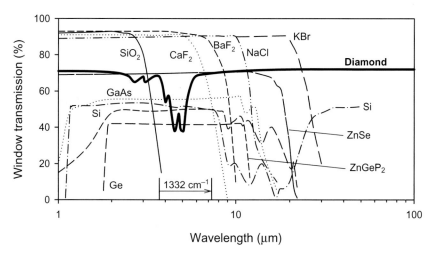

Figure 8.11 Transmission spectra of thin windows for prominent infrared materials. The diamond wavelength shift is also shown for the example of a pump wavelength at 3.7 μm.

Diamond may provide an alternative approach with advantages of robustness and, potentially, high average power.

The main absorption band between 4 and 6 μm is narrower than diamond's 1332 cm^{-1} shift when pumped in the mid-IR, and opens up the possibility of using mature pump sources of wavelengths less than 4 μm to Raman shift to the longside of the main absorption band. For example, for a pump wavelength situated in the local minimum of the multi-phonon absorption profile at 3.7 μm, the first Stokes wavelength is 7.3 μm as shown in Figure 8.11. The intrinsic multiphonon absorption and the diminishing Raman gain coefficient with increasing Stokes wavelength present significant challenges. Modeling as outlined in Section 8.1.1.2 can be used to estimate laser performance. Using a Raman gain coefficient of 2 cm GW^{-1} (see Figure 8.4), and pump and Stokes absorption coefficients of 0.5 cm^{-1} and 0.1 cm^{-1} (see Chapter 1), respectively, and an external cavity configuration similar to that shown in Figure 8.7, pump intensities in the vicinity of 1 GW cm^{-2} are required to reach the threshold for 10-ns pulses. Although this is less than the diamond damage threshold, it presents a major challenge for the dielectric optical coatings. Variants of this scheme, such as off-axis pumping and waveguiding in the diamond, may enable the parameter space between damage and efficient operation to be increased. It should be noted that, in the case of waveguides, the pump energies required to reach a threshold will be significantly reduced (e.g., several μJ instead of approximately 1 mJ), but due to the field enhancement of the waveguide no major reduction would be expected in the required intracavity power or energy density needed to achieve threshold. Hence, robust resonator reflectors present the major challenge currently for realizing diamond Raman lasers longer than 6 μm.

8.4.2
Deep Ultraviolet Generation

Diamond is unique among high-gain Raman crystals in that it has good transmission between 230 and 350 nm. This wavelength range is of interest for applications that include the stand-off detection of explosives, short-range covert communications, and environmental sensing, to name but a few. Several alternative approaches have been proposed that address this region, such as cerium-doped lasers and the second harmonic of yellow or red lasers. Nevertheless, it is of interest to investigate a directly pumped Raman laser in the UV range in order to understand the design issues for a high-gain Raman system and to address applications that may benefit from a cascaded, multiwavelength output.

The 266-nm synchronously pumped diamond Raman laser, as reported by Granados et al. [35], represents the first crystalline Raman laser pumped in the UV range. To the present authors' knowledge, the shortest pump wavelength to be used previously was 514 nm for an external cavity barium nitrate laser [79]. Here, the diamond crystal was pumped at the fourth harmonic of a Nd:YVO$_4$ laser in a resonator that was similar in design to the 573 nm synchronously pumped laser shown in Figure 8.10. A threshold of approximately 5 nJ (four times lower than a similar visible pumped system [18]) was obtained for the first Stokes output at 273 nm. By using a model to solve the rate equations for the evolution of pump phonon fields for the synchronously pumped system, a gain coefficient in the vicinity of 100 cm GW^{-1} was deduced (cf. 10–15 cm GW^{-1} at 1064 nm). The model also included the effect of two-photon absorption in the diamond which was found to lower the effective gain and cause rollover in slope efficiency at high pump energies. The conversion efficiency of 10% at maximum output, and the slope efficiency of 25%, were two- to threefold lower than would normally be expected from Raman lasers. The major cause of the low efficiency was attributed to non-ideal mirror coatings (this is a common problem with deep UV laser mirrors) and the effects of two-photon absorption.

A more critical concern for deep UV operation is damage to the diamond crystal facets. Only short-term operation was obtained, and the power was observed to decrease steadily over a 10-min period. An inspection of the facets using an optical surface profiler revealed pits with a diameter that was slightly smaller than the beam and a depth of approximately 50 nm. These were formed cumulatively during the laser beam exposure at an etch rate of approximately 300 atoms per pulse, on average. This type of damage was in contrast to the usual modes of damage for UV lasers, such as catastrophic surface damage (ablation) or the formation of color centers in the bulk. A similar "slow etching" was observed previously for a KrF excimer laser beam (248 nm) incident on single crystal diamond by Kononenko *et al.* [80].

In order to understand how facet etching might be prevented, further studies were undertaken to determine how the UV beam interacts with the surface and to identify factors that influence the etch rate [81]. These experiments were made using 10-ns, 266-nm pulses on single crystal and polycrystalline polished samples, after which the surfaces were analyzed using an optical profiler and surface-sensitive near-edge X-ray absorption fine-structure (NEXAFS) spectroscopy.

One of the most revealing results of the study was that the observed etch rate increased with the square of the incident fluence, up to the ablation threshold (Figure 8.12). The latter is defined here as the minimum fluence when using a burst of 60 pulses in 50 ms (the shortest burst available from the shuttered UV laser system) to produce the irregularly shaped pit that is characteristic of conventional diamond ablation in the nanosecond regime. For a fluence corresponding to 60% of the ablation threshold, pits 400 nm deep were created after a 30-s exposure (2.2×10^5 pulses). For fluences 10-fold lower, the time required to achieve the same depth was 100 times longer (1×10^8 pulses or 220 min), and so on for lower fluences. The lowest fluence investigated corresponded to a few milliwatts of incident power (two orders below the ablation threshold), and required an exposure time of approximately 24 h to create a similarly sized pit. In each case, the average number of atoms removed per pulse was less than an atomic layer. The results of the NEXAFS surface studies confirmed that the etched surfaces were oxygen-terminated and free from graphite [81]. Moreover, the results to date have suggested that the etch rate is proportional to the two-photon absorption rate, and is largely independent of the surface facet direction, pulse duration in the range 0.02 to 100 ns, and the pulse repetition rate.

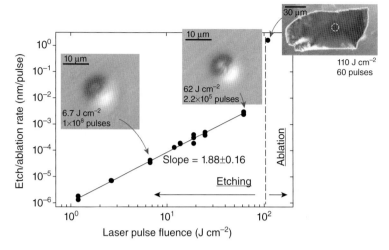

Figure 8.12 Etching rate (in terms of nm per pulse) as a function of the incident laser fluence. The images show the contrasting structure of the ablation and etching regimes. The dashed circle in the ablation image indicates the diameter of the incident beam.

The details of UV absorption in the surface layers and subsequent ejection kinetics are not well understood. Kononenko et al. [80] (see also Chapter 12) showed that the presence of oxygen is required for etching, which in turn indicates that surface oxidation is a crucial part of the process; thus, placing the crystal under a vacuum or inert atmosphere may enable a longer-term stable UV Raman laser operation. It has been found that the use of commercial anti-reflection coatings does not inhibit damage. An improved understanding of the mechanism is required to determine a practical method to prevent etching. The ejection of carbon species may be mediated by direct excitation of surface states or by electron–hole pairs created close to the surface by two-photon absorption events. The exciton energy (5.4 eV) is easily sufficient to exceed the activation energy for CO desorption (1.7 eV [82]). An increased knowledge of ejection kinetics would also be valuable in other fields when developing new methods for machining smooth, high-resolution surface structures in diamond. The square-dependence on the intensity is reproduced spatially, so etched features may be written smaller than the diffraction-limited spot-size of the writing beam. Although surface etching is a problem for UV laser operation, it may enable writing of submicron-sized surface features using a practical technique in air that is well suited to the rapid prototyping.

8.4.3
High Average Power

One of the most promising areas in which diamond may provide large increases in laser capability is in wavelength conversion at high average power. Based on the assumption that thermal gradients in the medium represent the main limit, much higher average powers compared to other crystals are foreshadowed in proportion to the higher thermal conductivity (by approximately two to three orders of magnitude). The exceptionally low thermal expansion coefficient also raises the amount of heat deposition that can be tolerated before fracture, or before end-face curvature and induced stress birefringence will need to be considered (see Section 8.2.2). Concepts such as Raman beam combination, in which multiple angularly displaced beams pump a Raman laser mode (see Refs [83, 84]) have been proposed for diamond. Indeed, a US defense-funded program has been initiated to demonstrate a combination of high-average power diamond Raman lasers and Raman beam combination at the multi-kW level [85]. In this study, the results of which were first made available in 2011 [19], involved an external cavity diamond Raman laser pumped using a 340 W cryogenic Q-switched Yb:YAG laser. The Raman laser cavity, which consisted of two identical mirrors with high transmission at the pump wavelength and a reflectance of 83% at the first Stokes, was not ideal for achieving the low thresholds afforded by double-pass pumping arrangements, nor for comprising a practical single-ended output device. Nevertheless, the configuration enabled the demonstration of a diamond Raman laser with a combined-end output of 24.5 W and a slope efficiency of 58% at the first Stokes. The laser was operated only up to 1.5 times above threshold due to damage to the anti-reflection coatings on the diamond, and consequently only modest overall conversion efficiencies were obtained (up to 13%). Although the system's performance would surely be improved by using a double pass of the pump beam and applying other optimizations, this result represents the highest average power obtained to date for a Raman laser based on any crystalline material. To the present authors' knowledge, the previous best was 11 W for barium nitrate [86], which has a very similar gain coefficient to diamond but with a thermal conductivity that is approximately 1500 times poorer. Clearly, temperature gradients are not limiting maximum powers to date; rather, damage represents the major limiting factor for the average power of this configuration.

In the present authors' laboratories, investigations are currently being made of an alternative power scaling approach based on CW Nd pump lasers. This approach has the practical advantage of building on mature CW high-power laser technology, and is able to address applications that require CW output. To the present authors' knowledge, the only CW pumped external cavity Raman laser reported to date used a 6.8 cm-long barium nitrate crystal and a pump laser in the visible range (514 nm) [79] where the gain is high (ca. 50 cm GW^{-1}). For this device, the efficiency was low (0.16 W for 5.5 W of pump power), yet even at this low power level thermal lensing effects were noted in the barium nitrate. Hence, low-gain materials of relatively low thermal conductivity face a fundamental problem that the output power is restricted to levels not much greater than the laser threshold. Diamond's combination of high gain and thermal conductivity presents an opportunity to investigate CW operation using NIR laser pumps, at output powers significantly larger than are possible with barium nitrate.

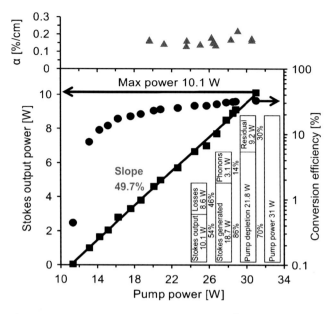

Figure 8.13 Output power (squares), conversion efficiency (circles) and combined scatter and absorption loss coefficient (α, triangles) of the diamond Raman laser as a function of incident pump power. The inset shows the results of power budget calculations.

Diamond has been used to demonstrate the first external cavity Raman laser pumped using a CW Nd (1064 nm) laser, and to generate CW Stokes output powers above 10 W [70]. An anti-reflection-coated diamond crystal (9 mm long) was placed in a confocal high-finesse resonator to create an intense Stokes waist within the crystal. The output coupling was approximately 0.4%, the laser threshold was 11 W, and a maximum output power of 10.1 W was obtained for 31 W of input power (33% conversion efficiency). This output power is approximately double the previous highest reported for a CW crystalline Raman laser, which was an intracavity diamond Raman laser generating 5.1 W at 1240 nm [29].

The output power increased linearly with input power, with a slope efficiency of 49.7% (see Figure 8.13), and the fact that no thermal effects were observed is highly promising for maintaining the slope efficiency with more powerful pump laser pumps. The external cavity approach has important benefits for operations at elevated power. The placement of a Raman crystal at the midpoint of a concentric resonator mitigates the effects of any thermally induced refractive index gradients. Moreover, in contrast to intracavity designs, the thermal lens effects in the pump laser and the Raman laser are decoupled, thereby greatly simplifying design at high powers.

The external cavity configuration is also convenient for investigating system losses, as good access is available to the unconverted pump laser beam. By noting the pump depletion, which represented the difference between the pump and the unconverted pump upon alignment of the Raman laser cavity, it was possible to determine the power coupled into the Stokes and phonon fields. The results are summarized in the inset of Figure 8.13 for operation at maximum power. It is found that although more than half of the pump power was converted to the Stokes, a significant fraction (ca. 8 W) was lost in the bulk diamond due to absorption and/or scatter. Thus, in contrast to the pulsed diamond Raman lasers described above, an improved crystal quality would be expected to significantly increase the conversion efficiency and output power. By using the known output coupler transmission (0.4 ± 0.1%) and Stokes power circulating in the Raman cavity, the combined absorption and scatter coefficient of the diamond was deduced as approximately $0.17 \pm 0.05\%\,cm^{-1}$ at 1240 nm. This was close to the absorption coefficient value ($\approx 0.1\%\,cm^{-1}$ at 1064 nm) quoted by the manufacturer's low-nitrogen ($\approx$20 ppb) material (see Ref. [87] and Chapter 2). Although crystal losses are the major factor limiting conversion efficiency, there is also scope for improvement through output coupling optimization.

This demonstration of CW conversion is a significant step towards a technology that is much more applicable than previously possible using other materials, and towards the creation of high-power CW lasers capable of addressing wavelength-specific applications across the IR range, and at the harmonics by use of intracavity and extracavity nonlinear frequency generation.

8.5
Conclusions and Outlook

Diamond is among the most highly performing of all crystalline Raman laser materials, having yielded the highest conversion efficiency – both in terms of energy and photons – for pulsed Q-switched external cavity Raman lasers, and the highest average power crystalline Raman laser in the categories of CW intracavity, and CW and pulsed external cavity designs. Diamond has also proven to be an excellent replacement for conventional Raman crystals, with advantages of small size, reduced thermal effects, and large Stokes shifts. Of greater importance, there is also considerable scope for diamond to extend the normal range of Raman laser performance. This review of diamond's properties pertinent to Raman laser design has confirmed that it has major advantages for thermal dissipation and the mitigation of thermally induced stress, as well as a high gain coefficient with transmission across a very broad spectrum. Already, diamond has enabled an initial demonstration of solid-state Raman lasers operating in the UV range, and of efficient CW IR conversion in a simple external cavity.

Further advances in performance are expected as research efforts are expanded, with increases in power and wavelength range expected in the short term. Parallel advances in crystal properties will also fuel progress. Indeed, with major breakthroughs in the growth of high-quality large single crystals occurring only during the past decade, it can still be considered the "early days" of synthesis. At present, reduced absorption and scatter, larger sizes and lower production costs are only some directions of development (see for example Chapter 2) that will assist in Raman laser development. Reduced bulk losses will enable a higher efficiency with highly resonant cavities such as those employed for CW and low-peak power lasers. Reproducibility and lower costs will also benefit research, and are crucial to the future transfer of this technology into the commercial arena. Currently, CVD growth is perceived to be the most promising technique, due to its unrivaled level of control of absorbers and residual stress-birefringence. However, there may be scope to develop devices based on other types of material (e.g., HPHT or polycrystalline diamond) in configurations where birefringence or scatter loss are less critical. Doping of diamond with laser ions represents a highly attractive avenue of development following the success of Nd-doped Raman crystals in simplifying intracavity Raman laser design ("self-Raman lasers"). The present challenges involved in incorporating a sufficient density of lasant ions into the closely packed diamond lattice are large [88], and a major advance is required before this possibility can be exploited. Diamond waveguides also represent an area of development that will assist device design by using the beam confinement to enhance the Raman nonlinearity and mitigate against thermal effects. Recently, several methods for creating waveguides have been investigated (see Chapter 10), including ridge waveguides that hold promise for realizing CW diamond Raman lasers of reduced threshold [89]. Such developments in materials engineering will surely create exciting future prospects for the field of diamond lasers.

The future development of diamond lasers will be driven by the ever-increasing demand for laser devices of enhanced performance range. The most obvious area is high-power Raman lasers, where the wider range of output wavelengths, good beam quality, and intrinsically narrow linewidth provide advantages over other high-power laser technologies such as fiber, thin disk, and slab lasers. Recent demonstrations of the high-power conversion of Nd/Yb lasers in diamond [19, 70] represent a major step towards high-power pulsed and CW or quasi-CW Raman lasers at IR and harmonic wavelengths. Moreover, diamond's exceptional thermal properties are likely to make it the most preferred solid-state material for the exploration of concepts such as brightness conversion or Raman beam combination.

Acknowledgments

A.S. and R.M. would like to thank Prof. Jim Piper for stimulating discussions in the areas of Raman gain coefficients and long-wavelength generation. A.S. would also like to thank Dr Olivia Samardzic and Dr John Haub for their insightful comments during the preparation of the manuscript and the Defence Science and Technology Organisation for supporting his research. The material contained in this chapter is based on research sponsored by the Australian Research Council Future Fellowship Scheme (FT0990622), and the Air Force Research Laboratory under agreement numbers AOARD-10-4078 and AOARD-12-4055. The OptoFab node of the Australian National Fabrication Facility is acknowledged for contributions in the ultraviolet etching of diamond.

Note Added in Proof

A Raman gain coefficient of $21\,\text{cm}\,\text{GW}^{-1}$ at a pump wavelength of 1064 nm was recently reported using a pump-probe method [90]; this value was substantially higher than the data presented in Figure 8.4 which are all in the range of 8.5–13.5 cm GW^{-1}. These results highlight the need for further studies on Raman gain coefficients, with careful specification of the experimental conditions employed.

References

1 Ammann, E.O. and Falk, J. (1975) Stimulated Raman scattering at kHz pulse repetition rates. *Appl. Phys. Lett.*, **27** (12), 662–664.

2 Ammann, E.O. (1978) High-average-power Raman oscillator employing a shared-resonator configuration. *Appl. Phys. Lett.*, **32** (1), 52–54.

3 Ammann, E.O. (1979) Simultaneous stimulated Raman scattering and optical frequency mixing in lithium iodate. *Appl. Phys. Lett.*, **34** (12), 838–840.

4 (a) Eremenko, A.S., Karpukhin, S.N., and Stepanov, A.I. (1980) Stimulated Raman scattering of the second harmonic of a neodymium laser in nitrate crystals. *Sov. J. Quantum Electron.*, **10** (1), 113–114; (b) Eremenko, A.S., Karpukhin, S.N., and Stepanov, A.I. (1986) Generation of radiation in a resonator under conditions of stimulated Raman scattering in Ba(NO$_3$)$_2$, NaNO$_3$, and CaCO$_3$ crystals. *Sov. J. Quantum Electron.*, **16** (8), 1027–1031.

5 Chen, X.H., Zhang, X.Y., Wang, Q.P., Li, P., Li, S.T., Cong, Z.H., Liu, Z.J., Fan, S.Z., and Zhang, H.J. (2009) Diode side-pumped actively Q-switched Nd:YAG/SrWO$_4$ Raman laser with high average output power of over 10 W at 1180 nm. *Laser Phys. Lett.*, **6** (5), 363–366.

6 Basiev, T.T. and Powell, R.C. (2003) Solid-state Raman lasers, in *Handbook of Laser Technology and Applications* (eds C.E. Webb and J.D.C. Jones), CRC Press, Boca Raton, FL, Chapter B1.7, pp. 469–497.

7 Rawle, C.B., Ter-Mikirtychev, V., McKinnie, I.T., and Sandle, W.J. (2001) High-energy solid-state Raman laser based on barium nitrate crystal for near IR and visible spectral range. Advanced Solid-State Lasers, January 28, 2001, Seattle, WA, Optical Society of America.

8 Pask, H.M. (2003) The design and operation of solid-state Raman lasers. *Prog. Quantum Electron.*, **27** (1), 3–56.

9 Piper, J.A. and Pask, H.M. (2007) Crystalline Raman lasers. *IEEE J. Sel. Top. Quantum Electron.*, **13** (3), 692–704.

10 Mildren, R.P., Pask, H.M., Ogilvy, H., and Piper, J.A. (2005) Discretely tunable, all-solid-state laser in the green, yellow, and red. *Opt. Lett.*, **30** (12), 1500–1502.

11 Lee, A.J., Spence, D.J., Piper, J.A., and Pask, H.M. (2010) A wavelength-versatile, continuous-wave, self-Raman solid-state laser operating in the visible. *Opt. Express*, **18** (19), 20013–20018.

12 McQuillan, A., Clements, W., and Stoicheff, B. (1970) Stimulated Raman emission in diamond: spectrum, gain, and angular distribution of intensity. *Phys. Rev. A*, **1** (3), 628–635.

13 Mildren, R.P., Butler, J.E., and Rabeau, J.R. (2008) CVD-diamond external cavity Raman laser at 573 nm. *Opt. Express*, **16** (23), 18950–18955.

14 Mildren, R.P. and Sabella, A. (2009) Highly efficient diamond Raman laser. *Opt. Lett.*, **34** (18), 2811–2813.

15 Lubeigt, W., Bonner, G.M., Hastie, J.E., Dawson, M.D., Burns, D., and Kemp, A.J. (2010) An intra-cavity Raman laser using synthetic single-crystal diamond. *Opt. Express*, **18** (16), 16765–16770.

16 Lubeigt, W., Bonner, G.M., Hastie, J.E., Dawson, M.D., Burns, D., and Kemp, A.J. (2010) Continuous-wave diamond Raman laser. *Opt. Lett.*, **35** (17), 2994–2996.

17 Sabella, A., Piper, J.A., and Mildren, R.P. (2010) 1240 nm diamond Raman laser operating near the quantum limit. *Opt. Lett.*, **35** (23), 3874–3876.

18 Spence, D.J., Granados, E., and Mildren, R.P. (2010) Mode-locked picosecond diamond Raman laser. *Opt. Lett.*, **35** (4), 556–558.

19 Feve, J.-P.M., Shortoff, K.E., Bohn, M.J., and Brasseur, J.K. (2011) High average power diamond Raman laser. *Opt. Express*, **19** (2), 913–922.

20 Lubeigt, W., Savitski, V.G., Bonner, G.M., Geoghegan, S.L., Friel, I., Hastie, J.E., Dawson, M.D., Burns, D., and Kemp, A.J. (2011) 1.6 W continuous-wave Raman laser using low-loss synthetic diamond. *Opt. Express*, **19** (7), 6938–6944.

21 Sabella, A., Piper, J.A., and Mildren, R.P. (2011) Efficient conversion of a 1.064 μm Nd:YAG laser to the eye-safe region using a diamond Raman laser. *Opt. Express*, **19** (23), 23554–23560.

22 Parrotta, D.C., Kemp, A.J., Dawson, M.D., and Hastie, J.E. (2011) Tunable continuous-wave diamond Raman laser. *Opt. Express*, **19** (24), 24165–24170.

23 Jelínek, M., Kitzler, O., Jelínková, H., Šulc, J., and Němec, M. (2012) CVD-diamond external cavity nanosecond Raman laser operating at 1.63 μm pumped by 1.34 μm Nd:YAP laser. *Laser Phys. Lett.*, **9** (1), 35–38.

24 Boyd, R.W. (2008) *Nonlinear Optics*, 3rd edn, Academic Press Limited, London.

25 Penzkofer, A., Laubereau, A., and Kaiser, W. (1979) High intensity Raman interactions. *Prog. Quantum Electron.*, **6**, 55–140.

26 Mildren, R.P., Coutts, D.W., and Spence, D.J. (2009) All-solid-state parametric Raman anti-Stokes laser at 508 nm. *Opt. Express*, **17** (2), 810–818.

27 Koch, K., Moore, G.T., and Dearborn, M.E. (1997) Raman oscillation with intracavity second harmonic generation. *IEEE J. Quantum Electron.*, **33** (10), 1743–1748.

28 Bienfang, J.C., Rudolf, W., Roos, P.A., Meng, L.S., and Carlsten, J.L. (2002) Steady-state thermo-optic model of a continuous-wave Raman laser. *J. Opt. Soc. Am.*, **19** (6), 1318–1325.

29 (a) Savitski, V., Hastie, J., Dawson, M.D., Burns, D., and Kemp, A.J. (2011) Low-loss synthetic single-crystal diamond: Raman gain measurement and high power Raman laser at 1240 nm. European Conference on Lasers and Electro-Optics, 22 May 2011, Munich, Germany. Optical Society of America; (b) Savitski, V., Burns, D., and Kemp, A.J. (2011) Multi-watt continuous-wave diamond Raman laser at 1217 nm. European Conference on Lasers and Electro-Optics, 22 May 2011, Munich, Germany. Optical Society of America.

30 Band, Y.B., Ackerhalt, J.R., Krasinski, J.S., and Heller, D.F. (1989) Intracavity Raman lasers. *IEEE J. Quantum Electron.*, **25** (2), 208–213.

31 Demidovich, A.A., Voitikov, S.V., Batay, L.E., Grabtchikov, A.S., Danailov, M.B., Lisinetskii, V.A., Kuzmin, A.N., and Orlovich, V.A. (2006) Modeling and experimental investigation of a short pulse Raman microchip laser. *Opt. Commun.*, **263**, 52–59.

32 Spence, D.J., Dekker P., and Pask, H.M. (2007) Modeling of continuous wave intracavity Raman lasers. *IEEE J. Sel. Top. Quantum Electron.*, **13** (3), 756–763.

33 Kananovich, A.A., Voitikov, S.V., Demidovich, A.A., Danailov, M.B., and Orlovich, V.A. (2012) Output power and intracavity intensity profiles of a quasi-continuous end-pumped Nd:YVO$_4$ self-Raman mini laser. *Appl. Phys. B*, **106** (106), 9–17.

34 Degnan, J.J. (1989) Theory of the optimally coupled Q-switched laser. *IEEE J. Quantum Electron.*, **25** (2), 214–220.

35 Granados, E., Spence, D.J., and Mildren, R.P. (2011) Deep ultraviolet diamond

Raman laser. *Opt. Express*, **19** (11), 10857–10863.

36 Granados, E. and Spence, D.J. (2010) Pulse compression in synchronously pumped mode locked Raman lasers. *Opt. Express*, **18** (19), 20422–20427.

37 Straka, P. and Rudolph, W. (2004) Numerical simulations of pulsed and quasi-CW regimes of synchronously pumped Raman oscillators. *Appl. Phys. B*, **79** (6), 707–712.

38 Jalali, B., Raghunathan, V., and Shori, R. (2006) Prospects for silicon mid-IR Raman lasers. *IEEE J. Sel. Top. Quantum Electron.*, **12** (6), 1618–1627.

39 Cleveland Crystals. Lithium Iodate. Available at: http://www.clevelandcrystals.com/liio3.htm (accessed 20 January 2012) and references therein.

40 Belevtseva, L.I., Voitsekhovskii, V.N., Nazarova, N.A., Romanova, G.I., Shvedova, M.V., and Yakobson, V.E. (1989) Main properties of optical crystals of barium nitrate. *Sov. J. Opt. Technol.*, **56**, 233.

41 Basiev, T.T., Sobol, A.A., Zverev, P.G., Osiko, V.V., and Powell, R.C. (1999) Comparative spontaneous Raman spectroscopy of crystals for Raman lasers. *Appl. Opt.*, **38** (3), 594–598.

42 Mochalov, I. (1997) Laser and nonlinear properties of the potassium gadolinium tungstate laser crystal KGd (WO): Nd-(KGW: Nd). *Opt. Eng.*, **36** (6), 1660–1669.

43 Basiev, T.T. (2005) New crystals for Raman lasers. *Phys. Solid State*, **47** (8), 1400–1405.

44 Ge, W., Zhang, H., Wang, J., Liu, J., Li, H., Cheng, X., Xu, H., Xu, X., Hu, X., and Jiang, M. (2005) The thermal and optical properties of $BaWO_4$ single crystal. *J. Cryst. Growth*, **276** (1–2), 208–214.

45 Bonner, G.M., Pask, H.M., Lee, A.J., Kemp, A.J., and Wang, J. (2012) Measurement of thermal lensing in a CW $BaWO_4$ intracavity Raman laser. *Opt. Express*, **20** (9), 365–372.

46 Kaminskii, A.A., Ueda, K., Eichler, H.J., Kuwano, Y., Kouta, H., Bagaev, S.N., Chyba, T.H., Barnes, J.C., Gad, G.M.A., Murai, T., and Lu, J. (2001) Tetragonal vanadates YVO_4 and $GdVO_4$ – new efficient chi-3 materials for Raman lasers. *Opt. Commun.*, **194**, 201–206.

47 Biswal, S., O'Connor, S., and Bowman, S.R. (2004) Thermo-optical parameters measured in potassium-gadolinium-tungstate. Technical Digest Conference on Lasers and Electro-optics, 16 May 2004, San Francisco, CA. Optical Society of America, Washington, DC, paper CThT62.

48 Ralston, J.M., and Chang, R.K. (1970) Spontaneous-Raman-scattering efficiency and stimulated scattering in silicon. *Phys. Rev. B*, **2** (6), 1858–1862.

49 Rhee, H., Lux, O., Meister, S., Woggon, U., Kaminskii, A.A., and Eichler, H.J. (2011) Operation of a Raman laser in bulk silicon. *Opt. Lett.*, **36** (9), 1644–1646.

50 Lisinetskii, V., Mishkel, I.I., Chulkov, R.V., Grabtchikov, A.S., Apanasevich, P.A., Eichler, H.-J., and Orlovich, V.A. (2005) Raman gain coefficient of barium nitrate measured for the spectral region of Ti: sapphire laser. *J. Nonlinear Opt. Phys. Mater.*, **14** (1), 107–114.

51 Lisinetskii, V., Rozhok, S., and Ko, D.B. (2005) Measurements of Raman gain coefficient for barium tungstate crystal. *Laser Phys.*, **2** (8), 396–400.

52 Basiev, T.T., Basieva, M.N., Doroshenko, M.E., Fedorov, V.V., Osiko, V.V., and Mirov, S.B. (2006) Stimulated Raman scattering in mid IR spectral range 2.31–2.75–3.7 μm in $BaWO_4$ crystal under 1.9 and 1.56 μm pumping. *Laser Phys. Lett.*, **3** (1), 17–20.

53 Levenson, M. and Bloembergen, N. (1974) Dispersion of the nonlinear optical susceptibility tensor in centrosymmetric media. *Phys. Rev. B*, **10** (10), 4447–4463.

54 Grimsditch, M. and Ramdas, A.K. (1975) Brillouin scattering in diamond. *Phys. Rev. B*, **11** (8), 3139–3148.

55 Cardona, M., Calleja, J.M., Meseguer, F., and Grimsditch, M. (1981) Resonance in the Raman scattering of CaF_2, SrF_2, BaF_2 and diamond. *J. Raman Spectrosc.*, **10** (1), 77–81.

56 Kaminskii, A.A., Hemley, R.J., Lai, J., Yan, C.S., Mao, H.K., Ralchenko, V.G., Eichler, H.J., and Rhee, H. (2007) High-order stimulated Raman scattering

in CVD single crystal diamond. *Laser Phys. Lett.*, **4** (5), 350–353.
57 Chandrasekharan, V. (1963) Scattering matrix for Raman effect in cubic crystals. *Z. Phys.*, **175** (1), 63–69.
58 Trunta, W.R., Park, Y.K., and Byer, R.L. (1979) The dependence of Raman gain on pump laser bandwidth. *IEEE J. Quantum Electron.*, **15** (7), 648–655.
59 Stappaerts, E.A., Long, W.H., and Komine, H. (1980) Gain enhancement in Raman amplifiers with broadband pumping. *Opt. Lett.*, **5** (1), 4–6.
60 Vodchits, A.I., Busko, D.N., Orlovich, V.A., Lisinetskii, V.A., Grabtchikov, A.S., Apanasevich, P.A., Kiefer, W., Eichler, H.J., and Turpin, P.-Y. (2007) Multi-frequency quasi-continuous wave solid-state Raman laser for the ultraviolet, visible and near infrared. *Opt. Commun.*, **272** (2), 467–475.
61 Kemp, A.J., Millar, P., Lubeigt, W., Hastie, J.E., Dawson, M.D., and Burns, D. (2009) Diamond in solid-state disk lasers: thermal management and CW Raman generation. Proceedings, Advanced Solid-State Photonics, February 1, 2009, Denver, CO. Optical Society of America, Washington, DC, paper WE7.
62 Tidwell, S.C., Seamans, J.F., Bowers, M.S., and Cousins, A.K. (1992) Scaling cw diode-end-pumped Nd:YAG laser to high average powers. *IEEE J. Quantum Electron.*, **28** (4), 997–1009.
63 Klein, C.A. and Cardinale, G. (1993) Young's modulus and Poisson's ratio of CVD diamond. *Diamond Relat. Mater.*, **2** (5–7), 918–923.
64 Koechner, W. (2006) *Solid-State Laser Engineering*, Sixth Revised and Updated Edition, Springer, New York.
65 Chen, Y.-F. (1999) Design criteria for concentration optimization in scaling diode end-pumped lasers to high powers: influence of thermal fracture. *IEEE J. Quantum Electron.*, **35** (2), 234–239.
66 Paci, J.T., Belytschko, T., and Schatz, G.C. (2005) The mechanical properties of single-crystal and ultrananocrystalline diamond: a theoretical study. *Chem. Phys. Lett.*, **414** (4–6), 351–358.
67 Klein, C.A. (1995) Laser-induced damage to diamond: dielectric breakdown and BHG scaling. *Proceedings of Conference: Laser-Induced Damage in Optical Materials, 24 October 1994, Boulder, CO* (eds H.E. Bennett, A.H. Guenther, M.R. Kozlowski, B.E. Newnam, and M.J. Soileau), SPIE, vol. 2428, pp. 517–530.
68 Bettis, J.R., House, R.A., II, and Guenther, A.H. (1976) Spot size and pulse duration dependence of laser-induced damage. *Natl Bur. Stand. (U.S.) Spec. Publ.*, **462**, 338.
69 Godfried, H., Kriele, P., Hall, C., Pickles, C., Coe, S., and Sussmann, R. (2000) Diamond optics for high power CO_2 lasers. Lasers and Electro-Optics Europe, Conference Digest, 10–15 September 2000. Nice Acropolis, Nice, France. European Physical Society, paper CMJ8.
70 Kitzler, O., McKay, A., and Mildren, R.P. (2012) Continuous wave wavelength conversion for high power applications using an external cavity diamond Raman laser. *Opt. Lett.*, **37** (14), 2790–2792.
71 Zhang, C., Zhang, X.Y., Wang, Q.P., Fan, S.Z., Chen, X.H., Cong, Z.H., Liu, Z.J., Zhang, Z., Zhang, H.J., and Su, F.F. (2009) Efficient extracavity Nd:YAG/$BaWO_4$ Raman laser. *Laser Phys. Lett.*, **6** (7), 505–508.
72 Zverev, P.G., Basiev, T.T., and Prokhorov, A.M. (1999) Stimulated Raman scattering of laser radiation in Raman crystals. *Opt. Mater.*, **11** (4), 335–352.
73 Benabid, F., Bouwmans, G., Knight, J., Russell, P., and Couny, F. (2004) Ultrahigh efficiency laser wavelength conversion in a gas-filled hollow core photonic crystal fiber by pure stimulated rotational Raman scattering in molecular hydrogen. *Phys. Rev. Lett.*, **93** (12), 123903–123906.
74 Dekker, P., Pask, H.M., Spence, D.J., and Piper, J.A. (2007) Continuous-wave, intracavity doubled, self-Raman laser operation in $Nd:GdVO_4$ at 586.5 nm. *Opt. Express*, **15** (11), 7038–7046.
75 Colles, M.J. (1971) Ultrashort pulse formation in a short-pulse-stimulated Raman oscillator. *Appl. Phys. Lett.*, **19**, 23–25.
76 Straka, P., Nicholson, J.W., and Rudolph, W. (2000) Synchronously pumped H_2 Raman laser. *Opt. Commun.*, **178** (1–3), 175–180.

77 Weitz, M., Theobald, C., Wallenstein, R., and L'Huillier, J.A. (2008) Passively mode-locked picosecond Nd: YVO$_4$ self-Raman laser. *Appl. Phys. Lett.*, **92**, 091122.

78 Granados, E., Pask, H.M., and Spence, D.J. (2009) Synchronously pumped continuous-wave mode-locked yellow Raman laser at 559 nm. *Opt. Express*, **17** (2), 569–574.

79 Grabtchikov, A.S., Lisinetskii, V.A., Orlovich, V.A., Schmitt, M., Maksimenka, R., and Kiefer, W. (2004) Multimode pumped continuous-wave solid-state Raman laser. *Opt. Lett.*, **29** (21), 2524–2526.

80 Kononenko, V.V., Komlenok, M.S., Pimenov, S.M., and Konov, V.I. (2007) Photoinduced laser etching of a diamond surface. *Quantum Electron.*, **37** (11), 1043–1046.

81 Mildren, R.P., Downes, J.E., Brown, J.D., Johnston, B.F., Granados, E., Spence, D.J., Lehmann, A., Weston, L., and Bramble, A. (2011) Characteristics of 2-photon ultraviolet laser etching of diamond. *Opt. Mater. Express*, **1** (4), 576–585.

82 Frenklach, M., Huang, D., Thomas, R.E., Rudder, R.A., and Markunas, R.J. (1993) Activation energy and mechanism of CO desorption from (100) diamond surface. *Appl. Phys. Lett.*, **63** (22), 3090–3092.

83 Hilfer, G. and Menyuk, C.R. (1992) Observation in wave-number space of pump-beam replication in stimulated Raman scattering. *Opt. Lett.*, **17** (13), 949–951.

84 Mildren, R.P. (2011) Side-pumped crystalline Raman laser. *Opt. Lett.*, **36** (2), 235–237.

85 DARPA. Black Diamond. Available at: http://www.darpa.mil/Our_Work/MTO/Programs/Black_Diamond.aspx (accessed December 2011).

86 Lisinetskii, V.A., Riesbeck, T., Rhee, H., Eichler, H.J., and Orlovich, V.A. (2010) High average power generation in barium nitrate Raman laser. *Appl. Phys. B*, **99** (1–2), 127–134.

87 Friel, I., Geoghegan, S.L., Twitchen, D.J., and Scarsbrook, G.A. (2010) Development of high quality single crystal diamond for novel laser applications. *Proceedings of SPIE*, vol. 7838, p. 19.

88 Zunger, A. (2003) Practical doping principles. *Appl. Phys. Lett.*, **83** (1), 57–59.

89 McKnight, L.J., Dawson, M.D., and Calvez, S. (2011) Diamond Raman waveguide lasers: completely analytical design optimization incorporating scattering losses. *IEEE J. Quantum Electron.*, **47** (8), 1069–1077.

90 Savitski, V., Friel, I., Hastie, J., Dawson, M.D., Burns, D., and Kemp, A.J. (2011) Characterization of single-crystal synthetic diamond for multi-watt continuous-wave Raman lasers. *IEEE J. Quantum Electron.*, **48** (3), 328–337.

9
Quantum Optical Diamond Technologies
Philipp Neumann and Jörg Wrachtrup

9.1
Introduction

9.1.1
Single Quantum Systems

The search for single quantum systems that can be deliberately fabricated and controlled while retaining their quantum features is not only motivated by quantum computation and communication. While miniaturization constantly decreases the size of electronic or electro-optic devices, the limit of single quantum building blocks might soon be reached. The operation of such devices does not need to exploit the full complexity of single quantum systems, as would a quantum computer; nonetheless, single quantum systems still need to be addressed and manipulated individually. In addition, in fields such as materials research or in the life sciences, instruments are required to make high spatial resolution measurements, where the actual sensor might very well be of the size of one atom or a small molecule. In that case, however, the sensitivity of quantum coherence to the local environment can indeed be exploited for measurement purposes.

The major challenges associated with all approaches towards single quantum system control are the individual addressing, the measurement of the quantum state, and the protection of (especially quantum) coherences from decoherence due to the environment.

Today, a variety of systems exists that allow control on the single quantum level. First, there are single photons which can be manipulated by linear optics elements [1–2]. In addition, single ions [3] or atoms [4] can be trapped, controlled and addressed individually, usually in a vacuum. The control of single molecules in the solid state has been demonstrated [5], and single spins in semiconductors can also be controlled coherently [6]. Eventually, mesoscopic systems may be developed that nevertheless exhibit mesoscopic quantum states which behave much like single quantum systems. Examples of this include superconducting devices [7], nanomechanical oscillators [8], or surface plasmon polaritons that carry a single quantum of excitation [9].

Optical Engineering of Diamond, First Edition. Edited by Richard P. Mildren and James R. Rabeau.
© 2013 Wiley-VCH Verlag GmbH & Co. KGaA. Published 2013 by Wiley-VCH Verlag GmbH & Co. KGaA.

9.1.2
The Nitrogen-Vacancy (NV) Center in Diamond

The NV center in diamond, as one of the major candidates for quantum information-processing applications, consists of a substitutional nitrogen atom next to a carbon vacancy in the diamond lattice [10]. Single NV centers can be addressed optically because of their high fluorescence yield upon optical excitation [11], which makes the NV center extremely interesting for applications as a room-temperature single photon source [12–15]. In addition, the electronic ground state is a spin triplet with exceptional coherence properties. Intersystem crossing (ISC) enables optically detected magnetic resonance (ODMR) of the electron spin state of a single NV center [11]. More precisely, the electron spin can be initialized into its $m_S = 0$ projection to a very high degree by optical pumping, and the fluorescence intensity will depend on the spin state even under ambient conditions. At cryogenic temperatures, spin-selective optical excitations are also allowed [16]. These exceptional properties of the diamond host lattice protect coherent spin states to a very high degree, which leads to the NV center serving as both a good quantum memory and a quantum sensor.

These promising features of the NV center in diamond have inspired many interesting experiments with regards to its applicability for quantum information processing (QIP). First, it was shown that the electron spin and associated nuclear spins could be controlled in a coherent fashion [17, 18]. Moreover, in addition to nuclear spins, neighboring electron spins–like that associated with nitrogen impurity atoms in the diamond lattice–can also be coupled to the NV center spin in either an incoherent [19] or coherent [20, 21] fashion. A deeper analysis of the coupling between electron and nuclear spins at the NV center [22–24] enabled the storage and retrieval of quantum information in and from a ^{13}C nuclear spin. Last, but not least, single NV centers can be deliberately created by ion implantation [25, 26].

9.1.3
The Present Situation

At this point, it can be shown how the optical access to single NV centers is used to address surrounding nuclear spins individually, and how these nuclear spins serve as qubits in quantum processors and memories. While the NV center electron spin acts as a bus qubit to mediate interactions among the nuclear spins, the latter are formidable quantum memories because of their weak interaction with the environment. In this way entanglement–which is the prime prerequisite for quantum registers–can be created in such systems. In addition, by tailoring the nuclear spin reservoir, the quantum register can be increased up to a certain degree without decreasing its performance.

9.2
The NV Center's Electron Spin as a Master Qubit

The NV center in diamond has several advantageous properties, which makes it a qubit candidate. Notably, it possesses an electron spin $S = 1$ in the ground state, which is the actual qubit or even a qutrit. This spin possesses both long transverse and longitudinal relaxation times under ambient conditions [27]. Intersystem crossing during optical pumping leads to polarization of that spin into its $m_S = 0$ state of the electronic ground state. Furthermore, the fluorescence of the NV center upon optical excitation depends on the spin state, and thus enables optical spin state readout. Eventually, the electron spin state can be manipulated (e.g., by resonant microwave radiation). The NV center is sufficiently stable under optical excitation – that is, it does not blink or bleach and, in particular, the nuclear spin states are preserved.

In these studies, the NV center spin is coupled to various other spins which usually are not directly accessible on the single level (see Figure 9.1) although, by using their coupling to the NV center, individual control is possible. For example, the NV center spin can be used for the initialization and readout of surrounding spin qubits. Apart from directly manipulating the quantum states of proximal spins, these states can also be transferred from and to the NV center; on this basis it may be referred to as the "master qubit," or occasionally a "bus qubit" [28, 29]. The coupling strength to other spins is switchable; for instance by changing into the $m_S = 0$ level, the interactions can be decreased. In addition to the

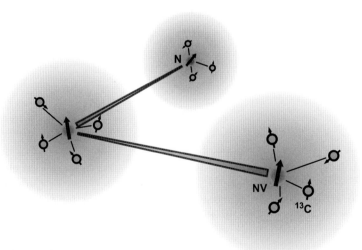

Figure 9.1 Schematic illustration of NV quantum register. ^{13}C or other nuclear spins coupled to NV center spins or other electron spins form local qubit clusters which are interconnected by the interaction of the electron spins among each other.

coupling to dark spins, several NV centers can also be coupled with each other (see Figure 9.1).

The NV center has additional promising applications in the field of metrology since, apart from coupling to other single spins, the magnetic dipole moment of the NV center electron spin is particularly susceptible to external magnetic fields. Due to its strong confinement to much less than $1\,\text{nm}^3$, the NV center allows magnetic field measurements with nanometer resolution and close proximity to the sample, irrespective of the environmental conditions [30, 31]. The external magnetic field which can be sensed by the NV spin could indeed also originate from a single external electron or nuclear spin [32]. In addition to magnetic fields, the NV center spin levels are susceptible to electric fields and crystal strain; thus, electric field sources (such as charges) in the vicinity of the NV center, as well as crystal strain around the NV position, can be detected [33]. The quantum registers examined in these studies also have very useful applications for future metrology devices based on the NV center and surrounding spins.

Although the NV center—with its above-mentioned properties—is quite unique at present, various strategies have been devised to search for defect centers with similar capabilities. However, the list of requirements for such a center reads like a "wanted" poster for the NV center [34]. Other materials where similar defect centers exist include SiC, MgO, or CaO [34, 35]. In diamond itself, other centers also exist that might be suitable for QIP, such as silicon and vacancy-related complexes [36].

9.3
Nuclear Spins as a Qubit Resource

In addition to the applicability of the electron spin of the NV center in diamond as a quantum bit, more sophisticated quantum effects can be studied if the electron spin qubit is coupled to other isolated quantum systems. Of these, single nuclear spins in the vicinity of the electron spin are probably the most obvious, and due to their extremely weak interaction with their environment they could serve as ideal storage qubits. Although the state of a coupled nuclear spin cannot be read out optically, as can the electron spin of the NV center, it can be measured indirectly by correlating its spin state with that of the NV center and reading out the latter.

Nuclear spins are also expected to be ideal candidates for QIP in other solid-state systems, an example being in the Kane proposal where the ^{31}P nuclear spin qubit is introduced [37]. Experiments using the phosphorus donor electron spin in silicon have recently achieved a single shot readout of that spin [38]. Other approaches in silicon are the use of ^{29}Si nuclear spin chains in otherwise spinless ^{28}Si [39]. In the case of diamond, other spins can also be used as qubits, such as electron and nuclear spins associated with single substitutional nitrogen impurities [20] or spins inside fullerenes [40].

9.3 Nuclear Spins as a Qubit Resource

In diamond, there are mainly two different nuclear spin species: (i) the nitrogen nuclear spin of isotopes ^{15}N and ^{14}N associated with the color center itself; and (ii) carbon nuclear spins of the ^{13}C isotopes spread over the whole lattice (see Table 9.1 and Figure 9.2). Coupling to these spins is mediated by a hyperfine interaction. For a coherent coupling between electronic and nuclear spins, the strength of that coupling must be stronger than the decoherence rates of both spins. Yet, despite this requirement, numerous nuclear spins in the vicinity of a NV center might serve as qubits. The interaction of NV centers with the electron spin is discussed in detail in the following subsection; furthermore, quantum gates such as the CNOT (controlled NOT) and the Toffoli gate are applied to generate highly

Table 9.1 Isotopes with nuclear spins. The important nuclear spins are those of certain isotopes of carbon and nitrogen; details of nuclear spin I, nuclear spin g_n-factor, and gyromagnetic ratio are provided, in addition to abundance.

Isotope	Natural abundance (%)	Nuclear spin I	g_n-factor and gyromagnetic ratio $\tilde{\gamma}_n$ (kHz mT^{-1})	
^{13}C	1.1	½	1.40482	10.7051
^{14}N	99.63	1	0.4037607	3.0766
^{15}N	0.37	½	−0.566380	−4.3156

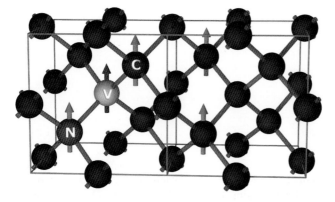

Figure 9.2 NV center and proximal nuclear spins two unit cells of the diamond lattice (C = carbon) containing an NV center (N = nitrogen, V = vacancy) and exemplary nuclear spins. The NV center electron spin is shown as a blue arrow, and the nuclear spins as orange arrows. From left to right the nuclear spins are that of nitrogen, ^{13}C in the first shell, and two ^{13}C in the third shell. For a color version of this figure, please see the color plate at the end of this book.

entangled states among electronic and nuclear spins. The results of these studies have been reported elsewhere [41, 42].

9.3.1
Interaction of a Single Electron Spin with nearby Nuclear Spins

9.3.1.1 Nuclear spin Hamiltonian

The Hamiltonian $\hat{H}_n$ of a nuclear spin surrounding an NV center in diamond can be divided into three terms [43]:

$$\hat{H}_n = \hat{H}_{hf} + \hat{H}_{nZ} + \hat{H}_Q \tag{9.1}$$

The first term is the hyperfine (hf) interaction with the NV electron spin, the second is the nuclear Zeeman (nZ) Hamiltonian depending on the static external magnetic field, while the final term describes the zero field energy for nuclear spins where $I > 1/2$. It should be noted, that the interaction with other nuclear spins is omitted here because it is negligible in strength due to their small concentration and, respectively, their large average distance. The effects of individual parts of the Hamiltonian are illustrated in Figure 9.3.

The hyperfine Hamiltonian $\hat{H}_{hf}$ comprises two contributions [43]; one is the isotropic Fermi contact interaction $\hat{H}_F$, and the other one is the anisotropic electron-nuclear magnetic dipole–dipole interaction $\hat{H}_{dd}$.

$$\hat{H}_{hf} = \hat{H}_F + \hat{H}_{dd} \tag{9.2a}$$

$$\hat{H}_F = -\frac{2\mu_0}{3}\tilde{\gamma}_e g_n \mu_n \left|\psi^e_{@n}\right|^2 \hat{\underline{S}} \cdot \hat{\underline{I}} = a_{iso} \hat{\underline{S}} \cdot \hat{\underline{I}} \tag{9.2b}$$

$$\hat{H}_{dd} = \frac{\mu_0}{4\pi}\tilde{\gamma}_e g_n \mu_n \frac{\hat{\underline{S}} \cdot \hat{\underline{I}} - 3(\hat{\underline{S}} \cdot \underline{e}_r)(\underline{e}_r \cdot \hat{\underline{I}})}{r^3} \tag{9.2c}$$

The Fermi contact interaction is proportional to the electron spin density at the location of the nucleus $\left(\left|\psi^e_{@n}\right|^2\right)$. Therefore, it is only of importance for nuclear

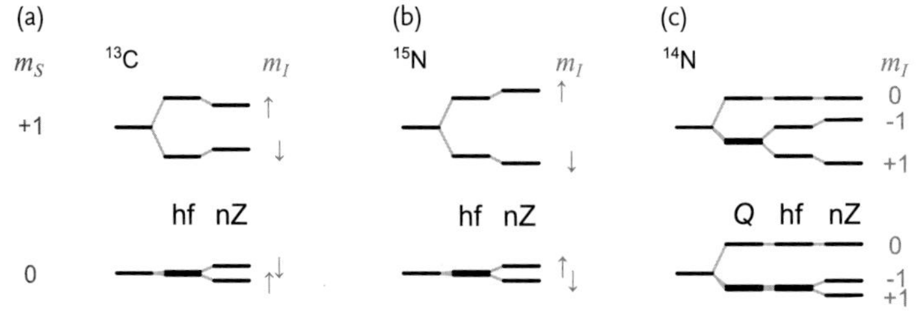

Figure 9.3 Nuclear spin Hamiltonian and energy levels. (a) Energy levels $m_I = \uparrow, \downarrow$ for a ^{13}C nuclear spin for two electron spin projections ($m_S = 0, +1$) are displayed. Subsequently, different parts of the Hamiltonian are switched on (see text); (b) Energy levels for ^{15}N; (c) Energy levels for ^{14}N exhibiting additional quadrupole splitting Q.

spins that are very close to the NV center, and thus can possess a reasonable amount of spin density. The magnetic dipole–dipole interaction, in contrast, is long range in nature, but this decreases very rapidly as $1/r^3$ with electron-nuclear spin distance r. As can be seen from Equation (9.2c), an anisotropy enters which depends on the orientation of the two spins with respect to each other. Here, $\underline{e}_r$ is the unit vector connecting them. For proximal nuclei, a point dipole approximation in Equation (9.2c) might no longer be valid, and it would be necessary to integrate $\hat{H}_{dd}$ for all r weighted by the electron spin density.

Often, the two contributions to the hyperfine interaction are summarized into one tensor **A** resulting in

$$\hat{H}_{hf} = \underline{\hat{S}} \cdot \mathbf{A} \cdot \underline{\hat{I}}. \tag{9.3}$$

In its eigensystem, the tensor is a diagonal of the form $A = \text{diag}(A_\perp, A_\perp, A_\parallel)$.[1]) The isotropic part of the interaction can be calculated as $a_{iso} = (A_\parallel + 2A_\perp)/3$ and the anisotropic one as $b = (A_\parallel - A_\perp)/3$. Conversely, $A_\perp = a_{iso} - b$ and $A_\parallel = a_{iso} + 2b$.

The nuclear Zeeman splitting is expressed in the following way

$$\hat{H}_{nZ} = -\tilde{\gamma}_n \underline{B} \cdot \underline{\hat{I}} \tag{9.4}$$

and the quadrupole splitting which, in the diamond lattice, applies only to the ^{14}N nuclear spin (see Table 9.1) can be written as

$$\hat{H}_Q = Q\hat{I}_z^2. \tag{9.5}$$

As illustrated in Figure 9.3c, Q for ^{14}N is negative and its value is −4.945 MHz [44].

9.3.1.2 Secular Approximation and Nonsecular Terms

In most cases, the electron Zeeman energy and the electron spin zero-field splitting are orders of magnitude larger than the hyperfine interaction (see Table 9.2). Thus, the electron spin states are almost unaffected by the nuclear spins, and this allows for the secular approximation of Equations (9.2) and (9.3). For a magnetic

Table 9.2 Hyperfine interaction strengths. Hyperfine splittings of NV center ESR lines due to nearby nuclear spins. For the carbon nuclear spins, several lattice positions with respect to the NV center are possible, leading to different values.

Nuclear spins	Observable hyperfine splitting (MHz)
^{13}C	0 . . . 126
^{14}N @ NV	−2.16
^{15}N @ NV	3.03

1) This is strictly true only for the point dipole approximation in Equation (9.2c). However, the hyperfine tensors of nuclei surrounding the NV are usually well approximated by $A_\perp$ and $A_\parallel$ [45, 46].

field applied parallel to the NV axis (z-direction), only the z-terms of the electron spin ($\hat{S}_z$) need to be taken into account. Hence, the eigenstates of the electron spin cannot be changed by the nuclear spins; only their energy can be altered by the nuclear spins. For the nuclear spins themselves, however, the situation is different. The electron spin state affects not only their energy levels but also their energy eigenstates by altering their quantization axes. An illustration of this would be the electron spin exerting a magnetic field on the nuclear spin in addition to the external field, and comparable in strength.

A consequence of the different nuclear spin quantization axes in different electron spin levels are nuclear spin Larmor precessions, which differ not only in frequency but also in their rotational axis. This behavior can be exploited to generate arbitrary controlled quantum gates on the nuclear spin [47].

In the case of the secular approximation, the sublevels of each electron spin projection differ only by their nuclear spin projection. Thus, magnetic dipole transitions between these levels can be induced by an oscillating magnetic field at the frequency of the respective energy splitting (i.e., the nuclear spin Larmor frequency). These transitions must obey the selection rules $\Delta m_I = \pm 1$. Usually, the Rabi frequencies achievable for these transitions are much smaller than for electron spin transitions, because of the much smaller nuclear gyromagnetic ratio compared to that of the electron spin. However, exceptions to this behavior can be observed where the nuclear Zeeman splitting as well as the nuclear spin Rabi oscillation frequency are larger than expected for a bare nuclear spin [22]. This whole effect is termed "hyperfine enhancement," and is caused by a tiny dressing of the nuclear spin transitions with electron spin character, which is a nonsecular effect. A detailed analysis of this is shown in Ref. [22]. In order to suppress the hyperfine enhancement, the magnetic field must be increased such that electron and nuclear Zeeman energies become increasingly off-resonant [48].

9.3.1.3 Examples of Nearby Nuclear Spins

Every NV center has at least a hyperfine interaction with one nuclear spin, namely that of the associated nitrogen atom. Typically, its nuclear spin is $I = 1$, but it might also be $I = \frac{1}{2}$, depending on the isotope (see Table 9.1). Due to a small electron spin density at the nitrogen nucleus, the hyperfine interaction is comparatively small. Depending on the nitrogen isotope present, the splitting of the ESR line is $\approx 2.2\,\text{MHz}$ (^{14}N) or $\approx 3\,\text{MHz}$ (^{15}N). For the ^{15}N nucleus, the nuclear magnetic resonance (NMR) resonance frequency can be deduced directly from the electron spin resonance (ESR) spectrum, and is the visible hyperfine splitting plus the nuclear Zeeman slitting. In the case of ^{14}N, the nuclear quadrupole splitting must also be taken into account. For moderate fields of up to 50 mT the observed nuclear Rabi frequencies will be higher than expected, due to hyperfine enhancement (see Section 9.3.1).

In addition to the nitrogen nuclear spin being present in every NV center, a special case also arises regarding symmetry. Typically, the nuclear spin lays on the NV axis (z-axis), and its hyperfine tensor therefore is rotationally symmetric and collinear with the NV axis. Thus, according to the secular approximation, the hyperfine field that the electron spin exerts on the nuclear spin also points in the z-direction (assuming small magnetic fields or fields in the z-direction). This has been exploited to demonstrate dynamic nuclear spin polarization [49] and quantum nondemolition measurements of single nuclear spins [48].

In addition to the nitrogen nuclear spin, every NV center is usually surrounded by numerous ^{13}C nuclear spins (see Figure 9.2). Depending on their position in the lattice, the hyperfine interaction strength varies drastically; for example, a ^{13}C atom on a lattice site next to the vacancy (i.e., in the first coordination shell) will generate a hyperfine splitting of 126 MHz [18]. Within this coordination shell, three indistinguishable positions have been identified. In general, those lattice sites around an NV center will form one coordination shell that can be converted into another by symmetry operations of the c_{3v} group with respect to the center. The next smallest observable hyperfine splitting is $\approx$14 MHz, followed by $\approx$13 MHz, $\approx$9 MHz, $\approx$6 MHz, $\approx$2 MHz, and $\approx$0.8 MHz [50]. Some corresponding ESR spectra exhibiting these splittings are shown in Figure 9.4. An assignment to distinct coordination shells or lattice sites is not straightforward, as the electron spin density is decreasing neither isotropically nor monotonically with distance from the vacancy. Some *ab initio* supercell calculation studies have been conducted of the electron spin density in the ground state of the NV$^-$ center, in addition to accompanying calculations of hyperfine interactions [45] which underline this behavior. Data from Ref. [45] are compared to the experimentally observed hyperfine splittings in Table 9.3, and a tentative assignment is given.

Due to the wide spread of nuclear spin resonance frequencies and hyperfine splittings, many different nuclei may be individually addressed at the same time.

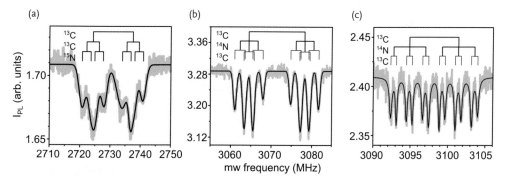

Figure 9.4 Optically detected magnetic resonance (ODMR) spectra with ^{13}C hyperfine splitting. (a) This spectrum shows the $m_S = 0 \leftrightarrow -1$ transition at $\approx$5 mT. The visible ^{13}C hyperfine interactions are 13.2 MHz and 4.2 MHz. The ^{15}N nuclear spin creates an additional splitting of $\approx$3 MHz. A concatenated stick spectrum illustrates the interactions; (b) This spectrum shows the $m_S = 0 \leftrightarrow +1$ transition at $\approx$11 mT. ^{13}C and ^{14}N hyperfine interaction is visible with the respective splittings 13.8 MHz, 2.2 MHz, and 2.6 MHz; (c) The same situation as in panel (b), but with hyperfine interactions 6.5 MHz, 2.2 MHz, and 0.8 MHz.

Table 9.3 Hyperfine interactions of various ^{13}C spins (experiment and theory). Experimentally observed ^{13}C hyperfine splittings (column 1) compared to results from *ab initio* supercell calculations (columns 2–3) (calculated results taken from Ref. [49]). The final column gives the distance of the corresponding nuclear spin to the vacancy. Question marks in the last row indicate that no calculated values have been available for such small coupling.

Experimental hf splitting (MHz)	Calculated hf splitting (MHz)	Calculated V–^{13}C distance (Å)
126	158.5	1.61
13.8	17.5	3.85
13.2	16.2	3.86
6.5	7.3	2.49
4.2	4.0	2.93
2.6	2.8	4.99
0.8	?	?

Therefore it is necessary that, apart from having different frequencies, the linewidths in ESR and NMR spectra must be smaller than the frequency difference of neighboring lines. For the example spectra in Figure 9.4, the lines are well separated, and this is true also for the respective nuclear spin frequencies because the linewidth there is usually much narrower. Taking into account the different nuclear spins in the list of Ref. [45], as well as experimentally observed spins, there are at least 12 different nuclear spin positions that can be addressed individually. If nuclear spins with even smaller hyperfine splittings are taken into account, then this number would increase; however, the distinguishability will also be reduced because the hyperfine splitting is then in the range of the ESR linewidth (cf. Figure 9.6a).

Apart from using nuclear spins in different coordination shells, more than one spin from the same coordination shell could be used. In this case, however, individual addressing might be difficult because especially nuclear spin transitions are almost equal. The reason for this is the small interaction of the nuclear spins among each other. However, if the interaction with the electron spin is sufficiently strong, a nuclear–nuclear interaction can be mediated. Especially in the case of two ^{13}C atoms, a level scheme arises in the first coordination shell where four distinct nuclear spin transitions are present (see Section 9.3.2).

In a natural diamond crystal, only 1.1% of all the carbon atoms possess a nuclear spin. Thus, the probability of identifying an NV center with a considerable amount of nuclear spins in close coordination shells is very low, and decreases exponentially with the number of desired spins. Although seeking such a center might pay dividends, other ways can be considered of increasing the probability. For example, molecules containing nitrogen atoms and a high amount of ^{13}C atoms might be implanted to produce a local enhancement of the spin density. An easier approach would be to fabricate diamond crystals with an enriched ^{13}C content, but in this case the nuclear spin bath would no longer be dilute and the effects on the NV electron spin would need to be considered (see Section 9.3.1.5).

9.3.1.4 Quantum Gates Using Nuclear Spins

As with the electron spin, every superposition of nuclear spin states can be created using radiofrequency (rf) pulses and phase shift gates that act directly on the nuclear spin. Unfortunately, these gates are usually not single-qubit in nature; rather, they are controlled two-qubit gates (i.e., they depend on the electron spin state). Because of the different hyperfine coupling strengths in m_S manifolds 0 and ±1, the rf pulses that are resonant in one manifold are off-resonant in the other (see Figure 9.5a). However, as long as only one electron spin manifold is being used, these gates can be considered as effective single-qubit gates. Again, many gates can be extracted from the nuclear Rabi oscillations [18] (see Figure 9.5b). Single-qubit phase shift gates can be realized in straightforward fashion by adjusting the phase of the driving rf field, or by using ancilla qubits.

A first demonstration of the "controlled" character of nuclear spin Rabi oscillations in the case of the NV center in diamond was demonstrated in Ref. [18], where a controlled rotation (CROT) gate was realized in a register out of an NV electron spin, and the nuclear spin of a ^{13}C atom in the first coordination shell (see Figure 9.5a, b). The large hyperfine interaction in this system means that it is easily possible to flip the electron spin conditional on the nuclear spin state, and vice versa. It should be noted that the corresponding ESR and NMR pulses therefore

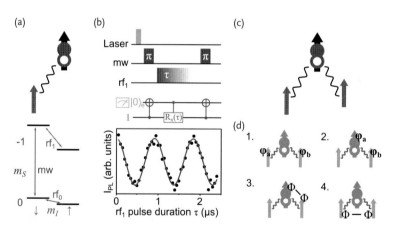

Figure 9.5 Nuclear spins and controlled qubit gates. (a) Illustration of a first shell ^{13}C nuclear spin coupled to an NV electron spin (top). The coupling is visualized by the wavy line. The corresponding energy level scheme is shown at the bottom. Blue and orange arrows illustrate EPR and NMR transitions, respectively; (b) Nuclear spin Rabi oscillation in the $m_S = -1$ electron spin sublevel (CROT gate, bottom) and corresponding pulse sequence (top, see text); (c) NV center spin coupled to two nuclear spins. The direct coupling among the nuclear spins is too weak to perform quantum gates; (d) Two qubit gates among nuclear spins exploiting the coupling to the common electron spin (see text). For a color version of this figure, please see the color plate at the end of this book.

comprise CNOT gates. To demonstrate this, the electron spin was first polarized into its $m_S = 0$ state after which, by a nuclear spin state-selective ESR π-pulse (CNOT gate), the nuclear spin state $|\downarrow\rangle_n$ was prepared at the $m_S = -1$ sublevel. An electron spin state-selective NMR pulse (CNOT gate) would then rotate the nuclear spin from $|\downarrow\rangle_n$ into $|\uparrow\rangle_n$ in the $m_S = -1$ sublevel, which would result in nuclear spin Rabi oscillations. This represented a controlled rotation, and therefore a CROT gate (see Figure 9.5b). As the fluorescence of the NV center usually depends only on the electron spin state, the nuclear spin Rabi oscillation performed in this way would not result in any oscillation of the fluorescence intensity. Consequently, the nuclear spin state of interest—namely, that in the $m_S = -1$ sublevel—must be correlated with the electron spin state, which is finally read out. Again, this correlation is achieved by performing the first CNOT gate on the electron spin.

In addition to driving the nuclear spin directly via rf radiation, the different hyperfine couplings in different m_S manifolds can be used to manipulate the spin state. In Section 9.3.1.2 it was shown how the quantization axis and the effective magnetic field at the nuclear spin depended on the m_S manifold and the external magnetic field. Larmor precessions around non-parallel quantization axes can finally be used to perform controlled quantum gates on the nuclear spin [47].

In principle, it is possible to compose a single-qubit rotation by applying two controlled rotations, each of which is controlled by another control qubit state.

To date, controlled gates have involved the electron spin and a nuclear spin, and this has been particularly easy because of the hyperfine coupling among them. The coupling among two nuclear spins, however, is usually far weaker and therefore not suitable to enable controlled gates (see Figure 9.5c). Controlled gates can, nevertheless, be achieved; one straightforward approach would be to use the electron spin as a bus qubit (see Figure 9.5d). Then, by swapping one nuclear spin qubit state onto the electron spin and performing the desired controlled operation with the second nuclear spin qubit and swapping it back onto the first nuclei, the interaction strength between the two qubits would be switched on virtually for computation, and switched off afterwards. A detailed theoretical analysis of how nuclear spin qubit gates can be realized in systems comprised of an NV electron spin and proximal nuclear spins is provided in Ref. [51].

9.3.1.5 The Nuclear Spin Bath

Although the use of nuclear spins as qubits has been discussed previously, in addition to being a qubit resource nuclear spins must first also form a spin bath that will affect the electron spin of the NV center. Because of the low natural abundance of the ^{13}C isotope (1.1%), spins in natural diamond possess favorable coherence properties. Apart from ^{13}C nuclear spins, electron spins associated with nitrogen defects in diamond represent another main source of decoherence [52]. However, the diamond samples used mainly for these studies contain only a very small fraction of nitrogen impurities (= 1 ppm, some even <1 ppb), and consequently the dilute nuclear spin bath may become the main source of decoherence [41]. The influence of the nuclear spin bath on the coherent superposition states of the NV electron spin will be detailed in the following subsection, and an analysis provided of the effects of nuclear spin baths with different densities.

Experimental Conditions The energy terms in most experiments on NV center spins are as follows. The ZFS parameter D is the strongest energy term for the electron spin Hamiltonian; second largest (and usually commuting with D) is the electron Zeeman energy, eZ; the smallest contribution is the coupling to other spins (electronic and nuclear). Usually, the latter is so small that only contributions commuting with D and eZ need to be taken into account, and thus a secular approximation is valid. In the case of nuclear spins the situation is different, however. Depending on the strength of the magnetic field and the distance to the NV center, either the hyperfine interaction (hf; see Table 9.2) with its electron spin or the nuclear Zeeman energy (nZ; see Table 9.1) is strongest, and usually both are not commuting. The interaction with other bath nuclear spins is much smaller (~10 Hz/%). As a consequence, switching the electron spin from one spin orientation m_S to another will change the quantization axis of the nuclear spin; this, in turn, will lead to an interesting behavior of the nuclear spin bath that affects the electron spin coherence of the NV center [22] (see Section 9.3.1.2).

The interaction of NV center spins with surrounding ^{13}C nuclear spins was first observed in ensemble experiments [53], while the case of a single NV electron spin was first observed and explained in Ref. [22]. In general, the interaction of a solid-state electron spin with a nuclear spin bath can be calculated using a cluster expansion method, as demonstrated for Si:P [54]. In this case, it could also be shown that this fully quantum mechanical approach would yield better results than a classical stochastic modeling. Here, the term "cluster" relates to clusters of more strongly interacting nuclear spins which are indeed allowed to interact with each other, whereas weaker interactions among nuclei would be neglected. It could also be shown that small cluster sizes would allow the decay of electron spin coherence (as measured by a Hahn-echo sequence) to be modeled, although larger cluster sizes are required to model echo modulations. The application of cluster expansion methods to the NV electron spin and the surrounding nuclear spin bath was first demonstrated in Ref. [55], where it transpired that some effects of this interaction could only be explained by quantum phenomena.

Electron Spin Decoherence Coherent evolution can be monitored in two main ways. The first approach involves a Ramsey-type experiment in which a coherent superposition between two electron spin states of the NV center is created, evolves freely, and is eventually read out. In this case, many runs are performed and an average time is obtained. During these many runs different statistically chosen nuclear spin bath settings will be present, whereas during one run the nuclear spin setting is hardly changed. Each setting creates a different local magnetic field at the NV center position, which can be explained as classical field fluctuation, and this leads to a Gaussian decay of the spin coherence (see Figure 9.6a) [56]. In particular, proximal nuclear spins have interaction strengths of up to ~1 MHz.

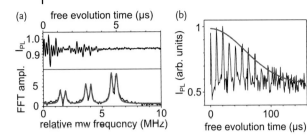
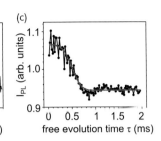

Figure 9.6 Ramsey and Hahn echo experiments affected by nuclear spins. (a) Ramsey oscillations and corresponding fast Fourier transform (FFT) spectrum showing the coupling to the ^{14}N nucleus and most likely a weakly coupled ^{13}C nuclear spin; (b) Electron spin echo envelope modulation (ESEEM) for Hahn echo of an NV center, showing collapses and revivals due to ^{13}C nuclear spin bath; (c) ESEEM for higher magnetic fields where collapses are suppressed. For a color version of this figure, please see the color plate at the end of this book.

Hence, the Ramsey signal decays on the time scale of a few microseconds (see Figure 9.6a), yielding the T_2^* time or the inhomogeneous linewidth:

$$\Delta v_{inh} = 2\sqrt{\ln 2} / (\pi T_2^*). \tag{9.6}$$

Experimentally, it emerges that the afore mentioned nuclear spin bath configurations are changed slowly from one to another (~2 kHz) compared to Δv_{inh}.

This situation is exploited, and thus made visible, by a second coherence monitoring method where, for a single measurement run, a free evolution due to a certain bath configuration can be refocused with a Hahn echo sequence. Hence, the coherence can be retained for up to $T_2 \approx 500\,\mu s$ (see Figure 9.6c). In order to monitor the coherence, the electron spin echo envelope modulation (ESEEM) is recorded. Consequently, a superposition of $m_S = 0$ and $m_S = -1$ usually evolves for time τ,[2] a microwave π-pulse on the corresponding transition acts as a time reversal for all static environmental influences, and after another evolution time τ the originally generated superposition is restored. For different values of τ the final state might differ from the original value, which is visible as a modulation of the echo amplitude (see Figure 9.6b). The modulation and the decay of the echo amplitude can no longer be explained by classical noise.

When the magnetic field is parallel to the NV center axis (i.e., when D and eZ commute) and is small such that nZ and hf are similar, the ESEEM will show distinct collapses and revivals (see Figure 9.6b). These can be explained by an entanglement and disentanglement of the electron spin with the bath nuclear spins [22]. If entanglement is present at the time of the microwave π-pulse, the ESEEM will show breakdowns. A revival of the electron spin coherence is caused by the common Larmor precession frequency ω_{L0} of the nuclear spins in the $m_S = 0$

---

2) Superpositions of states $m_S = 0$ and $+1$ or of $m_S = -1$ and $+1$ are also possible.

manifold. This leads to a common clock for the entanglement and disentanglement, which in turn leads to a lack of entanglement at times that are multiples of $2\pi/\omega_{L0}$.

On deviating from the usual scenario (as explained initially) and increasing the magnetic field, the nuclear Zeeman energy will become larger than the hyperfine interaction for almost all nuclear spins. This will reduce the possibility of entangling electron and nuclear spins, such that the collapses will disappear (see Figure 9.6c).

Finally, the ESEEM amplitude will decrease and eventually disappear. The reasons for this may include a nonuniform nuclear Larmor precession in the bath due to coupling to the NV spin, strong couplings of nearest-neighbor ^{13}C nuclei, or decoherence due to electron spins [55].

Nuclear Spin Bath of Varying Density Experiments on diamond samples with artificially tailored ^{13}C concentration c_{13} are summarized at this point [41]. Samples with both enriched and lowered ^{13}C concentrations were fabricated by collaboration partners (i.e., c_{13} = 00%[†], 20.7%, 8.4%, 1.1%, 0.35%, and 0.03%[†]).[3)] All samples exhibited a low nitrogen content such that the nuclear spin bath would be the dominant decoherence source for the NV electron spin.

Both, homogeneous and inhomogeneous linewidths (i.e., T_2 and T_2^*, respectively) have been determined for all samples; consequently, ODMR spectra were monitored, Ramsey experiments were conducted, and the ESEEM has been recorded. In the case of the 100% sample, no ESEEM was possible due to the rapid dephasing, whereas for samples with concentrations $c_{13} \leq 1.1\%$, T_2^* was derived from the ODMR linewidth [Equation (9.6)], and for the other samples by fitting the Ramsey decay.

On examining the inhomogeneous linewidth as a function of c_{13}, two regimes become visible (see Figure 9.7a). Whereas, in both regions the linewidth scales as $\sqrt{c_{13}}$ [41, 57, 58], there is an offset between both regimes of a factor of approximately 50. The reason for this behavior is the very localized electron spin density of the NV center. For concentrations $\leq 1.1\%$, it is unlikely that a ^{13}C atom is present in the closest shells around the NV center (i.e., within the high electron spin density region). Thus, the nuclear spins have a small interaction with the NV spin, due mainly to dipole–dipole coupling ($\approx 20\,\mathrm{kHz}\frac{\mathrm{nm}^3}{r^3}$). Hence, even if there is a ^{13}C in the closest shells, this will occasionally produce a visible splitting of the ODMR line that is not blurred by interactions of the other spins. If, however, the concentration exceeds 1.1%, an increasing number of nuclear spins will occupy the closest shells; in this case, the Fermi contact interaction is dominating over pure dipole–dipole interaction. Thus, if the number of nuclear spins in that region is sufficient to smear out the individual hyperfine split ODMR lines, an effective

---

3) The samples were provided by Element Six Ltd and Tokyo Gas Co., Ltd. All samples except those marked [†] were produced by microwave plasma-assisted homoepitaxial CVD growth, using different ratios of ^{12}CH$_4$ and ^{13}CH$_4$. The samples marked [†] were produced using the HPHT technique.

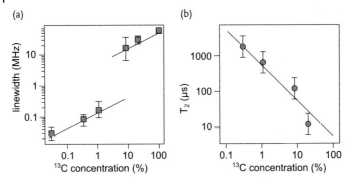

Figure 9.7 Dependence of NV center electron spin coherence properties on ^{13}C concentration. (a) Inhomogeneous linewidth of the NV electron spin EPR transition for various ^{13}C concentrations. Two regimes are clearly visible (see text); (b) Phase memory time T_2 of the NV center electron spin for various ^{13}C concentrations (see text).

broadening will occur with the same scaling as for the dipole–dipole case, but with a much larger prefactor.

The phase memory time T_2 will also increase with decreasing ^{13}C concentration. The mechanism of decoherence during a Hahn echo measurement can be simulated using the so-called disjoint cluster approach [55], which predicts a dependence of the T_2 time on the average nuclear spin–spin interaction within the bath C and the characteristic interaction strength between electron and nuclear spins A_c (i.e., $T_2 \propto 1/\sqrt{CA_c}$). Because A_c and C both depend on c_{13} (i.e., $A_c, C \propto c_{13}$), the overall concentration dependence of the phase memory time is $T_2 \propto 1/c_{13}$. This predicted behavior has been verified experimentally; in Figure 9.7b the concentration dependence of the T_2 time is displayed, and is nicely fitted by an inverse proportionality.

Although, T_2 decreases with increasing ^{13}C concentration, a working nuclear spin quantum register could be demonstrated in these samples, owing to the superior coherence properties of the nuclear spins [41].

The Frozen Core The frozen core around the NV center contains those nuclear spins that possess a stronger hyperfine coupling to the NV center electron spin than the width of the energy level splitting of the bath nuclear spins. In turn, these nuclear spins are decoupled from flip-flops with the rest of the bath spins, due to an energy mismatch. All nuclear spins that fall into the frozen core are potential qubit candidates.

In order to estimate the frozen core, some numbers must first be collected. The hyperfine coupling among the NV electron spin and a ^{13}C bath nuclear spin is on the order of 20 kHz for a distance of 1 nm (i.e., $E_{en} \approx 20\,\text{kHz}\,\text{nm}^{-3}$). The average distance of the nuclear spins for a concentration of 1% is almost 1 nm, and the corresponding dipole–dipole interaction among these is on the order of 10 Hz.

Indeed, experimental results have shown NMR linewidths of the ^{13}C spin transition of ≈ 7 Hz [59]. The average coupling among the bath spins scales in proportion to the concentration (i.e., $\tilde{E}_{nn} \sim 10$ Hz/% or $\tilde{E}_{nn} \sim 10$ Hz nm^{-3}). Now, it is possible to estimate the volume of the frozen core as $E_{en}/\tilde{E}_{nn} \sim 1000$ nm^3% (i.e., with an extension of approximately 10 nm and an inverse scaling with the concentration). Although the size of the frozen core changes, the number of nuclear spins inside will remain constant with E_{en}/E_{nn} equal to approximately 1000 spins.

The estimates for the frozen core are approximate, and neither do they take into account the full dipole–dipole interaction, with its angle dependence; however, the numbers do provide a feeling for the dimensions. With a demonstrated tailoring of the nuclear spin bath, it should be possible to extend the range of the frozen core further, from ~10 nm for 1% concentration to ~50 nm for 0.01%. This becomes interesting if distant nuclear spins become coupled to the NV spin, or even to nuclear spins outside the diamond lattice. In particular, it must be borne in mind that the frozen core exists only for electron spin projections $m_S = \pm 1$, whereas $m_S = 0$ does not possess a magnetic dipole moment that could produce a frozen core.

9.3.2
Nonlocal States: The Heart of a Quantum Processor

It has been shown that controlled qubit gates can be realized in a system comprising an NV electron spin and the nuclear spin of a ^{13}C atom in the first coordination shell [18]. Ultimately, however, the aim would be to create nonlocal states in such registers, and details of the generation and tomography of entangled states in a three-qubit quantum register are presented in the following section. The register comprises the NV electron spin plus two ^{13}C nuclear spins in the first coordination shell (i.e., one ^{13}C nuclear spin more than in the case described in Ref. [18]). This enables entanglements to be generated within a purely nuclear spin environment. Moreover, if the electron spin is also included, then a tripartite entanglement can be generated.

9.3.2.1 Two Nearest-Neighbor ^{13}C Nuclear Spins
Initially, the system is introduced focusing on interaction strength, energy levels, and eigenstates. The NV defect with two ^{13}C isotopes in the first coordination shell is depicted in Figure 9.8a, and in addition a mirror plane is highlighted to show the reduced symmetry when taking into account the nuclear spins. The NV spin exerts a large hyperfine field on both nuclei, and vice versa; this yields the ODMR spectrum shown in Figure 9.8c.

The spin Hamiltonian describing the system and finally the spectrum can be written in the following fashion:

$$\hat{H} = \hat{H}_{NV} + \hat{H}_{n1} + \hat{H}_{n2}$$
$$= \hat{H}_{NV} + \hat{H}_{nZ1} + \hat{H}_{nZ2} + \hat{\underline{S}} \cdot \sum_{i=1..2} \mathbf{A}_i \cdot \hat{\underline{I}}_i \qquad (9.7)$$

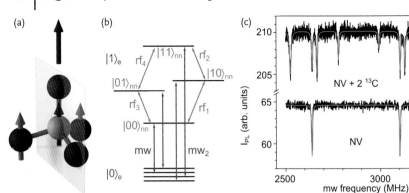

Figure 9.8 NV center with two ^{13}C nuclei in the first coordination shell. (a) NV center in the diamond lattice including the first coordination shell of carbon atoms, two of which are ^{13}C isotopes; (b) Spin energy level spectrum including electron spin $m_S = 0, -1$ ($|0\rangle_e, |1\rangle_e$) levels, times the four levels corresponding to the two nuclear spins with $I = ½$. For details of names, see the text; (c) Upper ODMR spectrum at 8.32 mT shows large hyperfine splitting due to the two nuclear spins. Both EPR transitions are shown, the red curve is a fit. The lower ODMR spectrum is a bare NV center at the same magnetic field, for comparison. For a color version of this figure, please see the color plate at the end of this book.

where the subscript i enumerates the ^{13}C nuclear spins. It should be noted that although both hyperfine tensors $\mathbf{A}_i$ possess the same constants $a_{iso} = 150.33$ MHz and $b = 27.33$ MHz [60], their symmetry axis is not parallel to the axis of the NV center. This implies that both tensors are also not parallel, and therefore have different reference frames and hence different representations in the laboratory frame.

According to secular approximation, the two central lines in Figure 9.8c would fall together but, as can be seen, the hyperfine constants are still smaller than D yet are sufficiently strong such that the nonsecular terms cannot be neglected. Thus, the hyperfine interaction between electron and nuclear spins leads to a certain mixing of the electron spin states $m_S = 0, \pm 1$. Second-order perturbation theory, taking into account also the nonsecular terms, leads to a splitting of the formerly degenerate central lines [43] (cf. Figure 9.8c). This provides a qualitative explanation of the spectrum, and the strength of the splitting is also quite well reproduced.

A further consequence of the perturbation approach is the new eigenstates. As the two nuclear spins are à priori not distinguishable, the single spin product basis is not appropriate. Indeed, the above-mentioned approximation assumes that the nuclear spin eigenstates are triplets and a singlet, whereas the electron spin states are only changed slightly. However, the different orientations of the hyperfine tensors of the two nuclei distort the singlet–triplet eigenfunctions. The indistinguishability of the two nuclear spins can be lifted by variations in the very local environment of the nuclei, or by magnetic or electric fields that do not possess the shown mirror symmetry. The correct eigenstates and eigenenergies can be obtained by diagonalizing the spin Hamiltonian of Equation (9.7).

In the energy level scheme obtained above, qubits can be encoded. In fact, the whole spin state $\Psi = \Psi_e \otimes \Psi_n$ can be written as a product state of a mainly electronic spin part Ψ_e and a mainly nuclear spin part Ψ_n. The electronic spin part represents one qubit, E. Starting in the $m_S = -1$ manifold which is labeled $\Psi_e = |E = 1\rangle_e$ ($m_S = 0 \rightarrow \Psi_e = |E = 0\rangle_e$), it is possible to encode two nuclear spin qubits N_1 and N_2 as $\Psi_n = |N_1 N_2\rangle_{nn}$ in the four nuclear spin energy levels. This is a straightforward procedure, according to the NMR transitions that can be driven (cf. Figure 9.8b). It should be noted that the qubit states are encoded in the given eigenstates, and can therefore not be identified with single nuclear spin product states (i.e., $|m_{I1}\rangle_{n1} \otimes |m_{I2}\rangle_{n2}$) in general [61, 62]. Whereas, this is possible for the states $|00\rangle_{nn} = |\uparrow\rangle_{n1} \otimes |\uparrow\rangle_{n2}$ and $|11\rangle_{nn} = |\downarrow\rangle_{n1} \otimes |\downarrow\rangle_{n2}$, it is not possible for states $|01\rangle_{nn}$ and $|10\rangle_{nn}$, where $|01\rangle_{nn}$ has a strong singlet character and $|01\rangle_{nn}$ has a triplet character.

9.3.2.2 Characterization of the Qubit System

The system comprised of qubits E, N_1, and N_2 must be characterized for use as a quantum register. Consequently, the energy level scheme calculated from the Hamiltonian must be verified, the basic qubit gates must be implemented, and an initialization procedure is required.

Following the application of a green laser pulse, the electron spin state is polarized into $\Psi_e = |0\rangle_e$ and the nuclear spins are in an incoherent mixture $\rho_{nn} = 1_{nn}$. The nuclear spins can be either initialized by an additional procedure [23, 49], or a probabilistic initialization of the nuclear spin in a certain electron spin sublevel can be performed by the application of a nuclear spin state-selective π-pulse on the electron spin (CNOT gate). Here, a CNOT gate is applied to the electron spin conditional on the nuclear spin state $\Psi_n = |00\rangle_{nn}$. Because of the incoherent mixture of nuclear spin states in the first place, this CNOT will result in $\Psi = |1\rangle_e \otimes |00\rangle_{nn}$, which occurs only in one-fourth of the time. Only then will the following operations have any effect on the final fluorescence signal.

In order to manipulate the nuclear spin qubits in the $|1\rangle_e$ level, it is essential to know the exact nuclear spin transition frequencies (see Figure 9.9a). As a starting point, the results of the diagonalized Hamiltonian are appropriate, but to refine the transition frequencies electron-nuclear double resonance (ENDOR) spectra are recorded (see Figure 9.9). Starting with the initialized state $\Psi = |1\rangle_e \otimes |00\rangle_{nn}$, an rf pulse almost in resonance with transition $|1\rangle_e \otimes |00\rangle_{nn} \leftrightarrow |1\rangle_e \otimes |01\rangle_{nn}$ or $|1\rangle_e \otimes |00\rangle_{nn} \leftrightarrow |1\rangle_e \otimes |10\rangle_{nn}$ is applied. It should be noted that this operation is a CROT gate, while the resultant nuclear spin state is measured by correlation with the electron spin state via a CNOT gate. Subsequently, the electron spin state is measured by a readout laser pulse. When the rf frequency is swept around the expected transition frequencies, the resonances appear (cf. Figure 9.9a, c). The rf frequencies for transitions

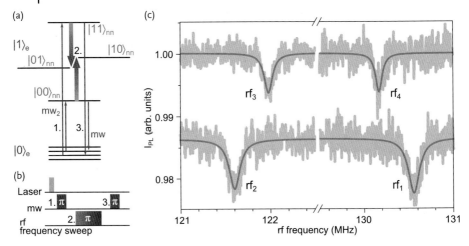

Figure 9.9 Electron-nuclear double resonance (ENDOR) spectra of ^{13}C nuclear spins. (a) Energy level scheme containing microwave (mw) and radiofrequency (rf) pulses to find rf resonance frequencies. The bold orange arrows with changing color indicate frequency sweeps; (b) Measurement sequence containing laser initialization, mw π-pulse (either mw or mw$_2$) and rf π-pulse with swept frequency. A spectrum is recorded as the rf frequency is swept while the sequence is running; (c) Actual spectra showing the correct transition frequencies for Figure 9.8b. For a color version of this figure, please see the color plate at the end of this book.

$|1\rangle_e \otimes |11\rangle_{nn} \leftrightarrow |1\rangle_e \otimes |01\rangle_{nn}$ and $|1\rangle_e \otimes |11\rangle_{nn} \leftrightarrow |1\rangle_e \otimes |10\rangle_{nn}$ are obtained in a similar fashion, but starting with state $|1\rangle_e \otimes |11\rangle_{nn}$ instead of $|1\rangle_e \otimes |00\rangle_{nn}$.

Due to the finite rf pulse lengths of a few microseconds (see Figure 9.9b), the linewidth is limited. Nuclear spin Ramsey measurements revealed a nuclear spin T_2^* of more than several hundred microseconds, and a T_2 time that was limited by the electron spin T_1 time.

By knowing the exact resonance frequencies, it is possible to start driving coherent Rabi oscillations on these transitions, from which $\pi/2$ and π pulse lengths can be deduced. As these Rabi oscillations are conditional rotations, the $\pi/2$ and π pulses are also conditional Hadamard and CNOT/Toffoli gates, respectively. The corresponding pulse sequences and Rabi oscillations are shown in Figure 9.10. Apparently, the Rabi frequency is very high for nuclear spins, a fact that can be attributed to hyperfine enhancement (compare Section 9.3.1.2). In addition, the Rabi oscillations exhibit a cosine form without any decay, thus underlining the superior coherence properties of nuclear spins as compared to electron spins.

9.3.2.3 Generation and Detection of Entanglement

Generation of Entangled States Although, until now, only superposition states of one qubit (e.g., $|0\rangle_1 + |1\rangle_1$) have been generated, when working with several qubits

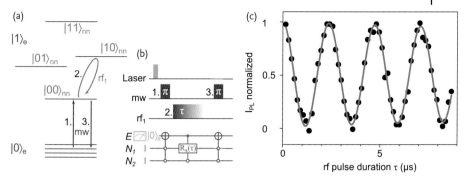

Figure 9.10 Coherent manipulation of ^{13}C nuclear spins. (a) Energy level scheme with arrows corresponding to microwave (mw) and radiofrequency (rf) transitions used for monitoring nuclear spin Rabi oscillations; (b) Pulse sequence for the experiment. A laser pulse initializes the electron spin into $|0\rangle_e$, after which a mw π-pulse flips the electron spin for nuclear spin state $|00\rangle_{nn}$ and a subsequent rf pulse resonant for the illustrated transition rotates the nuclear spin. Finally, spin state $|1\rangle_e |00\rangle_{nn}$ is probed by a second mw pulse and the laser pulse of the next sequence; (c) Measurement outcome. The fluorescence level oscillates while the nuclear spin is rotated. The vertical axis corresponds to the population that is found in state $|1\rangle_e \otimes |00\rangle_{nn}$. For a color version of this figure, please see the color plate at the end of this book.

it is also possible to create these superposition states on each qubit individually, leading to a separable or product state (e.g., $(|0\rangle_1 + |1\rangle_1) \otimes (|0\rangle_2 + |1\rangle_2)$). In addition, a new type of superposition states arises, namely the entangled state. Entanglement is present if a state $|\Psi\rangle$ is not separable; for example, it cannot be written as a product state. In order to access the whole Hilbert space spanned by the present qubits, it is also necessary to be able to generate all product states and all entangled states.

In the following subsection, the generation and detection of the most basic and most popular entangled states – namely the Bell states – are presented, together with details of entanglement among three qubits by the generation and detection of the so-called GHZ and W states.

The *Bell states*

$$\Psi^\pm = |01\rangle_{nn} \pm |10\rangle_{nn} \quad (9.8a)$$

$$\Phi^\pm = |00\rangle_{nn} \pm |11\rangle_{nn} \quad (9.8b)$$

are encoded among the nuclear spins in the electron state $\Psi_e = |1\rangle_e$. Starting with $|00\rangle_{nn}$, a conditional Hadamard gate, the first qubit generates $(|00\rangle_{nn} + |10\rangle_{nn})$, while a final CNOT gate on the second qubit leads to Φ^+. If the former CNOT gate is conditional on the state "0," the resulting state would be Ψ^+. However, by adjusting the Hadamard gate, Ψ^- and Φ^- can also be generated.

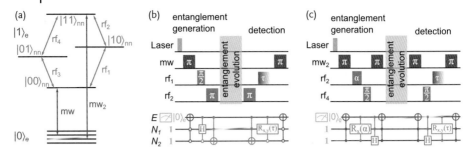

Figure 9.11 Sequence for entanglement generation and detection. (a) Energy level scheme showing contributing spin states and microwave (mw) and radiofrequency (rf) transitions used; (b) Exemplary sequence to generate and detect the Φ^- Bell state (see text). The other Bell states are created by changing the frequency of the rf$_2$ pulse and length and frequency of the rf$_1$ pulse. To record Ramsey oscillations, the free evolution time is varied and the final rf$_1$ pulse is a $\pi/2$-pulse. For state tomography, the final rf$_1$ pulse length is varied (Rabi oscillation) for two 90° phase-shifted rf fields; (c) Sequence to generate and detect the W state (analogous to Bell states; see text). For a color version of this figure, please see the color plate at the end of this book.

The GHZ state: Generation of the GHZ state starts with the Bell state Φ^+. Because this Bell state is present in the electronic spin state $\Psi_e = |1\rangle_e$, a Toffoli gate on the electron spin conditional on state $\Psi_n = |00\rangle_{nn}$ results in the GHZ state:

$$GHZ = |0\rangle_e \otimes |00\rangle_{nn} + |1\rangle_e \otimes |11\rangle_{nn}. \tag{9.9}$$

The W state: This is the following equal superposition state of three eigenstates,

$$W = |0\rangle_e \otimes |11\rangle_{nn} + |1\rangle_e \otimes |10\rangle_{nn} + |1\rangle_e \otimes |01\rangle_{nn}. \tag{9.10}$$

Its creation is a bit more complex. Starting with state $|1\rangle_e \otimes |11\rangle_{nn}$, a CROT gate on qubit N_2 with a rotation angle (α) of about 70.5° ($= 2\arccos\sqrt{2/3}$) results in $\sqrt{1/3}|1\rangle_e \otimes |10\rangle_{nn} + \sqrt{2/3}|1\rangle_e \otimes |11\rangle_{nn}$. A CROT gate on qubit N_1 with a rotation angle of $\pi/2$ then leads to $|1\rangle_e \otimes |10\rangle_{nn} + |1\rangle_e \otimes |11\rangle_{nn} + |1\rangle_e \otimes |01\rangle_{nn}$. Finally, a Toffoli gate on the electron spin conditional on the nuclear spins state $|11\rangle_{nn}$ leads to the W state.

The pulse sequences and the quantum circuit diagrams corresponding to the generation and detection of the above-mentioned entangled states are shown in Figure 9.11. After a free evolution period, the populations and coherences of the entangled states can be read out by similar pulse sequences.

Detection of Entangled States Entanglement can be measured in different ways. A first option is to monitor the coherent evolution of the phase of the entangled states (as detailed in Ref. [63]); a second option is to perform quantum state tomography.

The phase of an entangled state evolves characteristically with respect to the driving fields that are used for its generation. This evolution can be monitored in a Ramsey-type experiment, where the phase of the entangled state is read out after a free evolution time. In the case of the Bell states for generating entanglement, two rf driving fields are used (rf_i; see Figure 9.8b). The phases of the Bell states depend on the detunings Δv_i of these driving fields in the following way:

$$\Psi^+ : \varphi_\Psi^+ = 2\pi(\Delta v_1 - \Delta v_3)\tau \tag{9.11a}$$

$$\Psi^- : \varphi_\Psi^- = 2\pi(\Delta v_1 - \Delta v_3)\tau + \pi \tag{9.11b}$$

$$\Phi^+ : \varphi_\Phi^+ = 2\pi(\Delta v_1 + \Delta v_2)\tau \tag{9.11c}$$

$$\Phi^- : \varphi_\Phi^- = 2\pi(\Delta v_1 + \Delta v_2)\tau + \pi \tag{9.11d}$$

where τ is the free evolution period of the entangled state in the Ramsey experiment.

Figure 9.12 shows the measurement results for detunings Δv_i given in the caption. Clearly, the Ramsey frequencies and detunings obey Equations 9.11a–d. In addition, the "+" and "−" versions of the Ψ and Φ Bell states show Ramsey oscillations that are phase-shifted by π as compared to their counterparts (see Figure 9.12b, d).

A second possibility of unambiguously proving entanglement is quantum state tomography. For this, the entire density matrix of the entangled state must be reconstructed, including all populations and the quantum coherences among them. The populations and coherences can be extracted by performing Rabi oscillations on the respective transitions. As an example, the oscillation in Figure 9.13 reveals the coherence of the Φ^- Bell state.

The resulting density matrices for all four Bell states, which are shown in Figure 9.14, exhibit the characteristic features for entangled states. In order to evaluate

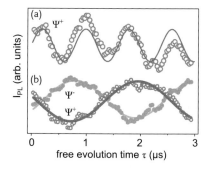
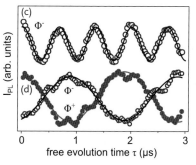

Figure 9.12 Ramsey interferometry with entangled states. (a) Ramsey oscillation of the Ψ^+ Bell state. The frequency should be $\Delta v_1 - \Delta v_3 = 1\,\text{MHz} - (-0.3\,\text{MHz}) = 1.3\,\text{MHz}$, and the fit yields 1.25 MHz; (b) Ramsey oscillations of Ψ^+ and Ψ^- reveal their π phase shift; (c) Ramsey oscillation of the Φ^- Bell state. The frequency should be $\Delta v_1 + \Delta v_2 = 0.5\,\text{MHz} + 1\,\text{MHz} = 1.5\,\text{MHz}$, and the fit yields 1.51 MHz; (d) Ramsey oscillations of Φ^+ and Φ^- reveal their π phase shift.

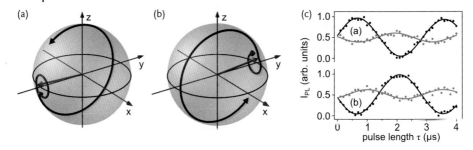

Figure 9.13 Quantum state tomography—Rabi signal. (a) Bloch sphere illustrating Rabi oscillations for quantum state tomography (see text). Here, the main density matrix entries of a Φ^- state (green arrow) have been transferred onto the working transition rf_1 (Bloch sphere), where the Rabi oscillations are performed around the x axis (black line) and the y axis (red line); (b) Tomography of Φ^+; (c) Actual measurement data of the Rabi oscillations (z-components of quantum states) for Φ^+ and Φ^- Bell states that lead to the state reconstruction shown in panel (a, upper part) and panel (b, lower part). For a color version of this figure, please see the color plate at the end of this book.

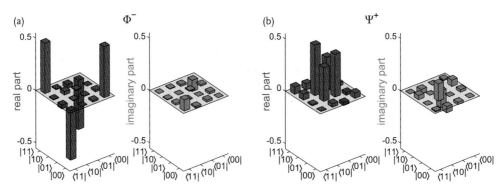

Figure 9.14 Results of quantum state tomography exemplary for (a) Φ^- Bell states and (b) Ψ^+ Bell states.

entanglement, different measures are used, namely the fidelity $\mathcal{F}$, the partial transpose $\mathcal{T}_p$, and the concurrence $\mathcal{C}$. The fidelity $\mathcal{F}$ is a measure of how well the actually generated state ρ matches the desired state σ. For a pure state σ (all Bell states are pure states), $\mathcal{F}$ can be calculated as follows:

$$\mathcal{F} = \text{Tr}(\sigma\rho) \tag{9.12}$$

where Tr is the trace of a matrix. The partial transpose, $\mathcal{T}_p$, is the smallest eigenvalue of a partially transposed density matrix of two qubits. It has been shown that two qubits are entangled if – and only if – this partially transposed density matrix

Table 9.4 Entanglement measures for the generated Bell states and the GHZ and W states. The partial transpose and the Concurrence are only defined for two qubit systems.

	Fidelity $\mathcal{F}$	Concurrence $\mathcal{C}$	Partial transpose $\mathcal{T}_p$
Ψ^+	0.80 ± 0.07	$0.65^{+0.15}_{-0.08}$	$-0.31^{+0.05}_{-0.06}$
Ψ^-	0.81 ± 0.06	0.59 ± 0.11	$-0.32^{+0.04}_{-0.05}$
Φ^+	0.98 ± 0.05	$0.96^{+0.04}_{-0.09}$	$-0.49^{+0.04}_{-0.02}$
Φ^-	$0.96 \pm .05$	$0.92^{+0.07}_{-0.08}$	$-0.47^{+0.05}_{-0.06}$
GHZ	0.87 ± 0.06	–	–
W	$0.85^{+0.05}_{-0.1}$	–	–

achieves a negative eigenvalue [64, 65]. Eventually, the Concurrence of a system of two qubits must be larger than zero in order for the entanglement of formation to be larger than zero [66, 67]. It should be noted that $\mathcal{C}$ and $\mathcal{T}_p$ are both defined for systems that comprise only two qubits. The details of all three above-mentioned entanglement measures for all four generated Bell states are presented in Table 9.4. As can be seen, the overlap of the generated state with the desired state is quite large and, in particular, values of $\mathcal{F}$ close to 1 are reached for the Φ Bell states. The Concurrence, as well as the partial transpose, also verify that entanglement has indeed been created.

Any deviation from the targeted Bell states can be explained by the rf pulses that were used for Bell state generation [68]; typically, these are square pulses with durations of a few microseconds. According to Ref. [68] the fidelity can be increased to values greater than 0.99 for all Bell states when specially designed robust rf pulses are used.

In addition to the degree of entanglement, the lifetime of the entangled states is also important. Usually, the coherence lifetime of an entangled state differs from that of a single qubit coherent superposition state; such behavior is expressed for instance by the phase dependence on the detuning of the driving fields, as in Equation (9.11). Such detuning can be produced by a magnetic field offset. Thus, the phase of $\Phi^\pm$ is twice as sensitive to magnetic field offset as would be the phase of a single qubit equal superposition state, while the phase of $\Psi^\pm$ would be unaffected. While the former states could be used for metrology, as they are more sensitive [69], the latter states may serve as ideal storage states for quantum information [70, 71].

In order to monitor the coherence lifetime of the entangled states, the amount of coherence for different free evolution times must be measured. It emerges that every Bell state has a coherence lifetime of ≈ 4 ms, which is similar to the electron spin T_1 time of the NV center. Thus, decoherence of the nuclear spin Bell states will be mainly limited by the longitudinal relaxation time of the electron spin. Differences between the decoherence rates of Φ and Ψ Bell states may have been noted for longer T_1 times of the electron spin.

The first quantum algorithms have been implemented in this register of two ^{13}C nuclear spins. In Ref. [72], entanglement was used to perform superdense coding and the Deutsch algorithm.

Three-Partite Entanglement As in the case of the Bell states, the density matrix of the GHZ has been partially reconstructed to verify its entanglement. Hence, all entries necessary for fidelity calculations have been measured in a similar fashion as in the two-qubit case (i.e., by Rabi oscillations of the corresponding density matrix elements on the working transition). The pulse sequences are shown in Figure 9.11c, while the acquired density matrix entries for the GHZ and W states are shown in Figure 9.15a, b (upper parts), and the resultant fidelities are listed in Table 9.4.

It should be noted that, whereas the GHZ state contains a three-qubit entanglement, the W state does not. Rather, the W state contains a two-qubit entanglement which manifests in the following fashion. If any qubit of the W state is measured projectively, then the other two will be either in an eigenstate or in a maximally

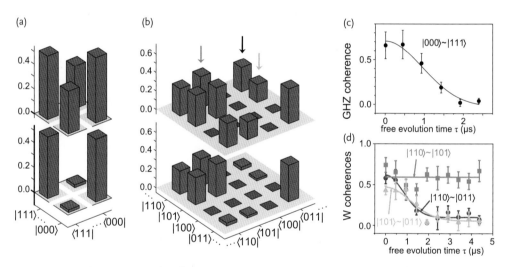

Figure 9.15 Results of quantum state tomography of GHZ and W states. (a) Partial tomography of the GHZ state, showing only the main density matrix entries. The upper part is for zero evolution time (see Figure 9.11), and the lower part shows the decayed coherences after 2.4 μs; (b) Partial tomography of the W state (gray entries have not been measured). The red, green, and black arrows mark the three different coherences. The upper topography is taken immediately after generation, and the lower tomography after an evolution time of 4.4 μs. The bare nuclear spin coherence (red arrow) survives longest; (c, d) Decay of the coherence of GHZ and W states, respectively, for increasing free evolution time. For the GHZ state, the decay of the single coherence (off-diagonal density matrix entry) is shown, whereas for the W state three coherences decay (for color coding, see panel b). For a color version of this figure, please see the color plate at the end of this book.

entangled Bell state, conditional on the result of the projective measurement. Thus, even when tracing out one qubit, the other two will have some degree of entanglement; this behavior becomes apparent when examining the time evolution of the coherences of the entangled states.

Such time evolution is monitored by performing a partial density matrix reconstruction after several free evolution times of the entangled states. For the GHZ state, only one coherence (namely that between $\Psi = |0\rangle_e \otimes |00\rangle_{nn}$ and $|1\rangle_e \otimes |11\rangle_{nn}$) needs to be monitored. However, because of the much shorter coherence lifetime of the electron spin (which now forms part of the entanglement), the coherence of the GHZ state will decay within 2 μs (see Figure 9.15c). As the measurement sequence is of the Ramsey type (see Figure 9.11), decay due to inhomogeneous broadening with lifetime T_2^* can be seen to occur. In addition to the partial density matrix reconstruction of the initial GHZ state, another reconstruction is shown for an evolution time $>T_2^*$ (see Figure 9.15a, lower part). As expected, the populations of eigenstates $\Psi = |0\rangle_e \otimes |00\rangle_{nn}$ and $|1\rangle_e \otimes |11\rangle_{nn}$ remained, but the coherence among them had decayed such that only classical correlations remained.

When monitoring the evolution of the W state, it is important to keep track of three coherences, namely those among $\Psi = |0\rangle_e \otimes |11\rangle_{nn}$, $|1\rangle_e \otimes |10\rangle_{nn}$, and $|1\rangle_e \otimes |01\rangle_{nn}$). On closer examination of these states, it is clear that a phase between states $|1\rangle_e \otimes |01\rangle_{nn}$ and $|1\rangle_e \otimes |10\rangle_{nn}$ does not possess any electron spin character, but that the other two phases (i.e., those between one of the former two states and $|0\rangle_e \otimes |11\rangle_{nn}$) do possess an electron spin character. Thus, different decay characteristics can be expected for these coherences. Indeed, it can be seen that two coherences of the W state decay within 2 μs, one coherence persists for at least 4 μs, but only the purely nuclear spin coherence remains (Figures 9.15b, d).

The rapid decay of electron spin coherence can be explained as a measurement of the electron spin state by the environment, but which occurs much faster for electron spins than for nuclear spins because of the much larger electron spin magnetic moment. These measurements have two possible outcomes: an electron spin state "0," or spin state "1." Whilst the "1" case delivers an entangled nuclear spin state, case "0" delivers a nuclear spin eigenstate, as verified by the density matrix reconstruction measured for a phase accumulation time $\tau > 2\,\mu s$ (Figure 9.15b, lower part).

9.4
Summary and Outlook

In addition to the storage of quantum information, a quantum register comprising the NV electron spin and two neighboring ^{13}C nuclear spins can also be used as a small quantum processor. The full Hilbert space of this three-qubit system was shown to be accessible, which is mandatory for quantum computation. As a consequence, up to three-partite entangled states with high fidelity were created [42], and the quantum register was then used either as a test bed for quantum algorithms in a solid state at room temperature (such as super-dense coding), or in the Deutsch algorithm [72]. Theoretical considerations regarding the use of several nuclear spin qubits associated with the NV center are provided in Ref. [51].

In other solid-state qubit systems, such as GaAs quantum dots, nuclear spins provide an excellent thread for quantum coherences [73]. In diamond, however, the nuclear spin concentration associated with ^{13}C is comparably low (1.1%), which results in other defects such as nitrogen causing a greater decoherence [52, 74]. With advances in artificial diamond production, however, concentrations of nitrogen and other defects might be reduced such that the nuclear spins may again cause the main part of the decoherence. This dependence was demonstrated by investigating isotopically tailored artificial diamond samples [41]. Indeed, the electron spin coherence properties of the NV center were improved when the ^{13}C concentration was reduced. In addition, it could be shown that even slightly higher than normal ^{13}C concentration would still allow working quantum registers. Hence, controlled quantum gates in a system of three ^{13}C and one ^{14}N nuclear spin, plus the NV electron spin, could be achieved with moderate coherence times [41].

With the availability of isotopically enriched ^{12}C diamond samples of reduced nuclear spin density, and accompanying increases in the coherence properties of the NV center, spins that were even further apart could be coupled coherently. In this way, it became possible to create a pair of coupled NV centers [75].

Until now, single NV centers together with proximal nuclear spins have been shown to provide highly versatile, small quantum registers that not only allow small algorithms to be performed [72, 76] but also provide protection for quantum information [48, 77, 78]. Further improvements, notably with regards to this protective role, can be achieved using special ^{13}C nuclear spins that lie on the NV center symmetry axis or ^{13}C dimers [79]. Both of these species should be capable of protecting quantum information to a very high degree. In addition, a scaling of the quantum register appears to be feasible by using arrays of coupled NV centers [75].

Ultimately, the challenge will arise to fabricate these registers in a more or less deterministic fashion, and this will require exceptionally pure diamond host crystals and advanced NV creation techniques, with nanometer positioning accuracy [80–83]. It would also be advantageous to produce spatially varying magnetic and electric fields on the same length scale for convenient and switchable spin selectivity [82, 84]. However, for selective readout and initialization, optical super-resolution techniques must first be adopted [85–88].

Progress in the above-mentioned diamond and NV production techniques should lead to the storage time for quantum information (especially on nuclear spins) exceeding seconds, while controlled quantum gates among an array of NV spins should be achievable on timescales of tens of microseconds. This ratio of coherence times versus quantum gate times is very favorable for potential quantum computers.

Today, with NV centers used as nanoscale magnetic field sensors having been shown feasible [30, 31, 89], the first use of an NV quantum register represents a clear improvement in the development of such (or similar) metrology devices. The proposed use of the NV center for quantum repeater networks [90] also requires a small quantum register that facilitates quantum error correction. Entanglement of the NV electron spin with an emitted photon has recently been demonstrated in a proof-of-principle experiment [91].

Apart from purely NV diamond-based quantum registers, hybrid approaches are also currently a major topic of discussion. In such quantum processor schemes, the advantages of several systems should be combined in a beneficial manner, an example being the coupling of NV centers and associated spins to superconducting resonators and qubits [92, 93]. Proposals such as the coupling of spins by nano-mechanical oscillators are also currently being discussed [94].

Although these studies are concerned with single NV centers, the fraction exploiting ^{14}N nuclear spins can also be employed in ensemble experiments, as almost every NV has its own ^{14}N nucleus. For example, bulk magnetometer applications would benefit greatly from readout enhancements enabled by nitrogen nuclear spins [89, 95, 96].

References

1 Knill, E., Laflamme, R., and Milburn, G.J. (2001) A scheme for efficient quantum computation with linear optics. *Nature*, **409** (6816), 46–52.
2 O'Brien, J.L. (2007) Optical quantum computing. *Science*, **318** (5856), 1567–1570.
3 Blatt, R. and Wineland, D. (2008) Entangled states of trapped atomic ions. *Nature*, **453** (7198), 1008–1015.
4 Bloch, I. (2008) Quantum coherence and entanglement with ultracold atoms in optical lattices. *Nature*, **453** (7198), 1016–1022.
5 Wrachtrup, J., von Borczyskowski, C., Bernard, J., Orritt, M., and Brown, R. (1993) Optical detection of magnetic resonance in a single molecule. *Nature*, **363** (6426), 244–245.
6 Hanson, R. and Awschalom, D.D. (2008) Coherent manipulation of single spins in semiconductors. *Nature*, **453** (7198), 1043–1049.
7 Clarke, J. and Wilhelm, F.K. (2008) Superconducting quantum bits. *Nature*, **453** (7198), 1031–1042.
8 Kippenberg, T.J. and Vahala, K.J. (2008) Cavity optomechanics: back-action at the mesoscale. *Science*, **321** (5893), 1172–1176.
9 Kolesov, R., Grotz, B., Balasubramanian, G., Stöhr, R.J., Nicolet, A.A.L., Hemmer, P.R., Jelezko, F., and Wrachtrup, J. (2009) Wave-particle duality of single surface plasmon polaritons. *Nat. Phys.*, **5** (7), 470–474.
10 Davies, G. and Hamer, M.F. (1976) Optical studies of the 1.945 eV vibronic band in diamond. *Proc. R. Soc. Lond., A, Math. Phys. Eng. Sci.*, **348** (1653), 285–298.
11 Gruber, A., Dräbenstedt, A., Tietz, C., Fleury, L., Wrachtrup, J., and Borczyskowski, C. (2012) 1997) Scanning confocal optical microscopy and magnetic resonance on single defect centers. *Science*, **276** (5321), 2014.
12 Kurtsiefer, C., Mayer, S., Zarda, P., and Weinfurter, H. (2000) Stable solid-state source of single photons. *Phys. Rev. Lett.*, **85** (2), 290–293.
13 Beveratos, A., Brouri, R., Gacoin, T., Poizat, J.P., and Grangier, P. (2001) Nonclassical radiation from diamond nanocrystals. *Phys. Rev. A*, **64**, 061–802.
14 Beveratos, A., Brouri, R., Poizat, J.P., and Grangier, P. (2002) Bunching and antibunching from single NV color centers in diamond, in *Quantum Communication, Computing, and*

Measurement 3 (eds P. Tombesi and O. Hirota), Springer, pp. 261–267.
15 Beveratos, A., Brouri, R., Gacoin, T., Villing, A., Poizat, J.P., and Grangier, P. (2002) Single photon quantum cryptography. *Phys. Rev. Lett.*, **89**, 187–901.
16 Jelezko, F., Popa, I., Gruber, A., Tietz, C., Wrachtrup, J., Nizovtsev, A., and Kilin, S. (2002) Single spin states in a defect center resolved by optical spectroscopy. *Appl. Phys. Lett.*, **81** (12), 2160–2162.
17 Jelezko, F., Gaebel, T., Popa, I., Gruber, A., and Wrachtrup, J. (2004) Observation of coherent oscillations in a single electron spin. *Phys. Rev. Lett.*, **92** (7), 076–401.
18 Jelezko, F., Gaebel, T., Popa, I., Domhan, M., Gruber, A., and Wrachtrup, J. (2004) Observation of coherent oscillation of a single nuclear spin and realization of a two-qubit conditional quantum gate. *Phys. Rev. Lett.*, **93** (13), 130–501.
19 Epstein, R.J., Mendoza, F.M., Kato, Y.K., and Awschalom, D.D. (2005) Anisotropic interactions of a single spin and dark-spin spectroscopy in diamond. *Nat. Phys.*, **1** (2), 94–98.
20 Gaebel, T., Domhan, M., Popa, I., Wittmann, C., Neumann, P., Jelezko, F., Rabeau, J.R., Stavrias, N., Greentree, A.D., Prawer, S., Meijer, J., Twamley, J., Hemmer, P.R., and Wrachtrup, J. (2006) Room-temperature coherent coupling of single spins in diamond. *Nat. Phys.*, **2** (6), 408–413.
21 Hanson, R., Mendoza, F.M., Epstein, R.J., and Awschalom, D.D. (2006) Polarization and readout of coupled single spins in diamond. *Phys. Rev. Lett.*, **97** (8), 087 601.
22 Childress, L., Dutt, M.V.G., Taylor, J.M., Zibrov, A.S., Jelezko, F., Wrachtrup, J., Hemmer, P.R., and Lukin, M.D. (2006) Coherent dynamics of coupled electron and nuclear spin qubits in diamond. *Science*, **314** (5797), 281–285.
23 Dutt, M.V.G., Childress, L., Jiang, L., Togan, E., Maze, J., Jelezko, F., Zibrov, A.S., Hemmer, P.R., and Lukin, M.D. (2007) Quantum register based on individual electronic and nuclear spin qubits in diamond. *Science*, **316** (5829), 1312–1316.
24 Jiang, L., Dutt, M.V.G., Togan, E., Childress, L., Cappellaro, P., Taylor, J.M., and Lukin, M.D. (2008) Coherence of an optically illuminated single nuclear spin qubit. *Phys. Rev. Lett.*, **100** (7), 073 001.
25 Meijer, J., Burchard, B., Domhan, M., Wittmann, C., Gaebel, T., Popa, I., Jelezko, F., and Wrachtrup, J. (2005) Generation of single color centers by focused nitrogen implantation. *Appl. Phys. Lett.*, **87** (26), 261–909.
26 Rabeau, J.R., Reichart, P., Tamanyan, G., Jamieson, D.N., Prawer, S., Jelezko, F., Gaebel, T., Popa, I., Domhan, M., and Wrachtrup, J. (2006) Implantation of labelled single nitrogen vacancy centers in diamond using N-15. *Appl. Phys. Lett.*, **88** (2), 023 113.
27 Balasubramanian, G., Neumann, P., Twitchen, D., Markham, M., Kolesov, R., Mizuochi, N., Isoya, J., Achard, J., Beck, J., Tissler, J., Jacques, V., Hemmer, P.R., Jelezko, F., and Wrachtrup, J. (2009) Ultralong spin coherence time in isotopically engineered diamond. *Nat. Mater.*, **8** (5), 383–387.
28 Mehring, M. and Mende, J. (2006) Spin-bus concept of spin quantum computing. *Phys. Rev. A*, **73**, 052 303.
29 Jiang, L., Taylor, J.M., Sørensen, A.S., and Lukin, M.D. (2007) Distributed quantum computation based on small quantum registers. *Phys. Rev. A*, **76** (6), 062 323.
30 Maze, J.R., Stanwix, P.L., Hodges, J.S., Hong, S., Taylor, J.M., Cappellaro, P., Jiang, L., Dutt, M.V.G., Togan, E., Zibrov, A.S., Yacoby, A., Walsworth, R.L., and Lukin, M.D. (2008) Nanoscale magnetic sensing with an individual electronic spin in diamond. *Nature*, **455** (7213), 644–647.
31 Balasubramanian, G., Chan, I.Y., Kolesov, R., Al-Hmoud, M., Tisler, J., Shin, C., Kim, C., Wojcik, A., Hemmer, P.R., Krueger, A., Hanke, T., Leitenstorfer, A., Bratschitsch, R., Jelezko, F., and Wrachtrup, J. (2008) Nanoscale imaging magnetometry with diamond spins under ambient conditions. *Nature*, **455** (7213), 648–651.
32 Grotz, B., Beck, J., Neumann, P., Naydenov, B., Reuter, R., Reinhard, F., Jelezko, F., Wrachtrup, J., Schweinfurth,

D., Sarkar, B., and Hemmer, P. (2011) Sensing external spins with nitrogen-vacancy diamond. *New J. Phys.*, **13** (5), 055 004.

33 Dolde, F., Fedder, H., Doherty, M.W., Nobauer, T., Rempp, F., Balasubramanian, G., Wolf, T., Reinhard, F., Hollenberg, L.C.L., Jelezko, F., and Wrachtrup, J. (2011) Electric-field sensing using single diamond spins. *Nat. Phys.*, **7** (6), 459–463.

34 Weber, J.R., Koehl, W.F., Varley, J.B., Janotti, A., Buckley, B.B., Van de Walle, C.G., and Awschalom, D.D. (2010) Quantum computing with defects. *Proc. Natl Acad. Sci. USA*, **107** (19), 8513–8518.

35 Weber, J.R., Koehl, W.F., Varley, J.B., Janotti, A., Buckley, B.B., Van de Walle, C.G.V., and Awschalom, D.D. (2011) Defects in sic for quantum computing. *Appl. Phys. Lett.*, **109** (10), 102–417.

36 D'Haenens-Johansson, U.F.S., Edmonds, A.M., Newton, M.E., Goss, J.P., Briddon, P.R., Baker, J.M., Martineau, P.M., Khan, R.U.A., Twitchen, D.J., and Williams, S.D. (2010) EPR of a defect in CVD diamond involving both silicon and hydrogen that shows preferential alignment. *Phys. Rev. B* **82**, 155205. doi: 10.1103/PhysRevB.82.155205.

37 Kane, B.E. (1998) A silicon-based nuclear spin quantum computer. *Nature*, **393** (6681), 133–137.

38 Morello, A., Pla, J.J., Zwanenburg, F.A., Chan, K.W., Tan, K.Y., Huebl, H., Mottonen, M., Nugroho, C.D., Yang, C., van Donkelaar, J.A., Alves, A.D.C., Jamieson, D.N., Escott, C.C., Hollenberg, L.C.L., Clark, R.G., and Dzurak, A.S. (2010) Single-shot readout of an electron spin in silicon. *Nature*, **467** (7316), 687–691.

39 Ladd, T.D., Goldman, J.R., Yamaguchi, F., Yamamoto, Y., Abe, E., and Itoh, K.M. (2002) All-silicon quantum computer. *Phys. Rev. Lett.*, **89**, 017 901.

40 Harneit, W. (2002) Fullerene-based electron-spin quantum computer. *Phys. Rev. A*, **65**, 032 322.

41 Mizuochi, N., Neumann, P., Rempp, F., Beck, J., Jacques, V., Siyushev, P., Nakamura, K., Twitchen, D.J., Watanabe, H., Yamasaki, S., Jelezko, F., and Wrachtrup, J. (2009) Coherence of single spins coupled to a nuclear spin bath of varying density. *Phys. Rev. B*, **80** (4), 041 201.

42 Neumann, P., Mizuochi, N., Rempp, F., Hemmer, P., Watanabe, H., Yamasaki, S., Jacques, V., Gaebel, T., Jelezko, F., and Wrachtrup, J. (2008) Multipartite entanglement among single spins in diamond. *Science*, **320** (5881), 1326–1329.

43 Weil, J.A., Bolton, J.R., and Wertz, J.E. (1994) *Electron Paramagnetic Resonance: Elementary Theory and Practical Applications*, John Wiley & Sons, Inc.

44 Steiner, M., Neumann, P., Beck, J., Jelezko, F., and Wrachtrup, J. (2010) Universal enhancement of the optical readout fidelity of single electron spins at nitrogen-vacancy centers in diamond. *Phys. Rev. B*, **81** (3), 035 205.

45 Gali, A., Fyta, M., and Kaxiras, E. (2008) Ab initio supercell calculations on nitrogen-vacancy center in diamond: electronic structure and hyperfine tensors. *Phys. Rev. B*, **77** (15), 155–206.

46 Gali, A. (2009) Identification of individual C^{13} isotopes of nitrogen-vacancy center in diamond by combining the polarization studies of nuclear spins and first-principles calculations. *Phys. Rev. B*, **80** (24), 241–204.

47 Hodges, J.S., Yang, J.C., Ramanathan, C., and Cory, D.G. (2008) Universal control of nuclear spins via anisotropic hyperfine interactions. *Phys. Rev. A* **78**, 010303. doi: 10.1103/PhysRevA.78.010303.

48 Neumann, P., Beck, J., Steiner, M., Rempp, F., Fedder, H., Hemmer, P.R., Wrachtrup, J., and Jelezko, F. (2010) Single-shot readout of a single nuclear spin. *Science*, **329** (5991), 542–544.

49 Jacques, V., Neumann, P., Beck, J., Markham, M., Twitchen, D., Meijer, J., Kaiser, F., Balasubramanian, G., Jelezko, F., and Wrachtrup, J. (2009) Dynamic polarization of single nuclear spins by optical pumping of nitrogen-vacancy color centers in diamond at room temperature. *Phys. Rev. Lett.*, **102** (5), 057 403.

50 Smeltzer, B., Childress, L., and Gali, A. (2011) ^{13}C hyperfine interactions in the nitrogen-vacancy centre in diamond. *New J. Phys.*, **13** (2), 025 021.

51 Cappellaro, P., Jiang, L., Hodges, J.S., and Lukin, M.D. (2009) Coherence and control of quantum registers based on electronic spin in a nuclear spin bath. *Phys. Rev. Lett.*, **102** (21), 210–502.

52 Takahashi, S., Hanson, R., van Tol, J., Sherwin, M.S., and Awschalom, D.D. (2008) Quenching spin decoherence in diamond through spin bath polarization. *Phys. Rev. Lett.*, **101** (4), 047 601.

53 van Oort, E. and Glasbeek, M. (1990) Optically detected low field electron spin echo envelope modulations of fluorescent N-V centers in diamond. *Chem. Phys.*, **143**, 131–140.

54 Witzel, W.M., de Sousa, R., and Das Sarma, S. (2005) Quantum theory of spectral-diffusion-induced electron spin decoherence. *Phys. Rev. B*, **72** (16), 161–306.

55 Maze, J.R., Taylor, J.M., and Lukin, M.D. (2008) Electron spin decoherence of single nitrogen-vacancy defects in diamond. *Phys. Rev. B*, **78** (9), 094 303.

56 Dobrovitski, V.V., Feiguin, A.E., Awschalom, D.D., and Hanson, R. (2008) Decoherence dynamics of a single spin versus spin ensemble. *Phys. Rev. B*, **77**, 245–212.

57 Kohn, W. (1957) *Solid State Physics*, vol. 5, Academic Press, New York.

58 Abe, E., Tyryshkin, A.M., Tojo, S., Morton, J.J.L., Witzel, W.M., Fujimoto, A., Ager, J.W., Haller, E.E., Isoya, J., Lyon, S.A., Thewalt, M.L.W., and Itoh, K.M. (2010) Electron spin coherence of phosphorus donors in silicon: effect of environmental nuclei. *Phys. Rev. B*, **82** (12), 121–201.

59 Zhou, J., Li, L., Hu, H., Yang, B., Dan, Z., Qiu, J., Guo, J., Chen, F., and Ye, C. (1994) Study of natural diamonds by dynamic nuclear polarization-enhanced ^{13}C nuclear magnetic resonance spectroscopy. *Solid State Nucl. Magn. Reson.*, **3** (6), 339–351.

60 Loubser, J.H.N. and van Wyk, J.A. (1977) Optical spin-polarization in a triplet state in irradiated and annealed type Ib diamonds. *Diamond Res.*, **11**, 4–8.

61 Lovett, B.W. and Benjamin, S.C. (2009) Comment on "multipartite entanglement among single spins in diamond". *Science*, **323** (5918), 1169.

62 Neumann, P., Mizuochi, N., Rempp, F., Hemmer, P., Watanabe, H., Yamasaki, S., Jacques, V., Gaebel, T., Jelezko, F., and Wrachtrup, J. (2009) Response to comment on "multipartite entanglement among single spins in diamond". *Science*, **323** (5918), 1169.

63 Mehring, M., Mende, J., and Scherer, W. (2003) Entanglement between an electron and a nuclear spin 1/2. *Phys. Rev. Lett.*, **90** (15), 153–001.

64 Peres, A. (1996) Separability criterion for density matrices. *Phys. Rev. Lett.*, **77** (8), 1413–1415.

65 Horodecki, M., Horodecki, P., and Horodecki, R. (1996) Separability of mixed states: necessary and sufficient conditions. *Phys. Lett.*, **223** (1–2), 1–8.

66 Hill, S. and Wootters, W.K. (1997) Entanglement of a pair of quantum bits. *Phys. Rev. Lett.*, **78** (26), 5022–5025.

67 Wootters, W.K. (1998) Entanglement of formation of an arbitrary state of two qubits. *Phys. Rev. Lett.*, **80** (10), 2245–2248.

68 Fisher, R. (2010) Optimal Control of Multi-Level Quantum Systems. PhD thesis, Technische Universität München, Department Chemie, Fachgebiet Organische Chemie.

69 Giovannetti, V., Lloyd, S., and Maccone, L. (2004) Quantum-enhanced measurements: beating the standard quantum limit. *Science*, **306** (5700), 1330–1336.

70 Kwiat, P.G., Berglund, A.J., Altepeter, J.B., and White, A.G. (2000) Experimental verification of decoherence-free subspaces. *Science*, **290** (5491), 498–501.

71 Kielpinski, D., Meyer, V., Rowe, M.A., Sackett, C.A., Itano, W.M., Monroe, C., and Wineland, D.J. (2001) A decoherence-free quantum memory using trapped ions. *Science*, **291** (5506), 1013–1015.

72 Beck, J. (2009) Quantenregister in Diamant: Experimentelle Realisierung von Mehrspinverschränkung. Diploma thesis, Universität Stuttgart, Physikalisches Institut.

73 Bluhm, H., Foletti, S., Neder, I., Rudner, M., Mahalu, D., Umansky, V., and Yacoby, A. (2011) Dephasing time of GaAs electron-spin qubits coupled to a

nuclear bath exceeding 200 μs. *Nat. Phys.*, **7** (2), 109–113.
74 Hanson, R., Dobrovitski, V.V., Feiguin, A.E., Gywat, O., and Awschalom, D.D. (2008) Coherent dynamics of a single spin interacting with an adjustable spin bath. *Science*, **320** (5874), 352–355.
75 Neumann, P., Kolesov, R., Naydenov, B., Beck, J., Rempp, F., Steiner, M., Jacques, V., Balasubramanian, G., Markham, M.L., Twitchen, D.J., Pezzagna, S., Meijer, J., Twamley, J., Jelezko, F., and Wrachtrup, J. (2010) Quantum register based on coupled electron spins in a room-temperature solid. *Nat. Phys.*, **6** (4), 249–253.
76 Shi, F., Rong, X., Xu, N., Wang, Y., Wu, J., Chong, B., Peng, X., Kniepert, J., Schoenfeld, R.S., Harneit, W., Feng, M., and Du, J. (2010) Room-temperature implementation of the Deutsch-Jozsa algorithm with a single electronic spin in diamond. *Phys. Rev. Lett.*, **105**, 040 504.
77 Jiang, L., Hodges, J.S., Maze, J.R., Maurer, P., Taylor, J.M., Cory, D.G., Hemmer, P.R., Walsworth, R.L., Yacoby, A., Zibrov, A.S., and Lukin, M.D. (2009) Repetitive readout of a single electronic spin via quantum logic with nuclear spin ancillae. *Science*, **326**, 267–271.
78 Steiner, M. (2009) Towards Diamond-Based Quantum Computers. Diploma thesis, Universität Stuttgart, Physikalisches Institut.
79 Zhao, N., Hu, J.L., Ho, S.W., Wan, J.T.K., and Liu, R.B. (2011) Atomic-scale magnetometry of distant nuclear spin clusters via nitrogen-vacancy spin in diamond. *Nat. Nanotechnol.*, **6** (4), 242–246.
80 Meijer, J., Pezzagna, S., Vogel, T., Burchard, B., Bukow, H., Rangelow, I., Sarov, Y., Wiggers, H., Plümel, I., Jelezko, F., Wrachtrup, J., Schmidt-Kaler, F., Schnitzler, W., and Singer, K. (2008) Towards the implanting of ions and positioning of nanoparticles with nm spatial resolution. *Appl. Phys. A, Mater. Sci. Process.*, **91**, 567–571.
81 Naydenov, B., Richter, V., Beck, J., Steiner, M., Neumann, P., Balasubramanian, G., Achard, J., Jelezko, F., Wrachtrup, J., and Kalish, R. (2010) Enhanced generation of single optically active spins in diamond by ion implantation. *Appl. Phys. Lett.*, **96** (16), 163–108.
82 Toyli, D.M., Weis, C.D., Fuchs, G.D., Schenkel, T., and Awschalom, D.D. (2010) Chip-scale nanofabrication of single spins and spin arrays in diamond. *Nano Lett.*, **10** (8), 3168–3172.
83 Schnitzler, W., Linke, N.M., Fickler, R., Meijer, J., Schmidt-Kaler, F., and Singer, K. (2009) Deterministic ultracold ion source targeting the Heisenberg limit. *Phys. Rev. Lett.*, **102** (7), 070 501.
84 Grinolds, M.S., Maletinsky, P., Hong, S., Lukin, M.D., Walsworth, R.L., and Yacoby, A. (2011) Quantum control of proximal spins using nanoscale magnetic resonance imaging. *Nat. Phys.*, **7** (9), 687–692.
85 Rittweger, E., Han, K.Y., Irvine, S.E., Eggeling, C., and Hell, S.W. (2009) STED microscopy reveals crystal colour centres with nanometric resolution. *Nat. Photonics*, **3** (3), 144–147.
86 Rittweger, E., Wildanger, D., and Hell, S.W. (2009) Far-field fluorescence nanoscopy of diamond color centers by ground state depletion. *Europhys. Lett.*, **86** (1), 14–001.
87 Han, K.Y., Kim, S.K., Eggeling, C., and Hell, S.W. (2010) Metastable dark states enable ground state depletion microscopy of nitrogen vacancy centers in diamond with diffraction-unlimited resolution. *Nano Lett.*, **10** (8), 3199–3203.
88 Maurer, P.C., Maze, J.R., Stanwix, P.L., Jiang, L., Gorshkov, A.V., Zibrov, A.A., Harke, B., Hodges, J.S., Zibrov, A.S., Yacoby, A., Twitchen, D., Hell, S.W., Walsworth, R.L., and Lukin, M.D. (2010) Far-field optical imaging and manipulation of individual spins with nanoscale resolution. *Nat. Phys.*, **6** (11), 912–918.
89 Steinert, S., Dolde, F., Neumann, P., Aird, A., Naydenov, B., Balasubramanian, G., Jelezko, F., and Wrachtrup, J. (2010) High sensitivity magnetic imaging using an array of spins in diamond. *Rev. Sci. Instrum.*, **81** (4), 043705.
90 Childress, L., Taylor, J.M., Sorensen, A.S., and Lukin, M.D. (2006) Fault-tolerant quantum communication based

on solid-state photon emitters. *Phys. Rev. Lett.*, **96** (7), 070 504.

91 Togan, E., Chu, Y., Trifonov, A.S., Jiang, L., Maze, J., Childress, L., Dutt, M.V.G., Sorensen, A.S., Hemmer, P.R., Zibrov, A.S., and Lukin, M.D. (2010) Quantum entanglement between an optical photon and a solid-state spin qubit. *Nature*, **466** (7307), 730–734.

92 Kubo, Y., Ong, F.R., Bertet, P., Vion, D., Jacques, V., Zheng, D., Dréau, A., Roch, J.F., Auffeves, A., Jelezko, F., Wrachtrup, J., Barthe, M.F., Bergonzo, P., and Esteve, D. (2010) Strong coupling of a spin ensemble to a superconducting resonator. *Phys. Rev. Lett.*, **105**, 140–502.

93 Amsüss, R., Koller, C., Nöbauer, T., Putz, S., Rotter, S., Sandner, K., Schneider, S., Schramböck, M., Steinhauser, G., Ritsch, H., Schmiedmayer, J., and Majer, J. (2011) Cavity QED with magnetically coupled collective spin states. *Phys. Rev. Lett.*, **107**, 060–502.

94 Rabl, P., Kolkowitz, S.J., Koppens, F.H.L., Harris, J.G.E., Zoller, P., and Lukin, M.D. (2010) A quantum spin transducer based on nanoelectromechanical resonator arrays. *Nat. Phys.*, **6** (8), 602–608.

95 Acosta, V.M., Bauch, E., Ledbetter, M.P., Santori, C., Fu, K.M.C., Barclay, P.E., Beausoleil, R.G., Linget, H., Roch, J.F., Treussart, F., Chemerisov, S., Gawlik, W., and Budker, D. (2009) Diamonds with a high density of nitrogen-vacancy centers for magnetometry applications. *Phys. Rev. B*, **80** (11), 115–202.

96 Acosta, V.M., Jarmola, A., Bauch, E., and Budker, D. (2010) Optical properties of the nitrogen-vacancy singlet levels in diamond. *Phys. Rev. B*, **82** (20), 201–202.

10
Diamond-Based Optical Waveguides, Cavities, and Other Microstructures

Snjezana Tomljenovic-Hanic, Timothy J. Karle, Andrew D. Greentree, Brant C. Gibson, Barbara A. Fairchild, Alastair Stacey, and Steven Prawer

10.1
Introduction

Color centers in diamond represent an outstanding opportunity for practical, solid-state quantum devices. The harnessing of their properties to the fullest requires optical structures, and exploring this space is one of the main aims of this chapter. The main question is, "What are the necessary conditions for diamond-compatible optical structures so as to enable practical diamond-based devices?" This is not an easy question to answer, given the relative immaturity of diamond photonic fabrication, and the ultimate solution must depend critically on the purpose to which the diamond devices will be put.

In this chapter, the creation of photonic structures–both in diamond and for diamond–are discussed. The main motivation for using diamond is to exploit the properties of the diamond color centers, and especially the negatively charged nitrogen-vacancy (NV) center. Diamond is introduced as a platform for quantum photonics, and the challenges and opportunities that result are outlined. First, the optical properties of three main forms of diamond are summarized, after which the design of diamond-based optical structures is reviewed. The machining of single-crystal diamond (SCD) is a difficult and developing process (see Section 10.4), while polycrystalline diamond offers its own possibilities and has proven useful in the creation of test structures (see Section 10.5). Finally, attention is turned to hybrid structures, in which diamond is combined with more conventional photonic platforms for the construction of quantum devices.

10.1.1
Motivation: Applications of Fluorescent Diamond Devices

Here, three applications are broadly considered that each benefit from–or are enabled by–optical confinement: (i) sensing (magnetometry); (ii) single-photon generation for quantum key distribution and quantum standards; and (iii) scalable quantum information processing based on NV diamond.

Optical Engineering of Diamond, First Edition. Edited by Richard P. Mildren and James R. Rabeau.
© 2013 Wiley-VCH Verlag GmbH & Co. KGaA. Published 2013 by Wiley-VCH Verlag GmbH & Co. KGaA.

10.1.1.1 Sensing

It is fair to say that most of the recent excitement in diamond derives from the fact that the NV center acts as an isolated, room-temperature qubit. As such, it can be used for a variety of protocols, including quantum information primitives and, perhaps most importantly, quantum sensing. There are other systems that can act as single-photon sources, for example quantum dots and trapped ions [1]; alternatives for solid-state room temperature quantum coherence, including ruby [2] and defects in silicon carbide [3]; and alternative systems for quantum memory [4]. The unique property of the NV center is that it offers *all* of these features, in addition to single-spin (electron or nuclear) single-shot readout in a robust and convenient biocompatible package. It is no exaggeration to say that NV diamond promises a disruptive platform for nanoscale sensing.

The room-temperature optically detected magnetic resonance (ODMR) is the real "engine" of the NV center. The ODMR provides the readout and spin-polarization of the NV, which in turn enables the NV to be used for readout and polarization of nearby spins [5–8]. The essential mechanism is that at zero strain, the transitions from the ground to excited state (^3E) are all spin-preserving. So, it is possible to excite the NV center with an off-resonant field, typically in the green, and the electronic spin will be unchanged through the excitation and fluorescence cycle, with the emission occurring in the red zero phonon line (ZPL) 637 nm or near infra-red (NIR) side band. This fluorescence cycle explains single-photon emission, but cannot explain the ODMR.

The ODMR can be understood due to the fact that the $m_s = \pm 1$ have a non-zero branching ratio to the singlet pathway ($^1A_1$ to ^1E, although the exact labeling is not completely proven). This pathway is considerably less allowed than the direct fluorescence route, with a total lifetime of around 300 ns [9] compared to the spontaneous emission lifetime of around 10 ns (this varies according to the NV environment) for the direct transition. The consequences of this change in lifetime are twofold. First, the change in emission time gives rise to a change in brightness conditional on the initial state of the NV center. The $m_s = \pm 1$ states are therefore less bright on average than the $m_s = 0$ state, enabling single spin readout. Second, the singlet de-excitation pathway is not spin preserving, and results in a net transfer of spin population into the $m_s = 0$ state. The branching ratio to the singlet pathway depends on the excited state properties, and in particular on the mixing between the spin states caused by either strain or electric field [10]. The degree of spin polarization and center brightness can vary considerably between individual NV centers, especially in nanodiamond NV, due to differences in the local environment. In many cases, a degree of either implicit or explicit post-selection is employed to determine the "optimal" center for a given application.

One important feature of the NV system is that of *photochromism*. It is common, even in single centers, to observe both the signature of NV$^-$ and NV^0; indeed, it is possible to photoionize the NV$^-$ [11]. Although electron paramagnetic resonance (EPR) has been observed from the excited state of the NV^0 [12], the NV^0 does not appear to be suitable for sensing applications, as it does not have a favorable ground-state spin structure. Hence, it is necessary to avoid this photochromism by a combination of sample preparation and selection, and the use of repump lasers.

Optical strategies for improving NV-based sensing schemes derive very naturally from the impetus to improve the efficiency of the readout of the ODMR. As mentioned above, one of the key ways is simply to post-select NV centers based on brightness, and only to use such centers for subsequent device integration [13–15]. Another approach is to note that the center emission varies with the substrate on which it is placed [16]. Alternatively, broadband collection techniques can be utilized, for example solid-immersion lenses [17], fiber integration [18–21], plasmonics [22, 23], or optical cavities [24–28]. In such systems, it is clear that *any* improvement in collection efficiency will lead to a corresponding improvement in readout, and hence reduce the integration time for sensing. In that sense there is no "threshold" for Q factors[1] required; instead, the best available Q should be utilized to improve device performance.

10.1.1.2 Single-Photon Sources for Quantum Key Distribution and Quantum Metrology

Fairly soon after single center imaging of, and single-photon collection from, NVs was demonstrated [29–31], diamond color centers were explored for applications in quantum key distribution [32]. At first glance, the strategies for the development of diamond-based single-photon sources would seem to follow the same steps that are required for sensing; that is, the collection of more photons using broadband techniques. In fact, although this is an important way to view the development of diamond single-photon sources, there is another criterion that we have overlooked until now, namely spectral linewidth. The most widely used diamond color center for single-photon applications is the NV center, but the NV is a relatively poor source of single photons in its natural state. The reason for this is that the emission from the center is typically broadened by around 200 nm due to the pronounced phonon sidebands, with only 2.7% of the total emission contained within the ZPL [33].

The broad linewidth of the NV center leads to several problems. The broad emission leads to serious dispersion for medium to long-range fiber-based single-photon distribution, while for free-space protocols the important metric is the intensity in a given spectral window above the background. By spreading the photon flux over such a broad window, the background levels must be correspondingly low, restricting NV-based free-space protocols to laboratory settings or nighttime operations.

Currently, there is considerable interest in developing single-photon sources that can be used as standards for single-photon counting modules [34] and, ultimately, to redefine the candela [35, 36]. In principle, diamond single-photon sources would be ideal, as a pulse train of single photons could be developed with reduced intensity fluctuations as compared to classical sources. However, for metrological consistency it is necessary to measure the intensity in a fashion

---

1) The quality factor (Q) is the ratio of power stored to power lost per optical cycle.

compatible with existing standards. This requires a photon flux that is large enough to give accurate readings on high-precision bolometers. Present sources suffer from photon fluxes that are too low to give accurate comparisons with existing sources, while the large fluorescence window for NV will give rise to a large spread in the energy deposited into the detector per photon, again limiting the precision of the metrology. However, these problems can be solved by narrow band sources with high collection efficiencies.

The above-discussed limitations apply to broadband (or alternatively incoherent) mechanisms for increasing the photon collection efficiency. Yet, considerable advantages can be obtained by considering *coherent* collection techniques instead. The requirement here is to take advantage of cavity quantum electrodynamic (CQED) effects to tailor the emission properties of the centers to be more compatible with device integration. This requires an exploration of the strong coupling regime to boost the fraction of light emitted into the ZPL, and to simultaneously increase the emission of the centers into the cavity mode, rather than into all space. It is important to stress that, when discussing coherent CQED processes with diamond emitters, there is an implicit restriction to cryogenic systems rather than to room-temperature operations. The reason for this is that mobile charges and phonon coupling all give rise to large spectral broadening of the excited state, and hence it is not possible at present to achieve a transform-limited emission of indistinguishable photons at room temperature. However, in a few cases using high-quality (low [N], few defects) samples, it is possible to achieve very stable, transform-limited emission [37], which is a necessary condition for coherent CQED.

To take full advantage of CQED requires a maximization of the coupling between the center and the optical mode of the cavity. This implies small cavity mode volumes and long photon storage times, Q-factors. As optical cavities are resonant structures, it is possible to optimize coupling for one emission line at the expense of other lines. In particular, Su *et al.* showed that it is possible to alter the fraction of light emitted into the ZPL, and to boost total emission rates from the NV center, by operating in the strong Purcell regime [38]. Assuming a cavity volume around λ^3, a cavity Q of about 10^4 was predicted to suffice to allow maximal emission into the ZPL, with minimal two-photon probability (see Figure 10.1). The emission of two photons per pulse becomes an issue when the system enters the strong coupling regime, due to the fact that when the coherent oscillation time is fast compared to the temporal width of the excitation pulse, then the cavity can store multiple excitations which have a probability of being ejected simultaneously from the cavity.

The cavity optimization of color center emission translates directly into improved device performance for quantum key distribution (QKD). Indeed, it has been shown that quantum key rates for the BB84 protocol [39] can actually be used as a metric to quantify the performance of single-photon sources [40]. Because the cavity can integrate photons, there is an optimal Q-factor to generate the highest key rate for any given system. A comparison of the BB84 QKD device performance for three cavity-enhanced diamond color centers is shown in Table 10.1 [40].

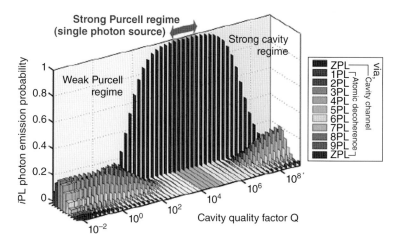

Figure 10.1 By ensuring that the ZPL emission is resonant with a cavity mode in the strong Purcell regime, all of the emission can be directed to be on the ZPL [38]. For a color version of this figure, please see the color plate at the end of this book.

10.1.1.3 Quantum Information Processing with NV Diamond

The two main approaches to scalable quantum information processing with diamond can be roughly classified as either nondeterministic or deterministic protocols. The requirements for each approach are quite different; nondeterministic protocols generally employ heralding approaches and can, in principle, operate with greatly reduced system requirements compared to deterministic schemes, although the nondeterministic protocols suffer from reduced rates of entanglement generation. The essential ideas and constraints for each scheme are discuss in the following subsections.

The nondeterministic approach is best explained with reference to the Barrett and Kok scheme [41], although there are strong similarities with the Lim, Beige, and Kwek [42] and brokered graph state approaches [43]. The essential idea underpinning all of the nondeterministic approaches is to entangle remote qubits by measurement of emitted photons in an entangled basis. In general, measurements are thought of as projecting out a single qubit, but they can equally well be used to project out the collective state of two qubits. By measuring the two qubits in an entangled basis, the measurement outcome forces the qubits to be projected into an entangled state. The details of the schemes vary, depending on the exact protocol and the use of memory, but in most cases a form of path erasure is used to ensure that the entangled basis measurement can be performed (see Figure 10.2a) [44]. The requirements on optical cavities to effect such a protocol are relatively modest. Notably, the emitted photons are required to be energetically indistinguishable, and hence a low temperature and a strong Purcell regime are needed. If photons are lost from the system, then the heralding and memory can usually compensate, albeit at the cost of a reduced gate time.

Table 10.1 Optimal cavity parameters to enhance single-photon emission from a variety of diamond color centers for mode volumes of one cubic wavelength.

Optical center	Dipole moment (10^{-29} cm)	Cavity parameters				Excitation pulse		Source performance			
		Wavelength λ_c (nm)	Freq ω_c (PHz)	Optimal Q-factor	Coupling Ω_0 (GHz)	Pump rate r (THz)	Pulse width T (ps)	Mean time (ps)	Spectral width $\Delta\lambda$ (nm)	Single-phase Prob. P_i	Repetition rate v (GHz)
NV	0.1[c]	637	2.95	64 000	9.3	1.3	2.3	125	0.007	0.95	2
	1[d]	–	–	–	–	1.3	6	1.16×10^4	120	0.74	0.086
NE8	2.1[c]	794	2.37	3700	127	20	0.16	10	0.15	0.954	30
	2.5[d]	–	–	–	–	9	1	1.15×10^4	2×10^{-5}	0.84	0.087
SiV	4.2[c]	738	2.55	1800	290	40	0.18	5	0.30	0.812	65
	4.7[d]	–	–	–	–	10	1	2.7×10^3	9×10^{-5}	0.05	0.37

a) bare emission properties;
b) optimal cavity enhanced performance;
c) estimated ZPL dipole moment;
d) total dipole moment over all phonon sidebands [40].

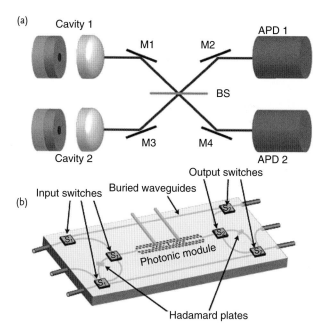

Figure 10.2 (a) Schematic of path erasure scheme for heralded, probabilistic gates for quantum information processing [44]; (b) Photonic module to realize multiphoton operator measurements for deterministic cluster-state quantum computing [45].

Deterministic approaches to quantum information processing with NV are far more demanding. Numerous schemes have been proposed, but the most powerful are those based on versions of cavity QED effects in the strong-coupling regime – that is, requiring single-photon–single-atom interactions. NV can be used for the Q-switching of cavities for nonclassical sources [46, 47], and this also leads to the possibility of fast quantum gates [48]. This immediately implies a requirement for small mode volume cavities and high Q-factors, around the 10^6 level. Although single cubic wavelength (λ^3) cavities with a Q-factor around 10^6 containing a single atom can enable conventional remote qubit interactions (e.g., via the Duan–Kimble method [49]), a more interesting approach is to use the atom-cavity system to measure out nondestructively the state of multiple photons passing through the cavity. This approach, which is taken with the photonic module scheme [45, 50] (see Figure 10.2b), is quite powerful as it allows access to three-dimensional (3-D) topological cluster state quantum computing, which has a high threshold, easy scalability, and fast logical qubit operations. Again, the required Q-factor for a λ^3 cavity is around 10^6 to 10^7.

Although, in most instances, concern relates to the λ^3 mode volume cavities, it is possible to construct cavities with mode volumes significantly less than λ^3, based on the slot waveguide approach [51]. It has been shown that the requirements on

Q-factors can be reduced markedly by combining slot waveguides with diamond emitters [52], and that arrays of slot waveguide cavities can also be used for two-dimensional (2-D) atom-cavity arrays [53].

10.2
Optical Properties

At one level, diamond appears to be an ideal material for photonic applications, as it has one of the highest optical bandgaps at 5.5 eV, with good transparency from the ultraviolet (UV) to the infrared (IR). Low-birefringence diamond is available [54] for applications such as diamond Raman lasers and, with a first-order Raman shift of 1332 cm^{-1}, diamond has been used for a number of key wavelengths of interest [55, 56]. Other important features of diamond are its very high thermal conductivity and low thermal expansion, which are great assets for high-power applications [57]. Finally, it should be noted that the hardness of diamond and its chemical resistance has meant it being employed in high-end window applications, the classic example being the use of diamond windows on some of the Venera landers to Venus. This combination of properties has led to significant waves of interest in diamond for both electronic and optical applications. There are, however, serious challenges to the broad application of diamond to photonics, and although many of these issues are currently being solved, further significant breakthroughs will be required before diamond can reach the level of market penetration enjoyed by silica and silicon.

Although, in principle, diamond is extremely transparent, until recently it was difficult to manufacture synthetic diamonds with reproducible defect concentrations. This is a significant point as the defects – especially nitrogen- and nickel-related defects – give rise to absorption bands that can limit optical absorption and scattering. To a large extent, this issue has been eliminated by the advanced homoepitaxial growth of diamond [58], though costs and wafer size are still unfavorable compared to high-quality silicon wafers. The creation of complicated multilayer heterostructures of diamond has been extremely problematic, and this has greatly impeded progress in the creation of efficient photonic structures (this point is discussed in Section 10.4).

To date, three main forms of diamond have been identified: SCD; polycrystalline films; and isolated nanodiamonds (NDs). The best candidate for diamond photonics is SCD. There are several reasons for this, but chief among them is that the optical properties of large monolithic SCD exceed those of isolated NDs and polycrystalline – either microcrystalline or ultra-nanocrystalline (UNCD) – diamond. The superior properties of bulk SCD arise due to the lack of scattering from grain boundaries, and the fact that color centers fabricated in bulk SCD witness more perfect environments and hence have are more desirable properties in terms of spectral diffusion and homogeneity. In a recent study of optical absorption conducted by Yoshitake *et al.*, an absorption coefficient in UNCD of 10^6 cm^{-1} was observed from around 3 to 5 eV [59]. Despite this, polycrystalline diamond films are significantly easier to fabricate and process than bulk single crystals, and many important test structures have been fabricated using these materials. The applications of polycrystalline diamond are described in Section 10.5.

10.3
Design of Diamond-Based Optical Structures

The refractive index of diamond ($n = 2.4$) falls between that of silica ($n = 1.45$) and silicon or high-index semiconductors ($n \approx 3.4$). Translating designs that will function at this different refractive index is straightforward for conventional optical structures such as waveguides and microdisks [60, 61]. However, diamond color centers emit light in the visible range (from 500 nm to 800 nm), and therefore attention is focused on creating structures of much smaller dimensions than those used at telecom wavelengths (ca. 1550 nm). For example, in order for a square-core diamond waveguide to be single-moded, the cross-section must be approximately 200 nm × 200 nm. In addition to the attendant fabrication challenge, coupling these waveguides to standard optical elements is also a major problem [60].

Theoretical investigations of more complex all-diamond-based optical structures have followed the designs developed in more commonly used photonics materials, such as silicon/high-refractive index semiconductors [62, 63]. Here, the main design constraints are due to the immature fabrication capabilities in diamond, compared to fabrication in more conventional optical materials. An improved design of structures in diamond necessarily follows the technical ability to fabricate interesting and exotic shapes. Recently, tremendous progress has been made in the fabrication of diamond-based photonics structures with cavity Q-factors reaching a few thousand [25]. The fabrication methods and the resultant optical structures in SCD are discussed in Section 10.4.

Hence, the major challenge is not only to design the best suitable structure, but also to design the most flexible structure that will enable a large fabrication tolerance. Since, 2006, many diamond-based optical structure designs have been proposed, including photonic crystal cavities [64–67], conventional waveguides [60], nanobeams [68], nanowires [69], triangular cavities [70], and slots [52] (see Figure 10.3a–c).

Optical microcavities based on photonic crystal slabs have attracted much attention recently [64–67]. Almost all of these studies have considered a 2-D photonic crystal slab (PhC) composed of an hexagonal array of cylindrical air holes in a dielectric slab. The structure is periodic in the plane and finite in the vertical direction, where it is surrounded by a low-index medium (usually air). The field is confined by the periodic dielectric structure in the plane, and by total internal reflection (TIR) out of the plane [62–64]. A cavity is usually formed in either of two ways: a point cavity (see Figure 10.3a); or a "double heterostructure" (DH) (see Figure 10.3c). Due to the smaller refractive index of diamond compared to that of high-index semiconductors, it was unclear if the diamond-based photonic band gap structure would have a sufficient photonic band gap. However, in 2006 it was

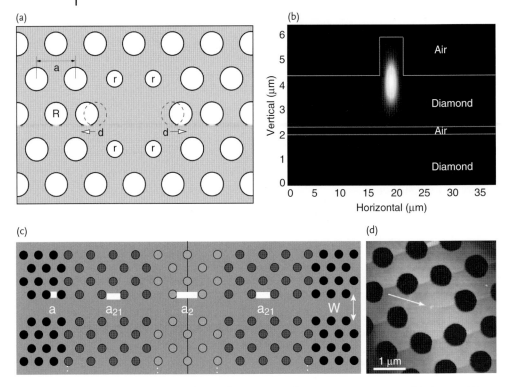

Figure 10.3 Schematics for: (a) the point-type photonic crystal cavity [64]; (b) rib waveguide [60]; (c) double-heterostructure photonic crystal cavity [67]; (d) photonic crystal fiber–diamond hybrid approach. The arrow indicates a nanodiamond [20].

shown that a high-Q PhC cavity ($Q = 30\,000$) could be designed in diamond [64], and this design has subsequently been improved with a fourfold higher quality factor [66]. In 2008, a DH PhC cavity with the ultrahigh-Q (10^6) was designed [67]. Unfortunately, in all of these examples the geometry of the ultrahigh-Q PhC cavities was finalized at the stage of fabrication, and required changes to be made with nanometer precision in the geometry, hole sizes, and positions [64–67].

There are alternative ways, however, to localize light within a PhC, without any geometric perturbation; notably, it is possible to form DH-type cavities simply by changing the refractive index of the background material [71] or holes [72]. The main advantage of this type of cavity is the possibility of it being post-processed around the region of interest. Unlike conventional DH, these designs do not require changes to be made in the geometry with nanometer precision; rather, the refractive index changes required for ultrahigh-Q cavities with $Q \sim 10^7$ are well within what can be achieved in diamond ($\Delta n \sim 0.02$) [73, 74]. The cavity modes have relatively small volumes $V < 2\,(\lambda/n)^3$, although absorption induced by damage may have a significant effect on the cavity [66, 71]. Despite this, waveguiding in the damaged region – induced by ion-implantation – has been demonstrated experimentally [75].

The carving of optical structures into SCD is a challenging task, and many attempts have been made to use thin nanocrystalline (NC) diamond films for this purpose. Unfortunately, the use of NC diamond film-based optical structures induces additional difficulties of absorption and/or scattering. Typically, NC chemical vapor deposition (CVD) -prepared diamond films have shown a strong absorption in the spectral range of optical center emission [66], due to the presence of sp^2-bonded carbon. The level of absorption depends heavily on the diamond grain size, and is generally highest for NC CVD diamond (grain sizes <100 nm) and smallest for polycrystalline CVD diamond (grain sizes >100 nm) and SCD films. On the other hand, increasing the grain size leads to an increase in scattering losses [66]. The properties of polycrystalline thin films and the fabrication of optical structures from such films are discussed in Section 10.5.

The combination of nanocrystals of ND with conventional optical media offers two alternative routes to integrated diamond nanophotonics:

- The manipulation of ND onto existing photonic structures consisting of more conventional materials such as GaP, Si_3N_4, polymer, and silica [15, 20, 76–78] (see Figure 10.3d). Recently, it has been shown numerically that a high-Q photonic crystal cavity can be induced by the presence of a ND on the air-hole side wall in an otherwise defect-free photonic crystal [72]. Experimental results on the combination of optical structures and NDs are discussed in Section 10.6.

- The incorporation of NDs into another photonic material such as silica, polymer, and soft glasses [21, 79] (see also Section 10.6).

10.4
Single-Crystal Diamond

Planar slab waveguides represent the starting point for optical integration, which in turn allows many functionalities including efficient sensing, confinement in cavities, and the routing of photons (flying qubits). With only tentative heteroepitaxial technologies, and no native oxide, diamond processing is devoid of material layers, which in Si/SiO_2 and Group III–V semiconductor processing are used as sacrificial layers in defining structures, for electronic confinement, and as low-index layers suitable as waveguide cladding materials. Typically, SCD is not susceptible to wet chemical processing; hence, thin films of diamond are prepared by either ion implantation, mechanical polishing or, occasionally, by ion beam etching. The etching and implantation techniques should avoid unwanted damage to the brittle diamond lattice.

Implantation has three main guises in diamond fabrication: (i) the implantation of light Me ions such as He, C, and O to create sacrificial layers; (ii) the implantation of N or similar-sized ions to add N or create vacancies, both of which are aimed at creating NV centers; and (iii) the unintentional implantation of heavy ions from focused ion beam (FIB) fabrication. Ion beam milling has a tendency to implant the heavy etchant ions (typically Ga) into the host lattice, which can induce absorption and hence optical loss [80].

Simple structures such as approximately 10 to 20 μm-thick slabs of SCD are formed by polishing to yield planar waveguides. The fabrication of thinner and more uniform slabs by ion implantation and lift-off techniques has been used to produce diamond membranes ranging from 200 nm to 5 μm thick, with side lengths of several hundreds of micrometers [81].

The dominant issues in the thin-layer growth of SCD are the method of preparation, the scalability of the technique, and the final layer quality. One of the most promising thin-layer growth techniques provides a route to the large-area heteroepitaxial growth of quasi-SCD on Si, using an iridium buffer on top of a layer of ytterbium-stabilized zirconia [82]. Full-width at half-maximum (FWHM) X-ray rocking curves for SCD are of order 0.003°, whereas current quasi-single-crystal material exhibits FHWMs of 0.1 to 1° [83]. The presence of Si in the growth chamber favors the inclusion of Si into the diamond lattice, and therefore the creation of bright SiV centers, the fluorescence signal of which may mask the NV side-band.

10.4.1
Lift-Off of Thin and Ultra-Thin Single-Crystal Diamond Films

10.4.1.1 Lift-Off

The lift-off method was first developed by Parikh *et al.* [84] for removing entire sheets of SCD from bulk samples. The technique is based on light MeV ion implantation, followed by annealing to convert a layer of end-of-range damaged [85] (amorphous material) to graphite. Graphite and diamond are chemically different, and graphite acts as a buried sacrificial layer that can be removed by galvanic etching or acid boiling, thus undercutting the film. Although both the lift-off method and FIB milling have been used to fabricate waveguides and cantilever structures [86], the single-energy lift-off method is limited to thick layers (>0.5 μm) due to strain effects in shallow implants. One approach to overcoming these difficulties was to grow a layer of CVD diamond on top of a shallow implanted region [87], but this still resulted in a film that was thicker than 0.5 μm and which could contain surface defects propagated through the CVD film [88]. In a revised method, a double energy implant was developed to create two damage layers with a single-crystal layer "sandwiched" between them [81]. This is shown in Figure 10.4, where a cross-sectional transmission electron microscopy (TEM) image is shown of a diamond with a double energy implantation, and a conceptual representation of 1.8 and 2.0 MeV helium ions damage profiles has been overlaid. By using this energy difference, the resulting sandwiched layer of diamond may be as thin as 210 nm.

10.4.1.2 Lift-Out

Following the implantation and etching process, the layer will be suitable for further fabrication, and can either be etched to form waveguides or cavities (see Sections 10.4.3 and 10.4.5), or the film can be lifted out. The technique for lift-out is shown in Figure 10.5, and the basic principle is the same for both single- and double-energy implantations. The free-standing film (post etch) can be lifted out

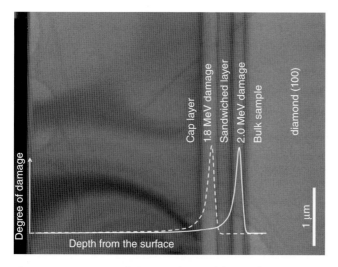

Figure 10.4 TEM image showing two layers of damage from 1.8 and 2.0 MeV helium ion implantations into diamond. Overlaid are two curves showing the resulting damage levels from the 1.8 (dashed) and 2.0 (solid) MeV helium ions [81].

from the bulk sample; this is performed *in situ* in the FIB by using a micromanipulator probe and a deposited weld, usually Pt, W, or C. When the microprobe has been attached the layer can be completely freed and transferred to another substrate [86, 89]. Recently, layers have been transferred directly, after galvanic etching using ultrapure water, to a SiO_2 substrate [90].

10.4.2
Fabrication: Lithography and Etching

In this section, the application of optical and electron beam (e-beam) lithography to pattern masks on diamond is first discussed, after which the use of FIB to directly write into diamond films or bulk material is examined. The various plasma etching techniques are used to transfer mask patterns into the diamond, and their suitability for the particular requirements of diamond photonic fabrication, are then outlined.

As noted in Section 10.3, the improved design of structures in diamond necessarily follows the technical ability to fabricate nanoscale shapes in reproducible fashion. Unfortunately, this requires an ability to work with typically small diamond samples that are not immediately conducive to common lithographic techniques.

The reliable fabrication of high-quality optical devices is dependent on the intrinsic material properties, the surface quality following processing, and the precision with which the design can be replicated. Common causes of optical device loss include material absorption, scattering due to roughness, bend loss, mode conversion due to sidewall angles, and leakage to the substrate caused by

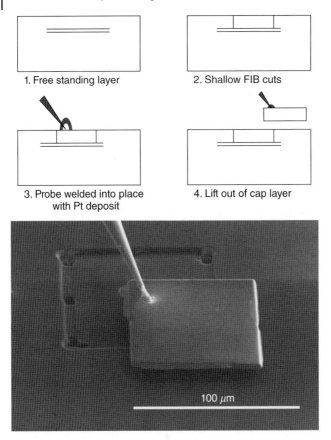

Figure 10.5 Schematic of the fabrication process. Step 1: this shows the sample after annealing and etching, with two air layers at depth within the sample. Step 2: this shows FIB cuts around ~80% of the edges of the cap layer, to enable liftout. Step 3: the probe is positioned and welded to the cap layer. Step 4: the final cut enables the cap layer to be removed, exposing the thin membrane beneath for further processing or milling. The SEM image shows the freshly exposed layer as the cap is lifted away [81].

insufficient cladding thickness. Absorption and scattering are dependent on the form of diamond (whether single crystal or polycrystalline film), and the operating wavelength. Side-wall roughness and side-wall angles depend primarily on mask transfer fidelity, etch chemistry, and mask selectivity; the latter as used here is the ratio of diamond removal rate to mask removal rate. It should be noted that, whenever diamond has been patterned into stripes to form rib or ridge waveguides, no propagation loss estimates are available.

Contact photolithography allows an efficient parallel patterning of large numbers of micron-scale features. In order to achieve a higher resolution (down to 10 nm), direct-write tools such as e-beam lithography and FIB milling are required; these serial write techniques are capable of dealing with relatively small samples.

10.4.2.1 Photolithography

A soft mask can be defined on the surface of the diamond by exposing a photosensitized resist to UV light through a photomask, although the resolution will be limited by diffraction to features on the order of the wavelength of the light used. Unfortunately, the move towards large-scale processing techniques is hampered by the very small size of the diamond platelets produced (low-cost samples for testing measure only a few square millimeters), and novel techniques are required to hold the samples in place. Indeed, in one study polydimethylsiloxane (PDMS) was used for this purpose, enabling the resist to be applied and the usable area of the diamond to be increased [91]. The subsequent inkjet printing of the photoresist [92] allowed millimeter-long, 20 μm-wide patterns to be defined, without any unwanted effects of edge beads (from resist spinning) hampering the contact lithography process that provided sharply defined 1 to 10 μm-scale features.

10.4.2.2 E-Beam Lithography

The use of an electron beam to write a pattern in an electron-sensitized resist can yield very high-resolution nanoscale patterns. Diamond has the highest atomic density of any material, but consists of a light element carbon, with a correspondingly low atomic number, Z. This results in a low backscatter coefficient (the ratio of backscattered to incident electrons) of a few percent, which is comparable with that of poly(methylmethacrylate) (PMMA), a popular high-resolution e-beam resist [93]. This allows the high-resolution patterning of diamond substrates and thin films (single [25] and polycrystalline [61]), with minimal influence from proximity effects [94].

10.4.2.3 FIB Milling

Having created a thin diamond film, it is possible to fabricate complex structures with the use of FIB milling [60, 81, 86, 89, 95], as shown in Figure 10.6. FIB has the advantage of being a direct write, maskless fabrication technique. In FIB, the sample can be mounted on a eucentric stage – for simultaneous milling and inspection by scanning electron microscopy (SEM) – at an angle with respect to the vertical, providing flexibility in patterning capability. FIB can be used to polish structures to trim their size to appropriate dimensions. The metal deposited in association with some of the FIB functionality – for example, the welding used in lift-out techniques used for TEM sample preparation, and metal sacrificial layers that help to avoid pattern rounding – can be detrimental to optical performance and requires careful stripping.

FIB milling is fundamentally limited by beam resolution, pattern reproducibility, and writing size. For example, in order to obtain 7 nm resolution with the instrument used to fabricate the structures in Figure 10.6a, a beam current of 1 pA is required. Likewise, to mill a similar structure in a 330 nm layer with a 30 keV gallium beam at 100 pA takes a few minutes. When writing times exceed 10 min, however, beam drift can become a problem; consequently larger structures must be written faster at higher currents, but this then leads to a fall in resolution (see Section 10.4.3). At higher currents, beam divergence and sputtering effects lead

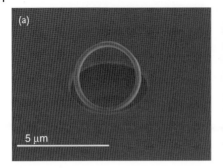 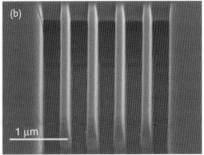

Figure 10.6 (a) A test ring resonator in a 330 nm layer of single crystal diamond; (b) Nanocantilevers in a 330 nm layer, ~100 nm wide and 3–4 µm long [81].

to the rounding of cuts and features [68], particularly at the tens of nanometers feature sizes shown in Figure 10.6a. A drop in resolution is acceptable for larger structures, but for structures smaller than 0.5 µm pattern stitching is required [96]. Because of these limitations, FIB milling is best suited to one-off or prototype devices, rather than to industrial fabrication.

Diamond is sensitive to low-level ion exposure [97], and milling leaves residual damage as well as redeposited amorphous carbon that must be removed if full optical functionality is to be obtained. Thin films are not always a prerequisite for FIB milling optical structures in SCD (as will be discussed in Section 10.4.4). Gallium ions accelerated at 30 keV at doses as low as $10^{14}\,\mathrm{m}^{-2}$ can cause amorphization of the diamond crystal [98, 99].

10.4.2.4 Ga Hard Mask

One radically different approach uses FIB not to mill, but rather to mask, the material. The FIB implantation of Ga in diamond results in an extremely resistant hard mask [100, 101]. Exposure of the diamond (either bulk or polycrystalline film) surface to Ga doses as low as $10^{16}\,\mathrm{cm}^{-2}$ leads to the formation of a hard mask that demonstrates resistance to both chemical (O) and physical (Ar) plasma dry etching. A selectivity of 9 (diamond : Ga mask) for a 50 nm hard mask has been demonstrated, with the added benefit of high resolution. High-pressure, high-temperature (HPHT) diamond and UNCD films have been structured in this way, and diamond etch rates of 100 nm min^{-1} have been achieved to depth of 450 nm in an oxygen plasma. This is an intriguing fabrication technique that merits further investigation.

10.4.2.5 Reactive Ion Etching: A Scalable Process

Reactive ion etching (RIE) is an industrial-scale process that is suitable for the simultaneous microfabrication of many components when combined with efficient masking techniques [102, 103]. The method employs an ionized gas that is accelerated by an electric field towards the sample surface; when the ions strike the surface energy is transferred such that material is either sputtered or chemically removed. Oxygen plasmas represent an obvious choice for etching diamond, as carbon will easily form volatile compounds with oxygen radicals. This causes the surface in regions that are not covered by the mask to be anisotropically etched, resulting in straight side walls. Diamond has been successfully etched, with a predominantly oxygen plasma, by a number of groups including those of Shiomi [104] and Leech *et al.* [105]. The results presented by both Shiomi *et al.* and Leech *et al.* suggest that the RIE process is ideally suited to the etching of large-area structures in SCD. In addition to capacitively coupled RIE, inductively coupled plasma (ICP) [106] and electron cyclotron resonance (ECR) [107] plasma etching processes have been used to offer more process control, by separating the physical and chemical etching components using different etchant species.

Importantly, RIE processes are capable of producing optical-quality surfaces that exhibit the low degree of roughness that is critical to micro-optical and photonic devices operating at short wavelengths in the visible spectrum. Materials used to mask diamond for dry etching include metal (Cr) [108] and dielectric resist films (SiO_2, photoresist, hydrogen silsesquioxane (HSQ)) [60, 106, 109] and also nanoparticles (Au, SiO_2, Al_2O_3) [109]. These masks need to demonstrate an appropriate selectivity [110] to allow controlled processing and the fabrication of high-aspect ratio structures.

10.4.2.6 Etching of Bulk Diamond: Microlenses

Etching recipes for diamond optical and photonic devices have been developed from larger-scale structures for applications in the mid- to long-IR range. By using a photoresist reflow method, microring- and cylindrical optical microlenses have each been fabricated using ICP etching with an Ar/O_2 plasma, to yield a low (3 nm) surface roughness. With an etch rate of 40 nm min^{-1}, long etches were required to achieve a ~1 μm vertical profile; thermal management was also needed to avoid overheating of the initially 7 μm-thick resists [106]. Etch parameter optimization with a high-density Ar/O_2 plasma achieved a higher etch rate of 230 nm min^{-1} of natural diamond and roughness of the order of 1.2 nm [111]. The development of etching recipes with a lower selectivity of approximately ~0.05 (as opposed to ~0.1 for oxygen etching) has led to the provision of smoother surfaces. This was achieved using an Ar/Cl_2 plasma, and resulted in an order of magnitude lower roughness of 0.19 nm for etching of natural Type IIa and HPHT diamond [112]. In this case, the volatile compounds were formed by the reaction of carbon atoms with chlorine ions to yield CCl_x compounds that could be readily pumped away. The chemical component of the reaction was shown to be responsible for the smooth etching, rather than the physical sputtering of the Ar. Although these slow etch rates allow control over the etch profile, the thicker masks limit the pattern resolution.

10.4.3
Optical Waveguiding in Single-Crystal Diamond

The fabrication of waveguides in SCD can be achieved by using combinations of the above-described three methods, namely direct ion-implantation, FIB lithography, and a photolithographic-RIE process. The latter method is especially important as it represents a scalable, industry-compatible technique.

10.4.3.1 Waveguides: FIB Lithography

By using techniques of lift-off and FIB (see Sections 10.4.1.1 and 10.4.2.3), it is possible to fabricate photonic structures in diamond. The optical waveguide fabricated by Olivero et al. was approximately 80 µm long, with a core cross-section of 3.5 µm height and 2 µm width [86]. The implant, anneal, and galvanic etching of the resultant graphitic material left an insufficient ~300 nm air gap below the waveguide. In this case, air acts as an out-of-plane waveguide cladding, while FIB-cut trenches are used to define the lateral confinement. Laser light (532 nm) was then coupled into and out of the short waveguide, using total internal reflection (TIR) mirrors. Light was then focused onto a micrometer spot on the 45° input mirror, using a 20× long-working-distance microscope objective that also collected the far-field modal intensity pattern from the output TIR mirror. A schematic representation of the coupling systems used to launch light into and out of the diamond waveguide is shown in Figure 10.7a, b.

Cleaving of the suspended SCD waveguide to provide access to optical quality facets is not viable, because the waveguide structure was manufactured in the center of a 3.5 mm^2 piece of SCD. Although the TIR mirror coupling technique (see Figure 10.7c) was highly suited for short waveguide lengths, the use of only one objective compromised on the coupling loss.

In Figure 10.7d–f the waveguide input is on the left of the image, and the multimode output is on the right. Magnified images of the waveguide output demonstrated the highly multimode nature of the propagating optical field; indeed, Olivero et al. estimated that more than 30 propagation modes were present. The input launch conditions were varied by moving the position of laser light relative to the input mirror, which led to the output modal intensity pattern being varied accordingly. The resulting pattern, which was attributed to the constructive and destructive interference of modes propagating along the waveguide, was the first demonstration of optical guidance in SCD.

The key limitation with regards to fabrication of the waveguide is the reliance on FIB to provide a transverse optical confinement of the propagating mode. FIB milling is not a scalable fabrication process, the main problem being that the field-of-view (FOV) of the instrument is limited to about 300 µm, which will be the upper length limit of the diamond photonic structures. Although the milled areas can be "stitched" together to create longer structures, this may unfortunately yield mismatched core geometries. There is some evidence that the use of the FIB may also damage the surrounding bulk diamond [68, 113, 114], which may also lead to additional optical losses.

Consequently, an alternative method to FIB that could be used to define the waveguides and which was also scalable was sought; the result of this quest was a combination of photolithography and RIE.

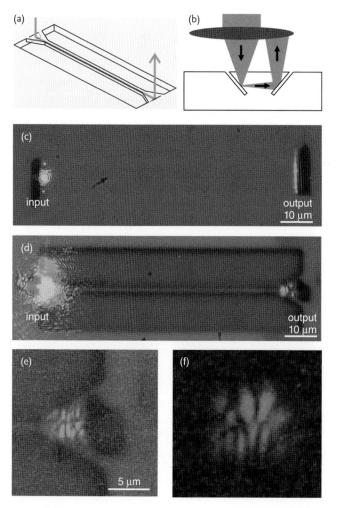

Figure 10.7 (a, b) Schematics of the (a) input/output light-collection system and (b) coupling to the waveguide structure; (c) Optical images of the incoming and outgoing 532 nm laser light between pairs of 45° mirrors facing each other in the bulk and (d) coupled through the bridge structure. In the former case, the outgoing light is spread along the output mirror, while in the latter case the propagating light is waveguided, which is indicated by the structured modal pattern at the output mirror; (e) Higher-magnification image of the output pattern; (f) same as (e), but without external illumination [86]. For a color version of this figure, please see the color plate at the end of this book.

10.4.3.2 RIE-Fabricated Waveguides

As RIE etching allows the precise control of etch depth, it is suitable for the fabrication of structures such as rib waveguides. The latter type of waveguide relaxes the constraints on waveguide dimensions to achieve a single-mode operation, compared to the channel waveguide (as described above). Hiscocks et al. fabricated 80 µm-long rib waveguides with dimensions of 4.5 µm rib width, 3.5 µm rib height, and 2 µm slab height, surrounded by an air-cladding [60, 115]. The fabricated waveguides were 1 µm wider than the single mode-designed waveguides, and resulted in a two-mode device. The coupling of light was similar to the method used in Ref. [86] (as discussed in Section 10.4.3.1). The propagation of light was investigated at $\lambda = 532$ nm, while the collection of transmitted light from the output mirror was achieved with the same objective (as shown schematically in Figure 10.7).

As alternative method of depositing photoresist on small diamond samples, that allowed the entire width of the sample to be used, was demonstrated by Zhang et al. [92]. In this case, inkjet printing, photolithographic patterning and ICP-RIE etching were used to fabricate large cross-section, edge-coupled waveguides on free-standing thin diamond substrates (see Figure 10.8). Well-defined photoresist microstripes without any edge and "coffee-stain" effects were produced, with the diamond ridge waveguides extending to the very edge of the substrate and enabling edge coupling and optical characterization. This new fabrication technique has demonstrated great potential for the production of long-length, high-quality diamond waveguides, and in the development of novel diamond microstructures.

10.4.3.3 Light Guiding in Ion-Implanted Diamond

In addition to the milling of material, ion implantation can be used to locally modify the refractive index of diamond [73, 74]. Lagomarsino et al. have demonstrated the fabrication of light-waveguiding microstructures in bulk SCD by means of direct ion implantation with a scanning microbeam. This resulted in a weak modulation of the refractive index of the ion-beam-damaged crystal [75]. Waveguiding through buried microchannels was also observed using a phase-shift micro-interferometric method, which allowed an examination of the multimoded structure of the propagating field. The possibility to define optical and photonic structures by direct ion writing opens a range of new possibilities in the design of quantum optical devices in bulk SCD [71].

10.4.4
Surface Emission: Solid Immersion Lenses and Nanopillars

The enhancement of collection efficiencies from otherwise isotropically emitting NV centers requires sculpting of the diamond surface; hence, the use of bulk diamonds will avoid the complexity of thin-film fabrication. Two different strategies have been applied to create solid immersion lenses (SILs) [116] or nanowires [69] to collect and redirect emission which otherwise would be lost into the diamond substrate.

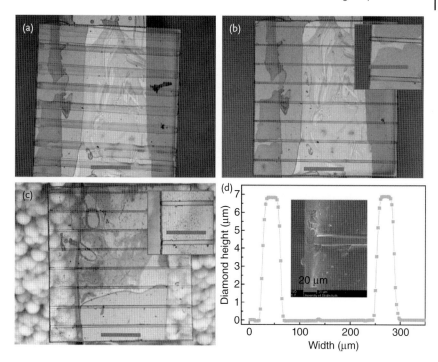

Figure 10.8 Optical microscopy images of the diamond sample during the process. (a) Inkjet printer-deposited photoresist stripes; (b) Photolithographic resist lines and inset with higher magnification; (c) Diamond waveguides etched by ICP and inset with higher magnification; (d) The surface profilometer measurements of the finished waveguide and the corresponding SEM image (inset). Scale bars: 500 μm for main figures; 200 μm for insets [92]. For a color version of this figure, please see the color plate at the end of this book.

10.4.4.1 Solid Immersion Lenses

A SIL is a hemispherical surface, the role of which is to redirect light. If centered on a color center, the SIL will reduce the light lost to TIR at the diamond–air interface. Polished bulk optical diamond SILs were first reported in 2006 [117], and are now available commercially [118]. Due to the high aspect ratio of SILs, their fabrication at the micron scale for use with quantum emitters necessitates FIB milling; this is achieved by milling a vertical series of concentric rings of increasing diameter [116]. Alignment to the individual centers is achieved by registering a confocal fluorescence microscopy image to a FIB-written ruler. These markers are referenced when milling the SIL [17], such that an eightfold increase in captured photons could be demonstrated. Similar SILs have been exploited to provide a higher count rate from two distinct NV emitters [119], allowing for quantum interference measurements between two photons.

10.4.4.2 Nanopillars

Anisotropic dry etching recipes have been developed for both polycrystalline and SCD surfaces, using a variety of masking techniques [109, 120]. These recipes were applied to form ~200 nm-diameter, 2 μm-high nanopillars in a Type Ib diamond [69]. Arrays of uniform diamond nanowires are shown in Figure 10.9a (with magnified images in Figure 10.9b, c). The masks included drop-cast particles of gold, silica, and sapphire, or films of gold that had been patterned using positive-resist and high-resolution negative-resist HSQ. The sapphire nanoparticles exhibited the greatest selectivity, but limited the control over particle deposition. The low etch selectivity of the silica nanoparticles resulted in their erosion by the end of the etching process. Nanopillar aspect ratios of 3 to ~10 were achieved with etch rates of approximately 200 nm min^{-1}. NVs were created within the single-crystal material by ion implantation, either before or after etching. A tenfold increase in photon flux was observed for pump powers at saturation, relative to representative NVs in bulk diamond [69]. The lifetimes of the emitters were also modified by the nanostructured environment.

The emission properties of NVs in nanopillars are sensitive to the environment surrounding the diamond. Choy *et al.* have shown that metallization of the surface of Type IIa diamond with a 500 nm silver film is sufficient to create plasmonic resonators [121]. Shorter, 180 nm-high, 130 nm-diameter pillars provide resonators with modal volumes of approximately 0.07 $(\lambda/n)^3$. The bulk NV lifetimes of 16.7 ns were increased to 37.17 ns in bare nanopillars, but when metallized the NV lifetimes were decreased to 5.65 ns. The spontaneous emission rate of the emitters could, in theory, be increased 30-fold for an optimally positioned radially polarized dipole. It should be noted that these metallic nanopillars are capable of almost isotropic collection, in contrast to all other structures presented in this chapter.

Figure 10.9 (a) Large arrays of 2 μm-tall, 200 nm-diameter pillars etched into the surface single crystal diamond; (b) Close-up of a group of pillars; (c) A single pillar [120].

10.4.5
Cavities

An optical cavity is a resonant structure that reinforces a narrow spectral emission band. As discussed above (see Section 10.1), a high Q and a small modal volume on the order of a cubic wavelength are desirable. Although, ideally, a single cavity would host a single color center at the peak of the modal antinode, this is statistically unlikely unless a targeted implantation or the preselection of color centers are considered. The resonance wavelength should also be simple to tune to overlap the ZPL. During the past decade, photonic crystal cavities have become the benchmark for high-Q, small-volume optical resonators, as their ability to reflect light over a wide range of incident angles allows them also to confine light efficiently in the plane. However, the relative simplicity of ring and disk resonators should not be overlooked, in terms of their processing and high-Q status, even though their modal volumes are significantly larger.

10.4.5.1 Nanobeams
One-dimensional (1-D) photonic crystal nanobeam cavities have been fabricated using FIB, at 30 kV, and selecting the beam current to produce a rapid patterning with good vertical profiles. The sizes of features, their side wall angles and their rounding, are functions of the beam current, focusing, and stigmation. Side milling has been used as a novel approach to undercut the diamond [68], which can be milled laterally and then vertically, or vice versa. The different parts of the pattern are milled with a beam current that is appropriate to the area of material to be removed, and the resolution required. A high Q of 10^6 design was presented. In 2008, Babinec's Stopping Range of Ions in Matter (SRIM) calculations predicted a 15 nm implantation of gallium ions into the diamond, which may not only account for the high absorption but also explain the problems encountered when measuring the cavity Q.

Bayn et al. [70] used angled undercutting to create a nanobeam cavity with rectangular holes (cut vertically). The designed cavity Q was <300, and the fabricated cavity Q was ~220, while subsequent secondary ion mass spectrometry (SIMS) analysis showed concentrations of Ga (maximum ca. 10^{22} ions cm^{-3}) implanted to a depth of 5 nm.

10.4.5.2 Two-Dimensional Photonic Crystals
An array of 2-D photonic crystal (PhC) defect cavities was patterned in a single-crystal membrane with the aid of lift-off and FIB milling [89]. Quality factors of ~8000 were predicted and subsequently measured on the order of ~30. The local separation to the substrate, at ~50 nm, was too small for optical isolation of the cavity, although the separation was seen to vary locally due to the contorted, swollen, and damaged membrane. In addition to the Ga ion implantation and damage, a 40 nm layer of chromium was deposited onto the surface to aid milling, and required acid cleaning post-fabrication. Together, these factors may play a role in Q-factor reduction.

Very recently, 2-D PhC defect cavities with predicted Q-factors of 11 800 were fabricated in quasi-SCD [122]. Before the PhC patterns were milled by FIB, the substrate and nucleation layers had to be removed and the diamond film thinned, by using a combination of ion milling and RIE processes. The samples were cleaned by annealing, acid-washing and oxidation processes to remove any damaged diamond, implanted Ga and graphitic material, respectively. The experimental Q-values for the 2-D PhCs were on the order of 450, and the impact of disorder and nonvertical 6° etched side walls was considered to account for the significant fall in Q. The intensity of the SiV ZPL was enhanced by a factor of 2.8 when it coincided with a cavity resonance. The resonances were tuned irreversibly by 3 nm, using repeated oxidation steps.

A 2-D PhC defect cavity has recently been demonstrated with an estimated Q-factor of ~3000 and modal volume, $V = 0.88\ (\lambda/n)^3$ [123]. Coupling of the ZPL of a single NV center to this cavity enhanced the spontaneous emission into the ZPL by a factor of approximately 70.

10.4.5.3 Ring Resonators

Multimode ring resonators were fabricated [25] in Type IIa diamond. In order to fabricate the cavities, a 5 μm-thick membrane was placed on a 2 μm-thick SiO_2 substrate and thinned to 280 nm by oxygen plasma etching. A plasma-enhanced CVD silicon nitride (Si_3N_4) hard mask was deposited on top, followed by a PMMA mask, which was patterned using e-beam lithography. An O plasma etch was then used to transfer the Si_3N_4 pattern into the diamond (see Figure 10.10a). Simulated Q-factors of 10^6 were predicted for the modal fields shown in Figure 10.10b, but the degree of roughness (as apparent in Figure 10.10c) limited the Q in this experiment. This was due mainly to the roughness of the top surface from the thinning process, and on the side walls due to the mask transfer (see Figure 10.10d, e). The cavity modes displayed Q-factors of about 4000, and modal volumes in the range of 17 to 32 $(\lambda/n)^3$.

Tuning a cavity resonance to coincide with the ZPL led to a significant enhancement of the spontaneous emission rate. The cavities were tuned by injecting the cryostat with xenon gas, which condensed on the microdisk surface and red-shifted the cavity resonance. Ideally, in this cavity environment an optimally positioned NV would experience a Purcell factor of between 10 and 19. The lifetimes of NVs resonantly coupled to the cavities were reduced from 11.1 ns to 8.3 ns. Including the branching ratio of the ZPL versus the sideband, an experimental Purcell factor of ~12 was deduced.

Recently, Hausmann *et al.* have demonstrated diamond ring resonators coupled to low-loss diamond waveguides [124]. Single-photon emission has been demonstrated coupled from integrated grating couplers.

With high-resolution patterning, highly selective masks and etch recipes capable of producing high-optical quality structures, the topic of monolithic diamond photonics is currently burgeoning. Nonetheless, challenges remain in preparing thin films of the highest quality diamond in suitably large areas, in positioning emitters within cavities, in tuning either color center or cavities to enhance lifetimes, and in interconnecting these quantum emitters.

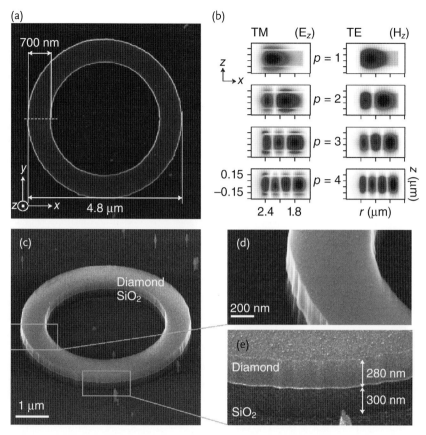

Figure 10.10 (a) Scanning electron microscopy image of a multimode ring resonator etched into a 280 nm-thick single crystal diamond membrane; (b) Simulated field profiles of 4 TM and 4 TE modes, with radial quantum numbers $p = 1$ to 4; (c) Side-view of the ring measuring 2.05 μm in radius and 700 nm in width, sitting on a SiO$_2$ pedestal that has been overetched by 300 nm; (d) and (e) Highlights of the surface roughness in the side and top walls of the structure [25]. For a color version of this figure, please see the color plate at the end of this book.

10.5 Polycrystalline Thin Films

10.5.1 Properties

Polycrystalline diamond films have been widely used for the fabrication of diamond-based optical structures, mainly because they can be grown with high nucleation densities [125] directly onto existing photonic substrates. This allows for the production of controlled thickness films, with smooth as-grown surfaces, submicron thicknesses, and large lateral dimensions, all of which are unachievable with homoepitaxial single-crystal growth methods. Both, nanocrystalline diamond (NCD) and UNCD are small grain-sized diamond films that can be deposited with submicron thicknesses on a wide range of substrates [126]. The successful production of optical structures with these films allows for readily scalable architectures and fabrication methods, thereby providing a significant boost to the outlook of commercial NV-based CQED systems. Yet, despite these advantages these materials also present unique challenges for the production of CQED-relevant optical structures, as discussed below.

First, it is not always trivial to calculate the precise impact of a material's scattering on the structure's photonic performance [127, 128], whereas absorption is relatively straightforward to include [66]. Consequently, the greatest obstacles to the successful production of optical structures in polycrystalline diamond remain their absorption [129] and scattering properties (see Section 10.3). In addition to surface roughness, internal scattering and nonuniform optical properties caused by graphitic inclusions combine to preclude the use of ordinary ellipsometry. For example, the refractive index of a diamond film may vary by as much as 50% among different areas of a 10 cm-diameter film [128].

Second, the existence of significant and often inhomogeneous strain fields within the grown material is an issue [130]. For NV color centers this crystal strain will typically break the symmetry of its excited state wavefunctions and introduce significant inhomogeneous shifts to the ZPLs of any fabricated centers. These variable ZPL wavelengths must then be accounted for by either a modification of the cavity's resonance wavelength, or by a tailored Stark-shift tuning of each NV center.

The quality of the color centers themselves can also vary significantly. Due to the lack of crystal orientation control during growth, even the highest-purity polycrystalline samples can house a variety of spin densities, which have been shown to vary even across single grains [131]. Since it appears to be difficult to produce transform-limited photon emission even from the purest single-crystal samples [132], these added materials quality issues are likely to prohibit the effective use of polycrystalline diamond material in CQED applications in the near future.

Despite these issues, significant progress has been made in the production and characterization of optical structures created in polycrystalline diamond films (see Section 10.5.2). Some unique properties also distinguish polycrystalline films from other forms of diamond. One such example is the observation of a new effect of reversible laser-induced change in the index of refraction of NCD membranes that appears as a shift in the interference fringes in the transmission and photoluminescence (PL) spectra [133]. Another example is the observation of second-order ($\chi^{(2)}$) nonlinear optical properties of nanocrystalline diamond that are most likely affected by the presence of grain boundaries [134]; indeed, $\chi^{(2)}$ processes cannot be observed in bulk diamond due to the SCD inversion symmetry. The observation of a second harmonic signal from NCD can be thus regarded as a probe of symmetry perturbation (e.g., the surface of the crystal) which, in NCD, is most likely due to the grain boundaries.

10.5.2
Optical Structures in Polycrystalline Films

Microdisk and PhC structures may be fabricated in NCD films grown directly on SiO_2/Si substrates, while 2-D PhC cavities may be fabricated in thin-film diamond.

10.5.2.1 Ring Resonators
Whispering gallery modes (WGMs) in cylindrical resonators exhibit high Q-factors with moderately small modal volumes. Microdisk cavities have been fabricated in films of ~350 nm-thick NCD, using standard e-beam lithography and ICP-RIE to define and transfer the pattern into the hard mask. In this case, oxygen-based RIE [61] was first used to create 10- to 15 µm-diameter diamond resonators; this was followed by a hydrofluoric acid wet etching step to undercut the structures, such that a central pedestal was left for mechanical support and the WGM would be surrounded by air. Subsequently, Q-factors on the order of 100 were measured by monitoring PL at 637 nm. As the WGMs are very sensitive to side wall roughness, a post-processing of the disks was performed with FIB to polish the side walls after the deposition of layers of SiO_2 and Ti, in order to protect the surface. Although this caused a lowering of PL emission and the optical modes vanished, measurements could still be made using the tapered optical fiber, and a degree of recovery was observed following a 30-min annealing at 700 °C in N_2. Such losses in microdisks are attributable to radiation from the curved interface, scattering at roughness, and absorption. The results of this experiment highlighted the absorption, as well as the surface- and grain-boundary scattering in the NCD material as being limiting factors to cavity Qs.

10.5.2.2 Two-Dimensional Photonic Crystals
Previously, 2-D PhC patterns were fabricated in NCD films [108], using both flowable oxide and metal masks, coupled with oxygen plasma RIE. The PhC defect cavities were subsequently fabricated in 130–160 nm-thick films, where grain sizes were of the order of 10–30 nm (Figure 10.11a). SiO_2 was used as a hard mask material, to transfer the triangular lattice of holes into the diamond by oxygen RIE. The PhC cavities also demonstrated relatively low Q-factors of 585, limited by the film roughness, defects, and absorption [65]. The top and side views in Figure 10.11b, c show the rough surface caused by the grains.

The patterning of a periodic or random structure into the surface of an NCD layer was studied in an attempt to enhance the PL emission from the surface [107]. For this, the NCD layers were grown on quartz substrates and patterned through Ni masks; the patterns were then etched into ~400 nm-thick diamond by CF_4/O_2 RIE plasma. When using excitation from a HeCd laser at 325 nm, the subgap luminescence from defects and impurities was strongly enhanced for the periodic pattern, with a sixfold enhancement in PL being observed at 600 nm [135].

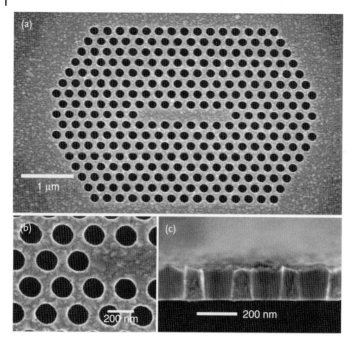

Figure 10.11 (a) L7 Photonic crystal defect cavity etched into 160 nm-thick nanocrystalline diamond membrane. The top of the rough polycrystalline surface is visible in panel (b) and in the cross-section (c). The lattice period is ~210 nm and the hole diameters are 160 nm. The silica substrate has been undercut [61].

10.6
Hybrid Approaches

Hybrid approaches include the nanopositioning of ND onto optical structures (these are discussed in greater detail in Section 10.6.1). The incorporation of NDs into optical materials transparent in the emission window of diamond color centers was recently demonstrated [21]. Specifically, optically active NDs were dispersed in tellurite glass, and a quantum emission was observed following a fiber-drawing process. This method provided a complete encapsulation of the emitter, in addition to a platform that could easily be integrated into fiber-based systems. This approach also allowed the possibility of coupling the emitter to a bound mode in hybrid photonic structures.

10.6.1
Nanomanipulation of Nanodiamond

The ability to readily fabricate single-defect centers in ND has, in recent times, enabled the commercialization of the first SPS based on NV defects in diamond [136]. The fabrication of these sources in desired locations, and on particular structures, remains difficult however due to the statistical nature of creating the atomic defect. Nonetheless, if opportunities for applications such as nanoscale magnetometry [137–140] and decoherence imaging [141] are to be fully explored, then an ability to manipulate single quantum systems and sources will be required. Some typical nanomanipulation methods are discussed in the following subsections.

The manipulation of nanoparticles has been reported by Barth et al., using atomic force microscopy (AFM) [13], and also by Balasubramanian et al., using a combined scanning confocal microscopy/AFM system [137]. The latter combined technique allows the position and optical characteristics of the emitter to be measured simultaneously, provided that the focused laser spot from the confocal microscope coincides with the AFM cantilever tip. Such systems have been shown to serve as an effective tool for attaching single quantum emitters onto the AFM tip, although the technique is limited to substrates that are transparent and optically thin. The technique of scanning near-field optical microscopy (SNOMs) has delivered similar outcomes, whereby the nanoparticle is attached to a tapered optical probe containing a thin layer of glue or PMMA [142], or by scanning the probe close to the substrate with the aim of attracting the particle to the tip [143]. Unfortunately, the manipulation success rate of this approach is low, and the ability to select a desired nanoparticle is a major challenge. Nevertheless, the manipulation of nanocrystals down to a size of ~25 nm has been demonstrated.

A fiber taper waveguide is a versatile, tool as shown by Barclay et al. [77] (see Figure 10.12). In addition to serving its usual role for probing the optical fields of a microcavity, the waveguide also functions as a positioning tool for NDs. This provides several benefits, notably the pre-selection of NDs with optimal spectral properties from a nominally lower-quality ensemble [144], as well as the controlled positioning of any number of NDs on photonic structures.

The nanomanipulation of NDs containing a single NV was also achieved by Ampem-Lassen et al. (Figure 10.13), using a custom-tapered optical fiber probe in

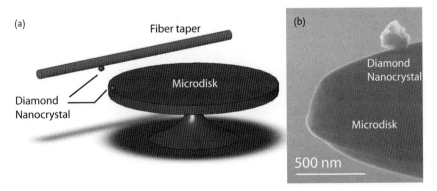

Figure 10.12 (a) Schematic of the nanocrystal fiber-to-microcavity positioning technique; (b) SEM image of a diamond nanocrystal positioned on the edge of a SiO$_2$ microdisk [77].

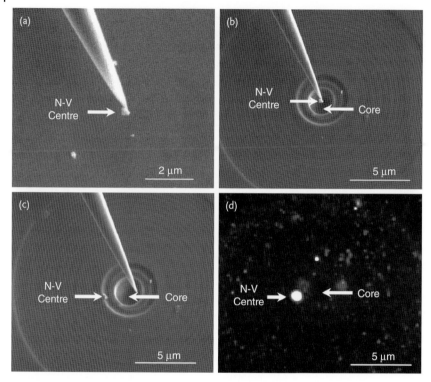

Figure 10.13 (a) SEM image of a single diamond nanocrystal (~300 nm) being removed from the marked silicon substrate; (b) SEM image of the tapered probe with the diamond nanocrystal attached and positioned above the optical fiber endface; (c) SEM image of the single emitting diamond nanocrystal positioned in the region outside the fiber core; (d) Confocal image of the fiber endface after manipulation and 1 h anneal at 650 °C in air [15].

a standard SEM system fitted with a nanomanipulator [15]. In similar fashion, van der Sar et al. (Figure 10.14) used a home-built nanomanipulator under real-time SEM imaging to manipulate NDs [145]. Each of these techniques offers a unique advantage of being able to inspect and manipulate a single ND onto varied substrates, while waveguiding devices with nanoscale precision.

10.7
Conclusions and Outlook

The fabrication and characterization of diamond-based optical structures is an emerging area that has faced a particularly challenging set of problems when compared to commonly used photonic materials. During the past decade, however,

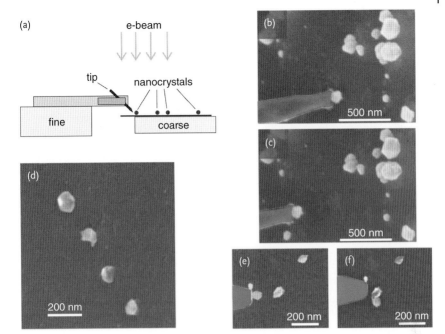

Figure 10.14 (a) Schematic of the nanomanipulator. The coarse stage can be moved in three dimensions by Attocube piezo steppers (range of 3 mm, smallest step size of 25 nm). A sharp, chemically etched tungsten tip is mounted on the piezo fine stage (range of 15 μm, smallest step size of subnanometer), which can also be moved in three dimensions. Both stages are controlled by an analog joystick, allowing for accurate adjustment of their positions; (b) and (c) SEM images showing a nanocrystal being attached to the tip and (c) subsequently being lifted from the substrate by retraction of the tip; (d) SEM image of four diamond nanocrystals that were subsequently picked up and positioned to form a line; (e) and (f) SEM images showing a nanocrystal (e) first attached to the tip (f) then being positioned onto another nanocrystal [145].

tremendous progress has been made in both the design and fabrication of optical structures in diamond. The unique properties of diamond have led to three broad applications: sensing; single-photon generation for quantum key distribution; and scalable quantum information processing that benefits from, or is enabled by, diamond-based optical devices. An examination of the optical properties of the three main forms of diamond has led to the conclusion that SCD has superior properties when compared to its counterparts. A review of the designs for diamond-based optical structures also concluded that practical devices require robust designs where fabrication tolerance is the first demand.

In this chapter, the three major platforms to create practical diamond photonic structures – namely single crystal, polycrystalline diamond, and hybrids – have been reviewed. The fabrication of SCD, as well as its processing and optical structures fabricated to date – including microlenses, rib waveguides, nanopillars, SILs, and 1-D and 2-D PhC cavities – have also been reviewed.

Whilst the carving of optical structures in SCD has been hampered by many challenges, it has recently re-emerged as a significant fabrication process. In particular, ring-resonator structures with Q-factors in excess of 12 000 [124] measured in the visible range, have marked a major milestone towards integrated diamond photonics.

As an alternative to monolithic diamond structures, the fabrication of optical structures in thin polycrystalline diamond films, such as PhC cavities and microdisks, has been proposed. Although these structures have pioneered the way towards diamond-based photonic devices, the optical properties of SCD are at present known to be superior to those of polycrystalline thin films. This is equally true regarding the quality of the light emitted from optical defects in these materials.

During recent years, while fabrication in SCD has been gaining momentum, many hybrid approaches that combine existing optical structures and NDs have also been pursued. In this respect, a variety of manipulation tools that include AFM, SNOM, fiber tapers have been developed, in addition to a number of optical structures that have been used as building blocks of the hybrid system, such as PhC slabs, microdisks, silica spheres, and single-mode fibers. Taken together, these techniques offer the unique advantage of allowing the manipulation of single NDs onto existing optical structures, with nanoscale precision, while typically employing evanescent fields to couple the emitters to the optical structures.

Recently, a hybrid material comprising a mixture of NDs in soft tellurite glass was developed, and single-photon emission from embedded NVs was observed [21]. This combination of advanced photonics and quantum emission heralds a new era of quantum sensors, and of more practical quantum information processing. Although plasmonic structures have not been discussed here in detail, the results of recent studies have indicated that it might be possible to avoid high absorption losses in metal [146], thus opening another promising avenue in diamond photonics.

At present, diamond photonics continues to face major challenges, due mainly to the difficulties encountered in the fabrication of photonic structures. The complexity of such processing is also testament to the lengths that research groups are forced to go to when processing this material [122]. Indeed, groups working with diamond photonics have long sought the ability to tune the resonant modes of an optical cavity to the wavelength of emitter emission, in real time. It should be noted that almost all of the presented optical structures are building blocks, and provide the basic elements for more complex structures and devices. It is believed that, within the next few years, significant progress will be made not only in building diamond-based optical structures individually, but also in coupling them together as integrated systems.

Acknowledgments

This chapter was produced with the assistance of the Australian Research Council (ARC) under the Discovery Project Scheme (project number DP1096288). S.T-H. is supported by the ARC Australian Research Fellowship (project number DP1096288), while A.D.G. acknowledges the support of an ARC QEII Fellowship (project number DP0880466).

References

1 Eisaman, M.D., Fan, J., Migdall, A., and Polyakov, S.V. (2011) Single-photon sources and detectors. *Rev. Sci. Instrum.*, **82**, 071101.

2 Bigelow, M.S., Lepeshkin, N.N., and Boyd, R.W. (2003) Observation of ultraslow light propagation in a ruby crystal at room temperature. *Phys. Rev. Lett.*, **90**, 113903.

3 Koehl, W.F., Buckley, B.B., Heremans, F.J., Calusine, G., and Awschalom, D.D. (2011) Room temperature coherent control of defect spin qubits in silicon carbide. *Nature*, **479**, 84.

4 Hedges, M.P., Longdell, J.J., Li, Y., and Sellars, M.J. (2010) Efficient quantum memory for light. *Nature*, **465**, 1052–1056.

5 Jelezko, F., Gaebel, T., Popa, I., Domhan, M., Gruber, A., and Wrachtrup, J. (2004) Observation of coherent oscillation of a single nuclear spin and realization of a two-qubit conditional quantum gate. *Phys. Rev. Lett.*, **93**, 130501.

6 Gaebel, T., Domhan, M., Popa, I., Wittmann, C., Neumann, P., Jelezko, F., Rabeau, J.R., Stavrias, N., Greentree, A.D., Prawer, S., Meijer, J., Twamley, J., Hemmer, P.R., and Wrachtrup, J. (2006) Room-temperature coherent coupling of single spins in diamond. *Nat. Phys.*, **2**, 408–413.

7 Hanson, R., Mendoza, F.M., Epstein, R.J., and Awschalom, D.D. (2006) Polarization and readout of coupled single spins in diamond. *Phys. Rev. Lett.*, **97**, 087601.

8 Dutt, G.M.V., Childress, L., Jiang, L., Togan, E., Maze, J., Jelezko, F., Zibrov, A.S., Hemmer, P.R., and Lukin, M.D. (2007) Quantum register based on individual electronic and nuclear spin qubits in diamond. *Science*, **316** (5829), 1312–1316.

9 Rogers, L.J., Armstrong, S., Sellars, M.J., and Manson, N.B. (2008) Infrared emission of the NV centre in diamond: Zeeman and uniaxial stress studies. *New J. Phys*, **10**, 103024.

10 Tamarat, Ph., Manson, N.B., Harrison, J.P., McMurtrie, R.L., Nizovtsev, A., Santori, C., Beausoleil, R.G., Neumann, P., Gaebel, T., Jelezko, F., Hemmer, P., and Wrachtrup, J. (2008) Spin-flip and spin-conserving optical transitions of the nitrogen-vacancy centre in diamond. *New J. Phys.*, **10**, 045004.

11 Gaebel, T., Domhan, M., Wittmann, C., Popa, I., Jelezko, F., Rabeau, J., Greentree, A.D., Prawer, S., Trajkov, E., Hemmer, P.R., and Wrachtrup, J. (2006) Photochromism in single nitrogen-vacancy defect in diamond. *Appl. Phys. B*, **82** (2), 243–246.

12 Felton, S., Edwards, A.M., Newton, M.E., Martineau, P.M., Fisher, D., and Twitchen, D. (2008) *Phys. Rev. B*, **77**, 081201(R).

13 Barth, M., Nuesse, N., Loechel, B., and Benson, O. (2009) Controlled coupling of a single-diamond nanocrystal to a photonic crystal cavity. *Opt. Lett.*, **34** (7), 1108–1110.

14 Gregor, M., Henze, R., Schroder, T., and Benson, O. (2009) On-demand positioning of a preselected quantum emitter on a fiber-coupled toroidal microresonator. *Appl. Phys. Lett.*, **95**, 153110.

15 Ampem-Lassen, E., Simpson, D.A., Gibson, B.C., Trpkovski, S., Hossain, F.M., Huntington, S.T., Ganesan, K., Hollenberg, L.C.L., and Prawer, S. (2009) Nano-manipulation of diamond-based single photon sources. *Opt. Express*, **17** (14), 11287–11293.

16 Inam, F.A., Gaebel, T., Bradac, C., Stewart, L., Withford, M.J., Dawes, J.M., Rabeau, J.R., and Steel, M.J. (2011) Modification of spontaneous emission from nanodiamond colour centres on a structured surface. *New J. Phys.*, **13**, 073012.

17 Marseglia, L., Hadden, J.P., Stanley-Clarke, A.C., Harrison, J.P., Patton, B., Ho, Y.-L.D., Naydenov, B., Jelezko, F., Meijer, J., Dolan, P.R., Smith, J.M., Rarity, J.G., and O'Brien, J.L. (2011) Nano-fabricated solid immersion lenses registered to single emitters in diamond. *Appl. Phys. Lett.*, **98**, 133107.

18 Kühn, S., Hettich, C., Schmitt, C., Poizat, J.-Ph., and Sandoghdar, V. (2001) Diamond colour centres as a nanoscopic light source for scanning near-field optical microscopy. *J. Microsc.*, **202** (1), 2–6.

19 Rabeau, J.R., Huntington, S.T., Greentree, A.D., and Prawer, S. (2005) Diamond chemical-vapor deposition on optical fibers for fluorescence waveguiding. *Appl. Phys. Lett.*, **86**, 134104.

20 Schroder, T., Schell, A.W., Kewes, G., Aichele, T., and Benson, O. (2011) Fiber-integrated diamond-based single photon source. *Nano Lett.*, **11**, 198–202.

21 Henderson, M.R., Gibson, B.C., Ebendorff-Heidepriem, H., Kuan, K., Afshar, V.S., Orwa, J.O., Aharonovich, I., Tomljenovic-Hanic, S., Greentree, A.D., Prawer, S., and Monro, T.M. (2011) Diamond in tellurite glass: a new medium for quantum information. *Adv. Mater.*, **23** (25), 2806–2810.

22 Schietinger, S., Schröder, T., and Benson, O. (2008) One-by-one coupling of single defect centers in nanodiamonds to high-Q modes of an optical microresonator. *Nano Lett.*, **8** (11), 3911–3915.

23 Kolesov, R., Grotz, B., Balasubramanian, G., Stöhr, R.J., Nicolet, A.A.L., Hemmer, P.R., Jelezko, F., and Wrachtrup, J. (2009) Wave-particle duality of single surface plasmon polaritons. *Nat. Phys.*, **5**, 470–474.

24 Barclay, P.E., Fu, K.-M.C., Santori, C., Faraon, A., and Beausoleil, R.G. (2011) Hybrid nanocavities for resonant enhancement of color center emission in diamond. *Phys. Rev. X*, **1**, 011007.

25 Faraon, A., Barclay, P.E., Santori, C., Fu, K.-M.C., and Beausoleil, R.G. (2011) Resonant enhancement of the zero-phonon emission from a color center in a diamond cavity. *Nat. Photonics*, **5**, 301–305.

26 Dumeige, Y., Alléaume, R., Grangier, P., Treussart, F., and Roch, J.-F. (2011) Controlling the single-diamond nitrogen-vacancy color center photoluminescence spectrum with a Fabry–Perot microcavity. *New J. Phys.*, **13**, 025015.

27 Wolters, J., Schell, A.W., Kewes, G., Nüsse, N., Schoengen, M., Döscher, H., Hannappel, T., Löchel, B., Barth, M., and Benson, O. (2010) Enhancement of the zero phonon line emission from a single nitrogen vacancy center in a nanodiamond via coupling to a photonic crystal cavity. *Appl. Phys. Lett.*, **97**, 141108.

28 van der Sar, T., Hagemeier, J., Pfaff, W., Heeres, E.C., Thon, S.M., Kim, H., Petroff, P.M., Oosterkamp, T.H., Bouwmeester, D., and Hanson, R. (2011) Deterministic nanoassembly of a coupled quantum emitter-photonic crystal cavity system. *Appl. Phys. Lett.*, **98**, 193103.

29 Gruber, A., Dräbenstedt, A., Tietz, C., Fleury, L., Wrachtrup, J., and von Borczyskowski, C. (1997) Scanning confocal optical microscopy and magnetic resonance on single defect centers. *Science*, **276**, 2012–2014.

30 Kurtsiefer, C., Mayer, S., Zarda, P., and Weinfurter, H. (2000) Stable solid-state source of single photons. *Phys. Rev. Lett.*, **85**, 290–293.

31 Beveratos, A., Brouri, R., Gacoin, T., Poizat, J.-P., and Grangier, P. (2001) Nonclassical radiation from diamond nanocrystals. *Phys. Rev. A*, **64**, 061802(R).

32 Beveratos, A., Brouri, R., Gacoin, T., Villing, A., Poizat, J.-P., and Grangier, P. (2002) Single photon quantum cryptography. *Phys. Rev. Lett.*, **89**, 187901.

33 Manson, N.B., Harrison, J.P., and Sellars, M.J. (2006) Nitrogen-vacancy center in diamond: model of the electronic structure and associated dynamics. *Phys. Rev. B*, **74**, 104303.

34 Schmunk, W., Rodenberger, M., Peters, S., Hofer, H., and Kück, S. (2011) Radiometric calibration of single photon detectors by a single photon source based on NV-centers in diamond. *J. Mod. Opt.*, **58** (14), 1252–1259.

35 Cheung, J.Y., Chunnilall, C.J., Woolliams, E.R., Fox, N.P., Mountford, J.R., Wang, J., and Thomas, P.J. (2007) The quantum candela: a re-definition of the standard units for optical radiation. *J. Mod. Opt.*, **54**, 373–396.

36 Zwinkels, J.C., Ikonen, E., Fox, N.P., Ulm, G., and Rastello, M.L. (2011) Photometry, radiometry and "the candela": evolution in the classical and quantum world. *Metrologia*, **47**, R15.

37 Tamarat, Ph., Gaebel, T., Rabeau, J.R., Khan, M., Greentree, A.D., Wilson, H., Hollenberg, L.C.L., Prawer, S., Hemmer, P., Jelezko, F., and Wrachtrup, J. (2006) Stark shift control of single optical centres in diamond. *Phys. Rev. Lett.*, **97**, 083002.

38 Su, C.-H., Greentree, A.D., and Hollenberg, L.C.L. (2008) Towards a picosecond transform-limited nitrogen-vacancy based single photon source. *Opt. Express*, **16** (9), 6240–6250.

39 Bennett, C.H. and Brassard, G. (1984) Quantum cryptography: public key distribution and coin tossing. Proceedings of the IEEE International Conference on Computers, Systems, and Signal Processing, Bangalore, India, pp. 175–179.

40 Su, C.-H., Greentree, A.D., and Hollenberg, L.C.L. (2009) High-performance diamond-based single-photon sources for quantum communication. *Phys. Rev. A*, **80**, 052308.

41 Barrett, S.D. and Kok, P. (2005) Efficient high-fidelity quantum computation using matter qubits and linear optics. *Phys. Rev. A*, **71**, 060310(R).

42 Lim, Y.L., Beige, A., and Kwek, L.C. (2005) Repeat-until-success linear optics distributed quantum computing. *Phys. Rev. Lett.*, **95**, 030505.

43 Benjamin, S.C., Browne, D.E., Fitzsimons, J., and Morton, J.J.L. (2006) Brokered graph-state quantum computation. *New J. Phys.*, **8**, 141.

44 Greentree, A.D., Fairchild, B.A., Hossain, F.M., and Prawer, S. (2008) Diamond integrated quantum photonics. *Mater. Today*, **11** (9), 22–31.

45 Stephens, A.M., Evans, Z.W.E., Devitt, S.J., Greentree, A.D., Fowler, A.G., Munro, W.J., O'Brien, J.L., Nemoto, K., and Hollenberg, L.C.L. (2008) Deterministic optical quantum computer using photonic modules. *Phys. Rev. A*, **78**, 032318.

46 Greentree, A.D., Salzman, J., Prawer, S., and Hollenberg, L.C.L. (2006) Quantum gate for Q switching in monolithic photonic bandgap cavities containing two-level atoms. *Phys. Rev. A*, **73**, 013818.

47 Su, C.-H., Greentree, A.D., Munro, W.J., Nemoto, K., and Hollenberg, L.C.L. (2009) Pulse shaping by coupled-cavities: single photons and qudits. *Phys. Rev. A*, **80**, 033811.

48 Su, C.-H., Greentree, A.D., Munro, W.J., Nemoto, K., and Hollenberg, L.C.L. (2008) High-speed quantum gates with cavity quantum electrodynamics. *Phys. Rev. A*, **78**, 062336.

49 Duan, L.-M. and Kimble, H.J. (2004) Scalable photonic quantum computation through cavity-assisted interactions. *Phys. Rev. Lett.*, **92**, 127902.

50 Devitt, S.J., Greentree, A.D., Ionicioiu, R., O'Brien, J.L., Munro, W.J., and Hollenberg, L.C.L. (2007) Photonic module: an on-demand resource for photonic entanglement. *Phys. Rev. A*, **76**, 052312.

51 Almeida, V.R., Xu, Q., Barrios, C.A., and Lipson, M. (2004) Guiding and confining light in void nanostructure. *Opt. Lett.*, **29** (11), 1209–1211.

52 Hiscocks, M.P., Su, C.-H., Gibson, B.C., Greentree, A.D., Hollenberg, L.C.L., and Ladouceur, F. (2009) Slot-waveguide

cavities for optical quantum information applications. *Opt. Express*, **17** (9), 7295–7303.

53 Su, C.-H., Hiscocks, M.P., Gibson, B.C., Greentree, A.D., Hollenberg, L.C.L., and Ladouceur, F. (2011) Coupling slot-waveguide cavities for large-scale quantum optical devices. *Opt. Express*, **19** (7), 6354–6365.

54 Martineau, P.M., Gaukroger, M.P., Guy, K.B., Lawson, S.C., Twitchen, D.J., Friel, I., Hansen, J.O., Summerton, G.C., Addison, T.P.G., and Burns, R. (2009) High crystalline quality single crystal chemical vapour deposition diamond. *J. Phys. Condens. Matter*, **21** (36), 364205–364212.

55 Spence, D.J., Granados, E., and Mildren, R.P. (2010) Mode-locked picosecond diamond Raman laser. *Opt. Lett.*, **35** (4), 556–558.

56 Sabella, A., Piper, J.A., and Mildren, R.P. (2010) 1240 nm diamond Raman laser operating near the quantum limit. *Opt. Lett.*, **35** (23), 3874–3876.

57 Lubeigt, W., Savitski, V.G., Bonner, G.M., Geoghegan, S.L., Friel, I., Hastie, J.E., Dawson, M.D., Burns, D., and Kemp, A.J. (2011) 1.6 W continuous-wave Raman laser using low-loss synthetic diamond. *Opt. Express*, **19** (7), 6938–6944.

58 Isberg, J., Hammersberg, J., Johansson, E., Wikström, T., Twitchen, D.J., Whitehead, A.J., Coe, S.E., and Scarsbrook, G.A. (2002) High carrier mobility in single-crystal plasma-deposited diamond. *Science*, **297**, 1670.

59 Yoshitake, T., Nagano, A., Itakura, M., Kuwano, N., Hara, T., and Nagayama, K. (2007) Spectral absorption properties of ultrananocrystalline diamond/amorphous carbon composite thin films prepared by pulsed laser deposition. *Jpn. J. Appl. Phys.*, **46** (38), L936–L938.

60 Hiscocks, M.P., Ganesan, K., Gibson, B.C., Huntington, S.T., Ladouceur, F., and Prawer, S. (2008) Diamond waveguides fabricated by reactive ion etching. *Opt. Express*, **16** (24), 19512–19519.

61 Wang, C.F., Choi, Y.-S., Lee, J.C., Hu, E.L., Yang, J., and Butler, J.E. (2007) Observation of whispering gallery modes in nanocrystalline diamond microdiscs. *Appl. Phys. Lett.*, **90** (8), 081110.

62 Kuramochi, E., Notomi, M., Mitsugi, S., Shinya, A., and Tanabe, T. (2006) Ultrahigh-Q photonic crystal nanocavities realized by the local width modulation of a line defect. *Appl. Phys. Lett.*, **88** (4), 041112.

63 Akahane, Y., Asano, T., Song, B.S., and Noda, S. (2003) High-Q photonic nanocavity in a two-dimensional photonic crystal. *Nature*, **425** (6961), 944–947.

64 Tomljenovic-Hanic, S., Steel, M.J., de Sterke, C.M., and Salzman, J. (2006) Diamond based photonic crystal microcavities. *Opt. Express*, **14** (8), 3556–3562.

65 Wang, C.F., Hanson, R., Awschalom, D.D., Hu, E.L., Feyetson, T., Yang, J., and Butler, J.E. (2007) Fabrication and characterization of two-dimensional photonic crystal microcavities in nanocrystalline diamond. *Appl. Phys. Lett.*, **91** (20), 201112.

66 Kreuzer, C., Riedrich-Moller, J., Neu, E., and Becher, C. (2008) Design of photonic crystal microcavities in diamond films. *Opt. Express*, **16** (3), 1632–1644.

67 Bayn, I. and Salzman, J. (2008) Ultra high-Q photonic crystal nanocavity design: the effect of a low-ε slab material. *Opt. Express*, **16** (7), 4972–4980.

68 Babinec, T.M., Choy, J.T., Smith, K.J.M., Khan, M., and Loncar, M. (2010) Design and focused ion beam fabrication of single crystal diamond nanobeam cavities. *J. Vac. Sci. Technol. B*, **29** (1), 010601.

69 Babinec, T.M., Hausmann, B.J.M., Khan, M., Zhang, Y.A., Maze, J.R., Hemmer, P.M., and Loncar, M. (2010) Diamond nanowire single-photon source. *Nat. Nanotechnol.*, **5** (3), 195–199.

70 Bayn, I., Meyler, B., Salzman, J., and Kalish, R. (2011) Triangular nanobeam photonic cavities in single crystal diamond. *New J. Phys.*, **13**, 025018.

71 Tomljenovic-Hanic, S., Greentree, A.D., de Sterke, C.M., and Prawer, S. (2009) Flexible design of ultrahigh-Q cavities in

diamond-based photonic crystal slabs. *Opt. Express*, **17** (8), 6465–6475.

72 Tomljenovic-Hanic, S., Greentree, A.D., Gibson, B.C., Karle, T.J., and Prawer, S. (2011) Nanodiamond induced high-Q resonances in defect-free photonic crystal slabs. *Opt. Express*, **19** (22), 22219–22226.

73 Bhatia, K.L., Fabian, S., Kalbitzer, S., Klatt, C., Krätschmer, W., Stoll, R., and Sellschop, J.F.P. (1998) Optical effects in carbon-ion irradiated diamond. *Thin Solid Films*, **324** (1–2), 11–18.

74 Draganski, M.A., Finkman, E., Gibson, B.C., Fairchild, B.A., Ganesan, K., Nabatova-Gabain, N., Tomljenovic-Hanic, S., Greentree, A.D., and Prawer, S. (2012) Tailoring the optical constant of diamond by ion implantation. *Opt. Mater. Express*, **2** (5), 644–649.

75 Lagomarsino, S., Olivero, P., Bosia, F., Vannoni, M., Calusi, S., Giuntini, L., and Massi, M. (2010) Evidence of light guiding in ion-implanted diamond. *Phys. Rev. Lett.*, **105** (23), 233903.

76 McCutcheon, M.W. and Loncar, M. (2008) Design of silicon nitride photonic crystal nanocavity with a quality factor of one million for coupling to a diamond crystal. *Opt. Express*, **16** (23), 19136–19145.

77 Barclay, P.E., Santori, C., Fu, K.-M., Beausoleil, R.G., and Painter, O. (2009) Coherent interference effects in a nano-assembled diamond NV center cavity-QED system. *Opt. Express*, **17** (10), 8081–8097.

78 Park, Y.-S., Cook, A., and Wang, H. (2006) Cavity QED with diamond nanocrystals and silica microspheres. *Nano Lett.*, **6** (9), 2075–2079.

79 Englund, D., Shields, B., Rivoire, K., Fariba Hatami, H., Vuckovic, J., Park, H., and Lukin, M.D. (2010) Deterministic coupling of a single nitrogen vacancy center to a photonic crystal cavity. *Nano Lett.*, **10** (10), 3922–3926.

80 de Ridder, R.M., Hopman, W.C.L., and Ay, F. (2007) Focused-ion-beam processing for photonics, 9th International Conference on Transparent Optical Networks, ICTON 2007, 1–5 July 2007, Rome Italy, pp. 212–215. Doi: 10.1109/ICTON.2007.4296183.

81 Fairchild, B.A., Olivero, P., Rubanov, S., Greentree, A.D., Waldermann, F., Taylor, R.A., Walmsley, I., Smith, J.M., Huntington, S., Gibson, B.C., Jamieson, D.N., and Prawer, S. (2008) Fabrication of ultrathin single-crystal diamond membranes. *Adv. Mater.*, **20**, 4793–4798.

82 Gsell, S., Bauer, T., Goldfuss, J., Schreck, M., and Stritzker, B. (2004) A route to diamond wafers by epitaxial deposition on silicon via iridium/yttria-stabilized zirconia buffer layers. *Appl. Phys. Lett.*, **84** (22), 4541–4543.

83 Yamada, H., Chayahara, A., Mokuno, Y., Umezawa, H., Shikata, S., and Fujimori, N. (2010) Fabrication of 1 inch mosaic crystal diamond wafers. *Appl. Phys. Express*, **3**, 051301.

84 Parikh, N.R., Hunn, J.D., McGucken, E., Swanson, M.L., White, C.W., Rudder, R.A., Malta, D.P., Posthill, J.B., and Markunas, R.J. (1992) Single-crystal diamond plate liftoff achieved by ion implantation and subsequent annealing. *Appl. Phys. Lett.*, **61** (26), 3124–3126.

85 Ziegler, J.F., Biersack, J.P., and Littmark, U. (eds) (1985) *The Stopping and Range of Ions in Solids*, Pergamon Press.

86 Olivero, P., Rubanov, S., Reichart, P., Gibson, B.C., Huntington, S.T., Rabeau, J., Greentree, A.D., Salzman, J., Moore, D., Jamieson, D.N., and Prawer, S. (2005) Ion-beam-assisted lift-off technique for three-dimensional micromachining of freestanding single-crystal diamond. *Adv. Mater.*, **17** (20), 2427–2430.

87 Wang, C.F., Hu, E.L., Yang, J., and Butler, J.E. (2007) Fabrication of suspended single crystal diamond devices by electrochemical etch. *J. Vac. Sci. Technol. B*, **25** (3), 730–733.

88 Friel, I., Clewes, S.L., Dhillon, H.K., Perkins, N., Twitchen, D.J., and Scarsbrook, G.A. (2009) Control of surface and bulk crystalline quality in single crystal diamond grown by chemical vapour deposition. *Diamond Relat. Mater.*, **18**, 808–815.

89 Bayn, I., Meyler, B., Lahav, A., Salzman, J., Kalish, R., Fairchild, B., Prawer, S.,

Barth, M., Benson, O., Wolf, T., Siyushev, P., Jelzko, F., and Wrachtrup, J. (2011) Processing of photonic crystal nanocavity for quantum information in diamond. *Diamond Relat. Mater.*, **20** (7), 937–943.

90 Lee, J.C., Aharonovich, I., Magyar, A.P., Rol, F., and Hu, E.L. (2012) Coupling of silicon-vacancy centers to a single crystal diamond cavity. *Opt. Express*, **20** (8), 8891–8897.

91 Hiscocks, M.P., Su, C.-H., Ganesan, K., Gibson, B.C., Greentree, A.D., Hollenberg, L.C.L., Ladouceur, F., and Prawer, S. (2010) Accessing diamond waveguides and future applications. *Proc. SPIE.*, **7604**, 760404.

92 Zhang, Y., McKnight, L., Tian, Z., Calvez, S., Gu, E., and Dawson, M.D. (2011) Large cross-section edge-coupled diamond waveguides. *Diamond Relat. Mater.*, **20**, 564–567.

93 Messina, G., Paoletti, A., Santangelo, S., and Tucciarone, A. (1997) Physical approximants to electron scattering. *Microelectron. Eng.*, **34**, 147–154.

94 Lister, K.A., Casey, B.G., Dobson, P.S., Thoms, S., Macintyre, D.S., Wilkinson, C.D.W., and Weaver, J.M.R. (2004) Pattern transfer of a 23 nm-period grating and sub 15 nm dots into CVD diamond. *Microelectron. Eng.*, **73–74**, 319–322.

95 Patton, B.R., Dolan, P.R., Grazioso, F., Wincott, M.B., Smith, J.M., Markham, M.L., Twitchen, D.J., Zhang, Y., Gu, E., Dawson, M.D., Fairchild, B.A., Greentree, A.D., and Prawer, S. (2012) Optical properties of single crystal diamond microfilms fabricated by ion implantation and lift-off processing. *Diamond-Relat. Mater.*, **21**, 16–23.

96 Imre, A., Ocola, L.E., Rich, L., and Klingfus, J. (2010) Large area direct-write focused ion-beam lithography with a dual-beam microscope. *J. Vac. Sci. Technol.*, **28**, 304.

97 Hamada, M., Teraji, T., and Ito, T. (2003) Field-induced effects of implanted Ga on high electric field diamond devices fabricated by focused ion beam. *Appl. Surf. Sci.*, **216** (1–4), 65–71.

98 McKenzie, W.R., Quadir, Md.Z., Gass, M.H., and Munro, P.R. (2011) Focused ion beam implantation of diamond. *Diamond Relat. Mater.*, **20** (8), 1125–1128.

99 Rubanov, S. and Suvorova, A. (2011) Ion implantation in diamond using 30 keV Ga+ focused ion beam. *Diamond Relat. Mater.*, **20**, 1160–1164.

100 McKenzie, W., Pethica, J., and Cross, G. (2011) A directwrite, resistless hard mask for rapid nanoscale patterning of diamond. *Diamond Relat. Mater.*, **20**, 707.

101 Hiscocks, M.P., Ganesan, K., McKenzie, W.R., Gibson, B.C., and Ladouceur, F. (2011) Towards characterisation of millimetre length waveguides and new fabrication method for nanoscale diamond photonic structure. *Diamond Relat. Mater.*, **20** (4), 556–559.

102 Sandhu, G. and Chu, W. (1989) Reactive ion etching of diamond. *Appl. Phys. Lett.*, **55** (5), 437–438.

103 Kiyohara, S., Mori, K., Miyamoto, I., and Taniguchi, J. (2001) Oxygen ion beam assisted etching of single crystal diamond chips using reactive oxygen gas. *J. Mater. Sci. Mater. Electron.*, **12** (8), 477–481.

104 Shiomi, H. (1997) Reactive ion etching of diamond in O_2 and CF_4 plasma, and fabrication of porous diamond for field emitter cathodes. *Jpn. J. Appl. Phys.*, Part 1 **36** (12B), 7745–7748.

105 Leech, P., Reeves, G., and Holland, A. (2001) Reactive ion etching of diamond in CF_4, O_2, O_2 and Ar-based mixtures. *J. Mater. Sci.*, **36** (14), 3453–3459.

106 Lee, C.L., Gu, E., and Dawson, M.D. (2007) Micro-cylindrical and micro-ring lenses in CVD diamond. *Diamond Relat. Mater.*, **16**, 944–948.

107 Ondič, L., Dohnalová, K., Ledinský, M., Kromka, A., Babchenko, O., and Rezek, B. (2011) Effective extraction of photoluminescence from a diamond layer with a photonic crystal. *ACS Nano*, **5**, 346–350.

108 Baldwin, J.W., Zalalutdinov, M., Feygelson, T., Butler, J.E., and Houston, B.H. (2006) Fabrication of short-wavelength photonic crystals in wide-band-gap nanocrystalline diamond films. *J. Vac. Sci. Technol. B*, **24**, 50.

109 Hausmann, B.J.M., Khan, M., Zhang, Y., Babinec, T.M., Martinick, K., McCutcheon, M., Hemmer, P.R., and Lončar, M. (2010) Fabrication of diamond nanowires for quantum information processing applications. *Diamond Relat. Mater.*, **19**, 621–629.

110 Tran, D.T., Fansler, C., Grotjohn, T.A., Reinhard, D.K., and Asmmusen, J. (2009) Investigation of mask selectivities and diamond etching using microwave plasma-assisted etching. *Diamond Relat. Mater.*, **19** (7–9), 778–782.

111 Gu, E., Cho, H.W., Liu, C., Griffin, C., Girkin, J.M., Watson, I.M., Dawson, M.D., McConnell, G., and Gurney, A.M. (2004) Reflection/transmission confocal microscopy characterization of single-crystal diamond microlens arrays. *Appl. Phys. Lett.*, **84** (15), 2754.

112 Lee, C.L., Gu, E., Dawson, M.D., Friel, I., and Scarsbrook, G.A. (2008) Etching and micro-optics fabrication in diamond using chlorine-based inductively-coupled plasma. *Diamond Relat. Mater.*, **17**, 1292–1296.

113 Castelletto, S., Harrison, J.P., Marseglia, L., Stanley-Clarke, A.C., Gibson, B.C., Fairchild, B.A., Hadden, J.P., Ho, Y.L.D., Hiscocks, M.P., Ganesan, K., Huntington, S.T., Ladouceur, F., Greentree, A.D., Prawer, S., O'Brien, J.L., and Rarity, J.G. (2011) Diamond-based. Structures to collect and guide light. *New. J. Phys.*, **13**, 02502.

114 Bayn, I., Bolker, A., Cytermann, C., Meyler, B., Richter, V., Salzman, J., and Kalish, R. (2011) Diamond processing by focused ion beam-surface damage and recovery. *Appl. Phys. Lett.*, **99** (18), 183109.

115 Hiscocks, M.P., Kaalund, C.J., Ladouceur, F., Huntington, S.T., Gibson, B.C., Trpkovski, S., Simpson, D., Ampem-Lassen, E., Prawer, S., and Butler, J.E. (2008) Reactive ion etching of waveguide structures in diamond. *Diamond Relat. Mater.*, **17**, 1831–1834.

116 Hadden, J.P., Harrison, J.P., Stanley-Clarke, A.C., Marseglia, L., Ho, Y.-L.D., Patton, B.R., O'Brien, J.L., and Rarity, J.G. (2010) Strongly enhanced photon collection from diamond defect centers under microfabricated integrated solid immersion lenses. *Appl. Phys. Lett.*, **97**, 241901.

117 Shinoda, M., Aito, K.S., Ondo, T.K., Akaoki, A.N., Furuki, M., Takeda, M., Yamamoto, M., Chaich, T.J.S., Van Oerle, B.M., Godfried, H.P., Kriele, P.A.C., Houwman, E.P., Elissen, W.H.M.N., Pels, G.J., and Spaaij, P.G.M. (2006) High-density near-field readout using diamond solid immersion lens. *Jpn. J. Appl. Phys.*, **45** (2B), 1311–1313.

118 Siyushev, P., Kaiser, F., Jacques, V., Gerhardt, I., Bischof, S., Fedder, H., Dodson, J., Markham, M., Twitchen, D., Jelezko, F., and Wrachtrup, J. (2010) Monolithic diamond optics for single photon detection. *Appl. Phys. Lett.*, **97**, 241901.

119 Bernien, H., Childress, L., Robledo, L., Markham, M., Twitchen, D., and Hanson, R. (2012) Two-photon quantum interference from separate nitrogen vacancy centers in diamond. *Phys. Rev. Lett.*, **108** (4), 043604.

120 Hausmann, B.M., Babinec, T.M., Choy, J.T., Hodges, J.S., Hong, S., Bulu, I., Lukin, M.D., and Lončar, M. (2011) Single-color centers implanted in diamond nanostructures. *New J. Phys.*, **13**, 045004.

121 Choy, J.T., Hausmann, B.J.M., Babinec, T.M., Bulu, I., Khan, M., Maletinsky, P., Yacoby, A., and Loncar, M. (2011) Enhanced single-photon emission from a diamond–silver aperture. *Nat. Photonics*, **5**, 738–743.

122 Riedrich-Möller, J., Kipfstuhl, L., Hepp, C., Neu, E., Pauly, C., Mücklich, F., Baur, A., Wandt, M., Wolff, S., Fischer, M., Gsell, S., Schreck, M., and Becher, C. (2011) One- and two-dimensional photonic crystal microcavities in single crystal diamond. *Nat. Nanotechnol.*, **5** (11), 1–6.

123 Faraon, A., Santori, C., Huang, Z., Acosta, V.M., and Beausoleil, R.G. (2012) Coupling of nitrogen-0 vacancy centers to photonic crystal cavities in monocrystalline diamond. *Phys. Rev. Lett.*, **109** (3), 033604.

124 Hausmann, B.J.M., Shields, B., Quan, Q., Maletinsky, P., McCutcheon, M., Choy, J.T., Babinec, T.M., Kubanek, A.,

Yacoby, A., Lukin, M.D., and Loncar, M. (2012) Integrated diamond networks for quantum nanophotonics. *Nano Lett.*, **12** (3), 1578–1582.

125. Akhvlediani, R., Lior, I., Michaelson, S., and Hoffman, A. (2002) Nanometer rough, sub-micrometer-thick and continuous diamond chemical vapor deposition film promoted by a synergetic ultrasonic effect. *Diamond Relat. Mater.*, **11**, 545–549.

126. Williams, O.A. and Nesládek, M. (2006) Growth and properties of nanocrystalline diamond films. *Phys. Status Solidi (A)*, **203** (13), 3375–3386.

127. Xu, T., Wheeler, M.S., Ruda, H.E., Mojahedi, M., and Aitchison, J.S. (2009) The influence of material absorption on the quality factor of photonic crystal cavities. *Opt. Express*, **17** (10), 8343–8348.

128. Plano, L. and Pinneo, M. (1993) *Non-electrical Applications of Diamond Films in Diamond Films and Coatings: Development, Properties and Applications* (ed. R.F. Davis), Noyes Publishers, pp. 339–380.

129. Nesládek, M., Meykens, K., Stals, L.M., Vaněček, M., and Rosa, J. (1996) Origin of characteristic subgap optical absorption in CVD diamond films. *Phys. Rev. B*, **54**, 5552–5561.

130. Golshan, M., Laundy, D., Fewster, P.F., Moore, M., Whitehead, A., Butler, J.E., and Konovalov, O. (2003) Measuring strain in polycrystalline CVD diamond films. *J. Phys. D Appl. Phys.*, **36**, A153–A156.

131. Markham, M.L., Dodson, J.M., Scarsbrook, G.A., Twitchen, D.J., Balasubramanian, G., Jelezko, F., and Wachtrup, J. (2011) CVD diamond for spintronics. *Diamond Relat. Mater.*, **20**, 134–139.

132. Fu, K.-M.C., Santori, C., Barclay, P.E., Rogers, L.J., Manson, N.B., and Beausoleil, R.G. (2009) Observation of the dynamic Jahn–Teller effect in the excited states of nitrogen-vacancy centers in diamond. *Phys. Rev. Lett.*, **103**, 256404.

133. Preclíková, J., Kromka, A., Rezek, B., and Malý, P. (2010) Laser-induced refractive index changes in nanocrystalline diamond membranes. *Opt. Lett.*, **35**, 577–579.

134. Trojánek, F., Žídek, K., Dzurňák, B., Kozák, M., and Malý, P. (2010) Nonlinear optical properties of nanocrystalline diamond. *Opt. Express*, **18** (2), 1349–1357.

135. Ondič, L., Kusová, K., Cibulka, O., Pelant, I., Dohnalová, K., Rezek, B., Babchenko, O., Kromka, A., and Ganesh, N. (2011) Enhanced photoluminescence extraction efficiency from a diamond photonic crystal via leaky modes. *New J. Phys.*, **13**, 063005.

136. Quantum Communications Victoria (QCV) (2005) SPS 1.01 product brochure.pdf. Available at: http://www.qcvictoria.com (accessed 19 September 2012).

137. Balasubramanian, G., Chan, I.Y., Kolesov, R., Al-Hmoud, M., Tisler, J., Shin, C., Kim, C., Wojcik, A., Hemmer, P.R., Krueger, A., Hanke, T., Leitenstorfer, A., Bratschitsch, R., Jelezko, F., and Wrachtrup, J. (2008) Nanoscale imaging magnetometry with diamond spins under ambient conditions. *Nature*, **455** (7213), 648–651.

138. Maze, J.R., Stanwix, P.L., Hodges, J.S., Hong, S., Taylor, J.M., Cappellaro, P., Jiang, L., Dutt, M.V.G., Togan, E., Zibrov, A.S., Yacoby, A., Walsworth, R.L., and Lukin, M.D. (2008) Nanoscale magnetic sensing with an individual electronic spin in diamond. *Nature*, **455** (7213), 644–647.

139. Taylor, J.M., Cappellaro, P., Childress, L., Jiang, L., Budker, D., Hemmer, P.R., Yacoby, A., Walsworth, R., and Lukin, M.D. (2008) High-sensitivity diamond magnetometer with nanoscale resolution. *Nat. Phys.*, **4** (10), 810–816.

140. Degen, C.L. (2008) Scanning magnetic field microscope with a diamond single-spin sensor. *Appl. Phys. Lett.*, **92** (24), 243111.

141. McGuinness, L.P., Yan, Y., Stacey, A., Simpson, D.A., Hall, L.T., Maclaurin, D., Prawer, S., Mulvaney, P., Wrachtrup, J., Caruso, F., Scholten, R.E., and Hollenberg, L.C.L. (2011) Quantum measurement and orientation tracking of fluorescent nanodiamonds inside

living cells. *Nat. Nanotechnol.*, **6** (6), 358–363.

142 Kuhn, S., Hettich, C., Schmitt, C., Poizat, J., and Sandoghdar, V. (2000) Diamond colour centres as a nanoscopic light source for scanning near-field optical microscopy. Proceedings of the 6th International Conference on Near-field Optics and Related Techniques, 27–31 August 2000, University Twente, Enschede, the Netherlands.

143 Cuche, A., Sonnefraud, Y., Faklaris, O., Garrot, D., Boudou, J., Sauvage, T., Roch, J.-F., Treussart, F., and Huant, S. (2008) Diamond nanoparticles as photoluminescent nanoprobes for biology and near-field optics. *J. Luminesc.*, **129** (12), 1475–1477.

144 Shen, Y., Sweeney, T.M., and Wang, H. (2008) Zero-phonon linewidth of single nitrogen vacancy centers in diamond nanocrystal. *Phys. Rev. B*, **77** (3), 033201.

145 van der Sar, T., Heeres, E.C., Dmochowski, G.M., de Lange, G., Robledo, L., Oosterkamp, T.H., and Hanson, R. (2009) Nanopositioning of a diamond nanocrystal containing a single nitrogen-vacancy defect center. *Appl. Phys. Lett.*, **94** (17), 173104.

146 Barth, M., Schietinger, S., Fischer, S., Becker, J., Nusse, N., Aichele, T., Lochel, B., Sonnichsen, C., and Benson, O. (2010) Nanoassembled plasmonic-photonic hybrid cavity for tailored light-matter coupling. *Nano Lett.*, **10** (3), 891–895.

11
Thermal Management of Lasers and LEDs Using Diamond

Alan J. Kemp, John-Mark Hopkins, Jennifer E. Hastie, Stephane Calvez, Yanfeng Zhang, Erdan Gu, Martin D. Dawson, and David Burns

11.1
Introduction

11.1.1
The Requirement for Thermal Management of Lasers and LEDs

The effective management of heat is the biggest hurdle to increasing the output power and efficiency of most lasers and light-emitting diodes (LEDs). This is particularly true if other desirable characteristics need to be maintained, notably the spectral or spatial quality of the output.

Energy is supplied to a laser medium via a process known as *pumping*, which usually is achieved either electrically or optically. The energy is then converted to laser output, but the conversion is not 100% efficient and some energy will always be dissipated as heat in the laser medium, which, typically, is either a semiconducting solid or dielectric solid doped with metal ions. The effect that the waste heat creates depends on the material employed; in semiconductor lasers, for example, the induced temperature rise directly reduces the efficiency. This type of laser is very widely used, ranging from DVD-players to medical devices.

Although lasers based on doped dielectric materials are less numerous due to their relative complexity, they are capable of the widest range of performance, from high-power systems to cut and weld steel to ultrashort pulses for biological imaging. In some of these lasers – such as the titanium-doped sapphire laser that underpins much of ultrafast science – the primary effect of the heat produced is to cause a direct reduction in efficiency, but in most doped-dielectric lasers the situation is more complex. Heat is typically deposited in the middle of the laser material and removed at the edges, but this results in temperature gradients, which can have a number of detrimental effects. First, differential thermal expansion results in stresses that lead to deformation and eventually to fracture as more pump power is used to produce more output power. The same temperature gradients also lead to gradients in the refractive index via the thermo-optic effect, such that so-called *thermal lensing* results. As the strength of this lens increases, the

---

Optical Engineering of Diamond, First Edition. Edited by Richard P. Mildren and James R. Rabeau.
© 2013 Wiley-VCH Verlag GmbH & Co. KGaA. Published 2013 by Wiley-VCH Verlag GmbH & Co. KGaA.

laser resonator can become unstable, preventing the generation of a laser beam. The thermal lens is typically also rather aberrated, and this leads to a deterioration in the spatial quality of the output beam. In LEDs – which today are used widely for illumination, displays, and instrumentation – heating has similar effects to those observed in semiconductor lasers, most notably a reduction in efficiency. Thus, for both lasers and LEDs, the efficient extraction of heat is vital – and this is where modern synthetic diamond, with its unrivaled thermal conductivity, is starting to play an important role.

11.1.2
The Advantages of Diamond: Thermal Conductivity and Rigidity

It is not simply the extraordinary thermal conductivity of diamond that makes it attractive for thermal management. Diamond is also a very stiff material with a low thermal expansion coefficient, and consequently it is being used increasingly to brace laser materials against thermal deformation. The advantages of diamond are illustrated in Table 11.1. The thermal conductivity of diamond is two orders of magnitude higher than the archetypal solid-state laser host material, yttrium aluminum garnate ($Y_3Al_5O_{12}$, commonly abbreviated to YAG). YAG is typically doped with either Nd^{3+} or Yb^{3+} ions to provide laser action, and is perhaps the most common host material for high-power doped-dielectric lasers. The Young's modulus of diamond (an indicator of rigidity) is almost fourfold higher, the thermal expansion coefficient is eightfold lower, and the tensile strength is an order of magnitude higher. Similar comparisons can be made with other materials, including host crystals for doped-dielectric lasers such as sapphire or YVO_4 or materials for semiconductor lasers and LEDs, such as GaAs and GaN. Exceptional thermal and mechanical properties – combined with transparency – make diamond a unique ancillary material for thermal management in lasers and LEDs.

Table 11.1 Comparison of the thermomechanical properties of typical optical materials in comparison to diamond. (Note: There is considerable variation in the reported values of some of these parameters.) [1–9].

Material	Thermal conductivity [W (m·K)$^{-1}$]	Young's modulus (GPa)	Thermal expansion coefficient ($\times 10^{-6}$ K^{-1})	Thermo-optical coefficient ($\times 10^{-6}$ K^{-1})	Tensile strength (MPa)
Diamond	2000	1100	1.0	9.6	2860
YAG	10.5	282	8	8.9	280
YVO$_4$	6.6/5.5	133	11.4/4.4	8.5/3	53
Sapphire	33	335	5	13	400
GaAs	44	86	5.7	267	–
GaN	130	181	3.2/5.6	–	–

11.2
The Use of Diamond in Lasers: A Brief Review

11.2.1
Diamond as a Sub-Mount for Lasers

The most common use of diamond to date has been as a sub-mount for semiconductor lasers to reduce thermal resistance, hence minimizing the operating temperature and maximizing the efficiency. In a semiconductor laser, the active region is very small, typically ~1000 µm^3. Diamond provides a low thermal resistance path between the small device and the larger heat sink by spreading the heat. A good example of this was reported by Bewley et al. [10], where diamond was pressure-bonded to a mid-infrared semiconductor laser; this provided effective thermal management while permitting optical pumping through the diamond, taking advantage of its transparency.

Diamond sub-mounts have also been used in a class of semiconductor lasers known as semiconductor disk lasers (see Section 11.4) [11]. These optically pumped lasers can produce high spatial quality output beams at much higher output powers than conventional semiconductor lasers. For example, Rudin and coworkers have reported an output power of 21 W with near diffraction-limited beam quality using a diamond sub-mount [12].

In doped-dielectric thin disk lasers, as pioneered by Giesen and coworkers [13, 14], diamond is increasingly used as a sub-mount to provide heat extraction and to reduce thermal deformation (see Section 11.5.1). The thin-disk approach to laser design relies on cooling a small volume through a large area by using a disk of laser gain material that is typically a few hundred micrometers thick but about a centimeter in diameter, while a diamond sub-mount prevents bowing of the structure under thermal load. This approach is seen as the route to developing kilowatt-class thin-disk lasers with substantially improved spatial quality of the output beam [15, 16].

Another class of high-power laser in which diamond has been exploited is the slab laser. Tzuk et al. demonstrated a 200 W Nd:YVO$_4$ slab laser that was bonded to a diamond window for cooling and through which the device was optically pumped [17]. However, in common with all the examples noted above, the generated laser beam within the laser cavity does not pass through the diamond, and thus the optical quality of the diamond is not critical. Although it may be advantageous to place the diamond within the laser cavity, this places tight constraints on its optical quality. These intracavity applications of diamond are discussed in the following subsection.

11.2.2
Intracavity Applications of Diamond for Thermal Management

Diamond is best used as close as possible to the heat source (see Section 11.3), and in some lasers this is best achieved by using diamond within the laser resonator, such that the intracavity laser beam passes through the diamond. For these applications it is vital that the diamond has excellent optical quality (see Section 11.3.2). Although, in the past, issues of loss and birefringence [18–20] have inhibited the intracavity use of diamond, recent rapid improvements in the optical quality of synthetic single-crystal diamond has led to radical change in this outlook (see Chapter 2 and Refs [1, 21, 22]).

The most widespread intracavity use of diamond to date has probably been in semiconductor disk lasers (see Section 11.4). Here, diamond is used as a heat spreader, to extract the heat from the intracavity surface of an optically pumped semiconductor laser material. Positioning the diamond within the laser resonator means that heat must travel through less semiconductor material [thermal conductivity of order a few tens of $W(mK)^{-1}$, or less] to reach the diamond [thermal conductivity of $2000 W(mK)^{-1}$]; this short extraction path means that the heat removal is less dependent on the semiconductor material properties. As a consequence, intracavity diamond heat spreaders allow watt-level output powers to be demonstrated from semiconductor disk lasers over an unprecedented range of semiconductor materials and hence wavelengths, from the red (0.67 μm) to the mid-infrared (2.3 μm) [23, 24]. The use of an intracavity diamond heat spreader in a semiconductor disk laser was first reported by Hopkins et al. [25], using natural diamond; however, the use of synthetic single-crystal diamond has since become standard as its optical quality has improved [26]. Notable demonstrations include 1 W of continuous-wave output power at 0.67 μm [27], 10 W at 1 μm [28], 11 W at 1.2 μm [29], 5 W at 1.5 μm [30], and 6 W at 2 μm [31]–all of which were enabled by the intracavity use of diamond.

The advantages of the intracavity diamond heat spreaders are most starkly illustrated by data recorded by the present authors from a 2 μm semiconductor disk laser, where the maximum output power was 5 W with diamond and 30 mW without; that is to say, the use of diamond improves the output power by two orders of magnitude. Although the performance improvements are less marked, intracavity diamond heat spreaders have also been used in doped-dielectric lasers. For example, Chou and coworkers [32] described a 50 W Nd:YAG laser, while Millar et al. [33] reported 25 W from a Nd:YVO$_4$ disk laser.

11.2.3
Direct Exploitation of Diamond as a Laser Gain Material

Diamond's ability to manage heat in solid-state lasers depends on it being used close to the heat source within the laser material. Using diamond directly as the laser gain material would be the ideal solution; however, obtaining laser gain from diamond is not straightforward because the compact nature of the diamond lattice makes it difficult to dope with metal ions in the manner of conventional laser materials such as Nd:YAG and Ti:sapphire. Diamond-based lasers have, however, been demonstrated in two ways: by using color centers, or by Raman scattering to provide gain.

In 1985, Rand and DeShazer demonstrated a laser based on H3 color centers in diamond [34]. This delivered output pulse energies of 0.17 mJ at 530 nm, based on an absorbed pump energy of 1.3 mJ in 4 ns pulses from a 480 nm dye-laser. While there has been little subsequent work on diamond color-center lasers [35, 36], there has been considerable recent interest in diamond Raman lasers (see Chapter 8). The potential of diamond as a material for Raman lasers has been known for many years (e.g., Refs [37, 38]), but it was typically assumed not to be viable for practical applications. The recent revival of interest was sparked by the demonstration of an intracavity pumped diamond Raman laser by Demidovich and coworkers [39], the work of Kaminskii and coworkers on stimulated Raman scattering in polycrystalline diamond [40], and the development of high-optical quality synthetic single-crystal diamond [1, 21]. This led to demonstrations which included a high-efficiency nanosecond-pulsed device [41], a picosecond-pulsed synchronously pumped device [42], and a multi-watt continuous-wave device [43].

In this chapter, the use of diamond as an ancillary material for thermal management in lasers and LEDs will be explored by examining some specific examples in greater detail, with particular emphasis placed on intracavity use enabled by high-optical quality material. The most effective means to exploit diamond for thermal management will first be discussed (Section 11.3), before examining its use in semiconductor disk lasers, doped-dielectric lasers, and LEDs (Sections 11.4–11.6).

11.3
Exploiting the Extreme Properties of Diamond

11.3.1
Intracavity versus Extracavity Use of Diamond

The thermal and mechanical properties of diamond are not merely marginally better than those of common laser materials – they are often orders of magnitude better (Table 11.1). How, then, might the best use be made of these properties when using diamond in concert with conventional laser materials? The archetypal geometry for a laser is the rod – a cylinder of laser material pumped along its axis and cooled on its barrel surface. A finite element simulation of the temperature rise in this simple scenario is shown in Figure 11.1a. A Ø5 × 10 mm rod of YAG is assumed (this could be doped with Nd^{3+} or Yb^{3+}) and 100 W of waste heat from the pumping process is deposited uniformly along the core of the rod in a cylinder 2 mm in diameter. The peak temperature rise at the center of the rod is calculated to be 215 K.

One approach that is used conventionally to control thermal effects in rod lasers is the addition of *end-caps*. These are most commonly made from the same material as the laser rod, but are not doped with the active metal ion that provides laser gain. The end-caps assist with heat removal and reduce the curvature of the end

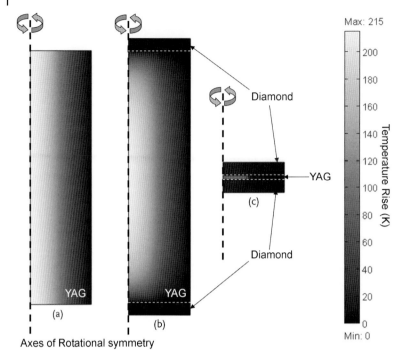

Figure 11.1 Finite element modeling of the use of diamond end-caps on a YAG laser rod. The temperature rise is plotted for 100 W of heat deposited: (a) with no end-caps on a 10 mm-long YAG rod; (b) with 0.5 mm-thick diamond end-caps on a 10 mm-long YAG rod; and (c) with 0.5 mm-thick diamond end-caps on a 200 μm-long YAG rod. The YAG rods are assumed to be 5 mm in diameter, with the heat deposited uniformly in a cylinder the length of the YAG rod and 2 mm in diameter. A fixed-temperature boundary condition is assumed on the barrel surfaces of the YAG and diamond. The end faces are assumed to be insulated.

faces that result from thermal expansion and contribute to the thermal lens. It is also possible to use end-caps of a different material; for example, Weber *et al.* examined the use of sapphire in a Nd:YAG rod laser [44]. A similar scenario is modeled in Figure 11.1b, but for 0.5 mm-thick end-caps of diamond, where the end-face curvature is considerably reduced, as it would be for the end-caps of YAG or sapphire [44]. However, the thermal conductivity of diamond–despite being 200-fold higher than that of YAG–only reduces the temperature rise near the ends of the rod; elsewhere the temperature is essentially unchanged. That is to say, the exceptional thermal conductivity of diamond is not well exploited in a rod laser geometry.

Alternatively, if the simplicity of cooling only on the barrel surface were to be retained and the rod length decreased to 200 μm, this 50-fold reduction in length would lead to a 50-fold increase in the heat load density. In the absence of diamond end-caps, the temperature rise would increase in proportion to this increase in

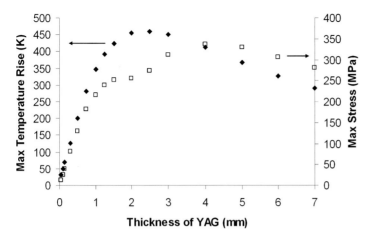

Figure 11.2 Finite element simulation of the maximum temperature rise and maximum stress in a YAG disk as a function of YAG thickness where both ends are bonded to 0.5 mm-thick diamond end-caps (as per Figure 11.1b, c). The disk is assumed to be 5 mm in radius, the pump spot is assumed to be 2 mm in diameter, and 100 W of heat is assumed to be deposited uniformly within the pumped region. A constant temperature boundary condition is assumed on the barrel surfaces of the YAG and the diamond. The end faces are assumed to be insulated. Note: The maximum temperature and stress plotted are those in the YAG, not in the diamond/YAG/diamond composite as a whole. In moving from thicker to thinner pieces of YAG, the locus of maximum stress in the diamond/YAG/diamond composite moves from the YAG to the diamond while the locus of maximum temperature remains in the YAG. As a result, the maximum stress in the YAG displays more complicated behavior than the maximum temperature in the YAG.

the heat load density, suggesting an unfeasible temperature rise in the excess of the melting point of YAG. With the diamond end-caps, however, the predicted maximum temperature rise would be 40% of that for a 10 mm Nd:YAG rod, despite the 50-fold increase in heat load density (Figure 11.1c). By contrast, if sapphire end-caps were to be used with the 200 μm-thick YAG crystal, the predicted temperature rise would be ninefold higher than for diamond – almost fourfold higher, indeed, than for the 10 mm-long YAG rod with no endcaps (Figure 11.1a).

The same scenario as for Figure 11.1 is considered in Figure 11.2, but here the thickness of the YAG material is varied. Again, the YAG is assumed to be sandwiched between two 0.5 mm diamond end-caps with 100 W of heat deposited, while cooling is through the barrel surface only and not through the large outer face of either diamond end-cap. The predicted maximum temperature rise and maximum tensile stress in the YAG laser material, plotted against the thickness of YAG used, are shown in Figure 11.2.

For large thicknesses, the temperature and stress in the YAG drop slowly as a result of the spreading of the fixed heat load over the length of what is now better thought of as a rod of YAG. A similar effect would be observed in the absence of diamond. For disks of YAG less than 1 mm thick – that is, thinner than the 1 mm radius of the region in which the heat is deposited – the temperature and stress in the YAG fall sharply as the disk is made thinner, despite the increase in the heat load density. This is a direct result of the excellent thermal and mechanical properties of diamond. For disk thicknesses less than 200 μm, the temperature and stress is predicted to be a factor of five or so less than for rods of a few millimeters in length, suggesting the potential for lasers that are simultaneously more powerful and more compact. Hence, it is in this regime of thin laser materials and short heat transport distances out of the material, that diamond is likely to be best exploited.

11.3.2
Material Requirements for Intracavity Applications

In the above-described scenarios the diamond was within the laser cavity, and intracavity use is indeed often important if the exceptional thermal and mechanical properties of diamond are to be best exploited. Yet, when a material is used intracavity its optical properties are also vital.

When using diamond in an intracavity role, two issues must be considered, namely birefringence and loss. Although natural diamond is, in principle, an isotropic material it often exhibits spatially varying birefringence due to internal stresses [18]; this may be problematic for intracavity use as it complicates the control of the laser's polarization [20, 45]. In addition, the absorption of natural diamond in the visible and near-infrared ranges varies considerably from sample to sample – an effect that is often related to the level of nitrogen (the most common impurity) within the diamond.

The grain size in polycrystalline diamond is such that its scatter loss is too high for many intracavity applications in the visible and near-infrared ranges [46]. Although the use of single-crystal diamond avoids this issue, it has proven difficult until recently to source synthetic single-crystal material that has simultaneously a low birefringence and low absorption [19, 20]. However, recent developments in the chemical vapor deposition (CVD) growth of diamond (see Chapter 2) have led to diamond with a low birefringence ($\Delta n < 1 \times 10^{-6}$) and low absorption (absorption coefficient ~0.001 cm^{-1} at 1064 nm) [1]. The applications of this material will be discussed in the following sections.

11.4
Current Uses of Diamond: Semiconductor Disk Lasers

11.4.1
Semiconductor Disk Lasers: Basic Principles

A semiconductor disk laser is a hybrid that combines some of the strengths of diode-pumped doped-dielectric lasers and a semiconductor laser. From the doped-

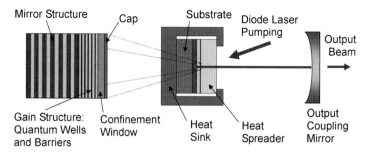

Figure 11.3 Schematic diagram of a semiconductor disk laser with an intracavity heat spreader. The inset on the left shows the details of the epitaxial layer structure of the semiconductor gain material.

dielectric laser the disk laser acquires the ability to produce higher powers whilst maintaining a good beam quality, while from semiconductor lasers it acquires the ability to design the gain material for a specific output wavelength. This combination of functionality is difficult to achieve with either conventional doped-dielectric lasers or conventional semiconductor lasers.

The layout of a typical semiconductor disk laser is shown schematically in Figure 11.3. The set-up, in which an optically pumped gain material is located in an extended cavity, is very similar to a conventional diode-pumped doped-dielectric laser, except that the gain material is an epitaxially grown multilayer semiconductor structure. Typically, a distributed Bragg reflector mirror of alternating high- and low-refractive index layers is grown on a semiconductor substrate. On top of this is grown a laser gain region consisting of quantum wells separated by higher band-gap barriers. The last two layers are a high band-gap material that confines the pump-generated carriers within the quantum well region, and a cap to protect the layers below. This structure usually forms one end mirror of an extended laser resonator that is completed with a discrete output coupling mirror. High-power variants also require some form of additional thermal management (as discussed in the next section), and this is often provided by the inclusion of diamond, either within the laser cavity (intracavity) or outside the cavity (extracavity).

Semiconductor disk lasers have demonstrated watt-level output powers at wavelengths ranging from 0.67 µm in the red to 2.35 µm in the mid-infrared [23, 24]. In the region around 1 µm, output powers of 106 W have been demonstrated for a single gain element [47]. The spectral coverage can be extended to the visible and ultraviolet by taking advantage of the extended laser resonator to enable intracavity frequency doubling [23]. Semiconductor disk lasers can also be frequency-narrowed to produce kHz linewidths [48], and modelocked to produce picosecond and femtosecond pulses [49]; this exceptional adaptability makes for a very practical and applicable class of laser. The use of diamond to improve thermal management – hence enabling higher output powers – is a cornerstone of this functionality, mainly because these lasers are particularly sensitive to the temperature rise in the quantum well region that is induced by waste heat from the pumping process. As the pump power is increased and the temperature of this region rises, the peak emission wavelength of the quantum wells will shift to longer wavelengths more quickly than does the laser's operating wavelength, which is typically dictated by resonances within the semiconductor multilayer structure [23]. This leads to a roll-over in output power and, eventually, to laser oscillation ceasing. In order to maximize the output power, the quantum well emission peak is typically designed to be at a shorter wavelength than the resonances when the device is unpumped. However, if this offset is made too large, the device will become too hot when the quantum well gain comes into alignment with the resonances, and the emission efficiency will be reduced. Without additional thermal management, the output power of a semiconductor disk laser is typically limited to a few tens or hundreds of milliwatts. Higher output powers require a smaller temperature rise per unit pump power, and this is where diamond becomes most beneficial, as its thermal conductivity [$2000\,\text{W}\,(\text{m}\,\text{K})^{-1}$] dwarfs that of typical semiconductor laser materials [typically a few tens of $\text{W}\,(\text{m}\,\text{K})^{-1}$ for binary alloys such as GaAs and GaSb, but lower for the ternary and quaternary alloys that are often required for certain layers within the semiconductor structure].

11.4.2
Thermal Management Strategies for Semiconductor Disk Lasers

11.4.2.1 Approaches to the Use of Diamond

The three approaches typically used to thermally manage semiconductor disk lasers are shown schematically in Figure 11.4. The simplest (Figure 11.4a) involves bonding a piece of the as-grown semiconductor structure to a suitable heat sink. Unfortunately, the output power must be limited to between a few tens and a few hundred milliwatts because of the large temperature rise that occurs when the waste heat from the pumping process is deposited into a very small volume (the quantum well and barrier region where the pump light is absorbed is only ~1 μm thick). The resultant heat density is three orders of magnitude greater than would be typical of a Nd:YAG laser at the same pump power.

Higher output powers require improved thermal management, and this can be achieved either by removing the substrate and bonding a high-thermal conductivity sub-mount beneath the mirror (Figure 11.4b), or by bonding a transparent, high-thermal conductivity heat spreader to the intracavity surface of the semiconductor structure (Figure 11.4c). These two set-ups are referred to as the "thinned device" and "heat spreader" approaches, respectively, and each has its own advantages and disadvantages. The thinned device approach requires considerable post-growth processing to remove the substrate, but the sub-mount is not within the laser cavity and so its optical quality is not important. In contrast, the heat spreader approach requires much less post processing, but the heat spreader is used intracavity so good optical quality is vital.

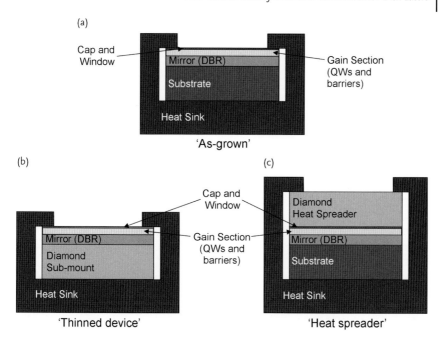

Figure 11.4 Schematic diagrams of the three approaches to thermal management of semiconductor disk lasers. (a) "As-grown"; (b) "Thinned-device"; (c) "Heat spreader." Note: the layers are not to scale. QW, quantum well; DBR, distributed Bragg reflector.

Table 11.2 Comparison of maximum temperature rise per unit pump power for semiconductor disk lasers designed for 0.98 µm and 2.35 µm, where diamond is used intracavity (heat spreader) and extracavity (thinned device on a diamond sub-mount). The 0.98 µm device is assumed to be pumped at 0.81 µm, and the 2.35 µm device at 0.98 µm. A pump spot radius of 100 µm is assumed [50, 51].

Approach to thermal management	Diagram	Location of diamond	Max. temperature rise per unit pump power (K W^{-1})	
			0.98 µm device	2.35 µm device
"As grown"	Figure 11.4a	No diamond	21	86
Thinned, diamond sub-mount	Figure 11.4b	Extracavity	2.7	26
Diamond heat spreader	Figure 11.4c	Intracavity	2.1	4.6

11.4.2.2 Intracavity and Extracavity Approaches to the Use of Diamond

The performance of the three approaches to thermal management (Figure 11.4) is summarized in Table 11.2. Finite element analysis has been used to model the maximum temperature rise in each case. Two semiconductor disk laser structures designed for emission at different wavelengths have been considered: (i) an

InGaAs-based device for 0.98 μm [52]; and (ii) a GaInAsSb-based device for 2.35 μm [50]. Assuming a pump spot radius of 100 μm, the temperature rise per unit pump power is large for the "as-grown" approach: $21\,\mathrm{K\,W^{-1}}$ for the 0.98 μm device and $86\,\mathrm{K\,W^{-1}}$ for the 2.35 μm device. As discussed in Section 11.4.1, the output power of a semiconductor disk laser rolls-over and oscillation ceases above a certain temperature (this is usually designed to be a few tens of Kelvin above ambient). As a result, the pump power that can usefully be used with the "as-grown" approach is limited to the watt-level, and hence output powers to the sub-watt level.

Both, the thinned device and heat spreader approaches permit much higher output powers, however. If a diamond sub-mount is assumed in the thinned device case, and a diamond heat spreader in the heat spreader case, then the maximum temperature rise per unit pump power is less than $3\,\mathrm{K\,W^{-1}}$ in both cases for an InGaAs-based device designed for 0.98 μm. This is almost an order of magnitude improvement on the as-grown scenario, and implies that pump powers of tens of watts and output powers of a few watts to a few tens of watts should be possible.

The situation for the GaInAsSb-based device for 2.35 μm is rather different. In this case, an order of magnitude reduction in the temperature rise per unit pump power is predicted only for the diamond heat spreader. This suggests that the intracavity diamond heat spreader approach will be favorable for the multi-watt operation of an GaInAsSb-based semiconductor disk laser.

This difference between the devices designed for 0.98 μm and for 2.35 μm is a result of the thermal properties of the semiconductor materials used, and the route by which the two approaches to thermal management extract heat. The integrated mirror in the 2.35 μm device is not only thicker, because of the longer operational wavelength, but also less thermally conductive, because of the materials that need to be used to enable a lattice-matched growth of the device structure. The combined effect is a sevenfold increase in the thermal resistance of the mirror structure in moving from the 0.98 μm device to the 2.35 μm device.

The heat flow path from the heat source in the gain region to the heat sink differs between the two approaches to thermal management. In the thinned device scenario, heat flow occurs predominantly through the mirror structure (vertically down in Figure 11.4b) and into the diamond sub-mount. The pumped region in the gain section is 100 μm in radius, whereas the gain and mirror structures are only a few micrometers thick, and hence there is little lateral heat spreading in these regions. The high thermal conductivity of the diamond sub-mount (assumed to be 0.5 mm thick) then promotes a lateral spreading of the heat, permitting an efficient extraction over an area much larger than the 100 μm radius pumped region. Hence, the thermal bottleneck in this approach derives from the thermal resistance of the mirror structure, which is much higher for the 2.35 μm device. The net result is a predicted temperature rise per unit pump power that is an order of magnitude higher for the 2.35 μm device than for the 0.98 μm device.

The scenario is rather different for an intracavity heat spreader. Here, heat flows out of the gain section into the diamond heatspreader through the relatively thin confinement window and cap layers with combined thickness of a few hundred nanometers (vertically up in Figure 11.4c). The heat is spread laterally by the diamond heat spreader (assumed to be 0.5 mm thick), and then extracted either directly into the heat sink or back down through the semiconductor structure (though over an area much larger than the 100 μm radius pumped region, and so with a much lower thermal resistance). These heat flow paths are much less dependent on the thermal properties of the semiconductor material, which is reflected in the low temperature rises predicted for both the 2.35 μm and 0.98 μm devices (Table 11.2). The heat load per unit pump power assumed in the gain region of the 2.35 μm device is 2.5-fold higher than for the 0.98 μm device, reflecting the greater disparity between the typical pump and laser photon energies. It is this higher heat load per unit pump power–not the lower thermal conductivity of the materials–that explains the twofold higher temperature rise in the case of the 2.35 μm device where a diamond heat spreader is used.

The region in which heat is deposited in a semiconductor disk laser is much thinner that it is broad; the gain section is typically a few micrometers thick while the pump spot radius is of the order of 100 μm. This disk-like geometry raises the question of whether power scalable behavior is possible. This is achieved–to first order at least–in doped-dielectric thin disk lasers by increasing the pump spot area in proportion to increases in the pump power, so as to keep the heat load density constant. In theory at least, this leaves detrimental thermal effects unchanged as the pump power is increased [14]. The results of finite element modeling of the expected temperature rise as a function of pump power if this strategy is used with a semiconductor disk laser are shown in Figure 11.5. Here, the data are presented for the 0.98 μm device (Figure 11.5a) and the 2.35 μm device (Figure 11.5b), and in each case for a thinned device and a heat spreader scenario. If an approach to thermal management is truly scalable, then the temperature rise should be independent of pump power, and the data of Figure 11.5 should fall on horizontal lines. However, this is not the case for any of the scenarios presented in Figure 11.5, where the temperature rise increases as a function of pump power, even though the pump intensity is held constant. This in turn indicates that the heat flow is not purely axial through the disk. The transverse profile of the pump is assumed to be Gaussian in these simulations to better reflect the typical experimental situation. Strictly, the thin-disk approach requires a "top-hat" pump profile; however, if such a profile is used in the simulations, the results are not significantly changed [50, 52]. Thus, the semiconductor disk laser is not an indefinitely scalable laser geometry using either a heat spreader or a thinned device approach.

Although the power scalability of semiconductor disk lasers will be limited–at least if only one pumped region is used–it is still important to assess whether the thinned device or the heat spreader approach has the greater potential to enable higher output powers. As illustrated in Figure 11.5, this depends on the semiconductor disk laser in question. For the 0.98 μm device (Figure 11.5a), the thinned device approach is better at large pump spot radii and thus higher pump powers, whereas the heat spreader approach is better for lower pump powers. This crossover occurs for pump spot radii of around 250 μm for the devices simulated; for

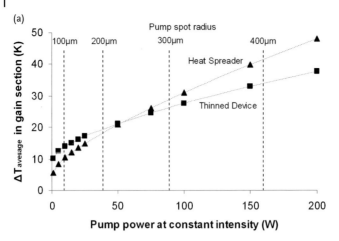

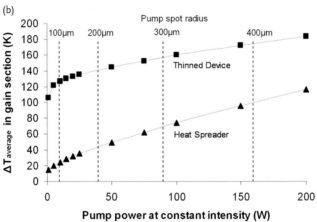

Figure 11.5 Averaged temperature rise as a function of pump power at constant pump intensity for (a) 0.98 μm and (b) 2.35 μm semiconductor disk lasers. Heat spreader and thinned-device approaches are considered assuming a diamond heat spreader and sub-mount, respectively. The pump radii for selected pump powers are indicted at the top of the graph. A weighted average of the temperature rise within the gain region has been taken over the Gaussian transverse profile of the pump [51].

larger pump spot radii, the effective thermal resistance of the distributed Bragg reflector mirror is reduced sufficiently, such that the thinned device approach outperforms the heat spreader approach. In contrast, for the 2.35 μm device (Figure 11.5b) the thermal resistivity of the materials comprising the distributed Bragg reflector is higher, and hence the temperature rise is too large to be viable in all cases for a thinned device. The heat spreader approach gives much lower temperature rises for all combinations of pump power and pump spot radius in Figure 11.5b, and so provides much greater potential for power scaling of the 2.35 μm device.

It is the lateral heat flow in the heat spreader or sub-mount that limits power scaling, because the effective area through which heat is extracted does not grow in proportion to the pumped area. The use of diamond is therefore crucial because its exceptionally high thermal conductivity minimizes the lateral thermal gradients that ultimately lead to excessive temperature rises. Hence, although the use of diamond does not lead to an indefinite scalability, it does enable considerably higher output powers than would otherwise be achievable.

11.4.2.3 Intracavity Diamond Heat Spreaders for Wavelength Diversity

One of the key advantages of the semiconductor disk laser is its potential to combine wavelength versatility with high brightness operation. It is important, therefore, to understand the performance of the thinned device and heat spreader approaches across the gamut of output wavelengths that can be generated. A plot of the ratio of the maximum temperature rise in the thinned device case to that in the heat spreader case as a function of operating wavelength is shown in Figure 11.6. This is based on finite element modeling of a number of different devices reported in the literature. The data are plotted for three pump spot radii, namely 100 μm, 200 μm, and 400 μm. Where the ratio is greater than one, the temperature rise is higher for the thinned device approach; where it is less than one, the temperature rise is higher for the heat spreader case. The thinned device approach only outperforms the heat spreader approach for a relatively narrow range of wavelengths around 1 μm, and then only for larger pump spot radii.

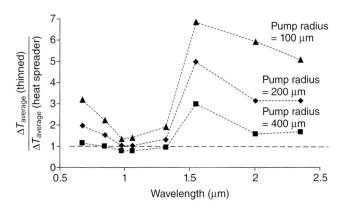

Figure 11.6 The ratio of the temperature rise in the heat spreader and thinned device cases for different wavelengths of operation and hence different semiconductor materials systems: a 0.67 μm device based on AlGaInP [27]; 0.85 μm, GaAs [53]; 0.98 μm, InGaAs [52]; 1.06 μm, InGaAs [45]; 1.32 μm, GaInNAs [25]; 1.55 μm, InGaAsP [54]; 2.0 μm, GaInSb [55]; and 2.35 μm, GaInAsSb [50]. A weighted average of the temperature has been taken over the Gaussian pump/laser mode (assumed to be the same size) within the gain section [50]. A diamond heat spreader or sub-mount is assumed. Above the horizontal dotted line, the temperature rise is higher for the thinned device case [51].

The spectral region where the thinned device functions better corresponds to those devices where appropriate gain regions can be lattice-matched to low thermal impedance mirror structures based on (Al)GaAs/AlAs layers. This region could, in principle, be widened by using wafer fusion techniques to bond a non-lattice-matched mirror to the gain structure; however, this requires additional post-processing and has so far only been demonstrated experimentally using the intracavity diamond heat spreader approach to thermal management [30].

The fact that the thinned device approach is only predicted to scale better for larger pump spot sizes arises from the need for axial rather than radial heat flow in the semiconductor structure to properly exploit this approach. However, large pump spot sizes are not in themselves a disadvantage as they enable the use of the highest pump powers (Figure 11.5) and hence the highest output powers – with the proviso that there must be sufficient gain to keep the laser's threshold reasonable.

By contrast, the heat spreader approach is predicted to work better for a wider range of wavelengths and, if the pump spot radius is less than 200 µm, for all the device designs modeled. This is because the intracavity heat spreader extracts heat through the thin window and cap layers, and not through the thicker mirror structure. This has two consequences: (i) the heat spreader approach is much less dependent on the thermal conductivity of the semiconductor materials; and (ii) transferring the heat into the diamond more directly leads to better heat spreading for smaller pump spot radii. Thus, the heat spreader approach can be expected to enable higher output powers over a broader range of wavelengths. As will be discussed in the next section, this is what is observed experimentally.

11.4.3
A Review of Progress to Date, and Future Prospects

At wavelengths below 0.92 µm and above 1.25 µm, semiconductor disk lasers with watt-level output powers have used a heat spreader approach to thermal management. Between 0.92 µm and 1.25 µm, both approaches have been used, but the highest powers have derived from thinned-device approaches. Notable demonstrations using the thinned device approach and a single gain element include 20 W at 0.92 µm [56], 30 W at 0.98 µm [56], and 106 W at 1.03 µm [47] – all of which utilized a diamond sub-mount. Notable demonstrations using an intracavity diamond heat spreader include 1 W at 0.67 µm [27], 0.7 W at 0.85 µm [57], 12 W at 0.92 µm [58], 11 W at 1.18 µm [29], 7 W at 1.31 µm [59], 5 W at 1.48 µm [30], 6 W at 1.99 µm [31], 0.9 W at 2.35 µm [60], and 0.6 W at 2.5 µm [61]. With only a few exceptions, watt-level demonstrations of semiconductor disk lasers have used diamond either as an intracavity heat spreader or as an extracavity sub-mount. Indeed, the remarkable spectral versatility of watt-level semiconductor disk lasers is built on the intracavity use of high-optical quality synthetic diamond.

The intracavity heat spreader can also have a secondary function. For example, Lindberg et al. demonstrated the use of a thin piece of diamond (50 µm) that served both as a heat spreader and an etalon to control the output wavelength [62]. In a

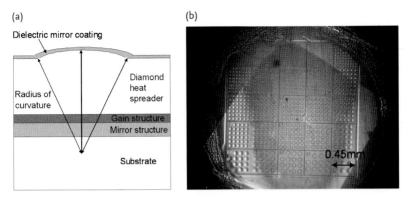

Figure 11.7 (a) Schematic diagram and (b) photographic image of a microlensed microchip semiconductor disk laser [64, 65].

similar vein, the present authors have used a dielectric coating on the heat spreader as the output coupling mirror for a microchip laser. This arrangement is both compact and robust because the requirement for discrete optics to complete the laser cavity is removed. The output coupling mirror can be coated directly onto the plane surface of the diamond, and the pump-induced thermal lens used to stabilize the resulting plane-plane cavity [63]. Alternatively, an array of microlenses can be etched into the diamond and the structured surface coated with the output coupling mirror to create an array of microchip lasers with intrinsically stable cavities (Figure 11.7) [64–66].

Future directions in semiconductor disk laser research are likely to include the use of quantum dot gain structures to reach new wavelengths [67], as well as the use of semiconductor disk lasers, which have low noise characteristics, as pump sources for other lasers (e.g., Ref. [31]) and nonlinear optical systems (e.g., Ref. [68]). In both cases, the use of diamond is likely to be to the fore in ensuring high-power operation. In addition, advances in the microfabrication of diamond are likely to lead to a further integration of thermal management and optical functionality (see Figure 11.8).

11.5
Current Uses of Diamond: Doped-Dielectric Disk Lasers

11.5.1
Doped-Dielectric Disk Lasers: Diamond Sub-Mounting for Mechanical Rigidity

The doped-dielectric disk laser, like the semiconductor disk laser, is predicated on the use of a thin piece of laser gain material. The doped dielectric crystal – which may be 15 mm in diameter and 150 μm thick – has a mirror coating applied to one of its large surfaces and is used as a cavity mirror. The crystal is cooled through

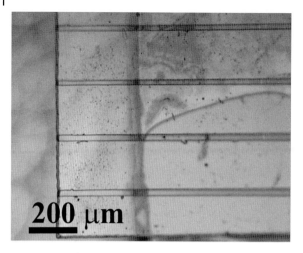

Figure 11.8 Ridge waveguides in diamond fabricated using inductively coupled plasma etching [69].

this mirrored surface by bonding it to a heat sink at the rear. The crystal is then pumped by the homogenized output of a high-power diode-laser stack. The pump spot diameter is usually of the order of a few millimeters, and hence is much wider than the disk is thick. This, in combination with cooling through the rear face, ensures that the heat flow in the laser material is axial to the laser cavity mode. The resulting gradients in temperature and refractive index are thus also axial, and so do not contribute to the thermal lens. In principle, this concept is scalable: if the pumped area is increased in proportion to an increase in pump power, the temperature rise remains constant. This thin-disk approach to solid-state laser design – as pioneered by Giesen and coworkers [13, 14] – has led to the development of kilowatt class lasers that are well suited to applications in manufacturing.

Although, the thin-disk approach largely eliminates contributions to the thermal lens in the laser medium due to radial temperature gradients, the axial temperature gradients lead to differential thermal expansion, and thus to a bowing of the disk and its sub-mount under high-density pumping [15]. This is illustrated in Figure 11.9a which shows a simplified finite element simulation of the temperature rise and deformation caused by a heat load of 100 W uniformly deposited in a 8 mm diameter region within a 150 μm-thick YAG disk (this could for example be Nd:YAG or Yb:YAG). The YAG disk is assumed to be bonded to a 1.5 mm-thick copper tungsten sub-mount, with the rear of the sub-mount held at a fixed temperature. (Cooling of the rear surface is often achieved by water jet impingement.) While there is very little radial change in the temperature within the pump region, there is a significant bowing of structure (the deformation is scaled 5000-fold to make it visible in Figure 11.9a). As the surface of the gain material bonded to the sub-mount typically carries a mirror coating, the result is a curved cavity mirror.

11.5 Current Uses of Diamond: Doped-Dielectric Disk Lasers

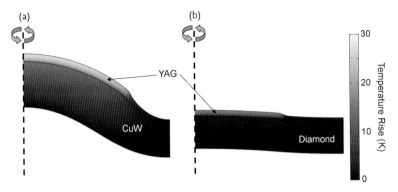

Figure 11.9 Finite element simulations of the temperature and deformation of a YAG disk mounted on (a) a copper tungsten (CuW) sub-mount and (b) a diamond sub-mount. A heat load of 100 W is assumed in an 8 mm-diameter pumped region. The YAG disk is assumed to be 150 μm thick × 14 mm in diameter (it could either Nd:YAG or Yb:YAG because the heat load rather than the incident pump power is set). The sub-mount is assumed to be 1.5 mm thick, and to be held at a fixed temperature on its lower surface (all other surfaces are taken to be thermally insulated). A radially symmetric simulation is assumed with the outside edge assumed to be mechanically constrained in radial dimension. The thermally induced deformation is scaled 5000-fold to make it visible on the scale of the diagram.

This curvature, and its associated aberrations, limit the potential to scale thin-disk lasers into the kilowatt range with diffraction-limited beam quality [15]. If the copper–tungsten sub-mount is replaced by diamond, the resulting deformation is considerably smaller (Figure 11.9b). For this reason, diamond sub-mounts are now increasingly used in high performance thin-disk lasers (e.g., Refs [15, 16]), including a recent demonstration of 500 W of output power with an M^2 beam quality factor of 1.6, which was much better than the M^2 of 20 or so typically associated with high-power thin-disk lasers [15].

11.5.2
Intracavity Use of Diamond in Doped-Dielectric Disk Lasers

Using diamond as a sub-mount does not require the use of high-optical quality material as the diamond is not used within the laser cavity. The intracavity use of diamond in doped-dielectric disk lasers requires high-optical quality diamond; hence, it is important to consider whether there are advantages over simpler extracavity use [26]. In Figure 11.10, the scenarios modeled in Figure 11.9 are modeled again but with the addition of a 0.5 mm-thick piece of diamond bonded to the intracavity surface of the YAG-based gain material. The same scalings for temperature and deformation are used in Figures 11.9 and 11.10 to allow for direct comparison.

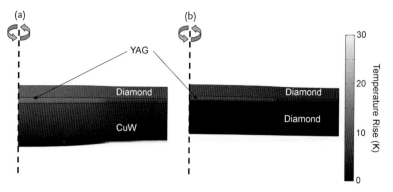

Figure 11.10 Finite element simulations of the temperature and deformation of a YAG disk with an intracavity diamond heat spreader. (a) With copper–tungsten sub-mount (CuW); (b) With a diamond sub-mount. The intracavity heat spreader is assumed to be 0.5 mm thick; otherwise the parameters are the same as Figure 11.9. The thermal-induced deformation is scaled 5000-fold to make it visible on the scale of the diagram.

The intracavity use of diamond reduces both the temperature rise and the deformation. For the particular cases shown in Figures 11.9 and 11.10, the temperature rise is reduced by a factor of two, and the deformation by almost an order of magnitude. Thus, there are potentially substantial benefits to the intracavity use of diamond – especially as mechanical deformation and not temperature-induced refractive index change is the primary source of thermal lensing in a thin-disk laser [70].

To date, however, there has been relatively little intracavity use of diamond in doped-dielectric disk lasers, certainly by comparison with semiconductor disk lasers (see Section 11.4). Examples include the demonstration of a 50 W Nd:YAG laser by Chou et al. [32], and of a 25 W Nd:YVO$_4$ laser by Millar et al. [33]. Neither was in the proper thin-disk regime examined in Figure 11.10: this is to say the pump spot radius was not significantly larger than the gain material thickness. Thus, it is likely that significant improvements in performance will be enabled by exploiting this regime.

One advantage of the approaches used by Chou et al. [32] and Millar et al. [33], when compared to a conventional thin-disk approach, is that complex multipassing of the pump was not used. One potential application of the intracavity diamond is to combine the thermal management advantages of the thin-disk laser with simpler pumping arrangements. This can be achieved by using a gain material that is strongly absorbing for the pump wavelength, so as to reduce the number of pump passes required for acceptable pump absorption efficiency in a thin slice of material. Without the use of diamond, Rivier et al. used this approach to demonstrate 9 W of output power from a Yb:KLu(WO$_4$)$_2$ thin-disk laser with only one double pass of the pump [71].

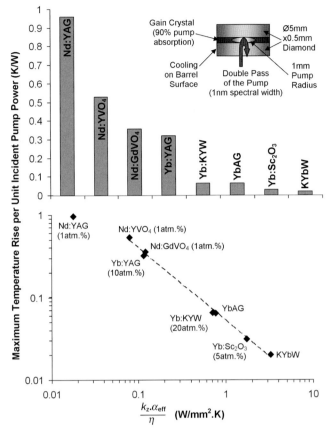

Figure 11.11 Temperature rise per unit pump power predicted from finite element modeling for gain materials sandwiched between diamond heat spreaders. The situation modeled is summarized in the inset on the upper graph. The thickness of the gain material is chosen to give 90% absorption on a double pass of the pump with the pump spectral width assumed to be 1 nm. The upper part shows a simple bar graph. The lower part shows the same data plotted on a log-log plot against the proposed figure of merit: $k_z.\alpha_{eff}/\eta$ (atm.% is the atomic percentage of Nd^{3+} or Yb^{3+} doped into the host crystal. Materials data are taken from Refs [1, 2, 72–75].

The results of finite element simulations that illustrate the potential of using materials with strong pump absorption in a diamond sandwich geometry are shown in Figure 11.11. Here, a thin piece of laser gain material is assumed to be sandwiched between two 0.5 mm-thick diamond plates. The thickness of the laser gain material is then chosen to give 90% absorption of the pump in a double pass (the pump spectral width is assumed to be 1 nm). The pump spot radius is 1 mm, and the combination of gain material and diamond is cooled only on the barrel surface (not through the large faces), which means that it can be used in very simple cavity configurations. The predicted maximum temperature rise per unit pump power for a variety of laser gain media is shown in the upper portion of Figure 11.11, where the lowest temperature rises are achieved with materials such as Yb:KYW (Yb:KY(WO$_4$)$_2$) and Yb:Sc$_2$O$_3$, which exhibit a strong pump absorption such that a thin slice of material can be used without compromising the absorption efficiency. This is because thin pieces of material minimize the heat transfer distance into diamond.

The pump absorption properties (and hence thickness) were shown to be a stronger determinant of performance in this geometry than thermal conductivity, although a material with a higher thermal conductivity would still perform better where the pump absorption and heat-loading characteristics are identical. With this in mind, a figure of merit for this geometry can be proposed:

$$\frac{k_z \cdot \alpha_{eff}}{\eta} \tag{11.1}$$

where k_z is the thermal conductivity perpendicular to the plane of the disk, α_{eff} is the effective pump absorption coefficient (accounting for the finite spectral width of the pump laser and the absorption feature), and η is the fraction of the absorbed pump power converted to heat. The data in the lower part of Figure 11.11 show that there is a strong correlation between this figure of merit and the temperature rise predicted by the finite element analysis. This indicates that the full exploitation of diamond in doped-dielectric lasers – at least in the format suggested here – will necessitate a move away from common gain media such as Yb:YAG towards materials with stronger pump absorption, such as Yb:KYW and Yb:Sc$_2$O$_3$.

11.5.3
Future Prospects

Recent studies on the scaling of high-brightness thin-disk lasers towards kilowatt output powers have suggested that the extracavity use of diamond sub-mounts will be crucial [15]. This is because the rigidity of diamond limits the optical aberrations that result from mechanical deformation of the laser gain material under large heat loads. Such extracavity use of diamond does not require high-optical quality material; however, as the finite element modeling outlined in Section 11.5.2 indicates, the intracavity use of diamond may allow for the simplification of thin-disk lasers by enabling the use of materials with stronger pump absorption, thus removing the need for complicated multi-pass pumping arrangements. This is likely to have a particular impact on the development of compact high-power lasers for applications that require portability or robustness. In a similar vein, Fork *et al.* [76] proposed and modeled a laser using interleaved disks of diamond and Ti:sapphire for applications in solar power delivery from space. Recent developments in diamond growth potentially bring the realization of such a laser a step closer.

11.6
Current Uses of Diamond: Light-Emitting Diodes

11.6.1
Introduction

Light-emitting diodes (LEDs) provide a very efficient and compact means of converting electrical energy to light, and are today widespread in applications from illumination to instrumentation, due mainly to their compactness, convenience, and ease of control. Typically, LEDs generate significantly less heat per unit optical output power than incandescent lamps although, as the light generation efficiency is typically only 15–30%, the majority of the injected electrical power is still converted to heat [77]. This is especially significant in the case of power LEDs with chip sizes on the order of a square millimeter. These produce hundreds of milliwatts of optical power at injection currents of around 1 A; however, unless this heat is managed properly, deleterious effects such as wavelength shift, high junction temperature, efficiency droop, and reduced lifetime will impair LED performance. Clearly, thermal management is a primary consideration in the design of modern high-power LEDs [78], and consequently many investigations have been conducted on optimized LED package design and the use of high thermal conductivity materials. This is exactly where diamond has the potential to play an important role.

11.6.2
Diamond as a Heat Spreader in LEDs

Chen *et al.* have reported the use of a 20 µm-thick diamond coating as a heat-spreading layer on the silicon substrate of a thin gallium nitride LED chip. This layer not only dissipates the heat effectively but also improves the temperature uniformity [79]. The same authors compared the temperature distribution over an LED chip bonded to either a conventional silicon substrate or a diamond-coated silicon substrate. For an injected current of 1 A, the junction temperature of the LED chip was 20 K lower when bonded to the diamond-coated substrate. In addition, Hsu and Lin have demonstrated that the thermal resistance of an LED on a diamond film is lower than on a graphite film [80].

An alternative use of diamond in this context is in diamond–metal composites that consist of diamond particles interspersed within a metal matrix. These materials are commercially available, with products including silver–diamond, copper–diamond and gold–diamond composites, and with various thermal conductivities. For instance, copper–diamond composites are available with thermal conductivities up to 800 W (m K)$^{-1}$. Recently, Horng *et al.* reported details of the use of a series of diamond–metal composites as heat spreaders for high-power gallium nitride LEDs [81, 82]. In addition, Fan and coworkers have reported on the optimization of diamond–copper composites and their use for the thermal management of high-power LEDs [83].

11.6.3
Monolithic Structures and the Epitaxy of Gallium Nitride on Diamond

The attractions of diamond as a heat spreader for gallium nitride LEDs and laser diodes naturally lead to the consideration of diamond as either a monolithic part of the device structure, or an epitaxial substrate for gallium nitride. To date, this has primarily been explored in the context of "high-added value" electronic devices, but there are also significant potential benefits in optoelectronics if the production and scalability costs of this technology can be addressed.

Currently, two main approaches are being pursued:

Polycrystalline diamond is deposited on silicon and then bonded to a device structure of gallium nitride on silicon. After the removal of both silicon substrates, this results in gallium nitride thin-film devices on diamond [84, 85].

Gallium nitride epitaxy is performed directly on diamond. Ultraviolet LEDs have been fabricated in this manner: for example, aluminum nitride was deposited onto diamond to create an aluminum nitride/diamond p-n junction [86]. Both (100) and (111) oriented substrates were explored.

Further recent success has been achieved in the use of single-crystal (111) diamond substrates for epitaxy of two-dimensional electron-gas (2DEG) structures and high electron mobility transistors (HEMTs) [87, 88], offering possibilities for optoelectronic structures in the future.

11.6.4
Diamond LEDs

Diamond itself is a semiconductor and, with appropriate doping, diamond p-n junctions and LEDs can be fabricated. The first diamond LED was reported by Koizumi *et al.* in 2001 [89], who used boron doping to grow p-type diamond on the {111} surface of single-crystal diamond, and then phosphorus doping to form n-type diamond. At a 20 V bias voltage, the diamond LED emitted ultraviolet light at 235 nm. Horiuchi *et al.* reported another diamond p-n diode in the same year [90], by using a sulfur-doped n-type diamond layer and also observing emission at 235 nm. Recently, Makino *et al.* developed a p-i-n structure diamond LED [91]; by optimizing the thickness of the intrinsic layer, these authors achieved an optical output power of 0.1 mW under pulsed-current injection [92].

11.7
Conclusions and Future Directions

The recent availability of high-optical quality diamond has already had a significant impact on solid-state laser engineering, and nowhere is this clearer than in the diamond-enabled wavelength versatility of semiconductor disk lasers with watt-level output powers. Here, the intracavity use of diamond has enabled an order of magnitude improvement in output power at wavelengths from the red (0.67 µm) to the mid-infrared (2.35 µm). While it is true that the highest output powers from semiconductor disk lasers have been achieved without the intracavity use of diamond, these demonstrations are nonetheless dependent on the extracavity use of diamond, and the wavelength coverage is restricted to the region around 1 µm. It is reasonable, therefore, to suggest that high-optical quality diamond is one of the key technologies that has enabled the semiconductor disk laser to become a truly wavelength engineerable laser.

The impact of diamond on doped-dielectric solid-state lasers has been less marked to date. Nonetheless, extracavity diamond is likely to be key to the development of high-brightness kilowatt-class thin-disk lasers, and the modeling introduced in Section 11.5.2 suggested that the intracavity use of high-optical quality diamond will enable simplified thin-disk lasers suited to mobile applications. In LEDs, the drive for ever higher powers may generate future applications for diamond, particularly polycrystalline diamond coating technologies.

The common factor between these approaches is the short distance that the heat has to flow out of the gain material and into the diamond, minimizing the thermal resistance of the structure as a whole. Taking this to its logical conclusion means using diamond itself as the laser gain material, and to date this has been achieved in two ways: (i) via color-center lasers [34]; and (ii) via Raman lasers [39]. The significant strides that have been made during recent years on diamond Raman lasers – including high-efficiency nanosecond-pulsed devices [93], synchronously pumped picosecond devices [42], and continuous-wave operation [94] – are detailed in Chapter 8. The combination of the high Raman gain of diamond with the developing ability to fabricate monolithic resonators in diamond potentially opens the way for compact, robust, monolithic Raman lasers. Although far fewer studies have been conducted on diamond color center lasers since the initial demonstration reported in 1985 [34], recent research into the physics and fabrication of color centers in diamond for applications in quantum optics (e.g., Ref. [95]) may provide a platform for renewed interest in diamond color-center lasers.

The motivation for the direct use of diamond as a laser gain material is illustrated in Figure 11.12, where finite element simulations of the temperature rise and mechanical deformation of two rod lasers are shown: one where the host matrix is YAG (Figure 11.12a), and one where it is diamond (Figure 11.12b). In both cases, the rod is assumed to be 2 mm in diameter and 5 mm long, with 100 W of heat deposited uniformly in a 0.5 mm-diameter cylinder along the axis. A fixed-temperature boundary condition is assumed on the barrel surface only. By replacing industry-standard YAG with diamond, the temperature rise and mechanical deformation are reduced by more than two and three orders of magnitude, respectively. Thus, approaches that can harness diamond directly as a laser gain material – enabled by high-optical quality synthetic diamond – have the potential to cause a step change in the engineering of compact, high-performance solid-state lasers.

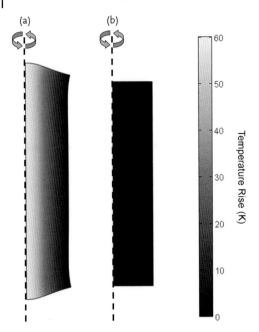

Figure 11.12 Finite element simulation of the temperature rise and deformation associated with 100 W of heat uniformly deposited in a 0.5 mm-diameter cylinder in Ø2 × 5 mm rods of (a) YAG and (b) diamond. The simulations are axisymmetric with a fixed-temperature boundary condition assumed on the barrel surface of the rod. Free surface boundary mechanical boundary conditions are assumed on all external boundaries.

Acknowledgments

The authors would like to thank Ian Friel, Harpreet Dhillon, Andrew Bennett, and Daniel Twitchen of Element Six Ltd for providing samples, and for their expert guidance on diamond materials science. The authors also gratefully acknowledge funding from the UK Engineering and Physical Sciences Research Council (EP/G00014X/1; EP/E004903/1; EP/E006000/1), the European Research Council (DiaL: 278389), and the Royal Society of Edinburgh/Scottish Government (personal research fellowship).

References

1 Friel, I., Geoghegan, S.L., Twitchen, D.J., and Scarsbrook, G.A. (2010) Development of high quality single crystal diamond for novel laser applications. Optics and Photonics for Counterterrorism and Crime Fighting VI and Optical Materials in Defence Systems Technology VII, September 2010, Toulouse, France, SPIE.

2 Didierjean, J., Herault, E., Balembois, F., and Georges, P. (2008) Thermal conductivity measurements of laser crystals by infrared thermography. Application to Nd:doped crystals. *Opt. Express*, **16**, 8995–9010.

3 Ioffe Physico-Technical Institute. (2006) Electronic Archive: New Semiconductor Materials, Characteristics and Properties. Available at: http://www.ioffe.ru/SVA/NSM/.

4 Adachi, S. (1983) Lattice thermal resistivity of III-V compound alloys. *J. Appl. Phys.*, **54**, 1844–1848.

5 Krupke, W.F., Shinn, M.D., Marion, J.E., Caird, J.A., and Stokowski, S.E. (1986) Spectroscopic, optical, and thermomechanical properties of neodymium-doped and chromium-doped gadolinium scandium gallium garnet. *J. Opt. Soc. Am. B*, **3**, 102–114.

6 Paschotta, R. (2008) *Encyclopedia of Laser Physics and Technology*, Wiley-VCH.

7 Peng, X.Y., Asundi, A., Chen, Y.H., and Xiong, Z.J. (2001) Study of the mechanical properties of Nd:YVO$_4$ crystal by use of laser interferometry and finite-element analysis. *Appl. Opt.*, **40**, 1396–1403.

8 Talghader, J. and Smith, J.S. (1995) Thermal-dependence of the refractive-index of GaAs and AlAs measured using semiconductor multilayer optical cavities. *Appl. Phys. Lett.*, **66**, 335–337.

9 VLOC Inc. (2003) YAG: Yttrium Aluminium Garnate Laser Materials Datasheet. Available at: http://www.vloc.com/PDFs/YAGBrochure.pdf.

10 Bewley, W.W., Felix, C.L., Aifer, E.H., Stokes, D.W., Vurgaftman, I., Olafsen, L.J., Meyer, J.R., Yang, M.J., and Lee, H. (1999) Thermal characterization of diamond-pressure-bond heat sinking for optically pumped mid-infrared lasers. *IEEE J. Quantum Electron.*, **35**, 1597–1601.

11 Kuznetsov, M., Hakimi, F., Sprague, R., and Mooradian, A. (1997) High-power (>0.5-W CW) diode-pumped vertical-external-cavity surface-emitting semiconductor lasers with circular TEM$_{00}$ beams. *IEEE Photonics Technol. Lett.*, **9**, 1063–1065.

12 Rudin, B., Rutz, A., Hoffmann, M., Maas, D.J.H.C., Bellancourt, A.R., Gini, E., Südmeyer, T., and Keller, U. (2008) Highly efficient optically pumped vertical-emitting semiconductor laser with more than 20W average output power in a fundamental transverse mode. *Opt. Lett.*, **33**, 2719–2721.

13 Giesen, A., Hugel, H., Voss, A., Wittig, K., Brauch, U., and Opower, H. (1994) Scalable concept for diode-pumped high-power solid-state lasers. *Appl. Phys. B*, **58**, 365–372.

14 Giesen, A. and Speiser, J. (2007) Fifteen years of work on thin-disk lasers: results and scaling laws. *IEEE J. Sel. Top. Quantum Electron.*, **13**, 598–609.

15 Mende, J., Schmid, E., Speiser, J., Spindler, G., and Giesen, A. (2009) Thin-disk laser–power scaling to the kW regime in fundamental mode operation, in *Solid State Lasers XVIII: Technology and Devices* (eds W.A. Clarkson, N. Hodgson, and R.K. Shori), SPIE Int. Soc. Optical Engineering, Bellingham, pp. 71931V-1–71931V-12.

16 Vretenar, N., Carson, T., Lobad, A., Peterson, P., Newell, T.C., and Latham, W.P. (2010) Thermal management investigations in ceramic thin disk lasers, in *Technologies for Optical Countermeasures VII* (eds D.H. Titterton and M.A. Richardson), SPIE Int. Soc. Optical Engineering, Bellingham, pp. 78360J-1–78360J-7.

17 Tzuk, Y., Tal, A., Goldring, S., Glick, Y., Lebiush, E., Kaufman, G., and Lavi, R. (2004) Diamond cooling of high-power diode-pumped solid-state lasers. *IEEE J. Quantum Electron.*, **40**, 262–269.

18 Lang, A.R. (1967) Causes of birefringence in diamond. *Nature*, **213**, 248–251.

19 Turri, G., Chen, Y., Bass, M., Orchard, D., Butler, J.E., Magana, S., Feygelson, T., Thiel, D., Fourspring, K., Dewees, R.V., et al. (2007) Optical absorption, depolarization, and scatter of epitaxial single-crystal chemical-vapor-deposited diamond at 1.064 μm. *Opt. Eng.*, **46**, 064002.

20 van Loon, F., Kemp, A.J., Maclean, A.J., Calvez, S., Hopkins, J.-M., Hastie, J.E., Dawson, M.D., and Burns, D. (2006) Intracavity diamond heatspreaders in lasers: the effects of birefringence. *Opt. Express*, **14**, 9250–9260.

21 Friel, I., Clewes, S.L., Dhillon, H.K., Perkins, N., Twitchen, D.J., and Scarsbrook, G.A. (2009) Control of surface and bulk crystalline quality in single crystal diamond grown by chemical vapour deposition. *Diamond Relat. Mater.*, **18**, 808–815.

22 Balmer, R.S., Brandon, J.R., Clewes, S.L., Dhillon, H.K., Dodson, J.M., Friel, I., Inglis, P.N., Madgwick, T.D., Markham, M.L., Mollart, T.P., et al. (2009) Chemical vapour deposition synthetic diamond: materials, technology and applications. *J. Phys. Condens. Matter*, **21**, 364221.

23 Calvez, S., Hastie, J.E., Guina, M., Okhotnikov, O.G., and Dawson, M.D. (2009) Semiconductor disk lasers for the generation of visible and ultraviolet radiation. *Laser Photon. Rev.*, **3**, 407–434.

24 Schulz, N., Hopkins, J.M., Rattunde, M., Burns, D., and Wagner, J. (2008) High-brightness long-wavelength semiconductor disk lasers. *Laser Photon. Rev.*, **2**, 160–181.

25 Hopkins, J.M., Smith, S.A., Jeon, C.W., Sun, H.D., Burns, D., Calvez, S., Dawson, M.D., Jouhti, T., and Pessa, M. (2004) 0.6 W CW GaInNAs vertical external-cavity surface emitting laser operating at 1.32 µm. *Electron. Lett.*, **40**, 30–31.

26 Millar, P., Birch, R.B., Kemp, A.J., and Burns, D. (2008) Synthetic diamond for intracavity thermal management in compact solid-state lasers. *IEEE J. Quantum Electron.*, **44**, 709–717.

27 Hastie, J.E., Morton, L.G., Kemp, A.J., Dawson, M.D., Krysa, A.B., and Roberts, J.S. (2006) Tunable ultraviolet output from an intracavity frequency-doubled red vertical-external-cavity surface-emitting laser. *Appl. Phys. Lett.*, **89**, 061114.

28 Kim, K.S., Yoo, J.R., Cho, S.H., Lee, S.M., Lim, S.J., Kim, J.Y., Lee, J.H., Kim, T., and Park, Y.J. (2006) 1060 nm vertical-external-cavity surface-emitting lasers with an optical-to-optical efficiency of 44% at room temperature. *Appl. Phys. Lett.*, **88**, 091107.

29 Korpijärvi, V.-M., Leinonen, T., Puustinen, J., Härkönen, A., and Guina, M.D. (2010) 11 W single gain-chip dilute nitride disk laser emitting around 1180 nm. *Opt. Express*, **18**, 25633–25641.

30 Lyytikainen, J., Rautiainen, J., Sirbu, A., Iakovlev, V., Laakso, A., Ranta, S., Tavast, M., Kapon, E., and Okhotnikov, O.G. (2011) High-power 1.48-µm wafer-fused optically pumped semiconductor disk laser. *IEEE Photonics Technol. Lett.*, **23**, 917–919.

31 Hempler, N., Hopkins, J.-M., Rösener, B., Rattunde, M., Wagner, J., Fedorov, V.V., Moskalev, I.S., Mirov, S.B., and Burns, D. (2009) Semiconductor disk laser pumped Cr^{2+}:ZnSe lasers. *Opt. Express*, **17**, 18136–18141.

32 Chou, H.P., Wang, Y.L., and Hasson, V. (2004) Compact and efficient DPSS laser using diamond-cooled technology. *Proc. SPIE*, **5448**, 550–560.

33 Millar, P., Kemp, A.J., and Burns, D. (2009) Power scaling of Nd:YVO$_4$ and Nd:GdVO$_4$ disk lasers using synthetic diamond as a heat spreader. *Opt. Lett.*, **34**, 782–784.

34 Rand, S.C. and DeShazer, L.G. (1985) Visible color-center laser in diamond. *Opt. Lett.*, **10**, 481–483.

35 Rand, S.C. (1994) Diamond lasers, in *Growth and Properties of Diamond* (ed. G. Davies), pp. 235–239.

36 Vins, V.G. and Pestryakov, E.V. (2006) Color centers in diamond crystals: their potential use in tunable and femtosecond lasers. *Diamond Relat. Mater.*, **15**, 569–571.

37 McQuillan, A.K., Clements, W.R.L., and Stoicheff, B.P. (1970) Stimulated Raman emission in diamond–spectrum, gain, and angular distribution of intensity. *Phys. Rev. A*, **1**, 628–635.

38 Eckhardt, G., Bortfeld, D.P., and Geller, M. (1963) Stimulated emission of stokes and anti-stokes Raman lines from diamond, calcite, and alpha-sulphur single crystals. *Appl. Phys. Lett.*, **3**, 137–138.

39 Demidovich, A.A., Grabtchikov, A.S., Orlovich, V.A., Danailov, M.B., and

Kiefer, W. (2005) Diode pumped diamond Raman microchip laser. Conference on Lasers and Electro-Optics Europe, June 12–17, Munich. Available at: http://www.cleoeurope.org.

40 Kaminskii, A.A., Ralchenko, V.G., and Konov, V.I. (2006) CVD-diamond – a novel χ^3-nonlinear active crystalline material for SRS generation in very wide spectral range. *Laser Phys. Lett.*, **3**, 171–177.

41 Mildren, R.P. and Sabella, A. (2009) Highly efficient diamond Raman laser. *Opt. Lett.*, **34**, 2811–2813.

42 Spence, D.J., Granados, E., and Mildren, R.P. (2010) Mode-locked picosecond diamond Raman laser. *Opt. Lett.*, **35**, 556–558.

43 Savitski, V.G., Friel, I., Hastie, J.E., Dawson, M.D., Burns, D., and Kemp, A.J. (2012) Characterization of single-crystal synthetic diamond for multi-watt continuous-wave Raman lasers. *IEEE J. Quantum Electron.*, **48**, 328–337.

44 Weber, R., Neuenschwander, B., MacDonald, M., Roos, M.B., and Weber, H.P. (1998) Cooling schemes for longitudinally diode laser-pumped Nd:YAG rods. *IEEE J. Quantum Electron.*, **34**, 1046–1053.

45 Maclean, A.J., Kemp, A.J., Calvez, S., Kim, J.Y., Kim, T., Dawson, M.D., and Burns, D. (2008) Continuous tuning and efficient intracavity second-harmonic generation in a semiconductor disk laser with an intracavity diamond heatspreader. *IEEE J. Quantum Electron.*, **44**, 216–225.

46 Sharda, T., Rahaman, M.M., Nukaya, Y., Soga, T., Jimbo, T., and Umeno, M. (2001) Structural and optical properties of diamond and nano-diamond films grown by microwave plasma chemical vapor deposition. *Diamond Relat. Mater.*, **10**, 561–567.

47 Heinen, B., Wang, T.L., Sparenberg, M., Weber, A., Kunert, B., Hader, J., Koch, S.W., Moloney, J.V., Koch, M., and Stolz, W. (2012) 106 W continuous-wave output power from vertical-external-cavity surface-emitting laser. *Electron. Lett.*, **48**, 516.

48 Holm, M.A., Burns, D., Ferguson, A.I., and Dawson, M.D. (1999) Actively stabilized single-frequency vertical-external-cavity AlGaAs laser. *IEEE Photonics Technol. Lett.*, **11**, 1551–1553.

49 Tropper, A.C. and Hoogland, S. (2006) Extended cavity surface-emitting semiconductor lasers. *Prog. Quantum Electron.*, **30**, 1–43.

50 Kemp, A.J., Hopkins, J.-M., Maclean, A.J., Schulz, N., Rattunde, M., Wagner, J., and Burns, D. (2008) Thermal management in 2.3 μm semiconductor disk lasers: a finite element analysis. *IEEE J. Quantum Electron.*, **44**, 125–135.

51 Calvez, S., Hastie, J.E., Kemp, A.J., Laurand, N., and Dawson, M.D. (2010) Thermal management, structure design and integration considerations for VECSELs, in *Semiconductor Disk Lasers* (ed. O.G. Okhotnikov), Wiley-VCH, pp. 73–117.

52 Kemp, A.J., Valentine, G.J., Hopkins, J.M., Hastie, J.E., Smith, S.A., Calvez, S., Dawson, M.D., and Burns, D. (2005) Thermal management in vertical-external-cavity surface-emitting lasers: finite-element analysis of a heatspreader approach. *IEEE J. Quantum Electron.*, **41**, 148–155.

53 Hastie, J.E., Hopkins, J.M., Calvez, S., Jeon, C.W., Burns, D., Abram, R., Riis, E., Ferguson, A.I., and Dawson, M.D. (2003) 0.5-W single transverse-mode operation of an 850-nm diode-pumped surface-emitting semiconductor laser. *IEEE Photonics Technol. Lett.*, **15**, 894–896.

54 Lindberg, H., Strassner, M., Gerster, E., Bengtsson, J., and Larsson, A. (2005) Thermal management of optically pumped long-wavelength InP-based semiconductor disk lasers. *IEEE J. Sel. Top. Quantum Electron.*, **11**, 1126–1134.

55 Harkonen, A., Guina, M., Okhotnikov, O., Rossner, K., Hummer, M., Lehnhardt, T., Muller, M., Forchel, A., and Fischer, M. (2006) 1-W antimonide-based vertical external cavity surface emitting laser operating at 2-mu m. *Opt. Express*, **14**, 6479–6484.

56 Chilla, J., Butterworth, S., Zeitschel, A., Charles, J., Caprara, A., Reed, M., and Spinelli, L. (2004) High power optically pumped semiconductor lasers. *Proc. SPIE*, **5332**, 143–150.

57 McGinily, S.J., Abram, R.H., Gardner, K.S., Erling, R., Ferguson, A.I., and Roberts, J.S. (2007) Novel gain medium design for short-wavelength vertical-external-cavity surface-emitting laser. *IEEE J. Quantum Electron.*, **43**, 445–450.

58 Kim, K.S., Yoo, J., Kim, G., Lee, S., Cho, S., Kim, J., Kim, T., and Park, Y. (2007) 920-nm vertical-external-cavity surface-emitting lasers with a slope efficiency of 58% at room temperature. *IEEE Photonics Technol. Lett.*, **19**, 1655–1657.

59 Rantamäki, A., Sirbu, A., Mereuta, A., Kapon, E., and Okhotnikov, O.G. (2010) 3 W of 650 nm red emission by frequency doubling of wafer-fused semiconductor disk laser. *Opt. Express*, **18**, 21645–21650.

60 Hopkins, J.M., Maclean, A.J., Burns, D., Riis, E., Schulz, N., Rattunde, M., Manz, C., Koehler, K., and Wagner, J. (2007) Tunable, single-frequency, diode-pumped 2.3 μm VECSEL. *Opt. Express*, **15**, 8212–8217.

61 Nikkinen, J., Paajaste, J., Koskinen, R., Suomalainen, S., and Okhotnikov, O.G. (2011) GaSb-based semiconductor disk laser with 130-nm tuning range at 2.5-μm. *IEEE Photonics Technol. Lett.*, **23**, 777–779.

62 Lindberg, H., Larsson, A., and Strassner, M. (2005) Single-frequency operation of a high-power, long-wavelength semiconductor disk laser. *Opt. Lett.*, **30**, 2260–2262.

63 Smith, S.A., Hopkins, J.M., Hastie, J.E., Burns, D., Calvez, S., Dawson, M.D., Jouhti, T., Kontinnen, J., and Pessa, M. (2004) Diamond-microchip GaInNAs vertical external-cavity surface-emitting laser operating CW at 1315 nm. *Electron. Lett.*, **40**, 935–937.

64 Laurand, N., Lee, C.L., Gu, E., Hastie, J.E., Calvez, S., and Dawson, M.D. (2007) Microlensed microchip VECSEL. *Opt. Express*, **15**, 9341–9346.

65 Laurand, N., Lee, C.L., Gu, E., Hastie, J.E., Kemp, A.J., Calvez, S., and Dawson, M.D. (2008) Array-format microchip semiconductor disk lasers. *IEEE J. Quantum Electron.*, **44**, 1096–1103.

66 Laurand, N., Lee, C.L., Gu, E., Calvez, S., and Dawson, M.D. (2009) Power-scaling of diamond microlensed microchip semiconductor disk lasers. *IEEE Photonics Technol. Lett.*, **21**, 152–154.

67 Schlosser, P.J., Hastie, J.E., Calvez, S., Krysa, A.B., and Dawson, M.D. (2009) InP/AlGaInP quantum dot semiconductor disk lasers for CW TEM_{00} emission at 716–755 nm. *Opt. Express*, **17**, 21782–21787.

68 Stothard, D.J., Hopkins, J.M., Burns, D., and Dunn, M.H. (2009) Stable, continuous-wave, intracavity, optical parametric oscillator pumped by a semiconductor disk laser (VECSEL). *Opt. Express*, **17**, 10648–10658.

69 Zhang, Y., McKnight, L., Tian, Z., Calvez, S., Gu, E., and Dawson, M.D. (2011) Large cross-section edge-coupled diamond waveguides. *Diamond Relat. Mater.*, **20**, 564–567.

70 Kemp, A.J., Valentine, G.J., and Burns, D. (2004) Progress towards high-power, high-brightness neodymium-based thin-disk lasers. *Prog. Quantum Electron.*, **28**, 305–344.

71 Rivier, S., Mateos, X., Silvestre, Ò., Petrov, V., Griebner, U., Pujol, M.C., Aguiló, M., Díaz, F., Vernay, S., and Rytz, D. (2008) Thin-disk Yb:KLu(WO$_4$)$_2$ laser with single-pass pumping. *Opt. Lett.*, **33**, 735–737.

72 Sudmeyer, T., Krankel, C., Baer, C.R.E., Heckl, O.H., Saraceno, C.J., Golling, M., Peters, R., Petermann, K., Huber, G., and Keller, U. (2009) High-power ultrafast thin disk laser oscillators and their potential for sub-100-femtosecond pulse generation. *Appl. Phys. B*, **97**, 281–295.

73 Jensen, T., Ostroumov, V.G., Meyn, J.P., Huber, G., Zagumennyi, A.I., and Shcherbakov, I.A. (1994) Spectroscopic characterization and laser performance of diode-laser-pumped Nd-GdVO$_4$. *Appl. Phys. B* **58**, 373–379.

74 Patel, F.D., Honea, E.C., Speth, J., Payne, S.A., Hutcheson, R., and Equall, R. (2001) Laser demonstration of Yb$_3$Al$_5$O$_{12}$ (YbAG) and materials properties of highly doped Yb:YAG. *IEEE J. Quantum Electron.*, **37**, 135–144.

75 Pujol, M.C., Bursukova, M.A., Guell, F., Mateos, X., Sole, R., Gavalda, J., Aguilo, M., Massons, J., Diaz, F., Klopp, P., *et al.*

(2002) Growth, optical characterization, and laser operation of a stoichiometric crystal KYb(WO4)$_2$. *Phys. Rev. B*, **65**, 165121.

76 Fork, R.L., Walker, W.W., Laycock, R.L., Green, J.J.A., and Cole, S.T. (2003) Integrated diamond sapphire laser. *Opt. Express*, **11**, 2532–2548.

77 U.S. Department of Energy. (2007) Thermal Management of White LEDs. Report no. PNNL-SA-51901. Available at: http://apps1.eere.energy.gov/buildings/publications/pdfs/ssl/thermal_led_feb07_2.pdf.

78 Arik, J.P.M. (2002) Thermal challenges in the future generation solid state lighting applications: light emitting diodes. ITHERM 2002: The Eighth Intersociety Conference on Thermal and Thermomechanical Phenomena in Electronic Systems, May 30–June 1, San Diego, CA, IEEE; American Society of Mechanical Engineers; and International Microelectronics and Packaging Society.

79 Chen, P.H., Lin, C.L., Liu, Y.K., Chung, T.Y., and Liu, C.Y. (2008) Diamond heat spreader layer for high-power thin-GaN light-emitting diodes. *IEEE Photonics Technol. Lett.*, **20**, 845–847.

80 Hsu, C.Y. and Lin, Y.L. (2010) Junction temperature of high-power LED packages with diamond film. 3rd International Nanoelectronics Conference, January 3–8, IEEE.

81 Horng, R.H., Hong, J.S., Tsai, Y.L., Wuu, D.S., Chen, C.M., and Chen, C.J. (2010) Optimized thermal management from a chip to a heat sink for high-power GaN-based light-emitting diodes. *IEEE Trans. Electron. Devices*, **57**, 2203–2207.

82 Horng, R.H., Hu, A.L., Lin, R.C., Peng, K.C., and Chiang, Y.C. (2011) Thermal behavior of sapphire-based InGaN light-emitting diodes with cap-shaped copper-diamond substrates. *Electrochem. Solid State Lett.*, **14**, H215–H217.

83 Fan, Y.M., Guo, H., Xu, J., Chu, K., Zhu, X.X., Jia, C.C., Yin, F.Z., and Zhang, X.M. (2011) Pressure infiltrated Cu/diamond composites for LED applications. *Rare Metals*, **30**, 206–210.

84 Chabak, K.D., Gillespie, J.K., Miller, V., Crespo, A., Roussos, J., Trejo, M., Walker, D.E., Via, G.D., Jessen, G.H., Wasserbauer, J., *et al.* (2010) Full-wafer characterization of AlGaN/GaN HEMTs on free-standing CVD diamond substrates. *IEEE Electron. Device Lett.*, **31**, 99–101.

85 Francis, D., Faili, F., Babic, D., Ejeckam, F., Nurmikko, A., and Maris, H. (2009) Formation and characterization of 4-inch GaN-on-diamond substrates. *Diamond Relat. Mater.*, **19**, 229–233.

86 Miskys, C.R., Garrido, J.A., Nebel, C.E., Hermann, M., Ambacher, O., Eickhoff, M., and Stutzmann, M. (2003) AlN/diamond heterojunction diodes. *Appl. Phys. Lett.*, **82**, 290–292.

87 Dussaigne, A., Gonschorek, M., Malinverni, M., Py, M.A., Martin, D., Mouti, A., Stadelmann, P., and Grandjean, N. (2010) High-mobility AlGaN/GaN two-dimensional electron gas heterostructure grown on (111) single crystal diamond substrate. *Jpn. J. Appl. Phys.*, **49**, 061001.

88 Hirama, K., Taniyasu, Y., and Kasu, M. (2011) AlGaN/GaN high-electron mobility transistors with low thermal resistance grown on single-crystal diamond (111) substrates by metalorganic vapor-phase epitaxy. *Appl. Phys. Lett.*, **98**, 162112.

89 Koizumi, S., Watanabe, K., Hasegawa, M., and Kanda, H. (2001) Ultraviolet emission from a diamond pn junction. *Science*, **292**, 1899–1901.

90 Horiuchi, K., Kawamura, A., Ide, T., Ishikura, T., Takamura, K., and Yamashita, S. (2001) Efficient free-exciton recombination emission from diamond diode at room temperature. *Jpn. J. Appl. Phys. Part 2 Lett.*, **40**, L275–L278.

91 Makino, T., Tokuda, N., Kato, H., Ogura, M., Watanabe, H., Ri, S.-G., Yamasaki, S., and Okushi, H. (2006) High-efficiency excitonic emission with deep-ultraviolet light from (001)-oriented diamond p-i-n junction. *Jpn. J. Appl. Phys. Part 2 Lett. Express Lett.*, **45**, L1042–L1044.

92 Makino, T., Yoshino, K., Sakai, N., Uchida, K., Koizumi, S., Kato, H., Takeuchi, D., Ogura, M., Oyama, K.,

Matsumoto, T., *et al.* (2011) Enhancement in emission efficiency of diamond deep-ultraviolet light emitting diode. *Appl. Phys. Lett.*, **99**, 061110.

93 Sabella, A., Piper, J.A., and Mildren, R.P. (2010) 1240 nm diamond Raman laser operating near the quantum limit. *Opt. Lett.*, **35**, 3874–3876.

94 Lubeigt, W., Bonner, G.M., Hastie, J.E., Dawson, M.D., Burns, D., and Kemp, A.J. (2010) Continuous-wave diamond Raman laser. *Opt. Lett.*, **35**, 2994–2996.

95 Greentree, A.D., Fairchild, B.A., Hossain, F.M., and Prawer, S. (2008) Diamond integrated quantum photonics. *Mater. Today*, **11**, 22–31.

12
Laser Micro- and Nanoprocessing of Diamond Materials[1]

Vitaly I. Konov, Taras V. Kononenko, and Vitali V. Kononenko

12.1
Introduction

A combination of the extreme properties of diamond – in particular its record hardness and thermal conductivity – creates major problems for the conventional (mechanical) processing of the material. Although, in a number of cases, chemically assisted diamond mechanical treatment may represent a solution to this problem, such an approach is not effective for the precise microfabrication of diamond which, at least in principle, can be achieved by scanning electron or ion beams, although vacuum conditions are required in order to conduct these techniques.

Ultimately, the laser emerged as a perfect diamond treatment tool. From a practical viewpoint, the first appearance of lasers during the early 1960s [1] indicated they could indeed be applied successfully to fabricate diamond draw plates. This success was followed by the laser cutting of diamond crystals for jewelry, as well as many other laser-based diamond technologies that began with polishing and shaping and ended with micro- and nanostructuring.

For light–diamond interactions and the laser processing of diamond materials, a variety of lasers was applied (see Table 12.1) that allowed a broad range of pulse durations to be covered, including femto-, pico-, nano-, and microsecond pulses, continuous-wave (CW) regimes, pulse repetition rate f (from single pulse up to $f \approx 100\,\text{kHz}$), and wavelengths in the infrared (IR), visible, and ultraviolet (UV) spectral regions.

Various diamond materials were used in laser experiments.

Single crystal material was represented by three types of diamond:

- natural gemstones
- clear crystals produced by high-pressure, high-temperature (HPHT) techniques [2]
- epitaxially grown chemical vapor deposition (CVD) monocrystalline diamond [3].

---

1) Parts of this article have been published before in Konov, V.I. (2012) Laser in micro and nanoprocessing of diamond materials, Laser Photonics Rev. 6, No. 6, 739–766/DOI 10.1002/Ipor.201100030. Published with permission.

Optical Engineering of Diamond, First Edition. Edited by Richard P. Mildren and James R. Rabeau.
© 2013 Wiley-VCH Verlag GmbH & Co. KGaA. Published 2013 by Wiley-VCH Verlag GmbH & Co. KGaA.

Table 12.1 Lasers used in diamond processing.

Type	Nd:YAG			Ti:Al$_2$O$_3$	Cu	Ar$^+$	XeCl	KrF	ArF	CO$_2$
λ (nm)	1064	532	266	800	510.5	488	308	248	193	10 600
$\hbar\omega$ (eV)	1.17	2.33	4,66	1.55	2.42	2.54	4,02	5.0	6.42	0.12
Regime	Pulsed, CW			Pulsed	Pulsed	CW	Pulsed	Pulsed	Pulsed	Pulsed, CW

CW, continuous-wave.

Other subjects of investigation were also very popular, and led to a broad range of diamond types, including CVD diamond films and plates [4, 5]. Depending on the growth conditions employed, these materials may demonstrate very different properties; for example, their thermal conductivity k may be several-fold higher for transparent (optical grade) than black (mechanical grade) samples, while difference in optical absorption may be even larger, varying by several orders of magnitude.

For comparison, a number of other well-known diamond materials have also been investigated, including so-called nanocrystalline diamond (NCD) films or ultra-nanocrystalline (UNCD) films [6, 7]. These coatings, the thicknesses of which vary between approximately 100 nm and 10 μm, consist mostly of nanograins of diamond (4–20 nm in size) in a matrix of nanographite and polyacetylene, and can be deposited at lower temperatures than can regular CVD polycrystalline diamond films. In particular, the UNCD films are very hard (hardness ≤80–90 GPa) and smooth (surface roughness ≈20–50 nm). If, during the CVD process, nitrogen is added to the gas mixture, the UNCD films produced will be based on diamond nanorods [8] rather than on nanocrystals of diamond, and contain more graphitic material than would regular UNCD films.

Conducting the CVD process in hydrocarbon gases, or the ablation of graphite in vacuum by laser or/and an electric arc, may result in the deposition on cold substrates of so-called diamond-like carbon (DLC) films [9]. These films, which were also used in the experiments described, are amorphous and consist of a mixture of tetrahedral (sp^3) diamond and hexagonal (sp^2) graphite-bonded nanostructures. The ratio of sp^3/sp^2 bonds may be quite high, being minimal when the films are hydrogenated (a-C:H) but reaching ≈70–90% for so-called a-C and ta-C coatings.

Despite all of the described materials being diamond-based, their properties can vary widely. For example, a key parameter that influences the laser heating of materials is thermal conductivity, k (see Table 12.2), which depends heavily on the type of diamond material. Typically, k is highest for natural single crystals, but is comparable though slightly lower for polycrystalline CVD diamond. In the latter case, the k-value is four- to fivefold higher for transparent CVD diamond plates than for black diamond samples. The thermal conductivity falls to only about 26 W(m·K)$^{-1}$ for UNCD films, and is even lower for diamond-like coatings [10–13].

Table 12.2 Thermal conductivity of diamond materials at room temperature.

Material	k [W (m·K)$^{-1}$]
Single crystal Type IIa	2000–2400
Polycrystalline CVD diamond	400–2100
Nanocrystalline CVD diamond	26
Diamond-like Type ta-C	2.2–3.5
Diamond-like Type a-C:H	0.3–0.6

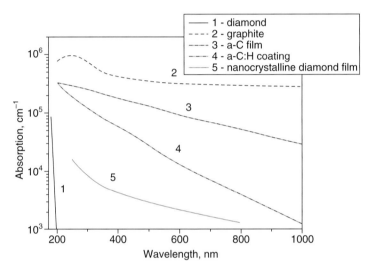

Figure 12.1 Dependence of the optical absorption coefficient on the radiation wavelength for different diamond materials and graphite [98]. Reproduced with permission.

In total, the variation among diamond material results in four orders of difference in thermal conductivity.

The light absorption coefficients α_0 for diamond materials in the visible and UV spectral ranges are displayed in Figure 12.1 [14–16]. It can be seen that α_0 varies with the laser wavelength λ, and depends heavily on the diamond material (for comparison, the data for graphite are shown). It should be noted that only the single-crystal and white CVD polycrystalline films are relatively transparent; all other diamond materials, that are contaminated with graphitic phases, absorb light to a high degree ($\alpha_0 > 10^3$–10^4 cm^{-1}). Yet, even within a single type of diamond the spread in α_0 values may be extremely large; for example, for photon energy $\hbar\omega \approx 1$ eV ($\lambda \approx 1000$ nm) with the highest quality polycrystalline CVD diamond, $\alpha_0 \approx 3 \times 10^{-3}$ cm^{-1}, whereas poor-quality samples with $k \approx 400$–500 W (m·K)$^{-1}$ (black CVD diamond) have $\alpha_0 \approx 100$ cm^{-1}.

In spite of such a wide scatter in their optical and thermal properties, all of the described materials have shown similar behaviors under the action of lasers. At the same time, a broad range of k and α_0 values, and combinations thereof, have greatly simplified the understanding of laser–diamond interactions.

The aim of this chapter is to detail the interaction of intense laser light with diamond materials, to determine how such interactions are governed, and identify which laser parameters are required to realize the different processing operations. The technologies of the laser fabrication of diamond elements and devices will also be outlined. It should be noted that the data presented are based mainly on results obtained during a more than 20-year period of research, carried out at the General Physics Institute, Russian Academy of Sciences, Moscow.

12.2
Laser-Induced Surface Graphitization

The specific feature of diamond as a material to be laser-processed is that its vaporization or chemical etching (e.g., oxidation) takes place not directly but rather via a diamond transformation into a graphitic phase; this carbon phase transition is a crucial point in the laser processing of diamond materials. Diamond is transparent, highly thermally conductive, and chemically inert, whereas graphitic phases are opaque, much less conductive, but chemically more reactive.

12.2.1
Mechanisms of Diamond Surface Graphitization

It is well known that at elevated temperatures ($T > T_g \approx 700\,°C$ in atmospheric air) diamond is transformed into graphite. The graphitization of a sp^3-bonded carbon atom in a diamond lattice takes place when the carbon atoms can obtain sufficient energy to undergo a diffusion jump and move into the sp^2 state, with a smaller binding energy, in a new lattice cell. During this graphitization process the mean distance between the adjacent carbon atoms is increased; consequently, graphitization will occur more easily when a free volume is available, for example, close to defects or on a free surface, and the graphitized material will have a lower density than diamond [17–22].

The thermal graphitization of monocrystalline diamond has been studied extensively to determine equilibrium heating conditions. Moreover, it was also shown that the graphitization rate may increase under nonequilibrium conditions when the sample is affected by high-energy particle beams, or when an electron-hole plasma is created inside the sample [22].

The specific features of laser-induced diamond surface graphitization are as follows. Laser radiation is first absorbed in diamond by multiphoton ionization and free electrons in elastic collisions with atoms and ions (the so-called *inverse bremsstrahlung* process); the initial concentration of electrons n_0 is then determined by multiphoton absorption, with electrons tunneling into the conduction band for radiation with high-intensity W or direct interband transitions for short laser wavelength λ. In addition, electron cloud initiation is also determined by many impurities and imperfections, which usually exist in real diamond samples and can absorb laser radiation better than can a perfect diamond crystal, and thus produce free electrons. Depending on W and the pulse duration τ, two different possibilities of sample heating exist. For low W and long pulses, electrons that absorb laser energy are in thermal equilibrium with the lattice and gradually raise its temperature to $T \approx T_g$. When W is sufficiently high, an electron "avalanche" may develop – that is, the electron temperature may substantially exceed the temperature of the lattice atoms, while the electron concentration can be much higher than that under equilibrium conditions. In this way, fast electrons can reach energy $\varepsilon > \varepsilon_a$ and thus initiate a carbon atom ionization by direct impact. In addition, laser pulse energy absorbed in the plasma will be transferred finally to the lattice, raising its temperature to $T \geq T_g$. Consequently, depending on the beam parameters, the result of the laser action may be quite different, ranging from the formation of micro (nano) regions with sp^2-dominated bonding up to a complete transformation of the diamond into a graphitic phase.

12.2.2
Experimental Data

Over a broad spectral range, the absorption coefficient in graphitic materials α_g $\approx 10^5\,\text{cm}^{-1}$ is substantially larger than in different forms of mono- and polycrystalline diamond ($\alpha_0 \ll \alpha_g$), and it is for this reason that the surface graphitization of such materials is accompanied by the fall in optical transmittance [23].

The dynamics of CVD polycrystalline diamond sample transmittance for intense picosecond laser pulses ($\lambda = 0.54\,\mu\text{m}$), that are capable of inducing surface graphitization, are illustrated graphically in Figure 12.2. Here, it can be seen that for a relatively high laser fluence $E = 10\,\text{J}\,\text{cm}^{-2}$ ($E = W\tau$, where W is intensity), the decrease in transmittance starts in the first pulse, whereas for $E = 4.3\,\text{J}\,\text{cm}^{-2}$ the graphitization is triggered only after pulse number $N \geq 20$ (sometimes, thousands of laser pulses are required to graphitize the diamond surface). This accumulation effect strongly supports the model that implies an influence of microdefects as starting absorbing centers in diamond–graphite transformation. Microscopy studies of the irradiation spot have shown that not only optical density but also the diameter of the graphitized zone grows with pulse number, because sections of the laser beam with a lower intensity induce graphitization for a larger N. The importance of defects for diamond graphitization was also supported experimentally [24] with an ArF-laser. In this case, i was found that a blackening of the isotopically enriched (concentration of ^{13}C was below 0.1%) diamond surface occurred at much larger laser fluencies than in the case of natural diamond crystals (when the concentration of ^{12}C was ca. 89.9%). This fact was also supported by the measurements of thermal conductivity when, even in the case of polycrystalline diamond, k was substantially higher for isotopically enriched samples [25]. Although accumulation effects can be quite important for diamond-like and

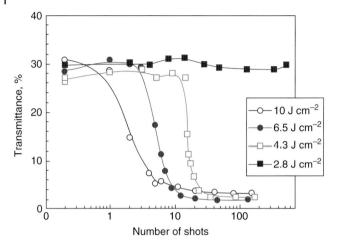

Figure 12.2 Variation of CVD diamond sample transmittance on laser ($\tau = 220$ ps, $\lambda = 539$ nm) pulse number at different fluencies. Reproduced from [23], copyright SPIE.

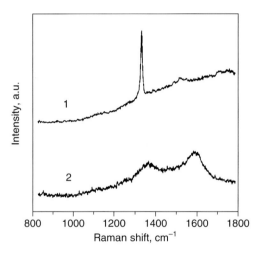

Figure 12.3 Raman spectra of the original (1) and KrF laser-graphitized (2) diamond film surfaces. Reproduced from [26], with permission from Elsevier.

nanocrystalline films, the optical monitoring may be complicated because the light-absorption coefficients of the films will be close to that of α_g.

Raman spectroscopy has strongly confirmed the results of these optical investigations. For a sufficiently high E and pulse number N, when defect-induced graphitization begins, the appearance of the material appears very similar to that of nanocrystalline graphite or glassy carbon. The Raman spectrum for original polycrystalline CVD diamond (see Figure 12.3) is characterized by the diamond peak at 1332.9 cm^{-1} and a weak band around 1550 cm^{-1} that is attributed to background amorphous sp^2-bonded carbon. Following intense KrF laser ($\lambda = 248$ nm) irradiation, two well-resolved broad peaks at 1350 cm^{-1} and 1580 cm^{-1} (D and G peaks, correspondingly) appear. Similar peaks are observed for all investigated diamond materials when graphitization takes place [26–32].

Near-edge X-ray absorption fine structure spectroscopy has been used [33] to confirm that the laser-produced nondiamond surface layer is nanocrystalline. Subsequent detailed Raman and transmission electron microscopy (TEM) studies of a-C:H DLC films that had been graphitized by KrF laser were also performed [34]. It was found that such amorphous films could undergo laser-induced recrystallization processes of two types: (i) amorphous-to-crystalline graphite; and (ii) amorphous-to-crystalline diamond. At fluencies below 100 mJ cm^{-2} an increase in the number of graphitic clusters and their ordering was observed, whereas at higher fluencies diamond particles of 2–7 nm in size appeared, embedded into the lower crystallized graphitic matrix, while a simultaneous and progressive growth of graphitic nanocrystals took place. In another study [18], transmission electron diffraction patterns of ArF laser-modified layers in Type IIa single crystal diamond (SCD) were obtained. These patterns highlighted the presence on the sample surface of a thin layer of graphite, below which was a further layer of graphite that was under a hydrostatic pressure of about 3 kBar. The planes of both nondiamond phases were perpendicular to the surface, the <001>-oriented diamond sample. Additional experiments with polycrystalline diamond have also shown that the graphitic layer thickness is dependent on the crystal orientation.

Laser-graphitized surface layers can also serve as good electrical conductors. For a high laser intensity and a sufficient number of laser pulses, the specific resistivity (σ) was found to fall from an initial value of $\approx 10^{14}\,\Omega \cdot$cm for polycrystalline CVD diamond to $\approx 10^{-3}\,\Omega \cdot$cm [31]. Similar results were obtained for Type IIa crystals [17], where the measured σ-value ranged from 10^{-3} to $10^{-4}\,\Omega \cdot$cm. The resistance was also found to be anisotropic; in fact, for <110> orientation it was about fivefold lower than for the <100> orientation.

The determination of σ requires a knowledge of the graphitized layer thickness, l_g. These measurements were made by the selective chemical etching of graphitic material. For example, oxidation in air at 450 °C over a 200 min period [35] led to the growth of resistance in the laser-graphitized strip on a CVD diamond surface by seven orders of magnitude, and it can be assumed that in this case most of the graphitized material was etched away. Further experiments [36] have shown that a much better precision in the graphitized layer depth detection can be obtained by oxidation at a higher temperature ($T \approx 600$ °C), although even these measurements proved to be incorrect because of the graphitized material expansion effect. Such an effect, which was observed both for CVD diamond and other diamond materials, manifests as the appearance of a bump with height l_b on an initially flat surface (see Figure 12.4). Consequently, in order to calculate the total thickness l_g of a laser-graphitized layer the following formula should be used:

$$l_g = \frac{\rho_d}{\rho_d - \rho_g} l_b \tag{12.1}$$

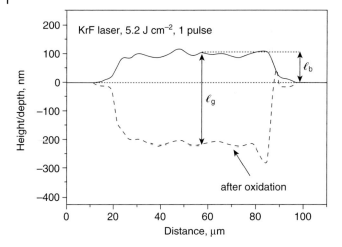

Figure 12.4 Profiles across the spot, as-irradiated (solid curve) and after oxidation in air at 600 °C (dashed curve) on the surface of a polished CVD diamond plate, graphitized by a single KrF laser pulse at $E = 5.2\,\mathrm{J\,cm^{-2}}$. Reproduced from [23], copyright SPIE.

where $\rho_g = 1.9\,\mathrm{g\,cm^3}$ and $\rho_d = 3.5\,\mathrm{g\,cm^3}$ are the mass densities of graphite and diamond, respectively.

The graphitized layer may be created not only as a result of diamond surface photothermal modification by a single laser pulse or a sequence of pulses; rather, if the laser fluence is sufficiently high, the surface graphitization can be induced by the leading part of the laser pulse. The energy of the remainder of the pulse is then absorbed into the sample much more efficiently, causing the sample surface to be heated up to sublimation temperature ($T = T_S$). As a result, an ablation of graphitic material will take place such that a crater is formed in the irradiation zone.

Investigations of the bottom of a crater with depth D_l revealed that it is always covered by a graphitized layer with thickness D_g, as shown in Figure 12.5 [23]. The properties and thickness of this layer are usually close to that of the graphitized layer produced without ablation. This has been confirmed by a combination of Raman spectroscopy studies, measurements of electrical conductivity, chemical etching of the material in the laser-produced crater, and by measurements of the crater depth before and after irradiation. In the latter case, it is important to again take into account any changes in the material mass density.

Among the several modes of diamond graphitization made possible by using a laser, in the first case the radiation is highly absorbed into the material. This can be achieved by selecting either UV or femtosecond laser pulses or (and) highly absorbing materials such as low-quality CVD diamond, NCD, and DLC films, and in this case even a single pulse can produce surface graphitization. Otherwise, it is necessary either to: (i) use a sequence of pulses that induce a gradual accumulation of absorbing defects; (ii) cover the surface with a highly absorbing thin (metal)

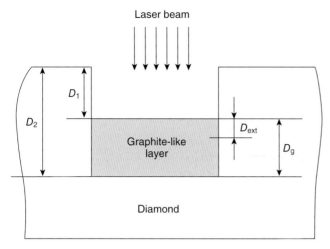

Figure 12.5 Diamond material cross-section as a result of laser ablation, formation of a graphitic layer on the crater bottom and its extension [98]. Reproduced with permission.

coating before the irradiation is carried out, thus inducing diamond vaporization and graphitization by laser; or (iii) apply a much more intensive laser pulse that can vaporize the diamond material. Whichever of these combinations is applied, the diamond surface will be covered by a graphitic layer following laser treatment.

Large differences in absorption (up to seven or eight orders of magnitude) between highly transparent diamond monocrystals or ultrapure CVD films and graphitic material lead to the fact that graphitization without ablation by a single pulse for such samples is practically impossible. Indeed, for surface graphitization the E levels required are so high that as soon as the graphitic layer appears it immediately triggers ablation during the following part of the laser pulse.

The data [37] for $l_g = D_g$ for different lasers, different laser wavelength and pulse durations, for a broad range of laser fluences (both below and above the ablation threshold), and multiple pulsed irradiations are summarized in Figure 12.6. Although the scattering of the data relating to the measured graphitized layer thickness was quite large, there were certain trends in $l_g(\tau)$ dependence. The thickness of the laser-graphitized layer could be characterized by the simple expression:

$$l_g = \max\{l_0, l_\tau\} \tag{12.2}$$

where $l_0 = 1/\alpha_0$ is the light penetration depth and $l_\tau = (\chi\tau)^{1/2}$ is the heat diffusion length ($\chi = k/c\rho$ is the material thermal diffusivity, where c and ρ are the material heat capacity and mass density, respectively).

The correlation between the experimental data in Figure 12.6 and the estimation (Equation 12.2) indicates that the process of diamond graphitization with high-intensity laser radiation is actually thermally stimulated. The thickness l_g of

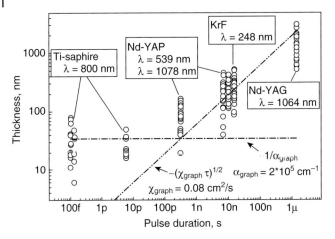

Figure 12.6 Dependence of the thickness of a laser-graphitized CVD diamond surface layer on pulse duration. Reproduced from [36], copyright MYUKK.

the modified layer is determined by the depth of the heat-affected zone. At long pulses ($\tau \geq 0.1$ ns) heat diffusion dominates and $l_g \propto (\chi\tau)^{1/2}$. The best fitting of the data was obtained for $\chi \approx 0.08$ cm^2 s^{-1}, which was close to that of nanographite, but 100-fold smaller than the diamond heat diffusivity. For short laser pulses ($\tau < 0.1$ ns), $l_g \approx 1/\alpha_o \approx 30$ nm and does not depend on the pulse duration. The graphite value $\alpha_o = \alpha_g = 2 \times 10^5$ cm^{-1} gives an excellent agreement between the estimations and experimental data. No definite correlation was found between the l_g value and laser wavelength in multiple-pulse action, when a graphitic layer that was stable in depth was formed. It should be noted that, for femtosecond laser pulses, in some cases the l_g is only a few nanometers. Indeed, the fact that such extremely thin graphitic layers are difficult to detect optically has led some to believe that diamond can be ablated, without graphitization, by applying intense, ultra-short laser pulses [38].

12.3
Laser Ablation

The results of numerous studies of the laser-induced ablation of diamond materials have revealed that, depending on the material, ambient atmosphere and laser parameters, two basic ablation modes can be realized:

- vaporization ablation (physical)
- chemical etching.

It was also found that, in the case of pulsed chemical etching, extremely thin (as low as 10^{-4} nm per pulse) layers of diamond can be removed; these regimes were termed *nanoablation*.

12.3.1
Vaporization Ablation

As soon as the surface graphitization is induced (either by the leading part of the laser pulse or by a sequence of pulses) and the laser intensity is sufficiently high to cause sample surface heating up to the carbon sublimation temperature (T_S) of $\approx 4000\,°K$), intense vaporization (ablation) will begin.

The transition from only surface graphitization to an ablation regime in a single-pulse action on turbostratic carbon (ta-C) films is shown graphically in Figure 12.7 [39]. At the fluence of KrF laser below the ablation threshold $E_a \approx 0.8\,J\,cm^{-2}$ and, as the result of graphitized material swelling, a bump appears the height of which may grow to $\approx 400\,nm$ as E is increased. Thus, it can be concluded that the graphitization takes place when $E_a > E > E_g \approx 0.1\,J\,cm^{-2}$. As soon as $E > E_a$, a detectable removal of the graphitized material begins; indeed in the case of $E \gg E_a$ the ablation rate became so high that the crater was produced in a single laser pulse.

For shallow craters, the depth is increased linearly [40] in line with the pulse number N (Figure 12.8). The dependence of the crater depth $D = D_1$ for different fluencies of 20 ns KrF-laser is shown in Figure 12.8a. In this case, the CVD films were covered by 5 nm-thick gold films, so that graphitization started during the first pulse. The slopes of the experimental curves define the average ablation rate V (nm per pulse). Accurate measurements of the $D(N)$ curves allowed the negative crossing of the ablative curves with the depth axis as $D = D_{ext} = 195\,nm$ to be identified. As the mass density of the graphitic phase is less than that of the diamond phase, the observed negative value of $D = D_{ext}$ and the further behavior of ablative curves can be explained by the effect of carbon material expansion. The difference between the measured crater depth D_1 and the total thickness of

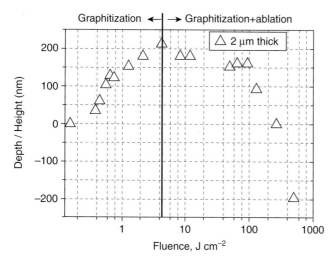

Figure 12.7 Dependence of the height of the extended graphitized surface layer or crater depth on the laser fluence (single-pulse irradiation of ta-C film with 2 μm thickness).

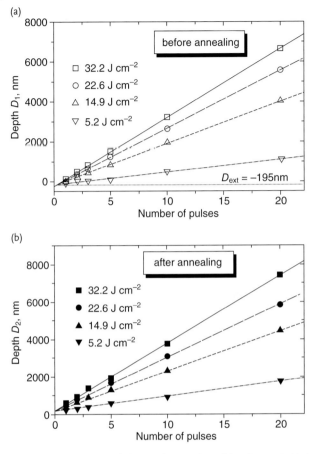

Figure 12.8 Variation with laser pulse number of the diamond ablation depth before (a) and after (b) thermal annealing. Reproduced from [40], copyright RIKEN.

removed material $D_1 + D_{ext}$, is indicated in Figure 12.5. In the second step, the laser-structured diamond plates were annealed in air at $T \approx 600\,°C$, and thus the graphitic layer with thickness l_g was removed. Then, by measuring D_2 (Figure 12.8b) it is possible to determine $l_g = D_2 - D_1 \approx 400\,nm$. It should be noted, that such an l_g value is comparable with D_1, and for precision surface machining should be taken into account. From these experiments the mass density of the laser produced graphitic phase can also be found:

$$\rho_g = \frac{D - D_{ext}}{D} \times \rho_d \tag{12.3}$$

For the diamond density $\rho_d = 3.5\,g\,cm^{-3}$, the calculated value $\rho_g = 1.86–1.97\,g\,cm^{-3}$ is in a good agreement with the mass density of nanocrystalline graphite.

In the case of multiple pulse action, when the graphitic layer is formed during the first few pulses, no influence of λ on material ablation was observed [41].

The vaporization ablation rate is determined by the laser fluence. Close to the ablation threshold, and depending on the laser parameters and target material, two models of ablation and, correspondingly, dependencies of $V(E)$ are in evidence. For long pulses, and highly thermally conductive materials such as diamond, when $(\chi\tau)^{1/2} \gg \alpha_o^{-1} \approx l_g$, the ablation depth is determined by the heat wave penetration into the sample. Likewise, if the sample surface temperature can be kept at $T = T_S$, then heat conduction and a material vaporization wave [19] is formed and the effective rate (depth per pulse) of ablation can be described as

$$V \approx \frac{AW\tau}{\rho_g L_b} \qquad (12.4)$$

where A is the sample absorptivity, W is beam intensity, and L_b is the specific heat of diamond transformation to graphite and further vaporization. Due to a lack of data relating to A and especially to L_b, calculations of V by means of Equation (12.4) are quite complicated. At the same time, some experimental results – for example, for the nanosecond-pulsed KrF laser ablation of CVD diamond – strongly support the $V \propto E$ dependence, following from Equation (12.4), as shown in Figure 12.9 [42]. It can be seen that for the three diamond materials, the ablation rate grows linearly with E, and only the inclination angles for these curves, determined by the $A/\rho_g L_b$ ratio, differ slightly from each other.

The type of material has a major influence on the ablation threshold E_a. For example, among the types of diamond represented in Figure 12.9, the nanocrystalline film has the lowest E_a. It follows from Figure 12.1 and Table 12.2 that this material also has the lowest thermal conductivity (and correspondingly, diffusivity) and the shortest penetration depth l_0, and so it should require less laser energy to be graphitized and vaporized. In contrast, the Type IIa single crystal is much more

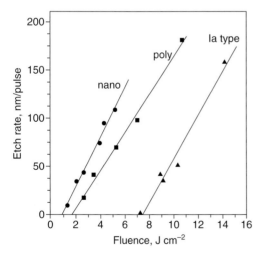

Figure 12.9 Ablation rate versus fluence of KrF laser for nanocrystalline, polycrystalline CVD, and single-crystal Type Ia natural diamond samples (the linear regime). Reproduced from [42], copyright NIST.

difficult to heat and graphitize by a laser beam, while the CVD polycrystalline film has intermediate k and l_0 parameters and, correspondingly, has an ablation threshold $E_a \approx 2\,\text{J}\,\text{cm}^{-2}$ in between that of nanocrystalline diamond ($E_a \approx 1\,\text{J}\,\text{cm}^{-2}$) and Type Ia single crystal diamond ($E_a = 7\,\text{J}\,\text{cm}^{-2}$). It should be noted that the thermal graphitization threshold temperature also depends on the diamond material; for example, the temperature is lower for DLC films than for diamond.

In the case of short pulses (when $(\chi\tau)^{1/2} \ll l_g$), the laser energy released in the graphitized layer is used only for material vaporization, and heat losses inside the sample are negligible. For such a condition, however, another well-known formula for ablation can be applied:

$$V \approx (1/\alpha)\ln(AE/E_\alpha) \qquad (12.5)$$

The results of femtosecond experiments [43], even with highly thermally conductive diamond, are in good agreement with the model of Equation (12.5), as follows from the dependence $V(E)$ presented in Figure 12.10a. Diamond-like films are much less conductive ($\chi \sim 10^{-2}\,\text{cm}^2\,\text{s}^{-1}$), and their ablation can be described [44] by Equation (12.5), even for nanosecond pulses (Figure 12.10b).

The models in Equations (12.4) and (12.5) usually function only near the ablation threshold, while at higher fluencies ablation rate saturation takes place. This effect, as well as some other ablation features, are illustrated by the $V(E)$ curves for different pulse durations presented in Figure 12.11 [37].

The ablation threshold E_a decreases with τ, but not as rapidly as predicted by the simple thermal model

$$AE_a = \rho c T_s (\chi\tau)^{1/2} \qquad (12.6)$$

This difference can be explained by the material phase transition that influences the laser heating rate. In addition, for very short laser pulses ($l_g \gg (\chi\tau)^{1/2}$), Equation (12.6) should be transformed into

$$AE_a = \rho c T_g l_g \qquad (12.7)$$

such that E_a no longer depends on τ, as follows also from pico- and femtosecond laser pulse experiments.

Above the threshold, the ablation rate initially grows according to Equation (12.4) or (12.5), but at high fluencies a saturation of the $V(E)$ curves definitely takes place. It is also important to mention that maximum values of $V(E)$ measured at $E \gg E_a$ increase for longer laser pulses, and for $\tau \approx 1\,\mu\text{s}$ can reach $V \approx 5\text{--}8\,\mu\text{m}$ per pulse, while for picosecond pulses V may be below 100 nm per pulse. This dependence can be explained in a similar fashion to the above, by estimating the thickness of the laser-graphitized layer. It should be noted that, for longer laser pulses, the heat-affected zone becomes deeper.

The saturation of $V(E)$ curves at high laser fluencies can be justified by the target material overheating and the sample surface screening by the vapor plasma flux. A particular situation takes place for femtosecond pulses, in which case the ejection of the target material begins only after the laser pulse has been turned off. This may provide the explanation for higher values of $V(E)$ in saturation regimes

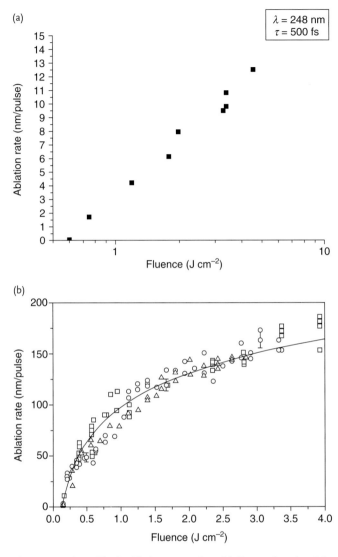

Figure 12.10 Logarithmic ablation rate regime. (a) Short pulses ($\tau = 0.5$ ps, $\lambda = 248$ nm), single-crystal diamond [43]; (b) Ablation of DLC film, where the experimental points are fitted by Equation (12.5). Reproduced from [44], with permission from Springer.

for femtosecond than for picosecond laser pulses. In the case of ultrashort laser pulses another problem arises however, especially if the focal plane is in front of the target surface [45]. Under such conditions, a gas optical breakdown plasma can be formed near the irradiation spot, and this will strongly limit the diamond ablation rates. The saturation of $V(E)$ curves for diamond and few other materials above the air breakdown threshold is illustrated in Figure 12.12. The threshold for

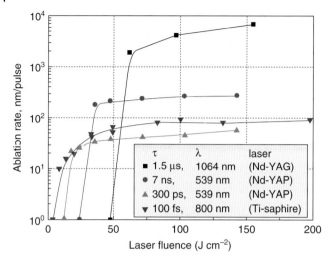

Figure 12.11 CVD diamond ablation rate as a function of the laser fluence for different pulse durations (multiple pulse irradiation). Reproduced from [37], copyright Turpion Ltd.

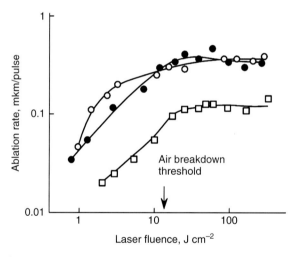

Figure 12.12 Rates of ablation in air of steel (○), Si_3N_4 (●), and diamond (□) by ultrashort ($\tau = 220$ fs, $\lambda = 532$ nm) laser pulses. Reproduced from [45], copyright Turpion Ltd.

air breakdown is $E = E_b \approx 10$–20 J cm^{-2}. It was found that, in the case of femtosecond laser pulses, such plasmas do not absorb the substantial part of the laser pulse energy, but rather especially transform it, such that an increase occurs in the minimum beam focal spot diameter and, even more importantly, there is a dramatic change in the distribution of energy inside the irradiation spot. As can be seen from Figure 12.13, when $E < E_b$ ($E = 5$ J cm^{-2}), a regular bell-shaped crater is

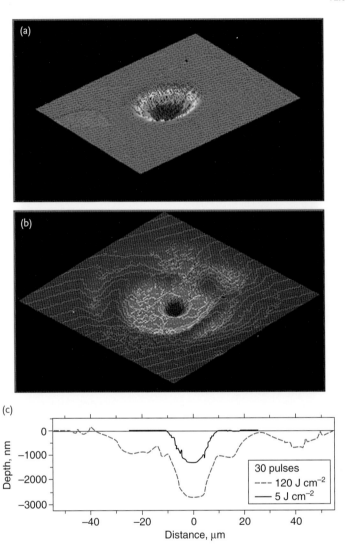

Figure 12.13 Influence of air ionization by 100 fs ($\lambda = 800$ nm) laser pulses on the quality and productivity of diamond ablation. (a) Image of the crater produced at $E = 5$ J cm^{-2} (low-intensity regime); (b) Image of the crater formed as a result of irradiation at $E = 120$ J cm^{-2} when air optical breakdown takes place; (c) Corresponding WLI crater profiles. For a color version of this figure, please see the color plate at the end of this book [98]. Reproduced with permission.

produced; otherwise, nonuniform laser energy deposition (not well-shaped central spot and outer rings) with growth of the spot size are observed for $E \gg E_b$ ($E = 120\,\mathrm{J\,cm^{-2}}$). Previous investigations [46] have shown that, as a result of gas breakdown, up to 70% of the pulse energy can be scattered outside the undisturbed laser beam. Notably, at high laser fluencies the air optical breakdown can also occur for longer laser pulses.

Although the most obvious route to suppress the gas breakdown effect would be to irradiate samples in a vacuum, from a technological standpoint material ablation in atmospheric air is much more preferable. The solution to this problem is to imitate conditions, close to that of a vacuum, near the sample surface. For example, with pico- and femtosecond pulses the transition from air to vacuum ablation regimes occurs at the air pressure $p \approx 100\text{--}200\,\mathrm{mBar}$ (see, e.g., ref. [45]). By using a specially designed nozzle [47] such pressures can be simulated close to the sample surface, thus avoiding the air breakdown for $E \leq 100\,\mathrm{J\,cm^{-2}}$. The growth of E_b is also possible by blowing the irradiation zone with a high-ionization-potential gas, such as helium, while the wavelength λ and laser pulse duration τ may also be optimized. In addition it was found that, as a result of multiple-pulsed ablation in a gas environment, charged microparticles could be formed that would remain within the beam caustic for long periods of time. Although these particles could ignite gas breakdown, they could be removed by applying the correct electrical field. Finally, an alternative approach to combining a high productivity with high-quality laser processing technology involves a diamond ablative treatment that can be performed in air by means of low energy (i.e., $E < E_b$) and ultra-high repetition pulse rates (up to 50–100 MHz).

Another influence of the gas environment was identified [48] for the ArF-laser ($\tau = 20\,\mathrm{ns}$) ablation of CVD diamond films of 5 µm thickness. When the dependence of ablation rate on pulse number was examined in different gases (see Figure 12.14), it became clear that in vacuum and light gases (He and H_2) the ablation rates were higher than in other gases, possibly as the result of gas breakdown suppression. In addition, the laser-produced vapor expanded radially more rapidly in helium and hydrogen (and more so in a vacuum) than in air, oxygen, or nitrogen at the same pressure. Correspondingly, in the latter case the vapor plasma should have a longer life and a higher temperature, thus shielding the target surface more efficiently. The higher ablation rates in oxygen than in air and nitrogen can be explained by the input of graphitic layer chemical etching by an oxidation reaction (this point is discussed in greater detail below).

12.3.2
Nanoablation

In the case of pulsed irradiation, specific regimes of diamond ablation were identified that are governed by oxidation etching and take place for $E < E_a$ (below the ablation threshold the temperature appears too low and vaporization is negligible) and are characterized by very low etching rates (less than a few nanometers per pulse). Hence, these regimes are referred to as *nanoablation*.

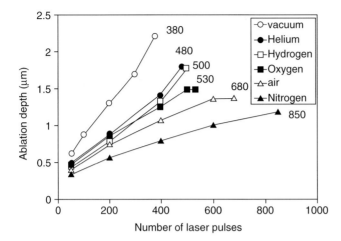

Figure 12.14 Ablation depth (per pulse) as a function of pulse number for various atmospheres. The number at each curve indicates the total number of pulses that can be applied before the CVD diamond film with 5 μm thickness is damaged. Reproduced from [48], with permission from Elsevier.

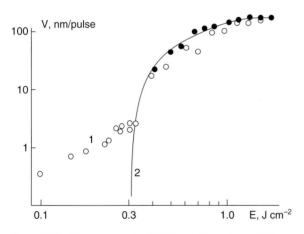

Figure 12.15 Dependence of a-C:H film etching rate on KrF laser pulse fluence in air (○) and vacuum (●). Reproduced from [49], copyright Turpion Ltd.

In one of the first observations of such a regime [49], diamond like a-C:H films were irradiated with 20 ns pulses of a KrF-laser; the obtained dependence $V(E)$ is shown in Figure 12.15. At a high laser fluence $E > E_a = 0.4\,\mathrm{J\,cm^{-2}}$, the vaporization of diamond material was induced, while the ablation rate was more than 10 nm per pulse and equal in air and vacuum. For smaller E-values (curve 1), the material removal rate fell by about an order of magnitude and occurred only in air. Taking into account that, in this experiment, the material graphitization threshold was as

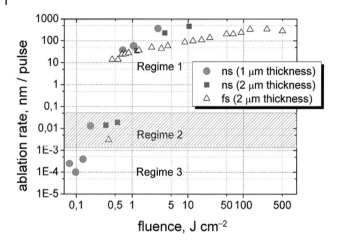

Figure 12.16 Different regimes of laser ablation for ta-C films. Reproduced from [39], copyright New Technologies.

low as $E_g = 0.07\,\mathrm{J\,cm^{-2}}$, it can be concluded that oxidation of the laser-graphitized layer in the temperature range $T_g < T < T_S$ determines the ablation process. The ablation (chemical etching) rates were quite low, at $V \leq 2\text{–}3\,\mathrm{nm}$ per pulse.

The specific feature of short-pulsed oxidation is that the chemical reaction takes place either during the laser pulse if the process has a photolithic nature, or at an elevated surface temperature for a pyrolytic case. For small pulse repetition rates ≤10 Hz (as are used in most experiments), the surface is cooled down between the pulses. As a consequence, only oxygen (water) molecules adsorbed at the surface between laser pulses can react with sp^2-bonded carbon atoms on the sample surface during a short period of laser-pulsed heating. It should be noted that similar peculiarities of pulsed-laser chemical reactions were observed in other experiments [32] on Si etching in Cl_2 by KrF-laser pulses.

Nanoablation was also studied for ta-C DLC films, using KrF and Ti:Al$_2$O$_3$ lasers [38]. The study results, summarized as $V(E)$ data, are presented in Figure 12.16, where three different ablation regimes can be distinguished:

- **Regime 1:** $E > E_a \approx 0.5\,\mathrm{J\,cm^{-2}}$ – vaporization regime.

- **Regime 2:** $E_g \approx 0.15\text{–}0.2\,\mathrm{J\,cm^{-2}} < E < E_a$. The graphitized surface layer is chemically etched at rates $V = 10^{-3}$ to $10^{-1}\,\mathrm{nm}$ per pulse. Surface swelling as a result of graphitization appears in the first pulse. However, for DLC films it was found that a noticeable decrease of the surface bump height can start after about $N_o \sim 10^3$ laser pulses. Such a long chemical reaction induction period can be related to variations in surface morphology; in other experiments [39] the surface roughness was almost doubled after this critical number N_o of the laser pulses.

- **Regime 3:** $E < E_g$. Surface graphitization was not detected optically. Surface etching takes place at ultra-low rates $V \geq 10^{-4}\,\mathrm{nm}$ per pulse. To measure such rates, about 10^4–10^5 laser pulses were required. To realize this, Regime 3 (and also Regime 2) requires an oxygen-containing atmosphere.

Similar results were obtained also for NCD films [50, 51], but the difference was in the E_g and E_a thresholds (both were shifted to higher values). In addition, a comparable study with NCD films that had been synthesized in a CVD reactor with a high nitrogen content (30%) of the gas mixture, was performed for excimer laser wavelengths $\lambda = 193$ and 248 nm (photon energy $hv = 6.4$ and 5 eV, respectively). It was found that, for a photon energy exceeding the diamond band gap of 5.4 eV, laser chemical etching in Regime 3 proceeded at higher rates. The results of this experiment may support the assumptions [49] that, in this regime, not only thermal but also photoinduced phenomena can play an important role.

To date, Regime 2 has not been definitively observed for single-crystal and high-quality CVD diamond samples. The reason for this is that, as soon as $E > E_g$, the absorption coefficient α_o in the sample surface layer rises dramatically (there is a large difference between α_o and α_g) and material vaporization always takes place.

Regime 3 was identified [52] when the KrF-laser irradiation of Type IIa diamond crystals was performed. The dependence of the diamond nanoablation rate on laser pulse fluence for $E < E_g \approx E_a = 14\,\mathrm{J\,cm^{-2}}$ is shown in Figure 12.17, and no etching was observed in vacuum. The threshold fluence for this effect was $\approx 2\,\mathrm{J\,cm^{-2}}$, and the maximum etching rate, V, at room temperature was $\approx 7 \times 10^{-4}$ nm per pulse. Additional heating of the sample resulted in a quite noticeable growth of the etching rate. For example, with $E = 5.4\,\mathrm{J\,cm^{-2}}$, V was increased from $\approx 2 \times 10^{-4}$ nm per pulse at room temperature to $\approx 10^{-3}$ nm per pulse at 600 °C. Estimations of the diamond sample temperature T during the laser pulse, based on the assumption that the temperature variation along the plate thickness is linear

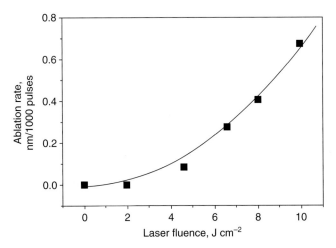

Figure 12.17 Dependence of the single-crystal diamond nanoablation rate on 20 ns KrF laser pulse fluence at room temperature. Reproduced from [39], copyright New Technologies.

and uniform, and applying experimental data for absorption and sample transmittance, showed $T < 100\,°C$. However, the diamond pulsed surface temperature could be higher because of a nonlinear growth in absorption by even minor material transformation close to the surface. In any case, calculations of the thermal diamond surface oxidation rate, using the data for a traditional constant-temperature process [53], and assuming that the etching time is equal to the laser surface heating period $N \cdot \tau$ ($N > 10^3$–10^4, $\tau = 20\,ns$), much smaller values of V were obtained than were measured in Regime 3 for laser repetitive-pulsed etching experiments.

Consequently, it was proposed that the Regime 3 oxidation mechanism was based on a photo-excitation of diamond by laser pulses (UV photon or multiphoton absorption). The injection of electrons into the conduction band should result in a diamond lattice deformation and the appearance on the sample surface of less strongly bonded carbon atoms or atoms and clusters with free carbon bonds that can easily react with oxygen. This mechanism may also explain why the observed rates of nanoablation were so small (much less than one monolayer of carbon atoms was removed per pulse).

12.3.3
Photoionization of Diamond

A number of optical methods have been applied to studies of the laser-induced ionization of high-quality CVD polycrystalline diamond and natural single crystals.

In one study, a transient photoconductivity technique was used to investigate the excitation and recombination of nonequilibrium free charge carriers [54], where picosecond ($\tau \approx 30$–$50\,ps$) pulses of a Nd:YAG laser at $\lambda = 355$ and $266\,nm$ were used as the source of excitation light. In order to record the photoconductivity signals, a diamond sample was placed inside a capacitor that was composed of two metallic plates to which an external pulsed voltage with amplitude $U_0 \leq 5\,kV$ was applied synchronously with the laser pulses. The induced photocurrent caused a voltage signal $U_{pc}(t)$ on a load resistor (a cable with wave impedance $R_0 = 50\,\Omega$) that was recorded with an oscilloscope using a temporal resolution of $\approx 1\,ns$. The transient voltage was proportional to the charge carrier concentration $n_e(t)$:

$$U_{ps}(t) = \alpha \frac{\mu_e U_0}{\varepsilon L^2 R_0} n_e(t) \qquad (12.8)$$

where $n_e(t)$, e, μ_e are the carrier concentration, charge, and mobility, respectively, L is the distance between electrodes, ε is the diamond dielectric constant, and $\alpha \approx 1$ is a geometric parameter.

Measurements [54] as well as with $\lambda = 430$–$690\,nm$ nanosecond pulses [55] have shown that photons with an energy smaller than the diamond band gap (5.4 eV) can efficiently ionize the material bulk by impurity and defect absorption, and in particular by nitrogen centers. The carrier lifetime was found to be about $\tau_R < 1$–$10\,ns$, depending on the diamond sample quality. Similar data on τ_R were obtained [56] when reflection and transmission of the microwave (frequency $\omega = 140\,\text{GHz}$) -probing radiation from the CVD diamond sample was measured during its laser-induced ionization. In this case, a strong variation of the diamond dielectric properties occurred when the carrier concentration reached a value of $n_{cr} = 2.5 \times 10^{13}\,\text{cm}^{-3}$ for $\omega = 140\,\text{GHz}$ (the resonant plasma frequency ω_{cr} is proportional to n_{cr}).

The sensitivity of the transient photoconductivity technique was very high, and the detected n_e value might be as low as $\leq 10^{11}\,\text{cm}^{-3}$. For small laser pulse fluences, the carrier concentration was shown to grow linearly with E until the applied electrical field became screened by a dense carrier cloud. This led to a saturation of $n_e(E)$ dependence, which made it difficult to obtain qualitative data on the photoinduced plasma parameters. However, estimations have shown that at picosecond laser intensities ($W \approx 10^9\,\text{W}\,\text{cm}^{-2}$), the carrier concentrations can reach $n_e \approx 10^{18}\text{--}10^{19}\,\text{cm}^{-3}$.

The pump-probe interferometric method (see, e.g. Ref. [57]) allows the investigation of bulk diamond ionization and plasma relaxation during the process and after the laser pulse action. Moreover, it also allows detection at the highest temporal resolution (100 fs or less).

The principle of the interferometric technique is as follows. First, pulsed-laser radiation is focused inside the diamond sample to induce its ionization or phase transformation (see Section 12.4) within the volume close to the focal plane. A probe beam can be created in either of two ways. For the first method, a pulse produced by another laser is fired at different time delays τ_D relative to the main laser pulse, whereas in the second method a single short (e.g., femtosecond) laser pulse is used. Part of the laser pulse energy passes through a delay line with variable length and corresponding τ_D. This probe beam illuminates the plasma region at the normal angle from the side sample surface.

In the traditional approach, a Michelson interferometer is used, with the sample under investigation being placed in one of its arms. The phase shift between undisturbed and passing through the excited area beams is then utilized to monitor changes in the optical path length. The latter is determined by variation of the refractive index n, which is expressed for the ionized solid material according to the Drude model as:

$$n = n_0 - \frac{n_e e^2}{2 m_e \omega^2} \tag{12.9}$$

where n_0 is the initial material refractive index, m_e is the electron mass, ω is the probe laser frequency, and n_e is the concentration of free electrons. Equation (12.9) shows that the presence of free electrons leads to a decrease in the refractive index. Hence, taking into account that the plasma is formed inside the main beam, the optical path change can be estimated as $\Delta \ell \approx \Delta n \cdot d$, where d is the spot diameter.

Unfortunately, when using the standard approach in time-resolved pump-probe interferometric studies, the precision of the phase shift measurements is about $\lambda/10$. At the same time, in laser-induced materials the refractive index variation, Δn, is usually very small (≤ 0.01); consequently, for typical focused-beam diameters,

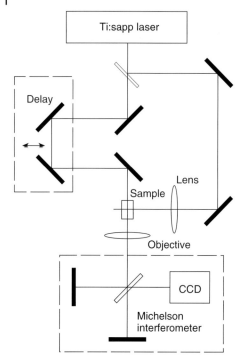

Figure 12.18 The set-up for the time-resolved study of laser pulse excitation of transparent matter. Reproduced from [59], with permission from Astro Ltd.

where $d \approx 1$–$10\,\mu m$, the detectable phase shift should be below $\lambda/100$. It is for this reason that more sophisticated interferometric techniques must be applied (see e.g., Refs [58, 59]).

An example of such a scheme is presented in Figure 12.18, where the probe beam passes through the interaction zone, "recording" the phase information. Inside the interferometer, the beam is divided into two parts that illuminate a charge-coupled device (CCD) array detector. The interferometer is adjusted specifically so that the beams are parallel and pass at some distance from each other; in this case, each beams will generate its own image, with one image being bright and the other dark. The distance between the images must be less than the size of the CCD, in order to provide a visualization of both images. The images contain a local phase perturbation that is related to any variation in refractive index in the irradiated area. If the interference fringe width is much greater than the image size, and the optical path difference between the interferometer arms is exactly zero, then image brightnesses of opposite sign variations are seen to appear. A subsequent digital treatment of the images leads to the interference fringes being practically eliminated, as shown in Figure 12.19.

If the absorption of the probe beam in the photoionized region is negligible, then the image intensity perturbations are determined only by variations in the

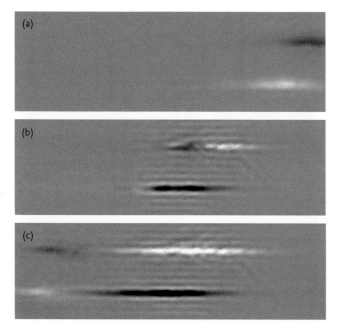

Figure 12.19 The interference image of the diamond refractivity dynamics in the laser waist at $\tau = 100$ fs and $\lambda = 800$ nm. The probe beam delays were (a) 0 fs, (b) 660 fs, and (c) 1320 fs. The pulse propagated from the right-hand side to the left-hand. The image length was 190 μm (for description, see the text).

refractive index, n. Opposite and equal-magnitude changes of both images are observed in this case. If light absorption by the plasma is notable, then both images will become darker simultaneously; correspondingly, measurements of the diamond's refractive indices before, during, and after pump laser irradiation are possible. Hence, a phase-amplitude analysis of the complete interference pattern will provide comprehensive information regarding the local polarizability of the laser-modified material, in terms of both its refractive index and absorption coefficient.

The interference images of 100 fs laser pulse propagation in the bulk of SCD are shown in Figure 12.19 (from the right-hand to the left-hand side). In order to avoid too much detail, all further comments here will concern only the upper (bright) interference image in the figure. The dark regions correspond to $\Delta n > 0$, and bright regions to the decrease in refractive index. The positive refractive index distortion is caused by the Kerr effect, and is observed only at the given point of the sample during the radiation action. Consequently, the dark cloud exactly coincides with a light wave packet, and it is possible to follow the motion of the laser pulse through the sample with a speed that is equal to the speed of light in diamond. In Figure 12.19a, the image was recorded before the pulse has reached the focal plane; in this beam cross-section the laser fluence is relatively small and a "pure" Kerr effect is detected. A more complicated image is observed, however, when the pulse reaches the focal plane zone where the beam intensity becomes maximal (Figure 12.19b). Here, both dark and light parts of the image can be seen that are determined, respectively, by the positive (Kerr effect) and negative (diamond ionization that occurs at times shorter than pulse duration) variations of the refractive index. Finally, as the light pulse passes the beam waist (Figure 12.19c) the laser fluence decreases and electron plasma formation ceases. At this point, the light wave packet can be seen to travel to the left side, while the electron–hole cloud persists for a long time close to the focal plane. The maximum measured value of refractive index change (Δn) due to plasma creation was about -0.02, while the corresponding carrier concentration, estimated with Equation (12.2), gives $n_e \approx 10^{20}\,\text{cm}^{-3}$.

It should be noted that the generation of carriers can most likely play a role in the development of submicron structures observed on the irradiated diamond surface [60, 61]. The period of these structures is $\approx 0.1\,\lambda$, and their orientation depends on beam polarization.

12.4
Bulk Graphitization of Diamond

It has been shown [62–67] that diamond–graphite phase transformation under intense laser radiation can occur within the bulk of a diamond sample. However, in order to achieve this, visible or near-IR laser radiation should be used to guarantee a negligible absorption while the intensity of the radiation remained at low level. A sharp focusing of the beam provides a huge rise in the laser intensity within a small focal volume that, in turn, initiates an efficient and nonlinear absorption of the radiation. However, when the laser intensity exceeds a certain threshold, local diamond graphitization becomes possible. Hence, by moving the focal volume inside the sample, it become possible to produce complex one-, two-, or three-dimensional (3-D) graphitic structures.

12.4.1
Threshold Conditions

It has been shown in various experiments that, when femtosecond pulses of a Ti:Al$_2$O$_3$ laser are tightly focused inside the body of an initially highly transparent diamond sample and pulse fluence $E > E_g^b$, the diamond can be graphitized inside the beam caustic. This effect has been confirmed by a Raman spectrum that proves the appearance of the graphitic D and G peaks, although it should be noted that the spectrum of the bulk-graphitized materials differs somewhat from that of the laser-graphitized layer on a diamond surface. This difference can be attributed to a confinement of the expanding graphitic material by the surrounding diamond lattice.

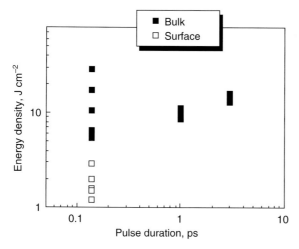

Figure 12.20 Surface and bulk graphitization thresholds for single-crystal diamond, depending on laser pulse duration ($\lambda = 800$ nm). Reproduced from [98], with permission.

The threshold fluencies for the surface E_g and bulk E_g^b graphitization of diamond for 130 fs, 1 ps, and 3 ps pulse lengths and multiple pulsed laser irradiation are shown in Figure 12.20. Two important conclusions can easily be made at this point: (i) $E_g^b \gg E_g$ (e.g., for femtosecond pulses $E_g^b \approx 7-30$ J cm^{-2}, while $E_g \approx 1-2$ J cm^{-2}); and (ii) the bulk optical breakdown effect has a highly statistical character. The weak dependence of E_g^b on the laser pulse duration should also be noted. Bulk graphitization has a pronounced "incubation" effect, similar to that observed for diamond surface graphitization. The reason for this effect is generally recognized as the appearance and accumulation of stable micro (nano) defects that absorb radiation much better than clear diamond, and can produce the electrons required for further electron avalanche, material heating, and phase transformation (optical breakdown).

When $E \approx E_g^b$ after a certain number of laser pulses, the initial graphitized volume appears in the focal plane. At such E levels, the size of this volume after the first breakdown is, as a rule, substantially smaller than the beam waist. Further pulsed irradiations cause a rapid expansion of the initial graphitized site until its size becomes consistent with the beam waist diameter, at which point an almost spherical (but slightly extended in the beam direction) graphitic drop is formed. Occasionally, several drops may appear in the neighborhood of the focal plane $(E \approx E_g^b)$. For higher fluencies $E > E_g^b$ (Figure 12.21, left side) [63], irregular and separate dark spots are formed inside the beam caustic; the first appears close to the focal plane, while others appear one after another at some distance from this plane in the laser direction, and increasing with the pulse number. The reason for this is that, at a larger distance from the focal plane, the beam intensity decreases and more laser pulses are needed for material optical breakdown. It should be

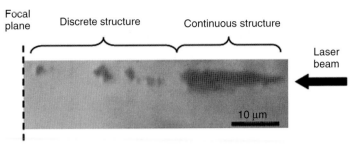

Figure 12.21 Optical microscopy image of laser-modified region inside diamond bulk created by multiple 120 fs pulses with energy of 320 nJ (spot size diameter 3 μm). The positions of the focal plane and the direction of laser beam propagation are shown. Reproduced from [63], with permission from Springer.

noted that the appearance of a new graphitic globule as a result of diamond optical breakdown leads to the conservation of all previously produced globules, because graphitic material is highly absorbing and the leading (i.e., towards the laser) graphitic site shields the remainder of the sites from the laser beam.

12.4.2
Laser-Induced Graphitization Waves

The interesting feature of bulk diamond graphitization is, that after initiation and at $E \leq E_g^b$, the graphitized region can extend continuously (from pulse to pulse) towards the laser beam (Figure 12.21, right side). The initial graphitic globule produced, on being irradiated by the next laser pulse, is heated and this can trigger graphitization of the adjacent (in the direction of the beam) layer of diamond at fluencies lower than the bulk diamond breakdown threshold E_g^b. As a result, the graphitization front can move inside a highly focused beam at rather long distances from the focal plane; this regime has been termed the laser-supported graphitization wave. Moreover, following the production of an initial graphitic site, the laser fluence can be lowered well below E_g^b but a continuous graphitic line will still be produced. However, if the focal plane position is changed (e.g., by moving the sample relative to the beam), the graphitized volume will follow the focal spot.

Here, it can be seen that the minimum (threshold) fluence value E_w needed to support the laser graphitization wave in diamond, is only $\approx 1\,\mathrm{J\,cm^{-2}}$, while $E_g^b \approx 10$–$50\,\mathrm{J\,cm^{-2}}$ (see Figure 12.20). Figure 12.22 demonstrates that the support of these waves requires much lower beam intensities than does the production of an initial material breakdown. This situation is well known for waves of radiation absorption and the heating of gases or solids (these are also termed optical discharges). The rate of graphitization grows strongly in line with E (from a few up to hundreds of nm per pulse) and saturates at $E \geq 10\,\mathrm{J\,cm^{-2}}$. Similar results were obtained for $\lambda = 400\,\mathrm{nm}$ and slightly longer laser pulses (1–3 ps).

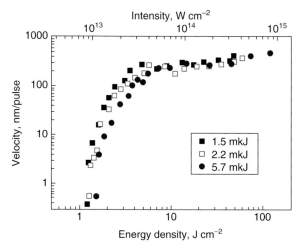

Figure 12.22 Continuous diamond graphitization wave velocity dependence on laser fluence [$\tau = 140$ fs, $\lambda = 800$ nm; pulse energy 1.5 (■), 2.2 (□), and 5.7 (●) µJ]. Reproduced from [98], with permission.

Previously, two mechanisms were proposed for the graphitization wave in diamond [63], both of which are based on laser heating of the initial graphitic site (thin layer) that is located at the graphitization front. For graphitic materials, the absorption depth is about 50 nm, and a high-intensity laser pulse can heat this layer to a very high temperature (up to $\approx 10^2$ eV). After termination of the laser pulse, about half of the absorbed energy is transmitted to the neighboring (i.e., towards the laser) diamond layer, causing it to be thermally graphitized. This hot graphitic layer is also an effective source of UV and X-ray radiation that can strongly activate the carbon phase transformation in diamond. Another mechanism is based on a fast electron ejection from the irradiated graphitic layer into diamond. Graphite has a zero-width band gap, and the Fermi level relates to the bottom of the conduction band. The velocity of electrons, excited by photons with energy $\hbar\omega = 1.55$ eV can reach a value of $V_e \approx \sqrt{2\hbar\omega/m_e} \approx 7.4 \times 10^7$ cm s^{-1}, where $m_e = 9.1 \times 10^{-28}$ g is the electron mass. As hot electrons move in the ballistic regime – that is, without large angle scattering – they can leave the graphite and penetrate inside the diamond body at a maximum distance $L_e \approx 90$ nm during the 120 fs pulse. As a consequence, diamond optical breakdown can be triggered in this layer. The limited propagation distance of external electrons in diamond restricts the thickness of the layer where the avalanche ionization and optical breakdown of diamond can be realized. It is important to note that the concentration of injected electrons inside the diamond layer on the wavefront decreases with the distance from the graphite–diamond interface. Consequently, it is important that the thickness of a newly graphitized layer (the V_w-value) should increase with the growth of the laser pulse intensity. The rise in V_w must decelerate and eventually stop when the modified diamond layer thickness (per pulse) becomes comparable with L_e, which is in reasonable agreement with the experimental data presented in Figure 12.22. An alternative explanation for the observed $V(E)$ saturation that occurs at fluences above the graphitization threshold is that, under these conditions, each laser pulse can act independently to form a new graphitized site that totally shields against radiation. Thus, the prehistory – namely graphitized globules produced by previous pulses – is of no importance in any further development of the graphitized diamond volume at a high laser fluence.

The use of intense short laser pulses for diamond bulk processing provokes a self-focusing of the laser beam, leading to a transformation of the spatial beam profile, or so-called *beam filamentation* (see Figure 12.23). Diamond luminescence in the visual spectral range provides a direct visualization of the appearing beam filaments (see Figure 12.23a). The evident consequence of beam profile transformation is a splitting of the laser-modified region in multiple thin wires at a large distance from the focal plane (see Figure 12.23b), under multipulse irradiation with a fixed focus position.

The self-focusing effect has a power-determined threshold; hence, the fluence threshold E_{sf} for self-focusing grows with the laser pulse duration (as noted in Table 12.3). The measured bulk graphitization thresholds are also listed for comparison, when it can be seen that, for femtosecond pulses, $E_{sf} \approx E_g^b$. Moreover, for the best quality diamond crystals and, correspondingly, the highest pulse energy (or fluence) thresholds, it is essentially impossible to avoid beam filamentation. As a result, the graphitized structures produced appear to be inhomogeneous, although for longer pulse durations the situation is substantially improved (for example, when $\tau \approx 3$ ps, $E_{sf} \gg E_g^b$).

12.4.3
Three-Dimensional (3-D) Laser Writing in Diamond

Movement of the diamond sample at constant speed along the beam outward of the laser makes possible the generation of straight graphitic wires of practically unlimited length [66]. The experimentally confirmed dependence of the wire diameter (d) on the sample scanning speed (V = 0.2–100 µm s^{-1}) is shown in Figure 12.24. It seems natural that the wire diameter decreases the smaller pulse energies, but that the effect of the scanning speed seems much more complicated and still has no adequate explanation. Variations in the wire diameter are accompanied by changes in its quality. For example, the filled symbols in Figure 12.24 correspond to high-quality homogeneous wires shown in Figure 12.25a, while open symbols indicate thicker wires split into a few thinner filaments, as shown in Figure 12.25b. This effect was observed at a minimal scanning speed (even for the lowest pulse energies examined), and cannot be explained by the self-focusing and filamentation of the laser beam (as discussed below).

By reducing the laser pulse energy and selecting a correct sample scanning speed, it is possible to reduce the wire diameter substantially below the beam waist size (2.2 µm in the experiments described). In particular, the measured diameter

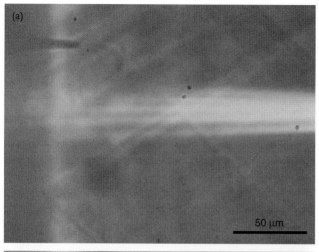
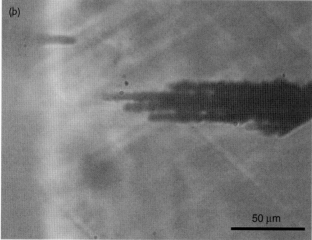

Figure 12.23 Filamentation of laser beam inside diamond due to self-focusing effect. (a) Filaments observed due to diamond luminescence; (b) Splitting of the laser-modified region (without sample scanning). Reproduced from [98], with permission.

Table 12.3 Thresholds for laser beam ($\lambda = 800\,\mu m$, $d = 2.8\,\mu m$) self-focusing and bulk graphitization in diamond.

Pulse length	Self-focusing threshold (μJ) (calculated)	Bulk graphitization pulse energy threshold (μJ) (experimental)
140 fs	0.4	0.15–0.9
1 ps	2.8	0.25–0.38
3 ps	8.5	0.4–0.5

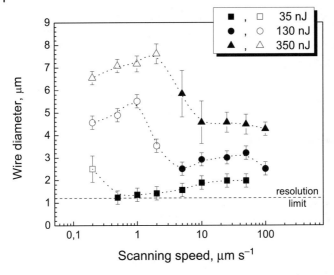

Figure 12.24 Diameter of straight graphitic wires as a function of scanning speed for different laser pulse energies (open symbols indicate wire splitting). Reproduced from [66], with permission from Elsevier.

could reach submicron values which, when close to a minimum of 1.2 μm, led to a progressive reduction in wire opacity, as shown using optical microscopy. This finding could be interpreted as a reduction in the actual wire diameter below the optical resolution limit of the microscope used. Moreover, it seems reasonable to assume that the wire diameter could be further decreased by sharper beam focusing and a transition to shorter laser wavelengths.

The fabrication of curved graphitic structures in diamond is also possible. The basic element of such structures is a graphitic wire inclined at an angle θ relative to the laser beam (axis Z). To create this, a continuous movement of the sample along the Z-axis must be combined with a movement in a perpendicular direction (in the XY plane). The inclination angle θ is determined by the relation between corresponding components of the sample scanning speed: V_Z and $(V_x^2 + V_y^2)^{1/2}$. The wire growth is the result of a graphitization wave movement initiated by laser irradiation of the graphitization front. Hence, the wire growth appears impossible under sliding incidence of the laser beam on the graphitization front that occurs when the angle between the laser beam and the wire becomes too large. The results of other experiments [66] have shown that the drawing of continuous graphitic filaments is restricted by the angles $\theta \leq 70$–$80°$. In addition, it is necessary to take into account that a notable portion of the laser pulse energy can penetrate behind the tip of the growing inclined wire and provoke an uncontrolled drawing of undesirable, additional wires. In order to achieve an efficient blocking of the laser beam by the growing wire tip, the pulse energy and all components of the sample scanning speed must be carefully chosen. This results in a reduction of the maximum inclination angle down to the value $\theta_{max} \sim 50°$. By taking this

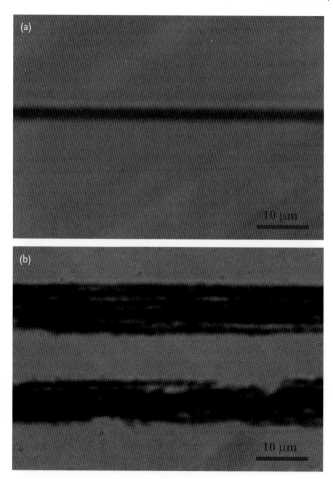

Figure 12.25 Examples of straight graphitic wires. (a) Homogeneous wire; (b) Split wire. Reproduced from [66], with permission from Elsevier.

account into restriction, complex filamentary structures such as the chain of hexagons shown in Figure 12.26 can be produced.

The 3-D movement of the focus inside diamond also allows the production of complex compact structures with a large cross-section, such as pillars and plates. The formation of such structures is associated with a specific challenge, and is the result of a substantially lower density of the graphitized material as compared to diamond. Indeed, decreases in the material density under phase transition causes an expansion of the modified material and the appearance of strong tensile stresses in the surrounding diamond. These stresses increase with the cross-sectional area of the structure, and can even reach the fracture limit. In experiments conducted to date [66], the maximum cross-sectional area for the compact structures has been restricted by a value of $\approx 100\,\mu m^2$, to avoid any visible cracking of the adjacent diamond material.

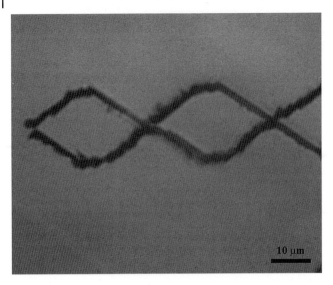

Figure 12.26 Fragment of the laser-produced graphitic hexagonal chain. Reproduced from [66], with permission from Elsevier.

The diamond damage considered is the result of a permanent material deformation and corresponding mechanical stresses that have remained on completion of the laser processing. A progressive rise in the cross-sectional area under pulse-periodic irradiation reinforces these steady-state stresses until the fracture occurs. It is also necessary to consider that the sharp change in material density induced by short laser pulses can generate intensive shock waves that are capable of destroying (inducing cracking in) the adjacent diamond.

When diamond damage due to transient mechanical stresses was investigated for the case of thin wire formation, visible diamond cracking was found to have occurred under certain combinations of two parameters, namely the rate of graphitization front movement V, and the local wire diameter d. The data obtained from numerous wires produced with different pulse energy and pulsewidths (from 100 fs to 3 ps) are summarized in Figure 12.27. In this case, the experimental points can be divided roughly into two regions separated by the curve $V = 80/(\pi d^2)$. Thus, nondestructive laser graphitization appears possible when the following requirement is fulfilled: $V(\pi d^2/4) < 20\,\mu\text{m}^3$ per laser shot. Whilst this condition can be interpreted as a restriction on the maximum diamond volume modified by one laser shot, in practice it represents a limitation of the productivity of the laser graphitization process.

Even slight (essentially invisible) diamond damage occurring around the graphitized region was found to affect the laser-structuring process. This conclusion was drawn on the basis of experiments investigating the formation of periodic arrays of straight wires [66]. These wires, which formed arrays with different periods (Λ) (see Figure 12.28) were drawn one after another, beginning from the upper wire. The correct choice of pulse fluence and scanning speed guaranteed the generation

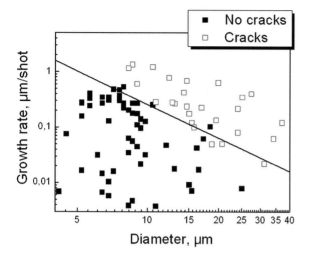

Figure 12.27 Diamond cracking dependence on the rate of graphitization front movement (V) and local wire diameter (d). The line relates to the expression $V = 80/(\pi d^2)$. Reproduced from [98], with permission.

of uniform and continuous wires when the array period exceeded the wire diameter (d) by a factor of more than two ($\Lambda > 2d$; see Figure 12.28a). Maintaining the same processing conditions but reducing the period until $\Lambda < 2d$ caused the wire to break (see Figure 12.28b), except for the wire which was first produced. In order to eliminate the possibility that the earlier-fabricated wires interfered with (screened) the propagation of radiation to the focal region under formation of the following wires, an alternative wire-writing strategy was tested. On this occasion, only short lengths ($\approx 10\,\mu m$) of all wires in the array were produced during the processing cycle, while successive cycles increased the sizes of all wires by the same step until the desired length was achieved. This strategy failed to provide any improvement in the array quality, however.

The most likely reason for this was the invisible diamond defects and microcracks that had been generated in close vicinity to each graphitic wire. Presumably, these defects had reduced the local diamond breakdown threshold and provoked diamond graphitization ahead of the continuous graphitization wave. The graphitic drop produced could be separated from the wire and screened against further irradiation; the graphitization wave then continued to move away from the new source until it was interrupted on meeting another strongly defective area.

12.5
Diamond Laser Processing Techniques

An understanding of laser light interactions with diamond materials resulted in the development of numerous techniques that allowed the modification

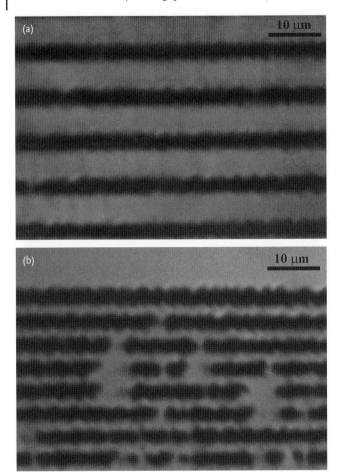

Figure 12.28 Regular arrays with different periods Λ. (a) $\Lambda = 8\,\mu m$; (b) $\Lambda = 4\,\mu m$. Reproduced from [66], with permission from Elsevier.

(graphitization) of diamond, and various processing operations to be performed. These included laser polishing, cutting, drilling, surface micro and nanostructuring, the formation of conductive features on the surface and in the bulk of diamond, the selective etching of laser-graphitized diamond materials, the encapsulation of conductive metal strips inside the diamond samples, and the fabrication of novel elements and devices for optics, electronics, and microsystems. The techniques employed, together with several examples of diamond-based and laser-fabricated elements, are detailed in the following subsections.

12.5.1
Laser Polishing

Polycrystalline CVD diamond is very attractive for many applications, but its growth surface is very rough (the surface roughness is about one-fifth or one-tenth

Table 12.4 Experimental data [34] of surface roughness R_a and etching depth as a function of the angle of laser beam incidence (KrF laser, $\tau = 15\,\text{ns}$). The roughness of a virgin diamond film surface was 0.65–0.8 µm. The value $\theta = 0$ corresponds to normal incidence.

Angle of incidence, θ (°)	0	48	60	68	74	77
Roughness R_a (µm)	0.38	0.4	0.37	0.38	0.25	0.15
Etching depth (µm)	4.0	4.0	5.3	4.4	2.7	2.0

of the plate thickness, which can reach 1–2 mm). As the mechanical polishing of this diamond is both complicated and time-consuming, the use of laser polishing would appear to be a promising alternative.

The effects of CVD polycrystalline diamond surface smoothening (polishing) were first described by Ageev et al. [19] and subsequently investigated by several groups [29, 35, 68–71]. It has been shown that at normal incidence of the laser beam, the bottom of a crater formed as the result of multiple pulsed ablation of CVD diamond may be much more smooth than the initial sample surface. It was also noted that, by increasing the angle of beam incidence θ from normal (0°) to 77°, the induced surface roughness fell to smaller values (to 0.15 µm), with less diamond material being removed to achieve the surface polishing (see Table 12.4).

Scanning electron microscopy (SEM) images of the surface areas treated at $\theta = 77°$, and also of a nonirradiated surface, are shown in Figure 12.29. The dramatic change in the polycrystalline CVD diamond surface morphology with growth of the laser-ablated depth was very clear. Notably, the tips of diamond crystallites were etched (cut) initially (Figure 12.29b), after which the plane areas could be ablated until the remaining pits were removed (Figure 12.29c).

The model of diamond pyramids removal at an oblique laser incidence has also been considered [72]. This technique is based on the fact that radiation absorption depends on the incident angle and beam polarization; as a consequence, the correct choice of irradiation parameters leads to a higher absorption on the pyramid slopes than on the flat diamond surface. In this case, for laser intensities slightly below the ablation threshold of a plane surface, diamond pyramids can be effectively removed. It should be noted that at an oblique incidence, the laser fluence on the plane surface $E_\theta = E*\cos\theta$ and more powerful lasers are required; similarly powerful lasers are also required to provide an effective surface polishing. Improvements in CVD diamond polishing can be achieved by sample rotation. An alternative approach to modeling the polishing mechanism has been considered [73] that is based on identifying the conditions of stability for different surface-profile harmonics.

Until now, the best results obtained for the laser polishing of polycrystalline films and plates has been a final surface roughness of approximately 0.1 µm; consequently, a more appropriate term for this technology might be laser "surface smoothing." Nonetheless, the high speed and simplicity of this technique may lead to its use as an important intermediate step in the fabrication of diamond elements and devices.

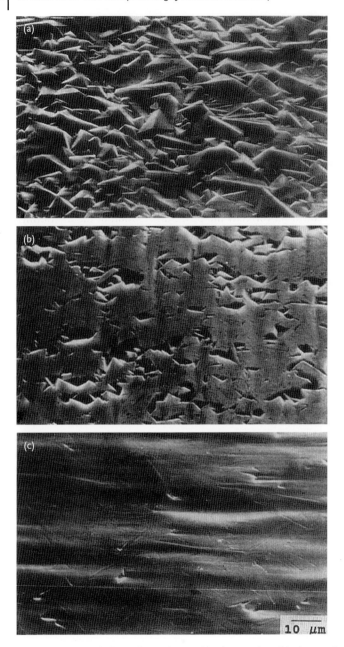

Figure 12.29 Morphology of CVD diamond surface irradiated by laser pulses under oblique incidence of the laser beam. (a) Original surface; (b) Angle of incidence $\theta = 77°$, etching depth 1 μm; (c) $\theta = 77°$, etching depth 2 μm. Reproduced from [35], with permission from Elsevier.

As a rule, as-grown CVD diamond films and plates have a nonuniform thickness, and the sample surface planarization – and possibly also shaping – of diamond is very important. To perform such a task using laser ablation and scanning the focused spot along the diamond surface will require the *in situ* control of the film thickness that will, in turn, be decreased during the laser processing stages.

The problem is relatively complicated because, in being laser-ablated, the surface will have a poor degree of smoothness and will be covered by a graphitized (highly absorbing) layer. Thus, standard interferometry employing light coherence cannot be used in this case. However, because the sample surface roughness was comparable with the probing laser wavelength, the observed interference pattern could be characterized by the development of "speckle" structures. Although it is practically impossible to reconstruct correctly the phase distribution of the interfering waves, a solution to the problem was found by applying low-coherence optical interferometry [74]. The technique used to measure the thickness of a transparent object is based on the fact that interference (the summation of electric field strengths) for partially coherent light beams occurs only when the optical path difference between the two beams does not exceed a certain value; this is termed the *coherence length* of the radiation source. In the experimental set-up (see Figure 12.30), a KrF laser was used for surface polishing, while the polycrystalline diamond samples were irradiated at $\theta = 70°$ to maximize the effect. The spot size was $200 \times 200 \mu m$, and the samples were optically transparent to the probing beam of a diode laser with $\lambda = 820 nm$ and a coherence length of only $\approx 15 \mu m$. The sample thickness was 150–200 μm, and the initial surface roughness 30 μm.

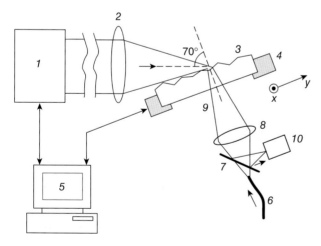

Figure 12.30 Experimental set-up for controlled laser ablation of diamond plates. (1) KrF laser; (2) Objective N 1; (3) CVD diamond plate; (4) X-Y table; (5) controller; (6) Optical fiber from the measuring block in the thickness control system; (7) Beam splitter; (8) Objective N 2; (9) Focused probing beam; (10) Photodetector. Reproduced from [74], copyright Turpion Ltd.

The rear surface was flat and was irradiated by the probe beam at 90° with a spot of 20–30 µm, which was smaller than (or at least comparable with) the size of the individual diamond crystals. The intensity of radiation reflected by the diamond plate surface was monitored using a photodetector such that a precision of about 1 µm (or better) was achieved for diamond sample thickness during the process of its ablation by KrF laser.

Finally, it is important to mention that laser polishing (smoothening) can be applied also for the treatment of localized areas of diamond samples. This technique can be important, for instance, in the tribological applications of diamond. Notably, it was found that the frictional coefficient of laser-polished polycrystalline diamond case could be as low as 0.05–0.1 [75, 76].

12.5.2
Formation of Conductive Structures in Diamond

As shown in Section 12.2.2, the laser-graphitized surface layer is highly conductive (the specific conductivity is usually smaller, but can be close to that of graphite and improved by additional thermal annealing); consequently, laser processing will allow the fabrication of large or selective area contacts on the diamond surface. Today, such structures appear especially attractive as electrodes for diamond detectors of radiation (e.g., UV light, X-rays) or high-energy particles.

The result of DLC film irradiation by single-pulse KrF laser ($E = 0.23\,\text{J cm}^{-2}$), using a mask projection technique, is shown in Figure 12.31, where the formation of an array of graphitic lines with a height of ≈70 nm and a period of ≈4 µm can be seen (Figure 12.31a). The line thickness of ≈200 nm under the sample surface plane can be determined by additional chemical etching (Figure 12.31b). Chemical etching of the laser-graphitized material also allows the production of micro- and nanostructures at the diamond surface. The graphitized layer thickness can be also be reduced by shortening the laser pulse or (and) using a radiation wavelength that is better absorbed in diamond material.

Continuous-wave lasers (e.g., Ar or Nd:YAG) can be also used to graphitize the surface of diamond materials, and thus to "laser-write" different structures by a controlled beam scanning of the sample surface [42, 77].

By using a tightly focused ultra-short pulsed laser beam for the bulk graphitization of diamond, it is possible to produce unique (with aspect ratios ≥1000) conducting vias. This can be achieved by starting the graphitization at the rear side of the transparent sample and moving a continuous graphitization wave to the front side. An example of a straight conductive via is shown in Figure 12.32. Potentially, more complicated conductive structures are possible applying the 3-D laser writing technique described in Section 12.4.3; however, it should be emphasized that none of the other beam techniques (e.g., ion beams) allow comparable results to be obtained due to the very low depth of penetration in solids.

Measurements of the specific resistivity, σ, of the graphitic columns inside diamond produced values down to $2-4\,\Omega\cdot\text{cm}$ [66, 78], which substantially exceeded the values for surface graphitization ($\sigma \approx 10^{-3}\,\Omega\cdot\text{cm}$; see Section 12.2.2). Accord-

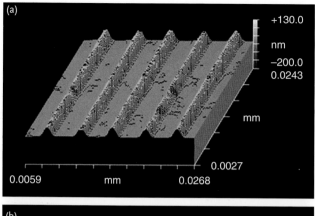

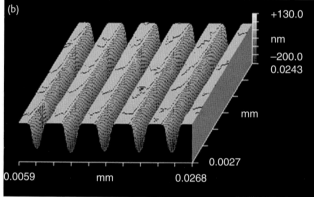

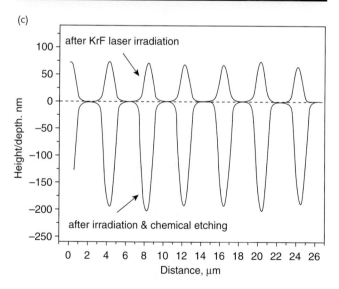

Figure 12.31 KrF laser-induced surface microstructuring of DLC ta-C sample. (a) Laser-produced graphitic strips; (b) Grooves formed after chemical etching of the laser-graphitized material; (c) Corresponding surface profiles. Reproduced from [98], with permission.

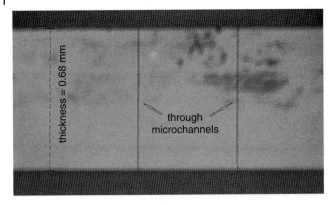

Figure 12.32 Optical microscopy image of the two graphitic microwires with diameter ≈10 μm produced throughout 0.68 mm-thick single-crystal diamond plate using 120 fs laser pulses. Reproduced from [64], with permission from Elsevier.

ing to Ref. [78], the resistivity of the graphitic columns depends heavily on the sample scanning velocity, and its value may be influenced by the homogeneity of the graphitic lines produced. One possible reason for the high resistivity of a graphitized material inside diamond bulk may be a partial transformation of the diamond into graphite, with the laser-modified material in fact consisting of a mixture of graphite nanocrystals and nanograins of disordered diamond.

An alternative technique to produce conductive metal structures encapsulated into diamond is based on a combination of laser processing of the diamond surface, wet chemical treatment, and CVD diamond deposition [26]. In the first step of the process (Figure 12.33a) the graphitized pathway is produced by KrF laser pulsed laser ablation, while in the second step (Figure 12.33b) the deposition of Ni or Cu is achieved via a conventional electroplating process. The metal coatings produced in this way were found to have good adhesion only on the laser-graphitized areas. The third stage (Figure 12.33c) involved a plasma CVD, where a polycrystalline diamond coating was deposited onto the diamond substrate with 2–3 μm-thick metal strips on its surface. In order to stimulate the nucleation process on a metal layer, the sample was first treated with an ultra-fine diamond powder suspended in alcohol, which resulted in a diamond coating of about 10 μm thickness that covered the diamond substrate with electroless copper-patterned films. The metal strips were then encapsulated into the diamond bulk. Notably, this operation could be repeated many times to produce multilayered metal structures inside the diamond.

An additional approach includes the well-known technique of high-energy ion implantation that leads to the formation of defective layers in diamond. As a result of ion implantation, diamond exhibits both a reduced optical transmission and a volume expansion (i.e., swelling of the implanted surface area), where such an effect is caused by defect formation and diamond graphitization similar to that occurring after laser irradiation. The main peculiarity of ion beam graphitization is that the defect layer is "buried" at a certain depth (ca. 1 μm for deuterium ions)

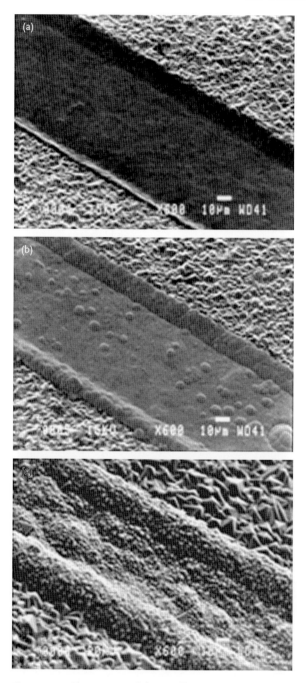

Figure 12.33 Three stages of the metallization process. (a) Activation of a diamond film surface by KrF excimer laser etching; (b) Metallization of a laser-etched trench with electroless copper; (c) Encapsulation of a copper line into diamond film. Reproduced from [26], with permission from Elsevier.

below the surface. It has been found [23, 79] that post-implantation pulsed laser treatment at low fluence ($E < E_g$) can lead to surface layer annealing, characterized by an increase in sample transmission, but this did not exceed 30% of the initial value and remained considerably lower than the transmission of the original nonimplanted diamond. For a diamond surface layer where $E > E_g$, complete graphitization is easily induced, triggered by laser irradiation of the buried defect layer.

12.5.3
Surface Structuring

Several experiments, performed with different lasers, have demonstrated a broad spectrum of irradiation regimes that allow the fabrication of micro- and nanostructures on diamond material surfaces.

The most widespread regime is based on vaporization of the graphitized surface layer; pulsed (periodic-pulsed) lasers are most appropriate for this purpose, and allow the production of both deep and shallow structures of various shapes. All diamond materials can be structured this way, and many examples of such a technique are known, including microholes, microgroves, and complex shape surface structures (e.g., triage teeth-like structures, as shown in Figure 12.34) [17, 23, 27, 28, 33, 41, 42, 49, 80–85]. Both, projection photolithography and laser writing irradiation schemes were employed, and cutting-edge sharpening of tungsten carbide–cobalt inserts covered by CVD diamond were also developed [42]. In addition, surface "pockets" with a controllable depth of only a few nanometers could be fabricated via the nanoablation of SCD (Figure 12.35) [20].

A purely chemical regime of diamond surface processing has also been realized [42]. In this case, a CVD diamond film (10 μm thick, highly absorptive, mechanical grade) was etched in air using a conventional CW Ar laser with 1 W power (focal

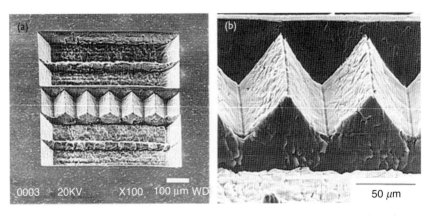

Figure 12.34 Triangular teeth-like and some other structures produced on the surface of a polished CVD diamond film by direct writing technique with ArF laser. (a) General view; (b) At higher magnification. Reproduced from [42], copyright NIST.

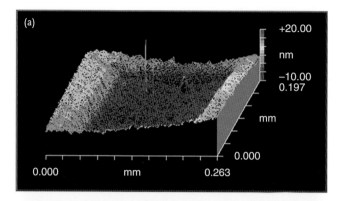

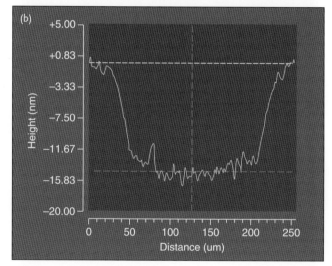

Figure 12.35 White light interferometric image (a) and profile (b) of the shallow crater formed as a result of KrF laser-induced nanoablation of IIa type diamond (irradiation in air by 3×10^5 laser pulses with $E = 15 \, \text{J cm}^{-2}$) [20]. For a color version of this figure, please see the color plate at the end of this book. Reproduced from [98], with permission.

spot diameter 2 μm). Because the thermal conductivity of DLC or nanocrystalline diamond films is much lower, a less-intensive laser radiation could be used for the chemical etching of these materials.

A third possibility would be to combine vaporization (physical) and chemical ablation regimes, and this was achieved using a pulse periodic Nd:YAG laser that combined a high average power (due to a high pulse repetition rate; $f = 10 \, \text{kHz}$) and a high pulse intensity (pulse duration 10 ns) [86]. The spot (diameter 30 μm) could be rapidly scanned along the sample surface, such that a variety of operations could be carried out, including cutting diamond plates, drilling holes, and shaping

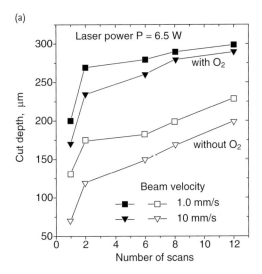

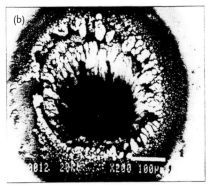 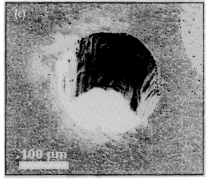

Figure 12.36 Groove depth produced by 6.5 W Nd:YAG laser power versus number of scans without (open points) and with O_2 injection (closed points). (a) The beam scan velocity was 1 mm s^{-1} (squares) and 10 mm s^{-1} (triangles); (b, c) Through-holes of 200 μm diameter drilled in 0.5 mm-thick diamond plate (b) without oxygen injection, and (c) with assistant O_2 stream. The laser power was 14 W. Note the improved shape of the hole and clean surface due to oxygen jet action. Reproduced from [86], copyright SPIE.

diamond plate edges for thermal management devices. This approach was substantially improved by the application of an oxygen gas jet; indeed, the injection of an O_2 stream into the interaction zone resulted in a large increase in etching rates for groove production on a CVD diamond surface (Figure 12.36a). The effects of the oxygen jet can be explained in two ways: (i) the jet supplies extra oxygen molecules to the chemical reaction zone and thus facilitates the etching rate; or (ii) the gas stream removes the reaction products and carbon microparticles ejected from the irradiation spot, and thus decreases their shielding action. The oxidation of carbon particles prevents their redeposition outside the laser spot. Improvements in processing quality as a result of particle oxidation are shown in Figure 12.36b and c.

12.5.4
Diamond Optics

The fact that diamond is transparent from UV to microwave spectral regions, combined with its record room temperature thermal conductivity and other unique properties, has led to diamond becoming a very attractive material as an optical element for high-power systems. At present, polished CVD diamond plates are considered to be the best windows for the 1 MW power submillimeter wave generators (gyrotrons) that are used for plasma heating in thermonuclear reactors. Moreover, lasers can also be efficiently utilized in the fabrication of such windows via the planarization and smoothening of CVD polycrystalline diamond disks.

Another important area of CVD diamond plate usage is in transmission optics for high-power CO_2-lasers [87]. The advantages of diamond optics over the best competitive material used in such optics (ZnSe) are listed in Table 12.5. Although there is one very important feature (optical absorption at 10.6 μm) where ZnSe has proven to be clearly superior, this difference is minimized by the ultra-high thermal conductivity and optical stability of diamond. It should also be noted that, if optical coatings are used that themselves have a substantial absorption, then the superiority of ZnSe in terms of its optical properties will be negated. Indeed, this is the main reason that the current use of diamond windows in high-power CO_2-lasers is rapidly expanding.

Surface and bulk laser microstructuring of diamond allows for the production of so-called diffractive optics elements (DOEs). One common use of a DOE is as a transparent plate on which a surface phase relief is produced; an example of this is the fully transmissive Fresnel lens (see Figure 12.37).

Table 12.5 Properties of CVD diamond and ZnSe.

Material	Diamond	ZnSe
Absorption coefficient at 10.6 μm (cm^{-1})	0.1–0.03	0.001–0.0001
Thermal conductivity [W (cm · K)$^{-1}$]	20–22	0.19
Thermal expansion coefficient (10^{-6}/K)	1.3	7.6
dn/dT (10^{-6}/K)	10	75
Refractive index at 10.6 μm	2.38	2.40
Microhardness (kg mm^{-2})	8300	137
Fracture strength (MPa)	500–1000	55

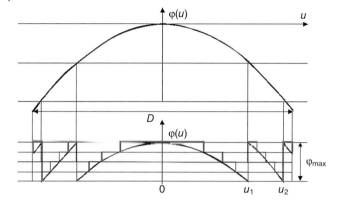

Figure 12.37 The profiles of a conventional plane-spherical lens and diffractive optical focusing element. Reproduced from [40], copyright RIKEN.

For comparison, a spherical-plane lens profile with the same diameter D and focal distance F, but with a much larger thickness is also shown. The geometric height of the DOE lens profile can be calculated as:

$$H_{max} = H(u = 0) = H(u_m) = \frac{\lambda}{n-1} \qquad (12.10)$$

where n is the material refractive index, u_m is the radius of the mth Fresnel zone ($m = 1, 2, 3, \ldots$), and for diamond and CO_2 laser the wavelength $\lambda = 10.6\,\mu m$. From Equation (12.10) it can be calculated that the thickness of the lens structure must be only 7.68 μm. This is why, for a diamond plate with a thickness exceeding 100 μm, the φ_{max} value is practically negligible and the DOE lens is much thinner than a spherical lens of equal focal length. Even for submillimeter waves, DOE elements can be laser-fabricated on thick (≥1 mm) diamond plates. The diameters of the feature structures of a DOE Fresnel lens are determined by

$$2U_m = (0.5\lambda F_m)^{1/2} \qquad (12.11)$$

In the instance where $\lambda = 10.6\,\mu m$, $F = 10\,cm$ and the circle number $m = 5$ [from Equation (12.11)], it follows that the lens diameter should be $2u_m = 3\,mm$. It can be seen that, for DOE fabrication, it is necessary to use a laser with a wavelength which is well absorbed in diamond and, correspondingly, can ablate thin layers of material by a single pulse. Alternatively, the laser wavelength should be much shorter than λ for which the DOE element is designed. This is why DOEs for CO_2 lasers or gyrotrons can be successfully produced using UV KrF lasers. Another problem is how to ensure that the laser-produced surface profile is close to the calculated version. The desired performance can be achieved by a multilayered step-type approximation of the DOE profile (in Figure 12.37, the number of layers is four).

The technology of laser fabrication of diamond DOEs was developed by several groups [40, 88–93]. By using a projection scheme, a square $50 \times 50\,\mu m$ irradiation

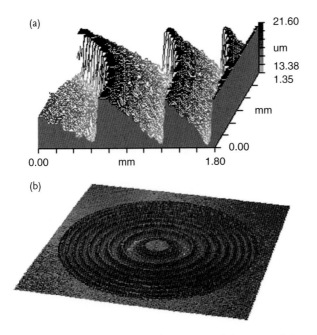

Figure 12.38 Magnified images of (a) a part and (b) a general view of the laser-produced diamond Fresnel lens (diameter 16.8 mm, surface relief depth ≈7.5 μm). Reproduced from [40], copyright RIKEN.

spot was first formed; a polished diamond sample, placed on a computer-driven X-Y stage, was then translated controllably (minimum step of 1 μm) so that a selected surface region with given coordinates could be irradiated with a certain number of laser shots in order to achieve a surface profile close to the calculated version. As the initial diamond plate absorption was small, surface graphitization and further ablation could be induced using either a very high laser intensity or by applying many pulses. In order to overcome this problem, a thin (≈5 nm) gold film was predeposited and, after patterning, the laser-graphitized layer was removed by surface oxidation in air at elevated temperatures (500–600 °C). Part of the microstructure and a general view of an eight-level diamond Fresnel lens produced by laser ablation is shown in Figure 12.38. The best samples produced demonstrated a high optical efficiency (>90%).

The diamond ablation technique is not only used to produce spherical or cylindrical DOE lenses; rather, it also allows the fabrication of elements that can transform the beam with a given aperture and cross-section intensity distribution into sought elements, at a given distance from the DOE. This type of DOE may also be referred to as a "beam shaper." As an example, the designed relief and results of tests with beam shapers that transform the CO_2-laser Gaussian beam into a contour at a distance of 100 mm are shown in Figure 12.39a and b, respectively. Based on a summary of the laser fabrication of diamond DOE (Table 12.6), it

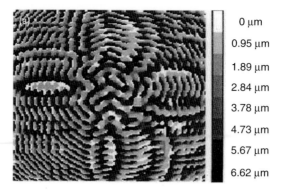

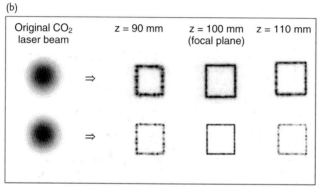

Figure 12.39 Diffractive optics element (DOE) diamond beam shaper transforming a parallel Gaussian beam into the contour at a distance of 100 mm. (a) The designed relief; (b) Experimentally measured (upper) and computer-simulated (bottom) intensity distributions in the original beam and at various distances from the DOE element. Reproduced from [40], copyright RIKEN.

Table 12.6 Parameters of CVD diamond DOEs.

	Lenses		Beam shaper into	
	Cylindrical	Spherical	Rectangle	Contour
Focal length (mm)	25	127	100	100
Quantization levels number	4	16	8	
Aperture	4 × 4 mm	Ø 16.8 mm	6.64 × 6.64 mm	
Experimental diffractive efficiency (%)	78	91	78	62
Theoretical estimation of diffractive efficiency (%)	80	95	98	75

is clear how well-matched are the results of the theoretical considerations and experiments. The difference is explained mostly by the need to resolve certain technological problems that lead to nonideal surface profiles that are produced by laser ablation. For example, the ablation rate depends on the diamond crystal orientation and its variation across the pixel structure, whereas the roughness of the pixel surface is increased for deeper profile levels. Surface roughness at the irradiation spot after laser ablation was increased from about $R_a \approx 20\text{--}30\,\text{nm}$ for the first level to $R_a \approx 200\,\text{nm}$ for the eighth level (corresponding to a depth of about $7.5\,\mu\text{m}$ on the processed diamond surface). The initial surface roughness of the mechanically polished diamond plate was 1–2 nm. Optical losses due to surface scattering depend heavily on R_a, and for $R_a \approx 100\,\text{nm}$ estimations yield a λ-value of $10.6\,\mu\text{m}$ and losses of $\approx 1\%$. Notably, the samples were always carefully oxidized after laser treatment in order to remove surface graphitized layer.

Among the key factors that determine whether an optical element can be used in high-power laser systems is the maximum level of radiation intensity that the element can withstand without destruction; this is termed the "optical damage threshold." In order to determine such data, tests were performed [40] using a CW multimode CO_2 laser with a maximum output power of 2 kW, with the laser beam focused on the CVD diamond diffractive optical element. When the peripheral flat surfaces of the diamond plate were fixed between water-cooled copper disks, the CVD diamond DOE was found to have remained unchanged (i.e., not optically damaged) at a CO_2 laser beam intensity of $\leq 50\,\text{kW}\,\text{cm}^{-2}$. This value proved superior to that of other IR-transparent optical elements.

It is also well known that the diffraction pattern of a grating with a period Λ less than λ/n (where n is the material's refractive index) contains only a zero-order maximum. The subwavelength relief is equivalent to a continuous surface layer of the material with a lower effective index of refraction $n_{eff} < n$. Such a structure will produce an antireflective effect, making it possible to reduce the reflectivity of a diamond sample without using easily damaged optical coatings. This technique has been investigated [94–97] via the laser production of two types of surface subwavelength relief: (i) a 1-D structure consisting of parallel microgrooves (Figure 12.40a, b); and (ii) a 2-D array of microcraters (Figure 12.40c). By means of such reliefs it was possible to increase the sample transmittance by $\approx 10\%$ for $\Lambda \approx 4\,\mu\text{m}$. Optimization of the processing regimes and a reduction of the structure period should also allow a further minimization of the diamond sample reflectivity.

As shown previously (in Section 12.4.3), the optimization of processing conditions allows for the formation of straight graphitic wires with diameters as low as $\leq 1\,\mu\text{m}$ and complex 3-D filamentary structures in the diamond bulk. This creates opportunities for the design of diamond-based photonic structures operating in the IR and THz spectral regions. A regular array of straight graphitic wires with the period of $\Lambda = 2\,\mu\text{m}$, that could be considered as a simplest example of such structures, is shown in Figure 12.41. Although, currently, investigations into the optimization and testing of diamond-based photonic crystals [78] and metamaterials are in the early stages, the prospects appear quite attractive. As an

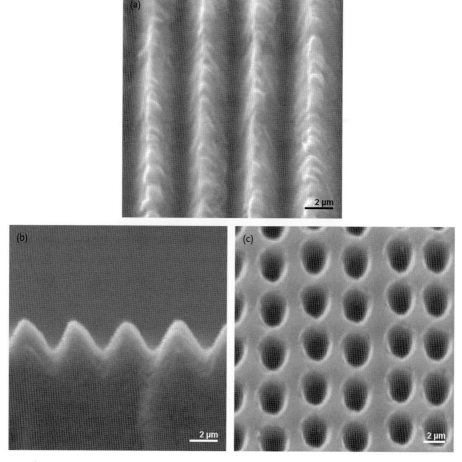

Figure 12.40 Diamond surface antireflective microstructures. Parallel grooves at (a) normal view and (b) cross-section; (c) Two-dimensional array of microoptics. Reproduced from [94], with permission from Springer.

additional opportunity, it should be noted that the modified material can be removed selectively by either oxidation or wet chemical etching (see Figure 12.42), if the structure produced touches the sample surface. Hence, the laser technique will allow the production of various 2-D and 3-D structures inside diamond samples, in the form of a combination of highly insulating and transparent diamond, conductive graphitic material, and hollow parts. Moreover, this can be achieved at relatively high rates (graphitization wave velocity up to ≈100 nm per pulse), taking into account that the pulse repetition rate of modern technological femto- and picosecond lasers can exceed 1 MHz.

12.5 Diamond Laser Processing Techniques | 437

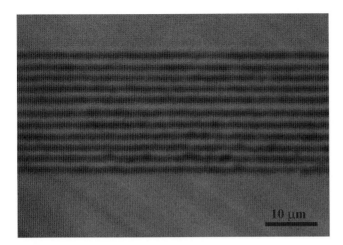

Figure 12.41 Regular array of straight graphitic wires with period of $\Lambda = 2\,\mu m$. Reproduced from [66], with permission from Elsevier.

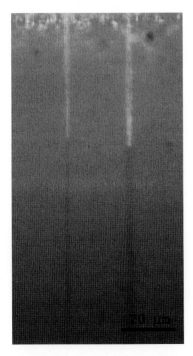

Figure 12.42 Empty microchannels in diamond fabricated by etching of the laser-graphitized material. Reproduced from [98], with permission.

12.6
Conclusions

The laser is an effective tool for the processing of various diamond materials, including SCD, polycrystalline diamond, NCD, and DLC. Moreover, the correct choice of laser parameters – such as temporal mode, pulse duration and repetition rate, intensity, and wavelength – permits the production of 1-D, 2-D, and 3-D micro- and nanostructures, both on the surface and in the bulk of diamond samples. These structures consist of the initial diamond material, a laser-produced graphitic phase, and hollow volumes. Other diamond-processing operations, such as cutting, drilling, polishing and shaping have been also demonstrated. In addition, laser radiation can be successfully applied to both the large-area or selective-area deposition of diamond materials, including combinations with nonlaser deposition techniques.

Acknowledgments

The authors are grateful to their colleagues at the General Physics Institute, notably S. Pimenov, V. Ralchenko, M. Komlenok, V. Pashinin, V. Tokarev, I. Vlasov, and to many others for important and determinative contributions to various parts of the described research activities, as well as valuable assistance in the preparation of the manuscript. They would also like to mention many years of fruitful collaboration with the teams from the Institute of Radioelectronics, Moscow, Russia, the Institute of Image Processing, Samara, Russia, the Institute for Laser Technologies, Stuttgart, Germany, the Institute of Applied Physics, Bern, Switzerland, and the Mediterranean University, Marseilles, France.

References

1 Epperson, I.P., Dyer, R.W., and Grzywa, I.C. (1966) *Laser Focus*, **19**, 26–33.
2 Burns, R.C., Hansen, J.O., Spits, R.A., Sibanda, M., Welborn, C.M., and Welch, D.L. (1999) Growth of high purity large synthetic diamond crystal. *Diamond Relat. Mater.*, **8**, 1433–1437.
3 Balmer, R., Brandon, J., Clewes, S., Dhillon, H., Dodson, J., Friel, I., Inglis, P., Madgwick, T., Mardham, M., Mollart, T., Perkins, N., Scarsbrook, G., Twitchen, D., Whitehead, A., Wilham, J., and Woollark, S. (2009) Chemical vapour deposition of synthetic diamond materials technology and applications. *J. Phys. Condens. Matter*, **21**, 364221–364244.
4 Prelas, M., Popovici, G., and Bigelow, L. (eds) (1997) *Handbook of Industrial Diamonds and Diamond Films*, Marcel Dekker, New York.
5 Spitsyn, B.V., Bouilov, L.L., and Derjaguin, B.V. (1981) Vapor growth of diamond on diamond and other surfaces. *J. Cryst. Growth*, **52** (Part 1), 219–226.
6 Konov, V.I., Smolin, A.A., Ralchenko, V.G., Pimenov, S.M., Obraztsova, E.D., Loubnin, E.N., Metev, S.M., and Sepold, G. (1995) D.c. arc plasma deposition of smooth nanocrystalline diamond films. *Diamond Relat. Mater.*, **4**, 1073–1078.
7 Gruen, D.M. (1999) Nanocrystalline diamond films. *Annu. Rev. Mater. Sci.*, **29**, 211–259.

8 Vlasov, I.I., Lebedev, O.I., Ralchenko, V.G., Goovaertz, E., Bertoni, G., Van Tendelo, G., and Konov, V.I. (2007) Hybrid diamond-graphite nanowires produced by microwave plasma chemical vapor deposition. *Adv. Mater. Weinheim*, **19**, 4058–4062.

9 Lifshitz, Y. (1999) Diamond-like carbon – present status. *Diamond Relat. Mater.*, **8**, 1659–1676.

10 Berman, R., Hudson, P.R.W., and Martinez, M.J. (1975) Nitrogen in diamond: evidence from thermal conductivity. *J. Phys. C Solid State Phys.*, **8**, L430.

11 Sukhadolau, A.V., Ivakin, E.V., Ralchenko, V.G., Khomich, A.V., Vlasov, A.V., and Popovich, A.F. (2005) Thermal conductivity of CVD diamond at elevated temperatures. *Diamond Relat. Mater.*, **14**, 589–593.

12 Liu, W.L., Shamsa, M., Calizo, I., Balandin, A.A., Ralchenko, V., Popovich, A., and Saveliev, A. (2006) Thermal conduction in nanocrystalline diamond films: effects of the grain boundary scattering and nitrogen doping. *Appl. Phys. Lett.*, **89**, 171915–171917.

13 Liu, W.L., Shamsa, M., Calizo, I., Balandin, A.A., Casiraghi, C., Milne, W.I., and Ferrari, A.C. (2006) Thermal conductivity of diamond-like carbon films. *Appl. Phys. Lett.*, **89**, 161921–1619123.

14 Chou, L.H. and Wang, H.W. (1993) On the microstructural, optical and thermal properties of hydrogenated amorphous carbon films prepared by plasma enhanced chemical vapour deposition. *J. Appl. Phys.*, **74**, 4673–4680.

15 Chen, Z.Y. and Zhao, J.P. (2000) Optical constants of tetrahedral amorphous carbon films in the infrared region and at a wavelength of 633 nm. *J. Appl. Phys.*, **87**, 4268–4273.

16 Ralchenko, V., Pimenov, S., Konov, V., Khomich, A., Saveliev, A., Popovich, A., Vlasov, I., Zavedeev, E., Bozhko, A., Loubnin, E., and Khmelnitskii, R. (2007) Nitrogenated nanocrystalline diamond films: thermal and optical properties. *Diamond Relat. Mater.*, **16**, 2067–2073.

17 Rothschild, M., Arnone, C., and Ehrich, D.J. (1986) Excimer-laser etching of diamond and hard carbon films by direct writing and optical projection. *J. Vac. Sci. Technol. B*, **4**, 310–314.

18 Geis, M.W., Rothschild, M., Kunz, R.R., Aggarwal, R.L., Wall, K.F., Parker, C.D., Mclntosh, K.A., Efremow, N.N., Zayhowski, J.J., Ehrlich, D.J., and Butler, J.E. (1989) Electrical, crystallographic, and optical properties of ArF laser modified diamond surfaces. *Appl. Phys. Lett.*, **55**, 2295–2297.

19 Ageev, V.P., Bouilov, L.L., Konov, V.I., Kuzmichov, A.V., Pimenov, S.M., Prokhorov, A.M., Ralchenko, V.G., Spitsyn, B.V., and Chapliev, N.I. (1988) Interaction of laser light with diamond films. *Sov. Phys. Dokl.*, **33**, 840–842.

20 Strekalov, V.N., Konov, V.I., Kononenko, V.V., and Pimenov, S.M. (2003) Early stages of laser graphitization of diamond. *Appl. Phys.*, **76**, 603–607.

21 Windholz, R. and Molian, P.A. (1998) Nanosecond pulsed excimer laser machining of chemically vapour-deposited diamond and graphite; Part II Analysis and modeling. *J. Mater. Sci.*, **33**, 523–528.

22 Strekalov, V.N. (2005) Graphitization of diamond stimulated by electron-hole recombination. *Appl. Phys. A*, **80**, 1061–1066.

23 Kononenko, V.V., Kononenko, T.V., Pimenov, S.M., Konov, V.I., Fischer, P., Romano, V., Weber, H.P., Khomich, A.V., Khmelnitskiy, R.A., and Strekalov, V.N. (2003) Laser induced structure transformations of diamonds. *Proc. SPIE*, **5121**, 259–270.

24 Desmukh, S., Hitchcock, L., Rothe, E., and Reck, G. (1994) Graphitization of synthetic diamond by 193 nm laser light: comparison of ^{12}C-enriched diamonds with those of natural isotopic composition. *Diamond Relat. Mater.*, **3**, 195–197.

25 Inyushkin, V.V., Ralchenko, V.G., Taldenkov, A.N., Artyukhov, A.A., Kravets, Ya.M., Gnidoi, I.P., Ustinov, A.L., Bolshakov, A.P., Popovich, A.F., Savelyev, A.V., Khomich, A.Vy., Panchenko, V.A., and Konov, V.I. (2007) Considerable increase in thermal conductivity of a polycrystalline CVD

diamond upon isotope enrichment. *Bull. Lebedev Phys. Inst.*, **34**, 329–333.

26 Pimenov, S.M., Shafeev, G.A., Konov, V.I., and Loubnin, E.N. (1996) Electroless metallization of diamond films. *Diamond Relat. Mater.*, **5**, 1042–1047.

27 Johnston, C., Chalker, P.R., Buckley-Golder, I.M., Marsden, P.J., and Williams, S.W. (1993) Diamond device delineation via excimer laser patterning. *Diamond Relat. Mater.*, **2**, 829–834.

28 Chan, S., Raybould, F., Arthur, G., Goodall, F., and Jackman, R. (1996) Laser projection patterning for the formation of thin film diamond microstructures. *Diamond Relat. Mater.*, **5**, 317–320.

29 Cappelli, E., Mattei, G., Orlando, S., Pinzari, F., and Ascarelli, P. (1999) Pulsed laser surface modifications of diamond thin films. *Diamond Relat. Mater.*, **8**, 257–261.

30 Chan, S., Whiffield, M., Jackman, R., Arthur, G., Goodall, F., and Lawes, R. (2003) The effect of excimer laser etching on thin film diamond. *Semicond. Sci. Technol.*, **18**, S47–S58.

31 Lade, R.J., Claeyssens, F., Rosser, K.N., and Ashfold, M.N. (1999) 193 nm laser ablation of CVD diamond and graphite in vacuum: plume analysis and film properties. *Appl. Phys. A*, **69**, 5935–5939.

32 Huang, B., Jou, S., Wu, Y., Chen, K., and Chen, L. (2010) Effect of XeF laser treatment on structure of nanocrystalline diamond films. *Diamond Relat. Mater.*, **19**, 445–448.

33 Smedly, J., Jaye, C., Bohon, J., Rao, T., and Fisher, D.A. (2009) Laser patterning of diamond. Part II. Surface nondiamond carbon formation and its removal. *J. Appl. Phys.*, **105**, 123108-1-5.

34 Nistor, L.C., Van Landuyt, J., Ralchenko, V.G., Kononenko, T.V., Obraztsova, E.D., and Strelnitsky, V.E. (1994) Direct observation of laser-induced crystallization of a-C:H films. *Appl. Phys. A*, **58**, 137–144.

35 Pimenov, S.M., Smolin, A.A., Ralchenko, V.G., Konov, V.I., Likhanski, S.V., Veselovski, I.A., Sokolina, G.A., Bantsekov, S.V., and Spitsyn, B.V. (1993) UV laser processing of diamond films: effects of irradiation conditions on the properties of laser-treated diamond film surfaces. *Diamond Relat. Mater.*, **2**, 291–297.

36 Pimenov, S.M., Konov, V.I., Kononenko, T.V., Garnov, S.V., Blatter, A., Maillat, M., Bogli, U., Romano, V., Weber, H.P., Hintermann, H.E., and Loubnin, E.N. (2000) Laser applications in tribology of hard materials and coatings. *New Diamond Front. Carbon Technol.*, **10**, 123–136.

37 Kononenko, V.V., Kononenko, T.V., Pimenov, S.M., Sinyavskii, M.N., Konov, V.I., and Dausinger, F. (2005) Effect of the pulse duration on graphitization of diamond during laser ablation. *Quantum Electron.*, **35**, 252–256.

38 Shirk, M.D., Molian, P.A., and Malshe, A.P. (1998) Ultrashort pulsed laser ablation of diamond. *J. Laser Appl.*, **10**, 64–70.

39 Komlenok, M.S., Pimenov, S.M., Kononenko, V.V., Konov, V.I., and Scheibe, H.-J. (2008) Laser microstructuring of the surface of ta-C films. *J. Nano Microsyst. Tech.*, **3**, 48–53.

40 Kononenko, V.V., Konov, V.I., Pimenov, S.M., Prokhorov, A.M., Bolshakov, A.P., Uglov, S.A., Pavelyev, V.S., Soifer, V.A., Ludge, B., and Duparre, M. (2002) Laser shaping of diamond for IR diffractive optical elements. *RIKEN Rev.*, **43**, 49–55.

41 Kononenko, T.V., Ralchenko, V.G., Vlasov, I.I., Garnov, S.V., and Konov, V.I. (1998) Ablation of CVD diamond with nanosecond laser pulses of UV-IR range. *Diamond Relat. Mater.*, **7**, 1623–1627.

42 Ralchenko, V.G., Pimenov, S.M., Kononenko, T.V., Korotoushenko, K.G., Smolin, A.A., Obraztsova, E.D., and Konov, V.I. (1995) Processing of CVD diamond with UV and green lasers. *Proceedings of the Third International Conference, NIST, Gaithersburg, MD, USA, NIST Special Publications*, vol. **885**, pp. 225–232.

43 Preuss, S. and Stuke, M. (1995) Subpicosecond ultraviolet laser ablation of diamond: nonlinear properties at 248 nm and time-resolved characterization of ablation dynamics. *Appl. Phys. Lett.*, **67**, 338–340.

44 Danyi, R. and Szoreny, T. (2004) KrF excimer laser processing of thick

diamond-like carbon films. *Appl. Phys. A*, **79**, 1373–1376.

45 Kononenko, T.V., Konov, V.I., Garnov, S.V., Danielius, R., Piskarskas, A., Tamoshauskas, G., and Dausinger, F. (1999) Comparative study of the ablation of materials by femtosecond and pico- or nanosecond laser pulses. *Quantum Electron.*, **29**, 724–728.

46 Lausinger, F., Lichtner, F., and Lubatschowski, H. (2004) Femtosecond technology for technical and medical applications, in *Topics in Applied Optics*, (eds F. Lausinger, F. Lichtner, and H. Lubatschowski) vol. 96, Springer, pp. 131–154.

47 Berger, P., Britling, D., Dausinger, F., Fohl, C., Hugel, H., Klimentov, S., Kononenko, T., and Konov, V. (2007) Vorrichtung zur bearbeitung eins werkstuckes mit einem laser strahl. Patent DE No.10203452, Bundesrepublik Deutschland.

48 Gloor, S., Pimenov, S.M., Obraztsova, E.D., Luthy, W., and Weber, H.P. (1998) Laser ablation of diamond films in various atmospheres. *Diamond Relat. Mater.*, **7**, 607–611.

49 Konov, V.I., Kononenko, T.V., Pimenov, S.M., Smolin, A.A., and Chapliev, N.I. (1991) Pulse-periodic laser etching of diamond-like carbon coatings. *Sov. J. Quantum Electron.*, **21**, 1112–1114.

50 Komlenok, M.S., Kononenko, V.V., Vlasov, I.I., Ralchenko, V.G., Arutyunyan, N.R., Obraztsova, E.D., and Konov, V.I. (2009) Laser "nano" ablation of ultrananocrystalline diamond films. *J. Nanoelectron. Optoelectron.*, **4**, 1–4.

51 Komlenok, M.S., Kononenko, V.V., Ralchenko, V.G., Pimenov, S.M., and Konov, V.I. (2011) Laser induced nanoablation of diamond materials. *Phys. Procedia*, **12**, 37–45.

52 Kononenko, V.V., Komlenok, M.S., Pimenov, S.M., and Konov, V.I. (2007) Photoinduced laser etching of a diamond surface. *Quantum Electron.*, **37**, 1043–1046.

53 Evans, T. and Qi, Z. (1979) The kinetics of the aggregation of nitrogen atoms in diamond, in *The Properties of Natural and Synthetic Diamond*, (ed. J.E. Field), Academic Press, London, Chapter 13.

54 Garnov, S.V., Klimentov, S.M., Pimenov, S.M., Konov, V.I., Kononenko, V.V., Tzarkova, O.G., Gloor, S., Luthy, W., and Weber, H.P. (1998) Ultrafast electronic processes in CVD diamond and GaAs: picosecond photoconductivity and high voltage switching. *Proc. SPIE*, **3287**, 67–77.

55 Klimentov, S.M., Garnov, S.V., Pimenov, S.M., Konov, V.I., Gloor, S., Luthy, W., and Weber, H. (2000) Transient photoconductivity spectroscopy of polycrystalline diamond films. *Quantum Electron.*, **30**, 459–461.

56 Garnov, S.V., Ritus, A.I., Klimentov, S.M., Pimenov, S.M., Konov, V.I., Gloor, S., Luthy, W., and Weber, H. (1999) Time-resolved microwave technique for ultrafast charge carrier recombination time measurements in diamonds and GaAs. *Appl. Phys. Lett.*, **74**, 1731–1733.

57 Guizard, S., D'Oliveira, P., Daguzan, P., Martin, P., Meynadier, P., and Petite, G. (1996) Time-resolved studies of carriers dynamics in wide band gap materials. *Nucl. Instrum. Methods Phys. Res. B*, **116** (1–4), 43–48.

58 Sarkisov, G.S. (1998) Refraction of an electromagnetic probe wave in a laser plasma. *Quantum Electron.*, **28**, 38–41.

59 Kononenko, V., Pashinin, V., Komlenok, M., and Konov, V. (2009) Laser-induced modification of bulk fused silica by femtosecond pulses. *Laser Phys.*, **19**, 1294–1299.

60 Yasumaru, N., Miyzaki, K., and Kiuchi, J. (2003) Femtosecond-laser-induced nanostructure formed on hard thin films of TiN and DLC. *Appl. Phys. A*, **76**, 983–985.

61 Shinoda, M., Gattass, R.R., and Mazur, E. (2009) Femtosecond laser-induced formation of nanometer-width grooves on synthetic single-crystal diamond surfaces. *J. Appl. Phys.*, **105**, 053102–053105.

62 Shimotsuma, Y., Sakakura, M., Kanehira, S., Qiu, J., Kazansky, P.G., Miura, K., Fujita, K., and Hirao, K. (2006) Three-dimensional nanostructuring of transparent materials by the femtosecond laser irradiation. *J. Laser Micro/Nanoeng.*, **1**, 181–184.

63 Kononenko, T.V., Meier, M., Komlenok, M.S., Pimenov, S.M., Romano, V.,

Pashinin, V.P., and Konov, V.I. (2008) Microstructuring of diamond bulk by IR femtosecond laser pulses. *Appl. Phys. A*, **90**, 645–651.

64 Kononenko, T.V., Komlenok, M.S., Pashinin, V.P., Pimenov, S.M., Konov, V.I., Neff, M., Romano, V., and Luthy, W. (2009) Femtosecond laser microstructuring in the bulk of diamond. *Diamond Relat. Mater.*, **18**, 196–199.

65 Neff, M., Kononenko, T.V., Pimenov, S.M., Romano, V., Luthy, W., and Konov, V.I. (2009) Femtosecond laser writing of buried graphitic structures in bulk diamond. *Appl. Phys. A*, **97**, 543–547.

66 Kononenko, T.V., Konov, V.I., Pimenov, S.M., Rossukanyi, N.M., Rukovishnikov, A.I., and Romano, V. (2011) Three-dimensional laser writing in diamond bulk. *Diamond Relat. Mater.*, **20**, 264–268.

67 Suzuki, H., Yoshikawa, M., and Tokura, H. (1994) Excimer laser processing of diamond films. Proceedings of Fourth International Conference on New Diamond Science and Technology Kobe, Japan, July 18–22, 1994 (eds S. Saito, N. Fujimory, O. Fukunaga, M. Kamo, K. Kobashi, and M. Yoshikawa), pp. 501–504.

68 Pimenov, S.M., Kononenko, V.V., Ralchenko, V.G., Konov, V.I., Gloor, S., Luthy, W., Weber, H.P., and Khomich, A.V. (1999) Laser polishing of diamond plates. *Appl. Phys. A*, **69**, 81–88.

69 Gloor, S., Luthy, W., Weber, H.P., Pimenov, S.M., Ralchenko, V.G., Konov, V.I., and Khomich, A.V. (1999) UV laser polishing of thick diamond films for IR windows. *Appl. Surf. Sci.*, **138–139**, 135–139.

70 Cremades, A. and Piqueras, J. (1996) Cathodoluminescence study of ArF excimer laser-induced planarization of large grain diamond films. *J. Appl. Phys.*, **79**, 8118–8120.

71 Tsai, H.Y., Ting, C.J., and Chou, C.P. (2007) Evaluation research of polishing methods for large area diamond films produced by chemical vapor deposition. *Diamond Relat. Mater.*, **16**, 253–261.

72 Tokarev, V.N., Wilson, J.I.B., Jubber, M.G., John, P., and Milne, D.K. (1995) Modelling of self-limiting laser ablation of rough surface: application to the polishing of diamond films. *Diamond Relat. Mater.*, **4**, 169–176.

73 Konov, V.I. and Tokarev, V.N. (1991) Light-induced polishing of an evaporating surface. *Sov. J. Quantum Electron.*, **21**, 442–444.

74 Kononenko, V.V., Konov, V.I., Pimenov, S.M., Volkov, P.V., Goryunov, V.A., Ivanov, V.V., Novikov, M.A., Markelov, V.A., Tertyshnik, A.D., and Ustavshikov, S.S. (2005) Control of laser machining of polycrystalline diamond plates by the method of low-coherence optical interferometry. *Quantum Electron.*, **35**, 622–626.

75 Pimenov, S.M., Smolin, A.A., Obraztsova, E.D., Konov, V.I., Bogli, U., Blatter, A., Loubnin, E.N., Maillat, M., Leijala, A., Burger, J., and Hintermann, H.E. (1996) Tribological properties of smooth diamond films. *Appl. Surf. Sci.*, **92**, 106–114.

76 Erdemir, A., Halter, M., Fenske, G.R., Krauss, A., Gruen, D.M., Pimenov, S.M., and Konov, V.I. (1997) Durability and tribological performance of smooth diamond films produced by Ar-C_{60} microwave plasmas and by laser polishing. *Surf. Coat. Technol.*, **94–95**, 537–542.

77 Armeyev, V.Yu., Chapliev, N.I., Loubnin, E.N., Mikhailov, V.I., Ralchenko, V.G., and Strelnitsky, V.E. (1991) Ar^+ laser annealing and etching of hydrogenated amorphous carbon films. *Surf. Coat. Technol.*, **47**, 279–286.

78 Shimizu, M., Shimotsuma, Y., Sakakura, M., Yunsa, T., Homma, H., Minowa, Y., Tanaka, K., Miura, K., and Hirao, K. (2009) Periodic metallo-dielectric structure in diamond. *Opt. Express*, **17**, 46–54.

79 Kononenko, V.V., Pimenov, S.M., Kononenko, T.V., Konov, V.I., Fisher, P., Romano, V., Weber, H.P., Khomich, A.V., and Khmelnitsky, R.A. (2003) Laser induced phase transitions in ion-implanted diamond. *Diamond Relat. Mater.*, **12**, 277–282.

80 Park, J.K., Ayres, V.M., Asmussen, J., and Mukherjer, K. (2000) Precision micromachining of CVD diamond films. *Diamond Relat. Mater.*, **9**, 1154–1158.

81 Smedley, J., Bohon, J., Wu, Q., and Rao, T. (2009) Laser patterning of diamond. Part I. Characterization of surface morphology. *J. Appl. Phys.*, **105**, 123107–123111.

82 Christensen, C.P. (1994) Micromachining of diamond substrates with waveguide excimer lasers. *Proc. SPIE*, **2062**, 14–21.

83 Windholz, R. and Molian, P.A. (1997) Nanosecond pulsed excimer laser machining of chemical vapor deposited diamond and highly oriented pyrolitic graphite. *J. Mater. Sci.*, **32**, 4295–4301.

84 Rothschild, M. and Ehrlich, D.J. (1987) Attainment of 0.13-μm lines and spaces by excimer-laser projection lithography in diamond-like carbon-resist. *J. Vac. Sci. Technol.*, **B5**, 389–390.

85 Ralchenko, V., Migulin, V., Kononenko, T., Kononenko, V., Konov, V., and Negodaev, M. (1999) Treatment of diamond films with lasers. *Adv. Sci. Technol.*, **21**, 109–118.

86 Migulin, V.V., Ralchenko, V.G., and Baik, Y.J. (1997) Oxygen assisted laser and drilling of CVD diamond. *Proc. SPIE*, **3484**, 175–179.

87 Douglas-Hamilton, D.H. and Hoag, E.D. (1974) Diamond as a high-power-laser window. *J. Opt. Soc. Am.*, **64**, 36–38.

88 Kononenko, V.V., Konov, V.I., Pimenov, S.M., Prokhorov, A.M., Pavelyev, V.S., and Soifer, V.A. (1999) Diamond diffraction optics for CO_2 lasers. *Quantum Electron.*, **29**, 9–10.

89 Kononenko, V.V., Konov, V.I., Pimenov, S.M., Prokhorov, A.M., Pavelyev, V.S., Soifer, V.A., Ludge, B., and Duparre, M. (2002) Laser shaping of diamond for IR diffractive optical elements. *Proc. SPIE*, **4426**, 128–134.

90 Konov, V.I., Kononenko, V.V., Pimenov, S.M., Prokhorov, A.M., Pavelyev, V.S., Soifer, V.A., Muys, P., and Vandamme, E. (2000) Excimer laser micromachining for fabrication of diamond diffractive optical elements. *Proc. SPIE*, **3933**, 322–331.

91 Konov, V., Kononenko, V., Pimenov, S., Prokhorov, A., Pavelyev, V., and Soifer, V. (1999) CVD diamond transmissive diffractive optics for CO_2 lasers. *Proc. SPIE*, **3822**, 2–5.

92 Kononenko, V.V., Konov, V.I., Pimenov, S.M., Prokhorov, A.M., Pavelyev, V.S., and Soifer, V.A. (2000) CVD diamond transmissive diffractive optics for CO_2 lasers. *New Diamond Front. Carbon Technol.*, **10**, 97–107.

93 Pavelyev, V.S., Soifer, V.A., Konov, V.I., Kononenko, V.V., and Volkov, A.V. (2009) Diffractive microoptics for technological IR-lasers, in *High power and femtosecond lasers: properties, materials and applications* (ed. P.-H. Barret and M. Palmer), Nova Science, Chapter 4, pp. 125–158.

94 Kononenko, T.V., Kononenko, V.V., Konov, V.I., Pimenov, S.M., Garnov, S.V., Tishenko, A.V., Prokhorov, A.M., and Khomich, A.V. (1998) Formation of antireflective surface structures on diamond films. *Appl. Phys. A*, **67**, 1–4.

95 Kononenko, T.V., Kononenko, V.V., Konov, V.I., Pimenov, S.M., Garnov, S.V., Tishenko, A.V., Prokhorov, A.M., and Khomich, A.V. (1999) Formation of antireflection microstructures on the surface of diamond plates by the laser patterning method. *Quantum Electron.*, **29**, 158–162.

96 Ralchenko, V.G., Khomich, A.V., Baranov, A.V., Vlasov, I.I., and Konov, V.I. (1999) Fabrication of CVD diamond optics with antireflective surface structures. *Phys. Status Solidi*, **174**, 171–176.

97 Gloor, S., Romano, V., Luthy, W., Weber, H.P., Kononenko, V.V., Pimenov, S.M., Konov, V.I., and Khomich, A.V. (2000) Antireflection structures written by excimer laser on CVD diamond. *Appl. Phys. A*, **70**, 547–550.

98 Konov, V.I. (2012) Laser in micro and nanoprocessing of diamond materials, Laser Photonics Rev. 6, No. 6, 739–766/ DOI 10.1002/Ipor.201100030.

13
Fluorescent Nanodiamonds and Their Prospects in Bioimaging
Nitin Mohan and Huan-Cheng Chang

13.1
Introduction

Diamond is an excellent material that is, by and large, desirable for jewelry and industrial applications owing to its unique optical, thermal, and electrical properties. During recent years, the synthesis of its nanosized particles bearing atomic fluorescent defects has enabled diamond to be used as a functional fluorescent material for probing biomolecules. Subsequently, diamond has become of interest to biomedical research, on the basis of its unmatched, stable, fluorescent characteristics. The visualization of biological specimens and biomolecular dynamics, aided by optical fluorescence microscopy, has helped to provide an understanding of biological systems in detail at the molecular, cellular, and organism levels. In order to complement such imaging technology, fluorescent biomarkers have been developed and used widely to provide a better image contrast, and also specifically to probe the molecules in question. Recent advances in nanotechnology have also enabled the evolution of novel fluorescent nanomaterials [1] and super-resolution fluorescence microscopy [2], since which time these cutting edge technologies have been applied in life sciences research to unravel intricate spatial-temporal biological phenomena. In particular, the development of semiconductor quantum dots (QDs) [3, 4], up-conversion fluorescent phosphors [5] and carbon materials [6] as bioimaging probes for *in vivo* applications is steadily improving. Such fluorescent nanoprobes are expected not only to be biologically inert and/or nontoxic but also to have excellent fluorescence properties.

The organic dye molecules that are used conventionally for bioimaging applications suffer from severe photobleaching effects, which limits their application to short-term imaging [7]. Although semiconductor QDs provide stable fluorescence with a tunable narrow emission spectrum [8], they have been found to be toxic such that their direct use in biology requires suitable surface treatments. Intermittency or "blinking" of fluorescence emission means that the material is undesirable for certain applications, such as the long-term dynamic tracking of a targeted biomolecule [9].

Optical Engineering of Diamond, First Edition. Edited by Richard P. Mildren and James R. Rabeau.
© 2013 Wiley-VCH Verlag GmbH & Co. KGaA. Published 2013 by Wiley-VCH Verlag GmbH & Co. KGaA.

Fluorescent proteins, on the other hand, have revolutionized the concept of "seeing is believing" in biotechnology [10]. Indeed, as biomolecules in their own right, fluorescent proteins can be easily tagged to proteins and other biomolecules to monitor protein expression, biochemical signaling pathways, and imaging [10]. One unfortunate point, however, is that fluorescent proteins also suffer from photobleaching, while their fluorescence may also occasionally be extremely sensitive to environmental conditions. In short, among the many fluorescent probes currently available, each will have its own advantages and disadvantages that must be critically selected before any particular application.

Fluorescent nanodiamond (FND), with its unique and impressive features, is a new addition to the list of ideal nanoprobes [11]. In fact, FNDs have arguably superlative photophysical characteristics in comparison to contemporary fluorescent nanoprobes. Moreover, the atom-like defect centers within the crystal enrich the material with its distinctive fluorescence properties. Although a number of defect centers may exist within the diamond, one in particular has attracted special attention, namely the negatively charged nitrogen-vacancy (NV) center, referred to as NV^-. This center provides FND with a far-red fluorescence emission that, in particular, enables better tissue penetration depths and a wide separation from the endogenous protein fluorescence that typically is associated with biological specimens. In addition, the fluorescence emission from FND does not photobleach or blink. The fluorescence quantum yield of the NV^- center is approximately 1 [12–15], while the surface of FND is easy to funtionalize and can be grafted to biomolecules for specific and sensitive labeling and targeting applications. An additional advantage is that FND is nontoxic [16–19].

In this chapter, attention will be focused on the production and optical characterization of FND, and on advancements in its application to biomedical research. Initially, the various defect centers, their properties, and the challenges involved in the mass production of FND will be discussed, followed by details of the material's impressive optical features. Issues relating to high-yield syntheses and the incorporation of maximum defect centers into smaller diamonds are also described. Finally, an account is provided of the imaging and biological applications of FND.

13.2
Color Centers

Pure diamonds possess the widest band gap of any natural material, and therefore are optically transparent to ultraviolet (UV) and/or infrared (IR) radiation. Defects or imperfections within the crystal lattice bring about interesting light–matter interactions, such as scattering and fluorescence, that contribute to the excellent photoluminescence of the material. To date, more than a 100 luminescent defects have been identified in diamond, and their charge, spin, and photophysical properties have been well characterized [20]. Among these defects, nitrogen-related defects are the most important as nitrogen is the most abundant and prominent

Table 13.1 Summary of the different vacancy-related defect centers within the diamond lattice.

Diamond	Defect center	ZPL (nm)	λ (nm)	τ (ns)	φ	References
Type IIa	V^0 (GR1)	741.2	898	20.55	0.014	[21–23]
Type Ib	NV^-	637.6	685	11.60	0.99	[21, 23]
Type Ib	NV^0	575.4	600	–	–	[21, 23]
Type IaA	NVN (H3)	503.5	531	16	0.95	[21–23]
Type IaB	N_3 + V (N3)	415.4	445	41	0.29	[21–23]

ZPL, Zero phonon line; λ, emission maxima; τ, emission lifetime; φ, quantum efficiency.

impurity in diamond. Diamonds are classified as different types, based on the concentration of nitrogen present: diamond with a high nitrogen concentration (200–4000 ppm) is classified as Type I, while that with a very low nitrogen content (<100 ppm) is referred to as Type II [21]. Type I diamond is further classified as Type Ia or Ib, based on whether the nitrogen impurities exist as atomically dispersed defects or as aggregates (Table 13.1).

Synthetic Type Ib diamonds produced by high-pressure, high-temperature (HPHT) processes contain nitrogen atoms at a concentration of 300 ppm or less by mass, dispersed throughout the crystal as single point defects [24]. These nitrogen impurities are responsible for the yellow color of the diamond material, as it absorbs light at wavelengths below 500 nm. High-energy (1–2 MeV) radiation damage with protons, electrons, neutrons or gamma rays knocks down the carbon atoms to interstitial positions and creates vacancies in the crystal [25, 26]. When annealed at high temperatures after radiation damage, however, these vacancies are mobilized and become trapped close to the nitrogen impurities to form NV centers. Depending on the type of diamond used for irradiation damage, different dominant color centers are formed in the crystal [22, 23, 27]. The various color centers produced by the different types of diamond, together with their optical absorption and emission wavelengths, are summarized in Table 13.1.

Among these defect centers, the NV^- color center exhibits distinguishable features that make it suitable for biological applications. The center has a broad absorption spectrum peaking at ~560 nm, while a dipole-allowed decay transition occurs from a triplet excited state to a triplet ground state, giving rise to a strong far-red emission with a zero phonon line (ZPL) at 637 nm. The quantum efficiency of the center is approximately 100%. The fluorescence is highly stable, without bleaching and blinking, even under continuous laser excitation in the range of 5 MW cm^{-2} [13]. The ZPL at 637 nm is accompanied by phonon side bands peaking at 560 nm, with an absorption cross-section of ~4 × 10^{-17} cm^{-2} per molecule [12]. The far-red fluorescence from the NV^- center also separates it from the endogenous autofluorescence of biological samples, which generally falls in the green–yellow region of the visible spectrum. Being the dominant end-product of an irradiation/annealing treatment, the NV^- center is therefore stable even at high

Table 13.2 Excitation maximum (λ_{ex}), emission maxima (λ_{em}), molar extinction coefficient (ε), and quantum yield (ø) of NV⁻ in diamond, and some red and far-red fluorescent proteins.

Fluorophore (acronym)	λ_{ex} (nm)	λ_{em} (nm)	ε (M⁻¹ cm⁻¹)	ø	In vivo structure	References
DsRed-monomer	556	586	35 000	0.10	Monomer	[10]
HcRed1	588	618	20 000	0.015	Dimer	[10]
AQ143	595	655	90 000	0.04	Tetramer	[10]
NV⁻	560	690	8 000	0.99	–	[10, 13]

temperatures around 1000 °C. Taken together, these unique photophysical features provide the NV⁻ center with great appeal as a fluorescent probe for biological applications.

The current trend in fluorescent probe technology is to expand the role of fluorophores such as NV⁻, which emit light in the far-red and near-IR regions, for biomedical use [10]. Hence, although other stable color centers are present in diamond (such as H3 defects which emit green fluorescence), attention is focused here only on the NV⁻ center. The optical properties of NV⁻ and two representative red and far-red fluorescent proteins are compared in Table 13.2. A simple calculation, based on molar extinction coefficient and fluorescence quantum yield, indicates that the brightness of a single NV⁻ center is twice as high as that of a DsRed-monomer red fluorescent protein.

13.3
Red Fluorescent Nanodiamonds

13.3.1
Mass Production

In this process, HPHT type 1b nanosized diamond powders (100 nm or 35 nm) containing high concentrations of nitrogen impurities (300 ppm) are subjected to radiation damage at room temperature by high-energy particles to create vacancies. When annealed at high temperatures (800 °C), the vacancies become mobile and are trapped at the nitrogen impurities to form NV⁻ fluorescent centers. Conventionally, an electron beam (2 MeV) from a van der Graaff accelerator is used to create the vacancies [11, 14]. Exposure to high-energy proton irradiation (3 MeV) [11, 13] may also create a comparatively large number of color centers, and thereby produce brighter FNDs.

Prior to the ion irradiation treatment, the commercially available 100 nm diamond powder was purified in a concentrated H_2SO_4–HNO_3 solution (3 : 1, v/v) at 100 °C in a microwave reactor for 3 h. Following this acid-washing procedure to remove the graphitic impurities from the diamond surface, the sample was rinsed extensively with deionized water and suspended in 95% ethanol. For proton irradiation, an aliquot of the diamond suspension containing 5 mg of the powders was deposited on a silicon wafer and air-dried to form a thin film with an area of 0.5 cm² and a thickness of 50 µm. The diamonds were then radiation-damaged with a 3 MeV proton beam generated from an NEC tandem accelerator at a dose of 1×10^{16} ions per cm², with a typical ion current in the range of 0.1 µA (or 6×10^{11} ions per second).

According to SRIM Monte Carlo simulations [28], a 3 MeV proton has a penetration depth of 50 µm, and would produce 12 vacancies at an estimated displacement energy of 35 eV for the carbon atom in the diamond. Wee *et al.* proved experimentally that the penetration depth of 3 MeV H⁺ was 50 µm by irradiating a diamond substrate, followed by annealing [12]. Under these conditions, the concentration of vacancies produced per 100 nm particle was 1×10^4, and gave rise to an NV defect concentration of 1×10^7 centers per µm³ after annealing, assuming minimal vacancy annihilation, divacancy formation, and recombination of vacancies and carbon interstitials [29]. Thus, homogeneously bright fluorescent nanodiamonds could be produced by matching the penetration depth of protons with the diamond film thickness.

These high-energy irradiation techniques demand sophisticated instrumentation, which in turn leads to high production costs. However, a fluorescent particle must be readily available in large quantities if its application for biological purposes is to be increased. Consequently, in order to circumvent this issue, the research group led by Chang developed a cost-effective technique for the mass production of FNDs, using medium-energy helium ion irradiation [15].

By taking correct safety measures, a device with a high-fluence, medium-energy (40 keV) He⁺ beam (Figure 13.1) was used to produce FNDs in large quantities in the laboratory [15, 30]. In this way, a diamond film of 0.2 µm thickness was prepared by spreading and drying a diamond suspension over a long copper tape of

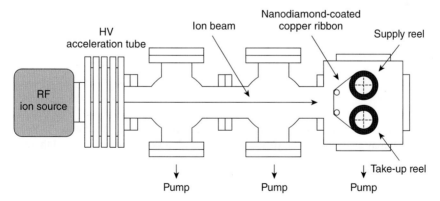

Figure 13.1 Schematic of the experimental set-up for mass production of FND using a medium-energy ion beam. Adapted from Ref. [30].

20 m length and 35 mm width, that had been exposed to He$^+$ ions generated by the discharge of pure helium from a radiofrequency (RF) positive ion source accelerated to 40 keV by using a high-voltage acceleration tube. The typical current of the unfocused He$^+$ beam without mass discrimination was 16 μA, and the flux at the sample target was 7×10^{12} He$^+$ cm^{-2} s^{-1}. Notably, one He$^+$ can create about 40 vacancies as it penetrates to about 200 nm thickness of the diamond film; this is in contrast to the 0.1 and 13 vacancies generated by 2 MeV e$^-$ and 3 MeV H$^+$, respectively [28, 30]. This high-level vacancy creation occurs because He$^+$ is heavier compared to e$^-$ and H$^+$. Such high damaging efficacy reduces the ion dosage required for irradiation, which has been estimated as 1×10^{13} He$^+$ cm^{-2} for Type Ib diamonds.

The main practical advantage of this mass production technique was that about 1 g of FND could be produced by irradiation with an He$^+$ beam in only 1 h, whereas more than 100 h irradiation with a 3 MeV H$^+$ beam would be required to prepare the same amount of FND. Whereas, irradiation with the high-energy H$^+$ beam would produce a greater number of NV centers compared to the He$^+$ beam, the fluorescence intensity would be 30% higher (see Figure 13.2).

Following ion irradiation and annealing at 800 °C for 2 h, the sample was oxidized at 490 °C for 2 h to remove the graphitic structures that had formed on the surface of the diamond powder. In addition, the FND was acid-washed for further purification and then rinsed extensively with deionized water. After undergoing

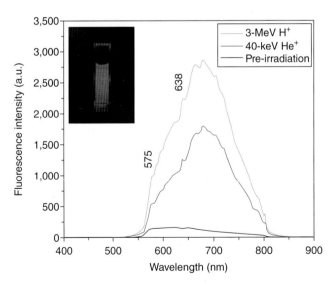

Figure 13.2 Fluorescence spectra of 35 nm FNDs prepared with either 40 keV He$^+$ or 3 meV H$^+$ irradiation. Inset: Fluorescence image of a cuvette containing FND suspension (1 mg ml^{-1}) excited by a 532 nm laser. Adapted from Ref. [15]. For a color version of this figure, please see the color plate at the end of this book.

these cleaning procedures, the FNDs were shown to be surface-functionalized with a variety of oxygen-containing groups, including the carboxyl moiety [31, 32]. They were also free from fluorescence quenching due to an absence of graphitic surface structures.

13.3.2
Fluorescence Spectra

The far-red fluorescence emission of FND, peaking at about 700 nm, is well separated from the ubiquitous endogenous fluorescence of biological specimens. By using appropriate optical filters, it is quite easy to isolate the FND signal from the autofluorescence background, and consequently a better signal-to-noise ratio (SNR), achieved by virtue of the far-red emission, proved to be an added advantage for FND as a biological nanoprobe in single-molecule studies and bioimaging applications. Fluorescence spectrum analysis at the single-particle level has been conducted by Fu *et al.* [13]. A confocal scanning image of 35 nm FNDs dispersed on a bare glass substrate with a density of about five particles per $10 \times 10\,\mu m^2$ is shown in Figure 13.3a. The full width at half-maximum (FWHM) of each peak in the image was 400–600 nm, which coincided with the diffraction limit of the optical microscope, and suggested that the peaks were derived from single isolated FND particles. This suggestion was confirmed using scanning electron microscopy, whereby each 35 nm FND exhibited a distinct spectrum in the wavelength region of 550–800 nm that was characteristic of single particle detection (see Figure 13.3b). While the observed spectra were highly heterogeneous, two sharp ZPLs from NV^0 at 576 nm and NV^- at 638 nm were identified for all of the 35 nm particles interrogated. Both ZPLs were accompanied by broad phonon sidebands that were red-shifted by ~50 nm. A close comparison revealed that the particles mainly differed in their ratio of NV^- to NV^0 content [13]. An additional benefit was that the good SNR guaranteed the imaging of single FNDs in complex biological environments.

13.3.3
Photostability

One outstanding photophysical feature that has led to FND being superior to contemporary fluorophores is its extreme photostability. The photostability of individual 35 nm FND particles was monitored over a long period of time (Figure 13.4a), and excellent stable fluorescence was observed from individual particles, without intermittency in emission. Under excitation with 532 nm light at a power density of $8 \times 10^3\,W\,cm^{-2}$, the fluorescence intensity of the particles investigated remained essentially unchanged over a period of 300 s, while no sign of photoblinking was detected within the time resolution of 1 ms for both 35 nm and 140 nm FNDs (Figure 13.4b). In contrast, single dye molecules such as Alexa Fluor 546 covalently linked to double-stranded DNA blinked randomly and photobleached rapidly within 12 s (Figure 13.4a). The exceptional photostability of the

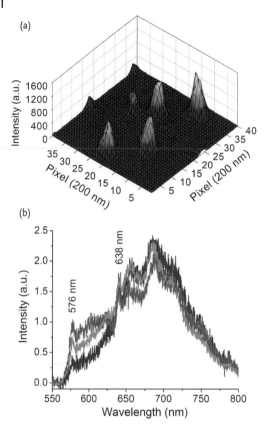

Figure 13.3 Fluorescence images and spectra of single 35 nm FNDs. (a) Confocal scanning image of 35 nm FNDs dispersed on a glass coverslip. The FWHM of each peak is 2–3 pixels, corresponding to a physical distance of 400–600 nm; (b) Fluorescence spectra of three different 35 nm FND particles. Adapted from Ref. [13]. For a color version of this figure, please see the color plate at the end of this book.

FNDs was also reflected in the power dependency measurement for the observed fluorescence intensity, which scaled linearly with the laser power density, from 1×10^2 to $1 \times 10^5 \, \text{W cm}^{-2}$, and gradually reached saturation at $1 \times 10^6 \, \text{W cm}^{-2}$ (Figure 13.4c).

It would be instructive to compare the fluorescence brightness of FNDs, dye molecules, and QDs under the same experimental conditions. Hence, from a measurement of 30 molecules or particles for each sample, it was estimated that the average fluorescence intensity of the 35 nm FNDs was higher than that of dye molecules such as Alexa Fluor 546 by roughly one order of magnitude, but was

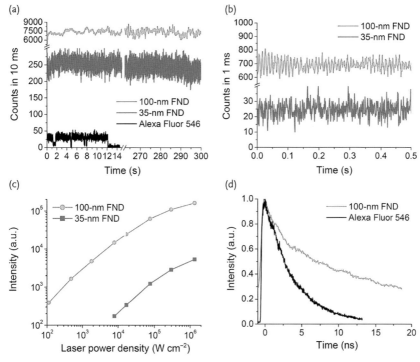

Figure 13.4 Spectroscopic characterization of 35 nm and 140 nm FNDs. (a) Typical time traces of fluorescence from a single 140 nm (green), 35 nm (red) and single Alexa Fluor 546 dye molecule attached to a single double-stranded DNA molecule (blue); (b) Time traces at a resolution of 1 ms, showing no photoblinking behavior; (c) Plot of fluorescence intensity as a function of laser power density; (d) Fluorescence lifetime measurements of 140 nm FNDs (green) and Alexa Fluor 546 dye molecule (blue). Fitting the time traces with two exponential decays for the FND revealed a fast component of 1.7 ns (4%) and a slow component of 17 ns (96%). The latter was fourfold longer than that (4 ns) of Alexa Fluor 546. Adapted from Ref. [13]. For a color version of this figure, please see the color plate at the end of this book.

about a factor of two lower than that of QDs such as CdSe/ZnS that emitted light in the same wavelength region.

13.3.4
Fluorescence Lifetimes

Fluorescence lifetime is another key parameter that could be utilized for improving the image contrast of fluorescent probes within a biological environment. When the fluorescence lifetime of 140-nm FNDs that had been spin-coated onto a glass substrate was measured (Figure 13.4d), the major component of decay had a lifetime of 17 ns, which was close to that (11.6 ns) measured for bulk diamond, and substantially longer than that of dye molecules (4 ns for Alexa Fluor 546) [13]. This distinct difference in fluorescence lifetime allowed for the easy isolation of the FND's emission from other backgrounds, such as cellular endogenous fluorescence, by using various time-gating techniques such as fluorescence lifetime imaging microscopy [33].

13.4
Smaller FNDs

Having overcome issues regarding the mass production of FNDs, the next important parameter to resolve for their enhanced biological application was the "size" of the FND particles. Clearly, a decrease in size would not only improve the mobility of the fluorescent nanoprobe inside a cell, but would also avoid altering the properties of the targeted molecules and promote translocation of the nanoprobe into the cell nuclei [34]. As the target proteins and other biomolecules are several nanometers in size, FND particles in the range of 10 nm would be suitable for biolabeling and for fluorescence resonance energy transfer (FRET) applications, without perturbing their function [35]. Whereas, the size-controlled, scaled-up, synthesis of smaller and brighter FNDs remains a major challenge, the production and isolation of smaller FND particles utilizing differential centrifugation [36], surface oxidative etching [36], and ball milling [37] have shown much promise.

Differential centrifugation is used basically to separate particles according to their size and density; larger and more dense particles sediment at lower centrifugal forces, while smaller particles remain in the supernatant and require more centrifugal force and longer times for their precipitation. This common procedure has been applied to the isolation of smaller FND particles from a homogenized FND solution containing particles with a broad size distribution [36, 38]. With serial variations of the speed and duration of centrifugation, larger particles with sizes of ~50 nm, ~35 nm, and ~25 nm were precipitated sequentially, while the remaining supernatants contained smaller particles. The dynamic light scattering (DLS) and transmission electron microscopy (TEM) data of particles in the final supernatant after differential centrifugation are shown in Figure 13.5a, b. The smallest particles obtained in this way were approximately 11 nm in size.

The signature photophysical properties of no photoblinking and no photobleaching were conserved in FNDs, irrespective of the particle size (Figure 13.6a). In addition, photon correlation spectroscopy [39, 40] was used to determine directly that each of the 11 nm FND particles contained three NV^- (Figure 13.6b).

The fluorescence intensity of single 11 nm FND particles, in comparison to that of single far-red fluorescent protein, Ds-Red, was determined using fluorescence correlation spectroscopy (FCS). A comparison of the normalized autocorrelation functions of the intensity fluctuations of DsRed-monomer (a 4 nm-sized monomeric red fluorescent protein) and the 11 nm FND excited under the same experi-

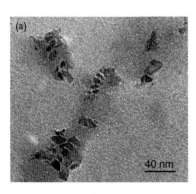 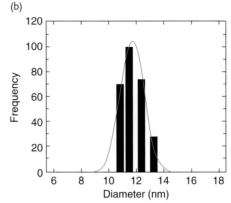

Figure 13.5 Particle size analysis of FNDs by TEM and DLS. (a) TEM image of 11 nm FNDs, showing sharp edges and irregular shape of the particles; (b) Size measurement with DLS for particles suspended in water (0.1 µg ml^{-1}). Adapted from Ref. [36].

mental conditions, is shown in Figure 13.6c. Typically, FCS is used to measure the size and fluorescence intensity of the particles, simultaneously [41, 42]. From a theoretical fitting of the autocorrelation function by a three-dimensional (3-D) diffusion model [42], the average particle size of the FND was given as ~8 nm, while the intensity distribution histogram (Figure 13.6d) of the particles showed an average fluorescence intensity of ~50 counts per millisecond, which is about eightfold higher than that for the DsRed-monomer. This eightfold difference in fluorescence intensity was in agreement with an estimation based on the molar extinction coefficient and quantum yield of each fluorophore (see Table 13.2), along with the measured number of NV$^-$ centers in the 11 nm FND particle. A reduction in the size of larger particles could also be achieved by the oxidative etching of sub-20 nm particles in air at 500 °C for 1–2 h. In this case, oxidation in air would cause the graphitic surface on the diamond to be burned away, thus reducing the size of the particles. The particles were first isolated by centrifugation, and then oxidized in air; iteration of these two processes led to the production of FNDs of a smaller size and with a sevenfold enhancement in fluorescence intensity [36].

Although isolation by centrifugation and oxidative etching processes may produce FNDs of smaller size with a satisfactory fluorescence efficiency, the final yield of production would be very low, at <1% of the initial quantity. Boudou et al. have developed a ball-milling technique for the high-yield fabrication of FNDs [37], where the yield is about 15% (w/w) of the microdiamond mass converted into 10 nm FNDs. Prior to milling, crystalline diamond powders of about 150 µm in size were radiation-damaged with an electron beam (10 MeV) and annealed at 800 °C in vacuum for 2 h. In the subsequent steps of milling, micrometer-sized diamond crystals were converted into 10 nm particles. The crystallinity and size of the particles thus produced were analyzed using high-resolution (HR) TEM

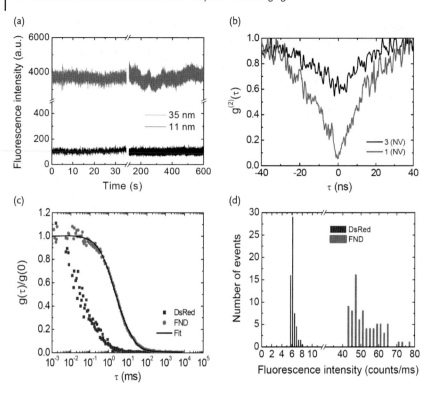

Figure 13.6 Optical characteristics of 11 nm FNDs produced by differential centrifugation. (a) Photostability tests of particles spin-coated on a bare glass substrate; (b) Typical photon correlation functions of the fluorescence intensities of two single FND particles containing one and three NV⁻ centers, respectively; (c) Spectroscopic characterization of DsRed-monomers and 11 nm FNDs by FCS; (d) Normalized autocorrelation functions of the intensity fluctuations and histograms of the fluorescence intensities of the particles. Adapted from Ref. [36]. For a color version of this figure, please see the color plate at the end of this book.

(Figure 13.7a, b), and the particle size was found to be ≤10 nm. In addition, it was found that the diamond structure was not altered by the milling procedures.

The FNDs thus obtained were characterized using atomic force microscopy (AFM) combined with optical microscopy (Figure 13.7c, d), such that the size and fluorescence of the nanoparticles were analyzed simultaneously. Based on the AFM image the particles were about 10 nm (Figure 13.7e), and from the fluorescence image they were photostable. The fluorescence spectrum (Figure 13.7f) also confirmed the characteristics of the NV⁻ center, but the ZPL line at 637 nm was absent due to the strain in small-sized crystals. Moreover, photon correlation spectroscopy demonstrated the presence of about 12 NV⁻ centers for a particular

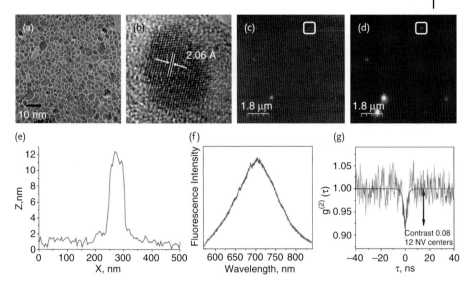

Figure 13.7 Particle size and optical characterization of 10 nm FNDs produced by milling. (a) TEM and (b) HRTEM images of the smallest particles obtained; (c) AFM and (d) fluorescence image of the same particles; (e) AFM dimensions of a 10 nm bright FND and (f) photoluminescence spectrum of the same particle; (g) Second-order autocorrelation function for the same particle. Adapted from Ref. [37]. For a color version of this figure, please see the color plate at the end of this book.

particle analyzed (Figure 13.7g). Thus, a high-yield production of smaller FNDs, with as many as 12 NV$^-$ centers, was realized via the ball-milling technique.

In order to improve the formation and number of NV centers, it is necessary to consider the vacancy annihilation effects along with the nitrogen concentrations in diamond nanocrystals. Smith et al. [43] performed a detailed study on the probability of vacancies forming NV centers in 5 nm nanodiamonds with a nitrogen concentration of 300 ppm. For this, the fluorescence from 55 nm HPHT diamonds and 5 nm single-digit nanodiamonds was compared after similar conditions of proton irradiation and annealing treatments. Estimates based on the fluorescence measurements suggested that the number of NV$^-$ centers per crystal was about 65 000-fold less for 5 nm particles than for 55 nm particles, despite the volume of a 5 nm particle being only 1300-fold less than its 55 nm counterpart. This clearly suggests the presence of a mechanism that hinders the formation of NV$^-$ centers within the smaller particles. Subsequent electron paramagnetic resonance (EPR) measurements revealed that the nitrogen concentrations in both types of particle were similar and that, in the smaller crystal, there was a greater possibility of the vacancy being lost during the annealing process. The probability that a vacancy forms the NV$^-$ center before reaching the surface scales up as the square of the crystal radius. The probability (based on Monte Carlo simulations) of one vacancy forming an NV center in various particle sizes of diamond with a nitrogen concentration of 300 ppm, predicts that the formation of NV^- centers in 55 nm FNDs would be 25-fold higher than in 5 nm crystals. It should be noted that for particles with a high nitrogen concentration or a large radius, the probability of a vacancy forming an NV^- center approaches one, and the dependence on radius is slowed.

So far in this chapter, the different approaches for the production of smaller FNDs have been described, and issues relating to the formation and stability of NV^- centers in smaller crystals have been discussed. The results of these studies have confirmed that there is still room to improve the fluorescence efficiency of smaller FNDs, by increasing the number of NV^- centers within the crystal. Indeed, i all of the nitrogen within a 10 nm crystal were to be converted into NV^- centers by preserving the nitrogen and vacancy from being exhumed during small particle production, then a single 10 nm FND should be as bright as a single QD.

13.5
Biological Applications

Having discussed the physical and chemical properties of FND that make it exceptional for bioimaging applications, attention is now focused on important real-time demonstrations of employing the advantages of FNDs as a bioprobe, both *in vitro* and *in vivo*.

13.5.1
Fluorescence Resonance Energy Transfer

Fluorescence resonance energy transfer (FRET) is an advanced spectroscopic technique that is used to study intermolecular interactions and conformational dynamics at the single-particle level. The development of 10 nm FNDs has enabled the use of FNDs for FRET applications. The possibilities of energy transfer between FND and a near-IR dye, IRDye-800CW, was demonstrated by H.C. Chang *et al.* [36]. In this case, a 19 nm FND particle was employed as a FRET donor and the near-IR dye as the FRET acceptor. These two materials formed an ideal FRET pair, as there was a considerable overlap between the emission spectrum of FND and the absorption spectrum of the IR dye (Figure 13.8a). A Förster distance of 5.6 nm was computed from the spectral overlap of the fluorophores. FRET signals could be detected by a direct coupling of the *N*-hydroxysuccinimide (NHS) ester group on the dye molecules to the amine-terminated FND surface. The FRET efficiency increased steadily with the ratio of IRDye to FND and levels of a 2:1 ratio and the highest FRET efficiency was recorded as ~7% (Figure 13.8b, c). This low FRET efficiency was attributed to the large size of the FND particles, and was in agreement with a theoretical simulation [44] which predicted a 6% efficiency for a 19 nm FND particle conjugated with two IRdye molecules. This was reasonable considering the low density of NV^- in the FND; in comparison, this efficiency was only

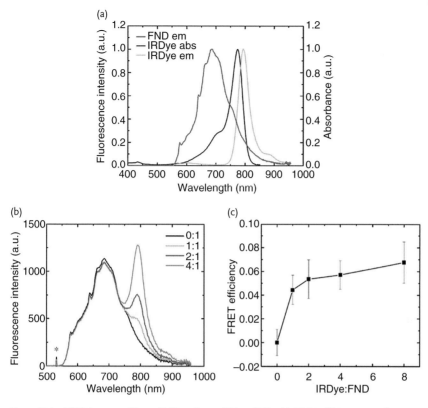

Figure 13.8 FRET between 19 nm FND and IRDye-800CW. (a) Overlap of the emission spectrum of FND and the absorption spectrum of IRDye; (b) Variation of the fluorescence spectra with the molar ratio of IRDye:FND; (c) FRET efficiency as a function of the IRDye:FND ratio. Adapted from Ref. [36]. For a color version of this figure, please see the color plate at the end of this book.

one order of magnitude lower than the FRET efficiency between QD donors and dye-labeled protein acceptors [45]. Thus, by employing further smaller particles with many NV⁻ centers to the surface of the crystal, there would always be a potential improvement in the FRET efficiency. Subsequently, this demonstration of FRET with FND-IRdye pair was extended to a real-time application by effectively conjugating the probes to the molecules under study.

13.5.2
Cellular Uptake and Fluorescence Imaging

The excellent fluorescence features of FNDs have led to their becoming promising candidates for *in vivo* imaging and diagnosis. To demonstrate this, 35 nm FNDs were incubated with HeLa cells (10^5 cells ml^{-1}) and cultured in DMEM supplemented with 10% fetal bovine serum. A well-sonicated aliquot of 0.2 ml of FND suspension was added to a chamber glass slide that contained the 24 h precultured HeLa cells; the slide was then incubated at 37 °C in 5% CO_2 for 5 h to facilitate FND uptake. The results of detailed studies performed by Vaijayanthimala et al. [19] and Faklaris et al. [14] revealed the uptake mechanism of FND by the cells to be clathrin-mediated endocytosis, and also verified the noncytotoxic properties of FNDs at the cellular level.

For HeLa cells, an intense auto-fluorescence was observed at 510–560 nm when they were exposed to blue light. Switching the excitation wavelength to 532 nm and collecting the emission at 650–720 nm, where the FND fluorescence resides, caused the background to be greatly reduced [13]. Thus, the extended red emission could be separated from the autofluorescence. The fluorescence images of a HeLa cell after incubation with 35 nm FNDs (see Figure 13.9a) confirmed that, while many nanoparticles formed aggregates in the cytoplasm, some isolated nanoparticles were also detected. Notably, particles were found *not* to enter the nucleus, as their size was greater than that of the nuclear membrane pores. The particles marked in the yellow boxes in Figure 13.9b were identified as single FNDs, as the spot sizes of their images were diffraction-limited (Figure 13.9c) and their fluorescence intensities were comparable to that of single FNDs that had been spin-coated onto the glass plate. These internalized FNDs proved to be photostable even after continuous excitation of the sample for 20 min at a laser power density of 100 W cm^{-2} (Figure 13.9d).

13.5.3
Two-Photon Excited Fluorescence Imaging

Two-photon excitation (TPE) microscopy is an advanced imaging technique that is employed for biomedical purposes, but which has an added advantage over one-photon excitation (OPE), namely that the autofluorescence and scattering signals from the specimen are reduced, so that a better contrast to the images is provided. The initial studies of NV centers using TPE were conducted by Wee et al. [12] in bulk diamonds and using a picosecond mode locked laser operating at 1064 nm. The two-photon absorption cross-section of NV^- at 1064 nm was determined as 0.45×10^{-50} $cm^4 s^{-1}$ per photon, based on the corresponding one-photon absorption cross-section of 3.1×10^{-17} cm^{-2} at 532 nm. In comparison to that of Rhodamine B, the TP absorption cross-section was about 30-fold smaller. Thus, FND particles of 40 nm containing more than 30 NV centers could be used for the TPE imaging of biological samples.

The first demonstrations of using FNDs for TPE imaging in cells were reported by Chang et al. [15], and their scanning confocal images obtained by one-photon excitation (532 nm) and two photon excitation (875 nm) for 140 nm FNDs in HeLa cells are compared in Figure 13.10. It is clear from these images that TPE was able to provide a better image contrast of FNDs by diminishing the cell autofluorescence. In addition, the excited volume was small due to the quadratic depend-

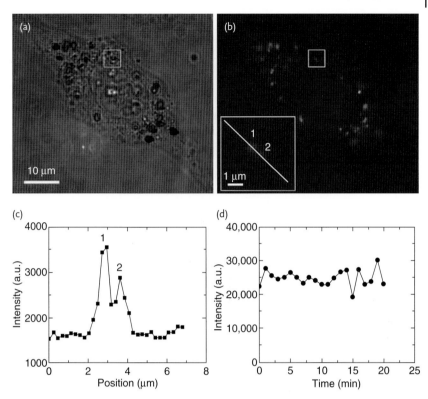

Figure 13.9 Observation of single 35 nm FNDs in HeLa cells. (a) Bright-field and epifluorescence merge images of a HeLa cells after uptake of 35 nm FNDs. Most of the particles are seen in the cytoplasm; (b) Epifluorescence image of a single HeLa cell after the FND uptake. Inset: An enlarged view of two fluorescence spots (denoted by 1 and 2) with diffraction-limited size (FWHM = 500 nm). The separation between the two particles is 1 μm; (c) Intensity profile of the fluorescence image along the line drawn in the inset of panel (b); (d) Integrated fluorescence intensity as a function of time for particle 1. The signal integration time was 0.1 s. Adapted from Ref. [13]. For a color version of this figure, please see the color plate at the end of this book.

ence of fluorescence intensity on the intensity of the incident light, which in turn led to a major reduction in the photo-mediated damage of the cells.

13.5.4
Fluorescence Lifetime Imaging

Fluorescence lifetime imaging, as the name indicates, is a time-resolved imaging technique where the difference in average fluorescence lifetime of different types

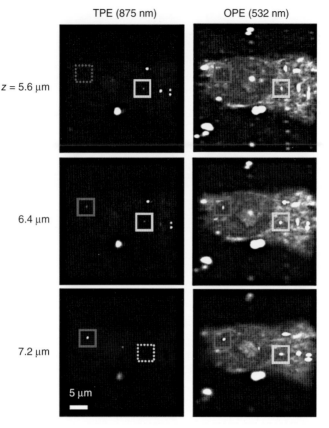

Figure 13.10 One-photon excited (OPE) and two-photon excited (TPE) fluorescence images of 140 nm FNDs in a fixed HeLa cell. Adapted from Ref. [15].

of fluorescent molecule is utilized to generate high-contrast images. Unlike FNDs, which possess relatively high fluorescence lifetimes of about 20 ns, fluorescent dye molecules or endogenous fluorescence from proteins within the specimen have very short fluorescence lifetimes (ca. 1–2 ns). Consequently, FND might be used for fluorescence lifetime imaging in biological samples.

Previously, Faklaris et al. [14] took advantage of this method to explore the possibilities of lifetime imaging of FND in HeLa cells. The time-resolved confocal images obtained of HeLa cells after FND uptake are shown in Figure 13.11. In this case, photons collected from the sample with a proper time delay reduces the background autofluorescence from the cells and provides an excellent SNR. The significant image contrast and high photostability of FND aids in identifying the precise location of particles inside the cells.

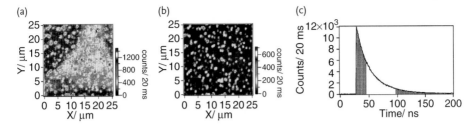

Figure 13.11 Time-resolved confocal raster scans of a fixed HeLa cell containing FND particles. (a) Raster scan obtained from the detection of all photons, displaying both FND and cell autofluorescence signals; (b) Time-gated raster scan constructed from photons detected at between 15 and 53 ns after the laser excitation pulses; (c) Fluorescence lifetime decay curve from the FNDs shown in panel (a). The gray region is the time-delay range, in which the photons are excluded to build the time-gated raster scan in panel (b). Adapted from Ref. [14]. For a color version of this figure, please see the color plate at the end of this book.

13.5.5
Super Resolution Imaging

With the advent of super resolution far-field fluorescence microscopy, it is possible to break the diffraction limit of conventional optical microscopy [2] and to obtain images of FND with a resolution of several tens of nanometers. For example, Hell, Eggeling and coworkers [46] have demonstrated a remarkably high resolution of approximately 6 nm for single NV$^-$ centers in bulk diamond by using stimulated emission depletion (STED) microscopy, which provided a sixfold improvement in resolution over confocal fluorescence microscopy for 35 nm FNDs dispersed on a cover glass [47]. The excellent stability and spectral features of FND make it ideal for STED imaging. When Tzeng et al. [48] explored the application of STED in imaging albumin-coated FND localization in HeLa cells, the covalent conjugation of bovine serum albumin (BSA) protein caused the FNDs to disperse better in phosphate-buffered saline and prevented the aggregation of FNDs when they were taken up by the cells. By using STED microscopy it can be shown clearly whether the particles have aggregated or have remained as single particles within the endosome of the cell.

In these experiments, a home-built stage scanning confocal and STED microscope was employed to record the images. For STED imaging, a continuous-wave tunable T-sapphire laser (3900S, Newport) running at 740 nm was used as the depletion STED beam for the FNDs. A 0-2π vortex-generating phase mask (VPP-1; RPC photonics) produced a helical phase retardation and a donut-shaped spot in the focal plane of the microscope objective, where the STED laser was overlapped with the excitation laser beam (532 nm). The undepleted signal was collected

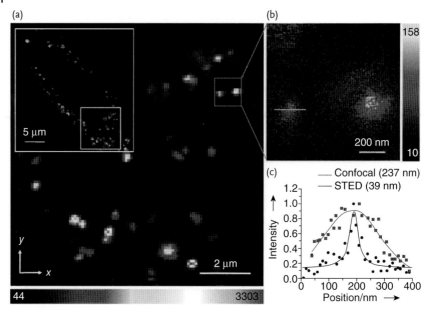

Figure 13.12 Confocal and STED imaging of HeLa cells labeled with BSA-conjugated FNDs by endocytosis. (a) Confocal image acquired by raster scanning of an FND-labeled cell. The fluorescence image of the entire cell is shown in the white box, and demonstrates fairly uniform cell labeling by BSA-conjugated FNDs; (b) STED image of single BSA-conjugated FND particles enclosed within the green box in panel (a); (c) Confocal and STED fluorescence intensity profiles of the particle indicated in panel (b) with a blue line. Solid curves are best fits to one-dimensional Gaussian (confocal) or Lorentzian (STED) functions. The corresponding full widths at half-maximum are given in parentheses. Adapted from Ref. [48]. For a color version of this figure, please see the color plate at the end of this book.

through a bandpass filter (FF01-685/40-25; Semrock) and detected by the avalanche photodiode. The confocal and STED images of HeLa cells labeled with BSA-conjugated FNDs by endocytosis are shown in Figure 13.12. Although a uniform cell labeling with BSA-conjugated FNDs was observed, also evident from the STED images of the endosome were about five FND particles that had not been observed in the confocal image.

13.5.6
Single Particle Tracking

In order to understand the vital functions of molecules and cellular organelles, it is first instructive to study their single-molecule behaviors, as this might help to unravel interesting dynamics and interactions that otherwise would go unnoticed

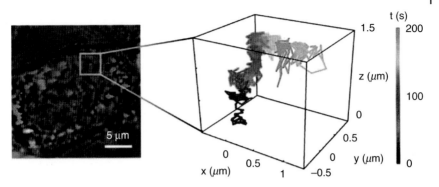

Figure 13.13 Three-dimensional tracking of a single 35 nm FND in a live cell. An overlay of bright-field and epifluorescence images (left) shows the existence of 35 nm FNDs in the cytoplasm of a live HeLa cell. The fluorescence intensity is sufficiently high and stable to allow tracking of a single FND in three dimensions over a time span of 200 s (right). Adapted from Ref. [15]. For a color version of this figure, please see the color plate at the end of this book.

in ensemble measurements. The excellent photostability and high brightness of FNDs makes it possible to conduct long-term, 3-D tracking of a single FND in a live cell [15]. This technique could be extended to reveal the details of cellular and intracellular activities, such as drug delivery and virus infection, in real time. An overlap of bright-field and fluorescence imaging of a live HeLa cell following the uptake of 35 nm FNDs at low concentration is shown in Figure 13.13. In this case, by using the home-built servo control system and operating the fluorescence microscope in widefield mode, it was possible to track a particular single particle over a time span of more than 200 s. From a mean square displacement analysis of the 3-D trajectories, a diffusion coefficient of $3.1 \times 10^{-3} \, \mu m^2 \, s^{-1}$ was determined for the internalized 35 nm particle. This value agreed well with reported diffusion coefficients for QDs confined within an endosome, and indicated that the particle was not moving freely within the cytoplasm.

13.5.7
In Vivo Imaging in *Caenorhabditis elegans*

Optical imaging represents an ideal noninvasive tool for visualizing biological processes, at both the molecular and cellular level, by probing specific biomolecules and their intermolecular interactions. *In vivo* imaging, in particular, enables the study of molecular dynamics in both real time and three dimensions, without any external perturbation of the whole systems. Unique photophysical features, combined with an ease of surface functionalization, have led to FNDs becoming potential candidates for long-term imaging *in vivo*. As an example, Mohan *et al.*

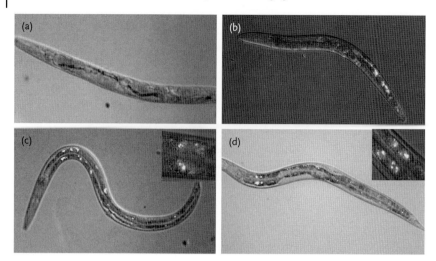

Figure 13.14 Epifluorescence/DIC-merged images of wild-type *C. elegans*. (a, b) Worms fed with bare FNDs for (a) 2 h and (b) 12 h; (c, d) Worms fed with (c) dextran-coated FNDs and (d) BSA-coated FNDs for 3 h. FNDs can be seen to be localized within in the intestinal cells. Adapted from Ref. [49]. For a color version of this figure, please see the color plate at the end of this book.

[49] reported an *in vivo* study of nano-biointeractions using an animal model, *Caenorhabditis elegans*, a free-living soil nematode with a simple and well-defined anatomy. The 1 mm long adult hermaphrodite worm consists of an invariable number of 959 cells, which are organized to form complex tissues including intestine, muscle, hypodermis, gonad, and nerve systems [50, 51]. The optically transparent body of *C. elegans* allows a precise observation of the biodistribution of FNDs within the organism.

The worm was grown at 20 °C in a nematode growth medium seeded with *Escherichia coli*. Prior to optical imaging, the worm was transferred onto an agar pad, anesthetized, and then sealed with a coverslip. Bare FNDs (~120 nm diameter) were incorporated into the worms by feeding in the absence of *E. coli*; this led to the appearance of an easily visible bright red fluorescence in the lumen of the digestive system, that extended from the pharynx to the anal region (Figure 13.14a). Interestingly, the particles were retained in the lumen even after 12 h of feeding (Figure 13.14b). Such retention may be caused, at least in part, by slow or no FND excretion, since it has been shown that the excretion of nanoparticles such as Y_2O_3:Yb,Er is inhibited when the worms are deprived of food [5]. The persistence of FNDs in the gut of the treated worms indicated that the nanomaterial was biostable and resistant to the worm's robust digestive enzymes (which can degrade intestinal bacteria in less than 2 min) [52]. Notably, there was no translocation of FNDs into the intestinal cells, though this may have been due to the bare FND particles forming large aggregates inside the intestinal lumen, which prevented their entry into the cells. On resuming feeding the worms with *E. coli*, the FNDs were forced down through the gut lumen and were excreted within an hour. It was determined from fluorescence intensity measurements that the amount of FNDs in the intestinal lumen after being fed with the particles for 2 h was in the range of ~3 ng, or ~1×10^6 particles, per worm.

The FND surface can be readily grafted with carboxyl groups for an ensuing conjugation with biomolecules such as carbohydrates, proteins, and nucleic acids, based on carbodiimide chemistry [31, 32], and subsequently conjugated covalently with carboxymethyldextran (CMDx) and BSA. When the bioconjugated FNDs were administered orally to the worms, most of the dextran- and BSA-coated FNDs were – somewhat surprisingly – absorbed into the intestinal cells, and very few remained in the gut lumen (Figures 13.14c, d). When the worms were transferred to a bacterial lawn for a 1-h recovery, the residual FNDs in the gut lumen were excreted whereas the FND bioconjugates that had been absorbed into the cells were retained (Figures 13.14c, d inset). It is likely that the surface modifications had permitted the FNDs to disperse more easily in solution, preventing their aggregation in the gut and helping their uptake via endocytosis by the intestinal cells. Bioconjugation may also have activated certain receptor-mediated endocytosis pathway(s) that facilitated the absorption of FNDs into the intestinal cells from the lumen [52, 53].

Microinjection is an alternative means of actively introducing FNDs to specific locations in *C. elegans*. In particular, FNDs can be delivered to the next generation if the particles are injected into the gonad of an adult hermaphrodite. Hence, a well-dispersed solution of bare FNDs was microinjected into the distal gonad of gravid hermaphrodites [54]; the injection site and dispersion of the injected FNDs in the germline syncytium, where the germline nuclei share the same cytoplasm, are shown in Figure 13.15a. The FNDs became incorporated into oocytes during oogenesis within 30 min after injection [55], and were retained in the fertilized zygotes (Figure 13.15b). Notably, their fluorescence signals persisted throughout embryogenesis (from the single-cell stage to the fully elongated pre-hatch stage), localized to the cytoplasm of the embryonic cells.

13.5.8
Other Imaging Techniques

Other than efficient fluorescence characteristics, the NV⁻ center possesses a single electronic spin which is ultrasensitive to its magnetic environment. Even a slight variation in magnetic field can lead to a shift in the spin resonance frequency that can be detected from the fluorescence signal. Balasubramanian *et al.* [56] demonstrated this effect by scanning a nanoscale magnetic tip over a single NV⁻ center containing nanodiamond, and precisely locating the position of the particle with nanometer resolution. It was found that two particles separated by 100 nm could be resolved with an accuracy of 20 nm. This was a remarkable finding, as the distance involved was far below the diffraction limit of optical microscopy. Hence, FNDs may be employed as magneto-fluorescent spin markers for bioimaging applications.

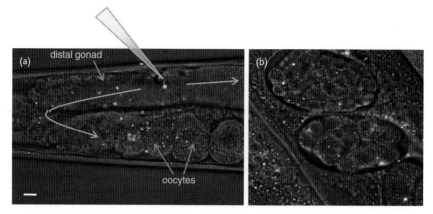

Figure 13.15 Microinjection of bare FNDs into C. elegans. Epifluorescence/DIC-merged images of an injected worm (a) and its progeny at the early (b) embryonic stage. The FNDs dispersed in the distal gonad and oocytes at approximately 30 min after injection (a). The green arrows indicate bulk streaming of FNDs with cytoplasmic materials, while the red triangle indicates the site of injection. Note that the injected FNDs are present in the cytoplasm of many cells in the early embryos (b). Scale bars = 10 μm. Adapted from Ref. [49]. For a color version of this figure, please see the color plate at the end of this book.

An additional superior optical property of diamond is that it has the highest refractive index of all transparent minerals ($n = 2.42$) [20, 21]. Thus, diamonds are excellent light scatterers and can be used as scattering probes for bioimaging. Smith et al. [57] estimated that the scattering from a single 55 nm ND particle was 300-fold brighter than that from a cell organelle of similar size. By using differential interference contrast microscopy, individual 55 nm nanodiamonds were imaged in mammalian epithelial cells. Typically, nanodiamonds cause less photodamage to cells than do metal scatters such as gold and silver nanoparticles, although in order to achieve a minimum contrast in scattering imaging, it is necessary to use particles of more than 40 nm in size.

Diamond also features a very distinct Raman peak, which makes it feasible to perform Raman mapping [58]. The Raman peak of diamond at 1332 cm^{-1} is extremely sharp and isolated, and can easily be distinguished from the signals emitted by a biological specimen [59, 60]. Subsequently, this property of diamond was exploited by Cheng et al. to study the interactions of protein lysozyme with E. coli bacteria. In this case, lysozymes were first adsorbed onto nanodiamonds, such that the nanodiamond Raman signal would serve as an indicator of protein–bacterium interaction sites. In future, with advances in both instrument design and confocal imaging schemes, Raman mapping may achieve a high spatial resolution in the range of 300 nm.

13.6
Conclusion

Fluorescent nanodiamond is a relatively new nanomaterial with unique characteristics that, in the future, will undoubtedly lead to extensive applications in biomedical research. The production, optical characterization and biological applications of FNDs, both *in vitro* and *in vivo*, and the issues and challenges related to all these aspects, have been discussed in detail in this chapter. As with any imaging probe, FND has its own disadvantages, notably that a reduction in the size of particles will reduce the number of NV centers and, in turn, reduce their fluorescence intensity. To improve the brightness of smaller FNDs represents a major challenge to be overcome, while problems associated with the aggregation of FNDs in cells and tissues must also be resolved. Recently, the issues of aggregation have been reduced to a great extent by performing surface functionalization, such as conjugation with BSA, though there is still room for improvement in this area.

The specific targeting of biomolecules and tissues and drug delivery represent future biological applications of FNDs that must be considered. In particular, the advantages of super resolution imaging and tracking may be exploited, while *in vivo* imaging applications may be extended to higher model organisms such as *Drosophila* and mouse. Clearly, many other biological questions remain to be explored and visualized in the future, for which FND is a potential candidate in this endeavor.

References

1 Alivisatos, P. (2004) *Nat. Biotechnol.*, **22**, 47.
2 Hell, S.W. (2009) *Nat. Methods*, **6**, 24.
3 Gao, X., Cui, Y., Levenson, R.M., Chung, L.W.K., and Nie, S. (2004) *Nat. Biotechnol.*, **22**, 969.
4 Park, J.-H., Gu, L., von Maltzahn, G., Ruoslahti, E., Bhatia, S.N., and Sailor, M.J. (2009) *Nat. Mater.*, **8**, 331.
5 Lim, S.F., Riehn, R., Ryu, W.S., Khanarian, N., Tung, C.-k., Tank, D., and Austin, R.H. (2006) *Nano Lett.*, **6**, 169.
6 Welsher, K., Liu, Z., Sherlock, S.P., Robinson, J.T., Chen, Z., Daranciang, D., and Dai, H. (2009) *Nat. Nanotechnol.*, **4**, 773.
7 Michalet, X., Kapanidis, A.N., Laurence, T., Pinaud, F., Doose, S., Pflughoefft, M., and Weiss, S. (2003) *Annu. Rev. Biophys. Biomol. Struct.*, **32**, 161.
8 Medintz, I.L., Uyeda, H.T., Goldman, E.R., and Mattoussi, H. (2005) *Nat. Mater.*, **4**, 435.
9 Derfus, A.M., Chan, W.C.W., and Bhatia, S.N. (2003) *Nano Lett.*, **4**, 11.
10 Piston, D.W., Lippincott-Schwartz, J., Patterson, G.H., Claxton, N.S., and Davidson, M.W. (2011) http://www.microscopyu.com/articles/livecellimaging/fpintro.html (accessed March 14, 2012).
11 Yu, S.-J., Kang, M.-W., Chang, H.-C., Chen, K.-M., and Yu, Y.-C. (2005) *J. Am. Chem. Soc.*, **127**, 17604.
12 Wee, T.-L., Tzeng, Y.-K., Han, C.-C., Chang, H.-C., Fann, W., Hsu, J.-H., Chen, K.-M., and Yu, Y.-C. (2007) *J. Phys. Chem. A*, **111**, 9379.
13 Fu, C.-C., Lee, H.-Y., Chen, K., Lim, T.-S., Wu, H.-Y., Lin, P.-K., Wei, P.-K., Tsao, P.-H., Chang, H.-C., and Fann, W. (2007) *Proc. Natl Acad. Sci. USA*, **104**, 727.
14 Faklaris, O., Garrot, D., Joshi, V., Druon, F., Boudou, J.-P., Sauvage, T., Georges, P., Curmi, P.A., and Treussart, F. (2008) *Small*, **4**, 2236.

15 Chang, Y.-R., Lee, H.-Y., Chen, K., Chang, C.-C., Tsai, D.-S., Fu, C.-C., Lim, T.-S., Tzeng, Y.-K., Fang, C.-Y., Han, C.-C., Chang, H.-C., and Fann, W. (2008) *Nat. Nanotechnol.*, **3**, 284.

16 Liu, K.-K., Cheng, C.-L., Chang, C.-C., and Chao, J.-I. (2007) *Nanotechnology*, **18**, 325102.

17 Schrand, A.M., Huang, H., Carlson, C., Schlager, J.J., Sawa, E., Hussain, S.M., and Dai, L. (2006) *J. Phys. Chem. B*, **111**, 2.

18 Vial, S., Mansuy, C., Sagan, S., Irinopoulou, T., Burlina, F., Boudou, J.-P., Chassaing, G., and Lavielle, S. *ChemBioChem*, 2008, **9**, 2113.

19 Vaijayanthimala, V., *et al.* (2009) *Nanotechnology*, **20**, 425103.

20 Rand, S.C. (1994) in *Properties and Growth of Diamond* (ed. G. Davies), INSPEC, Institute of Electrical Engineers, London, Ch. 7.4, pp. 221–240.

21 Collins, A.T. (1994) in *Properties and Growth of Diamond* (ed. G. Davies), INSPEC, Institute of Electrical Engineers, London, Ch. 3, pp. 81–125.

22 Davies, G. (1977) *Nature*, **269**, 498.

23 Davies, G., Lawson, S.C., Collins, A.T., Mainwood, A., and Sharp, S.J. (1992) *Phys. Rev. B*, **46**, 13157.

24 Hall, H.T. (1961) *J. Chem. Educ.*, **38**, 484.

25 Davies, G. and Hamer, M.F. (1976) *Proc. R. Soc. Lond., A*, **348**, 285.

26 Clark, C.D., Ditchburn, R.W., and Dyer, H.B. (1956) *Proc. R. Soc. Lond., A*, **234**, 363.

27 Lowther, J.E. (1984) *J. Phys. Chem. Solids*, **45**, 127.

28 Ziegler, J.F., Biersack, J.P., and Littmark, U. (2003) Free SRIM software (version 2003). Available at: http://www.srim.org/ (accessed January 10, 2007).

29 Prins, J.F. (2003) *Semicond. Sci. Technol.*, **18**, 27.

30 Wee, T.-L., Mau, Y.-W., Fang, C.-Y., Hsu, H.-L., Han, C.-C., and Chang, H.-C. (2009) *Diamond Relat. Mater.*, **18**, 567.

31 Huang, L.C.L. and Chang, H.-C. (2004) *Langmuir*, **20**, 5879.

32 Kruger, A., Liang, Y., Jarre, G., and Stegk, J. (2006) *J. Mater. Chem.*, **16**, 2322.

33 Lim, T.-S., Fu, C.-C., Lee, K.-C., Lee, H.-Y., Chen, K., Cheng, W.-F., Pai, W.W., Chang, H.-C., and Fann, W. (2009) *Phys. Chem. Chem. Phys.*, **11**, 1508.

34 Pante, N. and Kann, M. (2002) *Mol. Biol. Cell*, **13**, 425.

35 Periasamy, A. (2001) *J. Biomed. Opt.*, **6** (3), 287.

36 Mohan, N., Tzeng, Y.-K., Yang, L., Chen, Y.-Y., Hui, Y.Y., Fang, C.-Y., and Chang, H.-C. (2010) *Adv. Mater. Weinheim*, **22**, 843.

37 Boudou, J., *et al.* (2009) *Nanotechnology*, **20**, 235602.

38 Morita, Y., Takimoto, T., Yamanaka, H., Kumekawa, K., Morino, S., Aonuma, S., Kimura, T., and Komatsu, N. (2008) *Small*, **4**, 2154.

39 Hui, Y.Y., Chang, Y.-R., Lim, T.-S., Lee, H.-Y., Fann, W., and Chang, H.-C. (2009) *Appl. Phys. Lett.*, **94**, 013104.

40 Hui, Y.Y., Chang, Y.-R., Mohan, N., Lim, T.-S., Chen, Y.-Y., and Chang, H.-C. (2011) *J. Phys. Chem. A*, **115**, 1878.

41 Heikal, A.A., Hess, S.T., Baird, G.S., Tsien, R.Y., and Webb, W.W. (2000) *Proc. Natl Acad. Sci. USA*, **97**, 11996.

42 Qian, H., Sheetz, M.P., and Elson, E.L. (1991) *Biophys. J.*, **60**, 910.

43 Smith, B.R., Inglis, D.W., Sandnes, B., Rabeau, J.R., Zvyagin, A.V., Gruber, D., Noble, C.J., Vogel, R., Osawa, E., and Plakhotnik, T. (2009) *Small*, **5**, 1649.

44 Van der Meer, B.W., Coker, G., and Chen, S.-Y.S. (1994) *Resonance Energy Transfer: Theory and Data*, VCH Publishers, New York.

45 Clapp, A.R., Medintz, I.L., Mauro, J.M., Fisher, B.R., Bawendi, M.G., and Mattoussi, H. (2003) *J. Am. Chem. Soc.*, **126**, 301.

46 Rittweger, E., Han, K.Y., Irvine, S.E., Eggeling, C., and Hell, S.W. (2009) *Nat. Photon.*, **3**, 144.

47 Han, K.Y., Willig, K.I., Rittweger, E., Jelezko, F., Eggeling, C., and Hell, S.W. (2009) *Nano Lett.*, **9**, 3323.

48 Tzeng, Y.-K., Faklaris, O., Chang, B.-M., Kuo, Y., Hsu, J.-H., and Chang, H.-C. (2011) *Angew. Chem. Int. Ed. Engl.*, **50**, 2262.

49 Mohan, N., Chen, C.-S., Hsieh, H.-H., Wu, Y.-C., and Chang, H.-C. (2010) *Nano Lett.*, **10**, 3692.

50 Sulston, J.E. and Horvitz, H.R. (1977) *Dev. Biol.*, **56**, 110.

51 Sulston, J.E., Schierenberg, E., White, J.G., and Thomson, J.N. (1983) *Dev. Biol.*, **100**, 64.

52 McGhee, J.D. (2007) The *C. elegans* intestine, in *WormBook* (ed. The *C. elegans* Research Community), doi: 10.1895/wormbook.1.133.1. Available at: http://www.wormbook.org.

53 Karp, G. (1999) *Cell & Molecular Biology*, 2nd edn, John Wiley & Sons, Inc., New York.

54 Mello, C.C., Kramer, J.M., Stinchcomb, D., and Ambros, V. *EMBO J.*, 1991, 10.

55 Wolke, U., Jezuit, E.A., and Priess, J.R. (2007) *Development*, **134**, 2227.

56 Balasubramanian, G., Chan, I.Y., Kolesov, R., Al-Hmoud, M., Tisler, J., Shin, C., Kim, C., Wojcik, A., Hemmer, P.R., Krueger, A., Hanke, T., Leitenstorfer, A., Bratschitsch, R., Jelezko, F., and Wrachtrup, J. (2008) *Nature*, **455**, 648.

57 Smith, B.R., Niebert, M., Plakhotnik, T., and Zvyagin, A.V. (2007) *J. Lumin.*, **127**, 260.

58 Ferrari, A.C. and Robertson, J. (2004) *Philos. Trans. R. Soc. Lond., A*, **362**, 2477.

59 Chao, J.-I., Perevedentseva, E., Chung, P.-H., Liu, K.-K., Cheng, C.-Y., Chang, C.-C., and Cheng, C.-L. (2007) *Biophys. J.*, **93**, 2199.

60 Perevedentseva, E., *et al.* (2007) *Nanotechnology*, **18**, 315102.

Index

a

absorption coefficients 2–13
– diamond laser processing techniques 431
– integrated optical structures and microstructures 318
– laser micro- and nanoprocessing of diamond materials 387–390, 405, 409
– optical quality CVD diamond 44–48, 56, 60
– refractive and diffractive optics 134
– thermal management using diamond 360
– thermal stress fracture in Raman lasers 256
– two-photon absorption 29
AFM, see atomic force microscopy
AGPM, see annular groove phase mask
amorphous carbon
– integrated optical structures and microstructures 322
– laser micro- and nanoprocessing of diamond materials 390, 391
– polishing and shaping 86, 87, 91, 92, 97
anamorphic lenses 118, 119
annular groove phase mask (AGPM) coronagraphy 135, 136
antireflection (AR) coatings
– optics 122, 123, 133
– Raman lasers 241, 256, 257, 267
antireflective microstructures 122–124, 436
anti-Stokes scattering 17, 22, 26
apertures 112, 115, 118
AR, see antireflection
ArF laser ablation 428
astronomical applications 133
atomic force microscopy (AFM)
– CVD diamond films 41
– diamond optics 117, 118
– fluorescent nanodiamonds 456, 457
– integrated optical structures and microstructures 339
attenuated total reflectance (ATR) crystals 62
Auerbach's law 94, 95
autofluorescence 460, 463
axicons 118, 119

b

band bending model 210, 212–216, 225
barium nitrate
– Raman lasers 240, 241, 247, 249, 250, 253, 254, 260, 265, 268
– Raman linewidth 21
beam filamentation 414, 415
beam shapers 433–435
Bettis–House–Guenther scaling law 257
bioimaging 445, 446, 458–469
biolabeling applications
– fluorescent nanodiamonds 454
– surface doping 219
birefringence
– CVD diamond 49–52
– nitrogen-vacancy color centers 145
– Raman lasers 255, 256, 268, 271
– strain-induced 49–52
– thermal management using diamond 360
black spot microfeatures 46, 47, 53, 59
blazed gratings 119, 120, 122
blinking fluorescence 168, 233, 446
boron doping
– chemical vapor deposition of n-type diamond thin films 178
– light-emitting diodes 185–193, 197
– nitrogen-vacancy color centers 145
– surface doping 209
boron–deuterium (B–D) complex formation 178

Optical Engineering of Diamond, First Edition. Edited by Richard P. Mildren and James R. Rabeau.
© 2013 Wiley-VCH Verlag GmbH & Co. KGaA. Published 2013 by Wiley-VCH Verlag GmbH & Co. KGaA.

Bose–Einstein factors 8, 9, 15, 17
Bragg peaks 221, 222
Brewster-cut crystals 253, 258
bright-field fluorescence imaging 461, 465
Brillioun scattering 25–27
brittle-to-ductile transitions 87
brown diamonds 143
bruting diamond 73
bulk graphitization of diamond 410–419
– diamond laser processing techniques 424
– laser-induced graphitization waves 412–414, 424
– three-dimensional laser writing in diamond 414–419, 424–426, 435–437
– threshold conditions 410–412
bulk magnetometry 305
bus qubits 279
B–D, *see* boron–deuterium

c

Caenorhabditis elegans 465–468
CAMPP, *see* chemical-assisted mechanical polishing and planarization
capacitance–voltage measurements 186
carbon dioxide lasers
– CVD diamond windows 36, 47, 54, 56, 60
– diamond absorption coefficient 8
– laser micro- and nanoprocessing of diamond materials 432–435
– laser optics 111, 123, 125, 134
carbon phase diagram 37
cathodoluminescence (CL) 84, 189, 190
cavity quantum electrodynamic (CQED) effects 314, 317, 336
CCD, *see* charge-coupled device
cellular uptake 459, 460
charge-coupled device (CCD) array detectors 408
charged polymer interactions 231–235
chemical-assisted mechanical polishing and planarization (CAMPP) 126
chemical etching
– laser ablation 394, 404
– laser micro- and nanoprocessing of diamond materials 391, 392, 394, 404, 424, 429–431
– micromachining 128, 129, 131
– polishing and shaping 99–101
chemical inertness 110, 318
chemical–mechanical planarization 97, 98
chemical vapor deposition (CVD) diamond
– absorption characteristics 44–47

– diamond optics 110, 111, 113–118, 122, 123, 127, 134
– integrated optical structures and microstructures 321
– laser micro- and nanoprocessing of diamond materials 385–387, 389–407, 420–424, 426, 428–431, 433–435
– n-type doping 177–182, 185, 190
– nitrogen vacancy color centers 159, 160, 163, 166, 167
– nonoptical wavelength absorption 47–49
– optical applications 35, 36, 60–63
– polishing and shaping 82, 97, 101
– properties of optical quality CVD diamond 43–59
– Raman lasers 241, 242, 257, 258, 271
– scattering and depolarization losses 52–55
– strain-induced birefringence 49–52
– strength characteristics 56–59
– surface doping 216–219, 226, 228, 229
– thermal management using diamond 360
– thermal properties 54–56, 60–63
chemical vapor deposition (CVD) process 35–69
– atomic hydrogen role 37–39
– carbon phase diagram 37
– competitive grain growth 40, 41
– dopant impurities 40
– growth on refractory carbides 39, 40, 56
– isotopic content of diamond 49
– morphology and texture 42, 43
– point defects and defect complexes 44, 45
– polycrystalline diamond 35, 36, 38–43, 45–49, 52–55, 57, 58, 60, 61
– principles of diamond growth 36–43
– single-crystal diamond 35, 36, 38, 39, 42, 43, 49–54, 57, 61–63
chemiluminescence 89
chronometry 71, 72
CL, *see* cathodoluminescence
cleaving diamond 73, 98, 99
cluster-state quantum computing 289, 315, 317
CNOT, *see* controlled NOT
coatings, *see* antireflection coatings
coherence time 153, 157, 158, 164, 304
contact photolithography 324, 325
continuous-wave (CW) operation
– laser micro- and nanoprocessing of diamond materials 424

- Raman lasers   241, 244, 245, 255, 257, 258, 261, 262, 268–271
- thermal management using diamond   356, 357

controlled NOT (CNOT) gates   281, 282, 288, 295–297

controlled rotation (CROT) gates   287, 288, 295, 298

copper–diamond composites   375

copper encapsulation   426, 427

CQED, *see* cavity quantum electrodynamic

crack propagation   98, 99, 418, 419

critical point phonons   5–7

CROT, *see* controlled rotation

crystal morphology   42, 43

current density–voltage (J–V) characteristics   192–196, 198, 203, 204

current-injected light emission spectra   188–193, 195, 196, 198, 199

current–voltage (I-V) characteristics   186, 187, 191

curved diamond surfaces   71, 101, 102

curved graphitic structures   416–418

CVD, *see* chemical vapor deposition

CW, *see* continuous-wave

cyclotrons   220

cylindrical lenses   117

d

DBR, *see* distributed Bragg reflector

Debye temperature   15, 54, 246

deep-ultraviolet (DUV) generation   187–202, 205, 265–267

defects
- absorption in CVD diamond   44, 45
- applications of NV centers   163–168
- classification   144–158
- intrinsic and extrinsic   144, 145
- nitrogen-vacancy color centers   143–175
- polishing and shaping   74
- synthetic diamond   159–163

density functional theory (DFT) calculations   212, 214

density of states (DOS) calculations   5, 7, 216

deterministic QIP protocols   315

detonation nanodiamonds (DND)   159–161

DFT, *see* density functional theory

DH, *see* double heterostructures

diamond films   100

diamond laser processing techniques   419–437
- diamond optics   431–437
- formation of conductive structures in diamond   424–428
- laser polishing   420–424
- nanoablation   394, 402–406, 428
- surface structuring   428–431
- ultraviolet 2-photon etching   266, 267

diamond LEDs, *see* light-emitting diodes

diamond optics   109–142, 431–437
- applications   134–137
- CVD diamond   110, 111, 113–118, 122, 123, 127, 134
- diffractive optical elements   109, 119–125
- future directions   137
- lenses   112–118, 120, 121, 136, 137
- micromachining   126, 128–132
- novel devices   118, 119
- polishing and shaping   113, 118, 126–128
- polymer coatings   110, 113–115, 133
- prisms   118
- refractive optics   109, 112–119, 136, 137
- surface finishing   110, 113–115, 117, 122, 123, 128
- windows and domes   110–112

diamond sub-mounts for lasers   355, 363, 367, 369–371

diamond-like carbon (DLC) films   386, 392, 398, 399, 404, 424

diamond–diamond friction experiments   92–95

dielectric loss   61

differential interference contrast microscopy (Nomarski microscopy)   82, 84, 88, 98, 101, 468

diffractive optical elements (DOE)   109
- blazed gratings   119, 120, 122
- fan-out elements   124, 125
- Fresnel lenses   120–122
- laser micro- and nanoprocessing of diamond materials   431–437
- subwavelength gratings   122–124

directional dependence of polishing   75–79, 83, 84, 88, 91

disk lasers, *see* doped-dielectric disk lasers; semiconductor disk lasers

dislocations
- nitrogen-vacancy color centers   144
- polishing and shaping   84
- strain-induced birefringence   50–52

distributed Bragg reflector (DBR)   362, 363

DLC, *see* diamond-like carbon

DNA grafting   233, 234

DND, *see* detonation nanodiamonds

DOE, *see* diffractive optical elements

domes   110–112

doped-dielectric disk lasers
– diamond sub-mounts for mechanical rigidity 369–371
– future directions 374, 377
– intracavity applications of diamond 371–374
– thermal management using diamond 353–355, 360, 361, 369–374, 377
DOS, see density of states
double heterostructures (DH) 319, 320
doxorubicin 233, 234
drug delivery 165
dry chemical etching 99–101
DUV, see deep-ultraviolet
dynamic light scattering (DLS) 454, 455
dynamic nuclear spin polarization 285

e
EBIC, see electron beam-induced current
EBL, see electron beam lithography
ECR, see electron cyclotron resonance
edge dislocations 144
EELS, see electron energy loss spectroscopy
elastic strain tensor 25, 26
elaton materials 62, 63
electrical conductivity 143
electroluminescence 88, 89
electron beam lithography (EBL)
– diffractive optics 122
– integrated optical structures and microstructures 323, 325
– nanowires 132
electron beam-induced current (EBIC) analysis 186
electron cyclotron resonance (ECR) 129, 327
electron energy loss spectroscopy (EELS) 81, 89
electron-hole pair recombination 201, 202, 205
electron-nuclear double resonance (ENDOR) spectroscopy 295, 296
electron paramagnetic resonance (EPR)
– fluorescent nanodiamonds 457, 458
– integrated optical structures and microstructures 312, 313
– nitrogen-vacancy color centers 147, 151, 157
– quantum optical diamond technologies 287, 294
– surface doping 221
electron spin echo envelope modulation (ESEEM) 290–292

electron spin–nuclear spin interactions 282–293
electron spin resonance (ESR)
– n-type doping 182
– nitrogen-vacancy color centers 156
– quantum optical diamond technologies 283, 284, 286–288
electron spins
– dipole–dipole interactions 282, 283, 291–293
– Fermi contact interactions 282, 291–293
– generation and detection of entanglement 296–303
– gyromagnetic ratio 281, 284
– hyperfine interactions 281–289, 291–294, 296
– nuclear spin bath 289–292
– nuclear spins as a qubit resource 281
– Ramsey frequencies 289, 290, 296, 299
– single quantum systems 277
– Zeeman splitting 282–284, 289, 291
electronic nonlinearity 27–30
electroplating 117, 118
embedded grits 87, 88
end-caps 357–359
ENDOR, see electron-nuclear double resonance
endoscopic applications 136, 137
entanglement 296–303
– Bell states 297–303
– GHZ states 297, 298, 301–303
– W states 297, 298, 301–303
epifluorescence imaging 461, 465, 466, 468
EPR, see electron paramagnetic resonance
EQE, see external quantum efficiency
Er-doped lasers 260, 261
Escherichia coli inhibition 200, 205
ESEEM, see electron spin echo envelope modulation
ESR, see electron spin resonance
etch rate 266, 267, 397, 402–406, 430
etch selectivity 115, 131
excimer laser ablation 119, 121, 405, 427
external cavity Raman lasers 240, 258–261, 269, 270
external quantum efficiency (EQE) 199, 200
extrinsic defects 144, 145

f
Fabry–Perot interferometers 21
fan-out elements 124, 125
FCS, see fluorescence correlation spectroscopy

FEM, *see* finite element modeling
Fermi level engineering   209, 210, 214, 231, 236
FET, *see* field-effect transistors
FIB, *see* focused ion beam
fiber Raman lasers   240, 241
fiber taper waveguides   339
field-effect transistors (FET)   212
finite element modeling (FEM)   359, 371–374, 377, 378
fluorescence correlation spectroscopy (FCS)   454–456
fluorescence imaging   459–464
fluorescence lifetime imaging   453, 454, 461–463
fluorescence resonance energy transfer (FRET)   454, 458, 459
fluorescence spectra   450, 451, 457
fluorescent nanodiamonds   445–471
– autocorrelation functions   454–457
– biological applications   445, 446, 458–469
– cellular uptake and fluorescence imaging   459, 460
– fluorescence lifetime imaging   453, 454, 461–463
– fluorescence resonance energy transfer   454, 458, 459
– fluorescence spectra   450, 451
– *in vivo* imaging in *Caenorhabditis elegans*   465–468
– mass production   448–451
– nitrogen-vacancy color centers   165–167, 446–448, 456–458
– photostability   451–453, 456
– single particle tracking   464, 465
– super-resolution fluorescence microscopy   445, 463, 464
– surface doping   220, 228, 231–235
– two-photon excited fluorescence imaging   460–462
fluorescent proteins   446
fluorinated nanodiamonds   228, 229, 235, 236
FND, *see* fluorescent nanodiamonds
focused ion beam (FIB) fabrication
– diffractive optical elements   122
– integrated optical structures and microstructures   321–326, 328, 329, 333, 334
Förster resonance energy transfer (FRET)   234, 236
four-wave mixing   27, 28
Fourier transform infrared attenuated total reflectance (FTIR-ATR) spectroscopy   135

Fourier transform infrared (FTIR) spectroscopy   229
Fourier transform spectroscopy   14
fracture strength   56–59, 87, 256
freeform optical devices   119
Fresnel lenses   120–122, 433
Fresnel reflection   2, 3
FRET, *see* fluorescence resonance energy transfer; Förster resonance energy transfer
frozen core   292, 293
FTIR, *see* Fourier transform infrared
FTIR-ATR, *see* Fourier transform infrared attenuated total reflectance
fullerenes   281
fused silica   62, 63

g

GaInAsSb-based devices   364, 367, 368
gallium arsenide   304, 354
gallium hard mask   326, 333, 334
gallium nitride   354, 375, 376
gallium phosphide   117, 321
germanium sulfide   133
glide set   145
gold-doped silicon   61
gold–diamond composites   375
graphitic carbon
– bulk graphitization   410–419, 424
– integrated optical structures and microstructures   322
– laser micro- and nanoprocessing of diamond materials   388–395, 404, 405, 410–419
– polishing and shaping   91, 92, 97
– surface graphitization   388–395, 404, 405, 410, 411
graphitic wires   414–419, 424–426, 435–437
Griffiths model   57
gyrotrons   60, 61

h

H3 color centers   357
Hadamard gates   297, 298
Hahn-echo sequences   289, 290, 292
Hall-effect measurements   182–185, 190
heat spreaders   356, 361–369, 375
HeLa cells   460–463
HEMT, *see* high electron mobility transistors
Herzberger formula   14
heteroepitaxial growth   322
hexagonal diamond   85

HFCVD, *see* hot filament chemical vapor deposition
high electron mobility transistors (HEMT) 376
high-pressure, high-temperature (HPHT) synthetic diamond
– absorption characteristics 45
– carbon phase diagram 37
– diamond optics 118
– fluorescent nanodiamonds 447–451
– integrated optical structures and microstructures 326, 327
– laser micro- and nanoprocessing of diamond materials 385
– n-type doping 188, 192–193
– nitrogen-vacancy color centers 159, 160, 166, 167
– nonoptical wavelength absorption 48
– polishing and shaping 80, 82
– Raman lasers 242, 271
– strain-induced birefringence 50
– surface doping 215, 216, 219, 220, 226, 228, 229
high-resolution magnetometry 165
high-resolution transmission electron microscopy (HRTEM) 455–457
hole accumulation layers 212
homoepitaxial diamond junction devices 177–179, 188, 192, 194, 203, 204
hot filament chemical vapor deposition (HFCVD) 122
hot metal polishing 97
HPHT, *see* high-pressure, high-temperature
HRTEM, *see* high-resolution transmission electron microscopy
hydrogen bonded surface 82
hydrogenated nanodiamonds 225, 226, 230, 231, 235, 236

i

ICP, *see* inductively coupled plasma
in vivo imaging 459, 460, 465–468
inclusions 143
inductively coupled plasma (ICP) etching
– diamond optics 122–124, 129, 130, 136
– integrated optical structures and microstructures 327, 331
– n-type doping 191
– polishing and shaping 101
– thermal management using diamond 370
infrared (IR) absorption 3–11, 44–47, 201
infrared (IR) spectroscopy 61, 62
infrared (IR) transmission
– diamond optics 110, 111, 135
– Raman lasers 264, 265
– transmission spectra 2
infrared lattice absorption, *see* lattice absorption
InGaAs-based devices 364
integrated optical structures and microstructures 311–351
– applications of fluorescent diamond devices 311–318
– design of diamond-based optical structures 319–321
– fabrication techniques 323–327
– hybrid approaches 338–340
– lift-off/out methods 322–324, 328, 329
– microcavities 319, 320, 333–335
– nanobeams 333
– nanomanipulation of nanodiamond 338–340
– nanopillars 332, 333
– nitrogen-vacancy color centers 311–318
– optical properties of diamond 318, 319
– optical waveguiding in single-crystal diamond 327–330
– polycrystalline thin films 335–338
– quantum information processing 311, 315–318
– quantum key distribution 311, 313–316
– ring resonators 334, 335, 337
– sensing 311–313
– single-photon generation 311, 313–316, 342
– single-crystal diamond 311, 318, 319, 321–335
– solid immersion lenses 330, 331
– surface emission 330–333
– thin and ultra-thin single-crystal diamond films 322, 323
– two-dimensional photonic crystals 333, 334
interferometry 407–410, 429
intersystem crossing (ISC) 278
intervalley scattering 185
intracavity heat spreaders 56, 355–360, 363–368, 371–374, 377
intracavity Raman lasers 245, 261, 262, 269, 270
intrinsic defects 144, 145
inverse *bremsstrahlung* process 388, 389
ion beam-assisted etching 126, 128
ion implantation
– integrated optical structures and microstructures 330

– laser micro- and nanoprocessing of diamond materials 426–428
– nitrogen-vacancy color centers 162, 163
IR, *see* infrared
iron films 97
irradiative NV center formation 162
ISC, *see* intersystem crossing
isochronous cyclotron 220
isotopic content of diamond
– CVD diamond 49
– lattice absorption 9–11
– quantum optical diamond technologies 281, 286
– Raman scattering 23, 24
– UV edge absorption 13
I–V, *see* current–voltage

j

Jahn–Teller effect 151, 152
joint-density-of-states (JDS) calculations 5, 7
J–V, *see* current density–voltage

k

Kerr effect 409, 410
Klaus and Strawson's stereogram 78, 79
Klemens model 22, 23
Kramers–Kronig analysis 27, 29
KrF lasers 391, 392, 394, 395, 397, 403–405, 424, 425, 427, 429, 432

l

lanthanum–nickel eutectic alloys 97
larmor precession frequency 290, 291
laser ablation 394–410
– chemical etching 394, 404
– diamond laser processing techniques 421–424
– diamond optics 119, 121, 123
– nanoablation 394, 402–406
– photoionization of diamond 406–410
– vaporization ablation 394–402
laser damage threshold 240, 256, 257
laser etching
– diamond optics 119, 125, 126
– laser micro- and nanoprocessing of diamond materials 427
– nanoablation 402–406
– ultraviolet 2-photon etching 266, 267
laser micro- and nanoprocessing of diamond materials 385–443
– bulk graphitization of diamond 410–419, 424

– chemical etching 391, 392, 394, 404, 424, 429–431
– diamond laser processing techniques 419–437
– diamond optics 431–437
– formation of conductive structures in diamond 424–428
– graphitization waves 412–414, 424
– laser ablation 394–410, 421–424
– laser polishing 420–424
– nanoablation 394, 402–406
– photoionization of diamond 406–410
– surface graphitization 388–395, 404, 405, 410, 411, 424, 425
– surface structuring 428–431
– thermomechanical and optical properties of diamond 386–388
– three-dimensional laser writing in diamond 414–419, 424–426, 435–437
– threshold conditions for bulk/surface graphitization 410–412
– vaporization ablation 394–402, 429–431
lattice absorption 2–11
– CVD diamond 44–47
– isotopic content of diamond 9–11
– laser micro- and nanoprocessing of diamond materials 388, 389
– temperature dependence 8, 9
– two-phonon region 4–7
Laue geometry 76–78, 80
LEDs, *see* light-emitting diodes
lenses
– integrated optical structures and microstructures 327, 330
– laser micro- and nanoprocessing of diamond materials 431–435
– refractive optics 112–118, 120, 121, 136, 137
– *see also* microlenses; solid immersion lenses
life sciences applications
– diamond optics 136, 137
– fluorescent nanodiamonds 445–471
– n-type doping 200, 205
– nitrogen-vacancy color centers 164, 165
– quantum optical diamond technologies 277
– surface doping 215, 219, 228, 232–235
lift-off/out methods 322–324, 328, 329
light-emitting diodes (LEDs)
– diamond heat spreaders 375
– diamond LEDs 187–205, 376

– monolithic structures and GaN epitaxy on diamond 376
– n-type doping 187–202
– thermal management using diamond 353, 354, 375, 376
linear defects 144
linear optics, see diamond optics
linear variable displacement transducers (LVDT) 80, 81
lithium iodate Raman lasers 240, 241, 246, 247
lonsdaleite 85
luminescence
– laser micro- and nanoprocessing of diamond materials 414, 415
– nitrogen-vacancy color centers 166
– surface doping 210, 216–218, 222–236
– see also photoluminescence
LVDT, see linear variable displacement transducers

m

magnetic sensing 143, 149, 165
magneto-fluorescent spin markers 467
magnetometry 311–313
manganese powders 97
mechanical lapping 126
metallization of diamond 426, 427
microcavities 319, 320, 333–335
micro-crack features, see black spot microfeatures
microelectronics 111, 112, 114
microinjection 467, 468
microlenses
– integrated optical structures and microstructures 327
– refractive optics, fabrication 114–118
– thermal management using diamond 369
micromachining 126, 128–132
microstructures
– laser micro- and nanoprocessing of diamond materials 435–437
– polishing and shaping 74, 83, 84
– ultraviolet 2-photon etching 266, 267
– see also integrated optical structures and microstructures
microtexturing 128
microtrons 220
microwave plasma chemical vapor deposition (MPCVD)
– fundamentals of growth 37, 38
– n-type doping 180, 185

microwave plasma torch technique 167
microwave transmission 2, 48, 49, 61
missile domes 8, 61, 111, 112
molecular dynamics simulations 92, 93
Monte Carlo simulations 449, 457, 458
moth eye structures 122
MPCVD, see microwave plasma chemical vapor deposition
multiphonon absorption, see lattice absorption

n

n-type doping 177–208
– chemical vapor deposition of diamond thin films 177–182, 185, 190
– current theory and future directions 202–205
– deep-UV LEDs 187–202, 205
– electrical properties of diamond pn junctions 185–187
– electrical properties of n-type diamond 182–185
– homoepitaxial diamond junction devices 177–179, 188, 192, 194, 203, 204
– i-layer thickness 197–200
– light-emission mechanisms 200–202
– nitrogen-vacancy color centers 143
– pin junction LEDs 188, 192–200, 203, 204
– pn junction LEDs 188–192, 202–204
– surface doping 209
nanoablation 394, 402–406, 428
– see also ultraviolet 2-photon etching
nanobeams 333
nanocantilevers 326
nanocrystalline (NC) diamond films
– integrated optical structures and microstructures 321, 336, 337
– laser micro- and nanoprocessing of diamond materials 392, 397, 398, 402, 403
nanodiamond
– integrated optical structures and microstructures 318, 319, 338–340
– nanomanipulation of 338–340
– nitrogen-vacancy color centers 144, 159–161, 165–169
– surface doping 214, 219–236
– see also fluorescent nanodiamonds
nanogrooving 82–91
nanoimprint lithography (NIL) 132, 136
nanomechanical oscillators 305
nanopillars 132, 332, 333

natural diamond
- absorption characteristics   45
- carbon phase diagram   37
- diamond optics   110, 111
- high technology applications   35
- laser micro- and nanoprocessing of diamond materials   385
- nitrogen-vacancy color centers   143, 144
- quantum optical diamond technologies   286

NC, *see* nanocrystalline

Nd:YAG lasers
- design and performance of Raman lasers   239, 253–256, 261, 263
- laser micro- and nanoprocessing of diamond materials   386, 394, 406, 429, 430
- thermal management using diamond   354, 356–360, 370–373, 377, 378

Nd:YAP lasers   394

Nd:YVO$_4$ lasers
- design and performance of Raman lasers   261, 262, 265, 266, 271
- thermal management using diamond   354, 355, 372

Nd-doped lasers   239–241, 261, 268, 269

NEA, *see* negative electron affinity

near-edge X-ray absorption fine-structure (NEXAFS) spectroscopy   266, 267, 391

near-infrared (NIR) absorption   360

negative electron affinity (NEA)   202, 203

neutron scattering   5

NEXAFS, *see* near-edge X-ray absorption fine-structure

nickel encapsulation   426

NIL, *see* nanoimprint lithography

NIR, *see* near-infrared

nitrogen-doping
- laser micro- and nanoprocessing of diamond materials   405, 406
- n-type doping   177, 181, 182, 185
- Raman linewidth   21
- *see also* nitrogen-vacancy color centers

nitrogen-vacancy color centers   143–175
- A-centers   145–147
- applications   163–168
- B-centers   145–147
- bus qubits   279
- C-centers   145–147
- charge state of NV centers   150–152
- classification of defects in diamond   144–158
- controlled ground-state spin transitions   155, 156
- CVD diamond   159, 160, 163, 166, 167
- detonation nanodiamonds   159, 161
- electron spins   281–294, 296–303
- emission spectra   150
- energy level scheme for NV centers   152, 153
- enhancing surface emission   330–333
- fluorescent nanodiamonds   446–448, 456–458
- frozen core   292, 293
- high-pressure, high-temperature synthetic diamond   159, 160, 166, 167
- high-resolution magnetometry   165
- integrated optical structures and microstructures   311–318
- intersystem crossing   278
- intrinsic and extrinsic defects   144, 145
- ion implantation   163
- irradiation   162
- life sciences applications   165
- measurement of fluorescence intensity   156, 158
- nanotechnology applications   165–168
- nuclear spins   281–303
- optical and spin properties of NV centers   153–158
- optically detected magnetic resonance   278, 285, 291–294, 312, 313
- platelet defects   145, 147
- polarizability of NV centers   153–156
- quantum information technology   148, 149, 163, 164
- quantum optical diamond technologies   278–305
- relaxation time   156–158
- spin-selective optical excitations   278
- surface doping   209–236
- surface-related effects   167, 168
- synthetic diamond   159–163, 166

NMR, *see* nuclear magnetic resonance

Nomarski microscopy, *see* differential interference contrast microscopy

nondeterministic QIP protocols   315

nonlinear refractive index   29

nonlocal states   293–303

nonoptical wavelength absorption   47–49

nonplanar geometries   101, 102

nuclear magnetic resonance (NMR) spectroscopy
- quantum optical diamond technologies   286–288, 293, 295
- surface doping   228, 229

nuclear spins
– characterization of qubit systems 295, 296
– dipole–dipole interactions 282, 283, 291–293
– electron spin decoherence 281, 289–292, 296–303
– electron spin–nuclear spin interactions 282–293
– Fermi contact interactions 282, 291–293
– frozen core 292, 293
– gyromagnetic ratio 281, 284
– Hamiltonian operators 282, 283
– hyperfine interactions 281–289, 291–294, 296
– isotopes with nuclear spins 281
– nonlocal states 293–303
– nuclear spin baths 288–293
– quantum gates 287, 288, 304, 305
– Rabi frequency 296, 299, 300
– Ramsey frequencies 289, 290, 296, 299
– three-partite entanglement 302, 303
– two nearest-neighbour ^{13}C nuclear spins 293–295
– Zeeman splitting 282–284, 289, 291
nucleation and growth 58
NV, see nitrogen-vacancy

o

ODE, see ordinary differential equation
ODMR, see optically detected magnetic resonance
one-phonon absorption 3
one-photon excitation (OPE) microscopy 460, 462
optical absorption spectroscopy 221
optical coatings, see coatings
optical damage threshold 435
– see also laser damage threshold
optical microscopy
– fluorescent nanodiamonds 456, 457, 467, 468
– integrated optical structures and microstructures 331
– laser micro- and nanoprocessing of diamond materials 412, 416, 426
optical properties 1–34
– *ab initio* calculations 22
– absorption at wavelengths longer than 5 microns 7, 8
– Brillioun scattering 25–27
– critical point phonons 5–7
– electronic nonlinearity 27–30

– historical background 1, 2
– isotopic content of diamond 9–11, 13, 23, 24
– laser damage threshold 240, 256, 257
– lattice absorption 2, 3–11
– linewidth 20–22
– nonlinear refractive index 29
– Raman gain coefficient 24, 25, 247–254
– Raman scattering 16–24
– refractive index 13–16, 29
– temperature dependence 8, 9, 14–16, 22, 23
– transmission spectra 2–11
– two-phonon absorption 4–7, 29, 30
– UV edge absorption 2, 3, 11–13
– Verdet constant 16
optical pumping 153, 244, 245, 278
optically detected magnetic resonance (ODMR) 278, 285, 291–294, 312, 313
optics, diamond, see diamond optics
ordinary differential equation (ODE) solvers 244
oxidative etching 454–456
oxidized nanodiamonds 225, 226, 230, 235, 236

p

p-type doping
– homoepitaxial diamond junction devices 177, 178, 185–193
– nitrogen-vacancy color centers 143
– surface doping 209
particle induced X-ray emission (PIXE) 181, 182
particle size analysis 454, 455
PEI, see polyethylenimine
phonon scattering 185
phonon wave-vectors 27
phosphorus-doping
– homoepitaxial diamond junction devices 177–193, 195, 197
– surface doping 209
photochromism 312, 313
photoconductivity 145
photoelastic tensor 25, 26
photoionization of diamond 406–410
photolithography
– diamond optics 117, 121, 123
– integrated optical structures and microstructures 324, 325, 330, 331
– laser micro- and nanoprocessing of diamond materials 428
– n-type doping 191

photoluminescence (PL)
– fluorescent nanodiamonds   446, 457
– n-type doping   189, 190
– surface doping   211, 212, 215–218, 220–233
photonics applications   318–320, 333, 334, 336, 342
photoresist   114–118, 330, 331
photostability   451–453, 456, 465
photothermal ionization spectroscopy   201
pin junction LEDs   188, 192–200, 203, 204
PIXE, see particle induced X-ray emission
PL, see photoluminescence
planar slab waveguides   321, 322
plasma etching
– integrated optical structures and microstructures   323, 337
– refractive and diffractive optics   113–115, 117, 122–124, 128–130
– trenching   130
– see also inductively coupled plasma etching; reactive ion etching
plastic deformation
– nitrogen-vacancy color centers   143
– polishing and shaping   71, 83, 90, 91
– thermal management using diamond   354
platelet defects   145, 147
pn junctions
– homoepitaxial diamond junction devices   185–187, 188–192, 202–204
– surface doping   209, 210
point defects   44, 45, 144
Poisson equation   214
polarizability   17, 18, 153–156, 248, 250–253
polishing and shaping
– ab initio calculations   91
– apparatus and preparation   74, 75, 84–88
– atmosphere dependence   88, 96
– chemical–mechanical planarization   97–98
– diamond optics   113, 118, 126–128
– directional dependence of polishing   75–79, 83, 84, 88, 91
– dry chemical etching   99–101
– high-quality planar surfaces   98–101
– hot metal polishing   97
– laser polishing   420–424
– nanogrooving   82–91
– nonplanar and structured geometries   101, 102
– post-mechanical polishing treatment   99
– shaping techniques   73, 98, 99

– single-crystal diamond   71–107
– sliding diamond against other materials   95
– subsurface damage   84
– surface characteristics   82–84
– thermally induced wear   90, 97
– tribological behavior of diamond   92–95
– triboluminescence   88, 89, 102, 103
– velocity dependence   79–81
polycrystalline diamond
– absorption   45–49
– applications   35, 36, 60, 61
– diamond optics   111, 117, 118, 123, 131, 132, 134
– fracture strength   57, 58
– integrated optical structures and microstructures   318, 319, 335–338, 341, 342
– laser micro- and nanoprocessing of diamond materials   386, 387, 389–394, 397, 398, 406, 407, 420–424
– morphology and texture   42, 43
– n-type doping   177, 178
– polishing and shaping   97
– principles of diamond growth   38–41
– scattering and depolarization losses   52–54
– thermal conductivity   54, 55
– thermal management using diamond   360, 376, 377
– thin films   335–338
polyethylenimine (PEI)   233–235
polymer coating   113–115
post-mechanical polishing treatment   99
potassium gadolinium tungstate   247, 249, 250, 253, 254
prisms   118
proteins, fluorescent   446
Purcell regimes   314, 315

q

Q factor   313, 314, 317, 319, 320, 334, 337, 342
QD, see quantum dots
QIP, see quantum information processing
QKD, see quantum key distribution
quantum cascade lasers   264, 265
quantum computing applications   235, 236
quantum conversion efficiency   259, 260
quantum cryptography   164
quantum dots (QD)   304, 445, 452, 453, 465
quantum gates   304, 305
– CNOT gates   281, 282, 288, 295–297
– CROT gates   287, 288, 295, 298

– Hadamard gates   297, 298
– Toffoli gates   281, 282, 296
quantum information processing (QIP)
– integrated optical structures and microstructures   311, 315–318
– nitrogen-vacancy color centers   148, 149, 163, 164, 278, 280
– surface doping   215
quantum key distribution (QKD)   311, 313–316
quantum mechanical calculations   16, 17, 91
quantum optical diamond technologies   277–310
– *ab initio* calculations   285, 286
– characterization of qubit systems   295, 296
– dipole–dipole interactions   282, 283, 291–293
– electron spin decoherence   277, 281, 289–292, 296–303
– electron spin–nuclear spin interactions   282–293
– Hamiltonian operator   282, 283
– isotopes with nuclear spins   281
– nitrogen-vacancy color centers   164, 278, 279–305
– nonlocal states   293–303
– nuclear spin bath   288–293
– partial transpose   300, 301
– quantum gates   281, 282, 287, 288, 304, 305
– quantum state tomography   298–300
– qubits   279–305
– single quantum systems   277
– three-partite entanglement   302, 303
– two nearest-neighbor ^{13}C nuclear spins   293–295
quantum standards   311, 313–316
quarter-wave antireflection coatings   133
qubits   164, 279–305

r

Raman frequency   16
– isotopic dependence   23, 24
– temperature dependence   22
Raman gain coefficient   24, 25, 247–254, 377
– comparison with other materials   247, 248
– linewidth dependence   253, 254
– polarization dependence   250–252
Raman lasers   239–276
– deep ultraviolet generation   265–267
– design and modeling   245, 246

– development of diamond Raman lasers   258–263
– extending capability using diamond   264–270
– external cavity design   240, 258–261, 269, 270
– finite difference calculation   244
– future directions   270, 271
– gain coefficient   244, 245, 247–254, 377
– high average power   267–270, 271
– historical context   239–242
– integrated optical structures and microstructures   318
– intracavity design   245, 261, 262, 269, 270
– laser damage thresholds   256, 257
– linewidth dependence   253, 254
– long-wavelength generation   264, 265
– optical, thermal and physical properties of diamond   246–258
– polarizability   250–253
– principles   242–246
– stimulated Raman scattering   239, 241, 242, 243–245, 248, 249, 263
– synchronously pumped mode-locked lasers   262, 263
– target wavelengths for applications   239, 240
– thermal lens strength   255
– thermal management using diamond   356, 357, 377
– thermal properties   254–256, 268
– thermal stress fracture   256
– thermally induced stress birefringence   255, 256, 268
Raman mapping   468
Raman scattering   16–24
– isotopic content of diamond   23, 24
– linewidth   20–22
– second-order Raman scattering   5
– stimulated   24, 25
– temperature dependence   22, 23
– wavelength dependence   20, 21
Raman spectroscopy
– laser micro- and nanoprocessing of diamond materials   390–392, 410
– polishing and shaping   82, 85, 86, 89, 92
– surface doping   217, 218, 221, 222, 226
Raman tensors   17, 18
RBS, *see* Rutherford backscattering
reactive ion etching (RIE)
– integrated optical structures and microstructures   326, 327, 330, 337
– n-type doping   186

– surface processing of diamond optics 126, 128, 129
reciprocal polishing 85–88, 90
red fluorescence 446–454
reflection electron microscopy 82
refractive index 13–16
– CVD diamond 54
– diamond optics 110, 113
– integrated optical structures and microstructures 319, 320
– laser micro- and nanoprocessing of diamond materials 409, 410, 435
– nonlinear 29
– temperature dependence 14–16
refractive optics 109, 112–119
– lenses 112–118, 136, 137
– novel devices 118, 119
– prisms 118
refractory carbides 39, 40, 56
relaxation time 156–158
remote sensing applications
– diamond optics 133
– quantum optical diamond technologies 304, 305
– Raman lasers 260, 264, 265
rheological properties
– fracture strength 56–59, 87
– laser micro- and nanoprocessing of diamond materials 385
– polishing and shaping 71, 83, 87, 90, 91
– thermal management using diamond 354, 369–371
ridge waveguides 370
RIE, see reactive ion etching
ring resonators 334, 335, 337
ring-shaped lenses 117
RNA grafting 233, 234
Rutherford backscattering (RBS) analysis 181, 182

S

sapphire 54, 61, 358, see also titanium-doped sapphire lasers
sawing diamond 73
SBD, see Schottky barrier diodes
scaife plates 74, 75, 84–88
scanning confocal microscopy
– fluorescent nanodiamonds 451, 462–464
– integrated optical structures and microstructures 339, 340
scanning electron microscopy (SEM) 41
– diamond optics 115, 116, 122–125, 136
– integrated optical structures and microstructures 324–326, 331, 335, 339–341
– laser micro- and nanoprocessing of diamond materials 421, 422
– n-type doping 178, 179
– polishing and shaping 82
scanning near-field optical microscopy (SNOM) 339
scanning probe microscopy 82
scatter 52–54, 270
scattering efficiency
– Brillioun 26
– Raman 18, 19
SCD, see single-crystal diamond
Schottky barrier diodes (SBD) 202, 203
Schottky-pn diodes (SPND) 202, 203
secondary ion mass spectrometry (SIMS)
– integrated optical structures and microstructures 333
– n-type doping 181, 182, 192, 193
secular approximation 283–285
self-focusing of laser beams 27, 414, 415
self-interstitial carbon 144
self-Raman lasers 261
Sellmeier formula 14
SEM, see scanning electron microscopy
semiconducting characteristics, see n-type doping; nitrogen-vacancy color centers; p-type doping; surface doping
semiconductor disk lasers
– basic principles 360–362
– future directions 368, 369
– intracavity and extracavity applications of diamond 363–368
– quantum well 363
– strategies for thermal management 362–368
– thermal management using diamond 353–356, 360–369
– wavelength diversity 367, 368
sensing 311–313, see also remote sensing
shaping techniques 73, 98, 99, see also polishing and shaping
Shockley–Read–Hall-type recombination 201
shuffle set 145
signal-to-noise ratio (SNR) 451
SIL, see solid immersion lenses
silicon oxide 133
silicon Raman lasers 246–247, 250
silicon wafers 318
silicone oil 87
silver–diamond composites 375

SIMS, *see* secondary ion mass spectrometry
single-crystal diamond (SCD)
– applications   35, 36, 61–63
– diamond optics   111, 117, 118, 132, 134
– fabrication techniques   323–327, 339–342
– fracture strength   57
– integrated optical structures and microstructures   311, 318, 319, 321–335, 339–342
– laser micro- and nanoprocessing of diamond materials   385, 386, 391, 409–411
– lift-off/out methods   322–324, 328–329
– microcavities   333–335
– morphology and texture   42, 43
– optical waveguiding   327–330
– polishing and shaping   71–107
– principles of diamond growth   38, 39
– Raman lasers   242
– scattering and depolarization losses   52–54
– strain-induced birefringence   49–52
– surface doping   214–218, 221, 223–226
– surface emission   330–333
– thermal management using diamond   356, 360, 376
single particle tracking   464–465
single-photon generation   311, 313–316, 342
single-point diamond turning tools   119
single quantum systems   277
sintered diamond materials   75, 87
slab lasers   355
slot waveguides   317, 318
slow speed diamond–diamond sliding   92–95
SNOM, *see* scanning near-field optical microscopy
SNR, *see* signal-to-noise ratio
solid immersion lenses (SIL)   61, 62, 330, 331
solid-state quantum technologies   164
spin-flip relaxation time   156–158
spin-selective optical excitations   278
spin-spin relaxation time   157, 158
SPND, *see* Schottky-pn diodes
SRS, *see* stimulated Raman scattering
steady-state Raman gain coefficient, *see* Raman gain coefficient
STED, *see* stimulated emission depletion
sterilization applications   200, 205
stimulated emission depletion (STED) microscopy   463, 464

stimulated Raman scattering (SRS)
– gain coefficient   24, 25, 247–254
– Raman lasers   239, 241, 242, 243–245, 248, 249, 263
strain-induced birefringence   49–52
strength characteristics   56–59, 87, 110, 111, 354
stress fracture   256
structured geometries   101, 102
sub-mounts for lasers   355, 363, 367, 369–371
subsurface damage   84
subwavelength gratings   122–124
sulfur-doping   178
super-resolution fluorescence microscopy   445, 463, 464
surface doping   209–238
– annealing techniques for formation of NV centers   220, 221, 223–225
– band bending model   210, 212–216, 225
– creation of GR1 vs. ND1 vacancies   220, 221, 223
– depth of NV centers   216, 217
– fluorescent nanodiamonds   220, 228, 231–235
– fluorinated nanodiamonds   228, 229, 235, 236
– formation of variously charged NV centers   217–222
– interactions with charged polymers   231–235
– introduction of vacancies to diamond lattice   220–222
– irradiation strategies   220–222
– nitrogen-vacancy color centers   209–236
– photoluminescence   211, 212, 215–218, 220–222
– quenching of NV-luminescence   226–228
– size dependence of NV luminescence   229–231
– surface functional changes on luminescence   216–218
– surface and size effects   228–231
– theoretical considerations   212
– transfection monitoring by nanodiamond carriers   232–235
– transfer doping   210, 212, 225–236
surface finishing   110, 113–115, 117, 122, 123, 128
surface graphitization of diamond   388–394
– diamond laser processing techniques   424, 425
– experimental data   389–394
– laser ablation   395, 404, 405

– mechanisms 388, 389
– threshold conditions 410, 411
surface-related effects of NV color centers 168
surface structuring 428–431
surface topologies 112
synchronously pumped mode-locked lasers 262, 263

t

TEM, *see* transmission electron microscopy
temperature dependent properties
– lattice absorption 8, 9
– n-type doping 183–185
– Raman lasers 254–256
– Raman scattering 22, 23
– refractive index 14–16
– UV edge absorption 12, 13
tensile strength 354
terahertz windows 48
texture evolution 42, 43
thermal annealing 396
thermal conductivity
– CVD diamond 54, 55, 62, 63
– diamond optics 110
– laser micro- and nanoprocessing of diamond materials 385–387
– polishing and shaping 71
– Raman lasers 268
– thermal management using diamond 354, 365–367
thermal expansion 80, 246–247, 255, 256, 268, 354
thermal lens strength 255
thermal lensing 254, 255, 269, 353, 354
thermal management using diamond 353–384
– advantages of diamond 354
– diamond LEDs 376
– diamond sub-mounts for lasers 355, 363, 367, 369–371
– direct exploitation of diamond for laser gain 356, 357, 377, 378
– doped-dielectric disk lasers 353–355, 360, 361, 369–374, 377
– exploiting extreme properties of diamond 357–360
– extracavity applications of diamond 357–360, 363–367
– future directions 368, 369, 374, 376–378
– intracavity applications of diamond 355, 356, 357–360, 363–368, 371–374, 377
– lasers 353–374
– light-emitting diodes 353, 354, 375, 376

– material requirements for intracavity applications 360
– monolithic structures and GaN epitaxy on diamond 376
– requirement for 353, 354
– semiconductor disk lasers 353–356, 360–369
– thermal conductivity and rigidity 354
– thinned device approach 355, 362–363, 365–370, 372–374
– wavelength diversity 367–368
thermal resistance 55, 56, 60, 61
thermal stress fracture 256
thermally induced stress birefringence 255, 256, 268
thermomechanical polishing 126
thermo-optic coefficients
– comparison with other crystals 247
– diamond 15, 16, 254, 255, 354
thinned device approach 355, 362, 363, 365–370, 372–374
thorium fluoride 133
three-dimensional laser writing in diamond 414–419, 424–426, 435–437
three-partite entanglement 302, 303
time-resolved confocal imaging 462, 463
TIR, *see* total internal reflection
titanium-doped sapphire lasers 353, 354, 356, 394
Tm-doped lasers 239–240
TO, *see* transverse optical
Toffoli gates 281, 282, 296
total internal reflection (TIR) 319, 328
TPE, *see* two-photon excited
transfection monitoring 232–235
transfer doping 210, 212, 225–236
– fluorescent nanodiamonds 220, 228, 231–235
– fluorinated nanodiamonds 228, 229, 235–236
– interactions with charged polymers 231–235
– quenching of NV-luminescence 226–228
– size dependence of NV luminescence 229–231
– surface and size effects 228–231
– transfection monitoring by nanodiamond carriers 232–235
– variously terminated nanodiamonds 225–236
transient photoconductivity technique 406, 407
transition metal oxide Raman lasers 240, 241, 246, 247, 249, 250, 260

transmission
- diamond optics   110, 111, 135
- infrared Raman lasers   264, 265
- lattice absorption   2, 3–11
- transmission spectra   2, 3
- UV edge absorption   2, 3, 11–13
transmission electron microscopy (TEM)
- fluorescent nanodiamonds   454–457
- integrated optical structures and microstructures   322, 323, 325
- laser micro- and nanoprocessing of diamond materials   391
- polishing and shaping   82
transverse optical (TO) phonons   188
transverse relaxation time   157, 158
tribological behavior of diamond   92–95
triboluminescence   88, 89, 102, 103
TRIM simulations   215, 217
trying rings   75, 76
tungsten carbide   95
twinned crystal   49, 50
two-dimensional photonic crystals   333, 334, 337
two-phonon absorption   4–7
two-photon absorption   29, 30
two-photon excited (TPE) fluorescence imaging   460–462
two-photon UV laser etching   119, 125, 266, 267
- see also nanoablation

u

ultrafast coherent phonon spectroscopy   21
ultra-nanocrystalline diamond (UNCD) films
- integrated optical structures and microstructures   318, 319, 326, 336
- laser micro- and nanoprocessing of diamond materials   386–387
ultraviolet 2-photon etching   119, 266, 267
- see also nanoablation
UNCD, see ultra-nanocrystalline
up-conversion fluorescent phosphors   445
UV edge absorption   2, 3, 11–13
UV-visible absorption spectra   45, 46

v

vacancies, see nitrogen-vacancy color centers; surface doping

van der Drift model   40, 41
vaporization ablation   394, 395–402, 429–431
VECSEL, see vertical-external cavity surface emitting lasers
Verdet constant   16
vertical-external cavity surface emitting lasers (VECSEL)   262

w

Wallner lines   99
wavefront aberrations   112
waveguides, see integrated optical structures and microstructures
wavelength division multiplexing   62
wear anisotropy   75–77, 79, 89, 94
wear debris   80, 81, 89, 90, 92, 94
Weibull modulus   57, 58
whispering gallery mode (WGM)   337
windows   110–112, 135

x

X-ray absorption   47, 48
X-ray diffraction   76, 77, 80, 81
X-ray reflectivity   135
X-ray topography   50, 52

y

Yb:KYW lasers   373, 374
Yb:YAG lasers   370–373
Yb-doped lasers   239–240
Young's modulus   56, 57, 354

z

Z-scan technique   27, 28
zeoten   83, 99
zero phonon line (ZPL)
- fluorescent nanodiamonds   447, 451, 456, 457
- integrated optical structures and microstructures   312, 314, 315, 334, 336
- nitrogen-vacancy color centers   151–153
- surface doping   211, 217, 225, 226, 230
zinc selenide   54, 60, 431
zinc sulfide   133
zone center phonons   17, 21, 24
ZPL, see zero phonon line

Color Plates

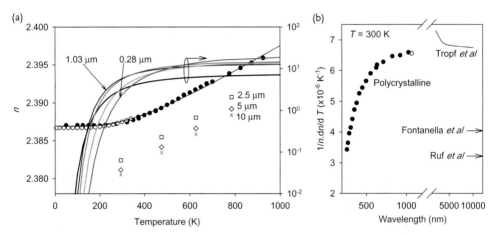

Figure 1.10 (a) Refractive index and thermo-optic coefficient versus temperature. Index data includes $\lambda = 100\,\mu m$ (solid circles [26]), $\lambda \to \infty$ (open circles [27]) and $\lambda = 2.5, 5,$ and $10\,\mu m$ [28]. The solid black curves for n and $1/n \times dn/dT$ are obtained using Equation (1.7). The colored curves correspond to Equation (1.7) using values ω_{eff} and K obtained for polycrystalline material at wavelengths 1.03, 0.62, 0.41, and 0.28 μm (brown, orange, light green, and dark green) [29]; (b) Values of $1/n \times dn/dT$ at 300 K as a function of wavelength [26–28]. The solid circles correspond to polycrystalline material [29].

Optical Engineering of Diamond, First Edition. Edited by Rich P. Mildren and James R. Rabeau.
© 2013 Wiley-VCH Verlag GmbH & Co. KGaA. Published 2013 by Wiley-VCH Verlag GmbH & Co. KGaA.

490 | Color Plates

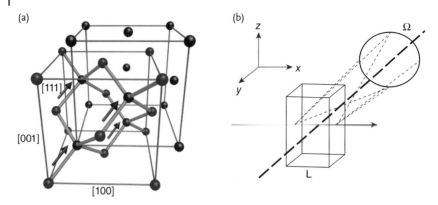

Figure 1.11 (a) The diamond lattice showing the two interpenetrating face-centered cubic lattices and the direction of relative movement for the zero wave-vector optical phonon involved in first-order Raman scattering; (b) Configuration for perpendicular Raman scattering.

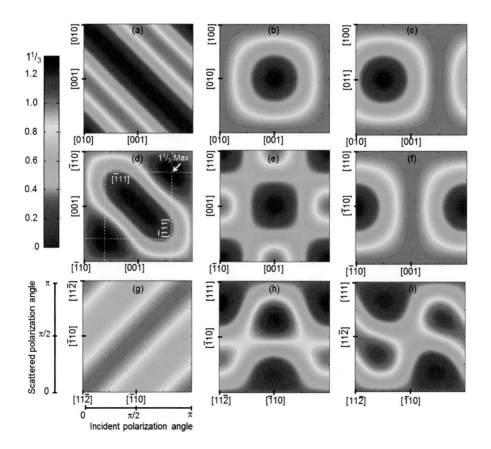

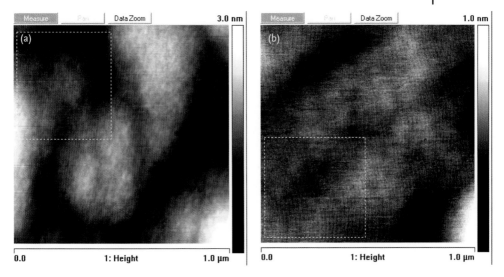

Figure 2.4 Atomic force microscopy images over a $1 \times 1\,\mu m^2$ area of a {100} as-grown CVD diamond film grown under conditions of (a) 2.0% CH_4 in H_2 and substrate temperature <800°C and (b) 0.8% CH_4 in H_2 and substrate temperature >900°C. Values of the rms roughness measured were 0.43 and 0.08 nm, respectively.

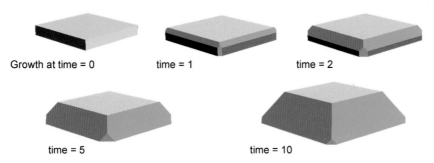

Figure 2.6 Theoretical morphology development under conditions of $\alpha = 4$, with low β and γ from a {100}-oriented, <110>-sided substrate (time = 0). At time > 0 {100} facet growth is shown in white, {111} in green, and {110} in pink. Morphology model courtesy of DTC Research Center, Maidenhead [2].

Figure 1.12 Scattering efficiency $\left(\sum_j |e_s R_j e_i|^2\right)/d^2$ as a function of input and Stokes beam polarizations for several beam directions. (a–c) For an incident beam along a <100> axis for axially directed scattering, perpendicular scattering in direction <100>, and perpendicular scattering in the direction <110>, respectively; (d–f) For an incident beam along a <110> axis for axially directed scattering, perpendicular scattering in direction <110>, and perpendicular scattering in the direction <100>, respectively; (g–i) For an incident beam along a <111> axis for axially directed scattering, perpendicular scattering in direction <112>, and perpendicular scattering in the direction <110>, respectively.

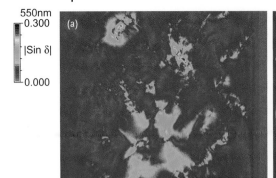

Figure 2.13 (a) |sin δ| false color image of a $2 \times 2 \times 2\,mm^3$ free-standing {100} CVD diamond plate illuminated along the <100> growth direction with $\lambda = 550\,nm$ light; (b) The same image overlaid with a grid of vectors representing the relative magnitude and direction of the local slow axis.

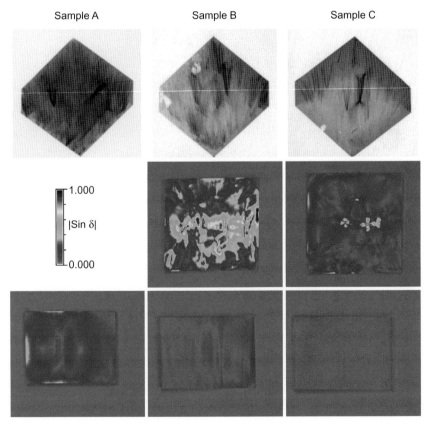

Figure 2.14 Free-standing, {100} CVD blocks imaged using {111} X-ray projection topography (top row) and birefringence microscopy as viewed parallel (middle row) and perpendicular (bottom row) to the direction of growth. |sin δ| > 1 parallel to growth direction for sample A (not shown).

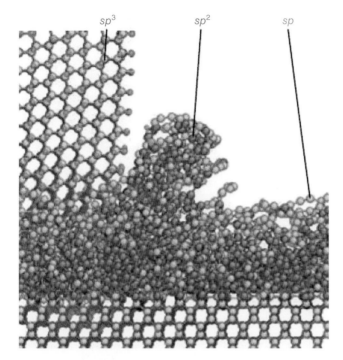

Figure 3.9 An illustration from a molecular dynamics study of diamond polishing by Pastewka *et al.* [73] This image shows the planing action of a sharp diamond grit particle on the amorphous sp^2 (green), sp (beige) layer on a diamond surface as it is polished on a scaife. Such simulations can now be carried out using realistic speeds and loads.

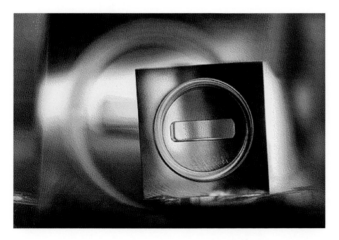

Figure 4.11 A diamond outcoupling window for CO_2 lasers. Courtesy of Element Six Ltd, UK.

Figure 4.12 (a) SEM image of a diamond annular groove phase mask (AGPM) component; (b) The Very Large Telescope in Chile, as operated by the European Southern Observatory. Courtesy of the European Southern Observatory.

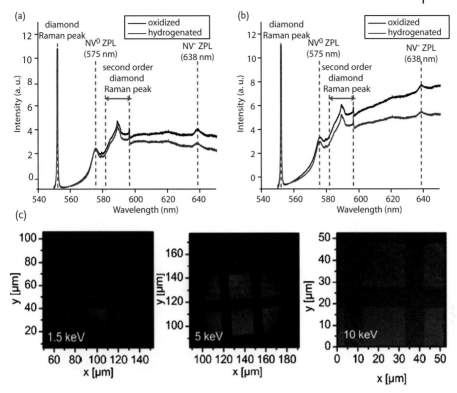

Figure 7.6 (a, b) 514 nm-pumped Raman and PL spectra of treated SCD, showing the influence of different surface termination on NV⁰ and NV⁻ related luminescence. NV⁰ ZPL (575 nm) peaks and NV⁻ ZPL (638 nm) peaks are visible for oxidized SCD. For SCD with hydrogenated surface, NV⁻-related luminescence is decreased with an increase in NV⁰-related luminescence, as shown in the difference spectra. All spectra were normalized to the diamond Raman peak (1332 cm^{-2}) and background-corrected. (a) 1 keV and (b) 6 keV N^+ ion implantation energy. When normalized to second-order diamond Raman peak, the number of counts (i.e., intensity) of NV-related luminescence was significantly lower for 1 keV implantation energy compared to 6 keV implantation energy. This suggests a different efficiency of creation of NV centers, depending on the implantation energy for the same total dose. This effect is discussed in the text and evaluated by mathematical modeling; (c) Fluorescence of NV defects implanted with three different energies. The dark stripes mark areas where the diamond surface is hydrogenated, while bright areas indicate regions where the surface was oxygen-terminated [12].

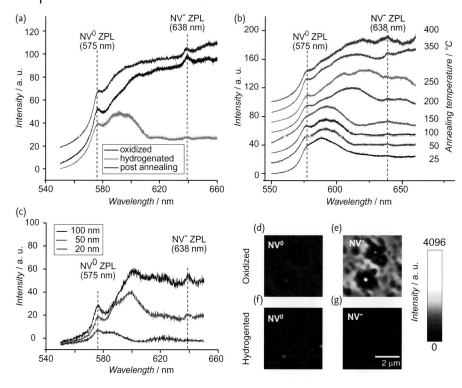

Figure 7.11 (a) Changes in NV⁻ and NV⁰ luminescence induced by various terminations of 40 nm-sized ND particles. Spectra of oxidized, hydrogenated, and annealed surfaces at 400 °C (leading to restoration of original surface termination). The hydrogenation of larger particles (~40 nm) resulted in a luminescence shift towards NV⁰ luminescence (hydrogenated 1), for the smaller particles (<20 nm), the luminescence quenched completely (hydrogenated 2); (b) Luminescence changes of hydrogenated ND (40 nm) upon annealing in air at different temperatures. Samples were heated to the target temperature (using a ramp of 25 °C per min), maintained at the set temperature for 30 min, and then cooled to room temperature. All spectra were recorded at room temperature and baseline-corrected; (c) PL spectra of hydrogenated ND of various sizes (20–100 nm) measured in physiological solution (pH 7), showing the ND size effect on the PL changes in hydrogenated/oxidized diamond and the possibility to monitor PL changes in liquid, which is essential for biosensor usage. The NV⁻ to NV⁰ ratio could be tuned with the size of ND. All spectra were normalized to diamond Raman peak, and the water Raman background was subtracted; (d–g) Confocal microscopy images of single ND particles upon various treatments. For oxidized NDs, both NV⁰ (d) and NV⁻ (e) luminescence is clearly visible in contrast to hydrogenated ND, where NV⁰ luminescence (f) is dominant and NV⁻ luminescence (g) is barely visible. The confocal PL images correspond to luminescence collected from a 570 nm to 610 nm spectral range for NV⁰, and from a 630 nm to 750 nm range for NV⁻.

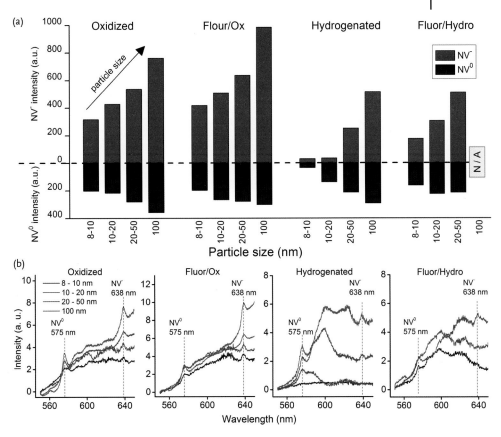

Figure 7.12 Luminescence intensity of variously terminated NDs as a function of size. Fluor/Ox represents NDs that were fluorinated after oxidization, leading to partially oxidized/partially fluorinated surface. Fluor/Hydro represents NDs with hydrogenated surfaces. (a) Peak intensity of the NV- and NV⁰ ZPL; (b) The corresponding photoluminescence spectra.

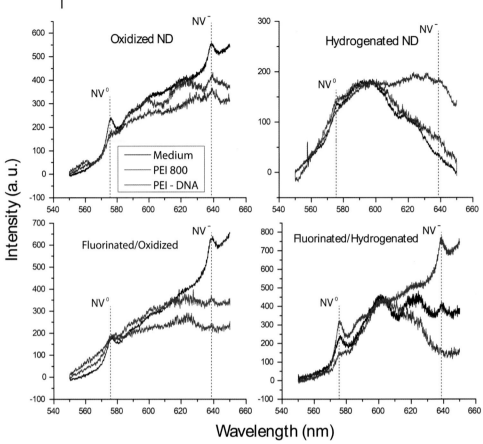

Figure 7.16 Photoluminsecence spectra of fluorescent NDs with different surface terminations. Large differences (e.g., the NV-fluorescence of the fluorinated/ hydrogenated nanodiamonds coated with PEI) could be used to monitor the interaction of the particles with strongly negatively charged biopolymers, such as DNA inside cells.

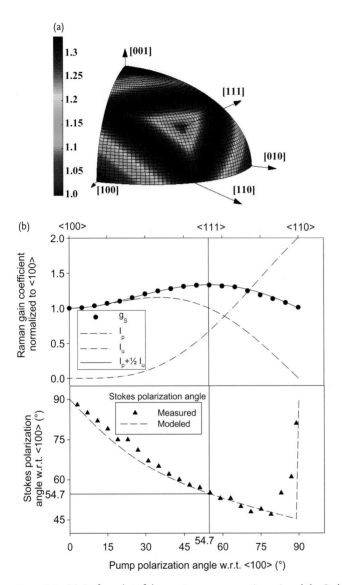

Figure 8.5 (a) Surface plot of the maximum Raman gain coefficient as a function of the propagation direction; (b) Measured polarization dependence of the gain coefficient g_S for propagation along <110> axes (upper) and the Stokes polarization (lower). The angle 54.7° corresponds to pump and Stokes alignment along the <111> direction.

(a)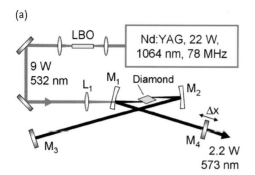

(b)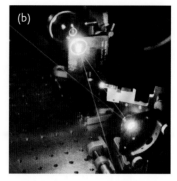

Figure 8.10 (a) Schematic diagram of the 573 nm synchronously pumped diamond Raman laser; (b) Photograph (courtesy of Eduardo Granados) of the center of the resonator, showing the diamond crystal and the intracavity yellow beam.

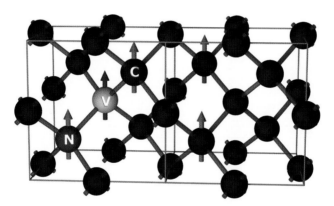

Figure 9.2 NV center and proximal nuclear spins Two unit cells of the diamond lattice (C = carbon) containing an NV center (N = nitrogen, V = vacancy) and exemplary nuclear spins. The NV center electron spin is shown as a blue arrow, and the nuclear spins as orange arrows. From left to right the nuclear spins are that of nitrogen, ^{13}C in the first shell, and two ^{13}C in the third shell.

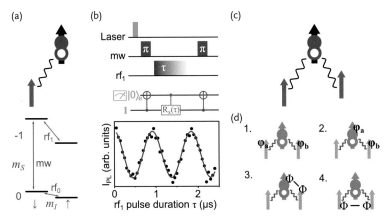

Figure 9.5 Nuclear spins and controlled qubit gates. (a) Illustration of a first shell ^{13}C nuclear spin coupled to an NV electron spin (top). The coupling is visualized by the wavy line. The corresponding energy level scheme is shown at the bottom. Blue and orange arrows illustrate EPR and NMR transitions, respectively; (b) Nuclear spin Rabi oscillation in the $m_S = -1$ electron spin sublevel (CROT gate, bottom) and corresponding pulse sequence (top, see text); (c) NV center spin coupled to two nuclear spins. The direct coupling among the nuclear spins is too weak to perform quantum gates; (d) Two qubit gates among nuclear spins exploiting the coupling to the common electron spin (see text).

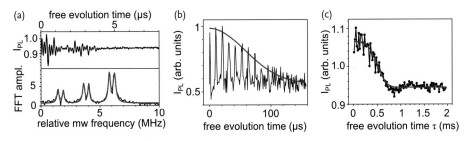

Figure 9.6 Ramsey and Hahn echo experiments affected by nuclear spins. (a) Ramsey oscillations and corresponding fast Fourier transform (FFT) spectrum showing the coupling to the ^{14}N nucleus and most likely a weakly coupled ^{13}C nuclear spin; (b) Electron spin echo envelope modulation (ESEEM) for Hahn echo of an NV center, showing collapses and revivals due to ^{13}C nuclear spin bath; (c) ESEEM for higher magnetic fields where collapses are suppressed.

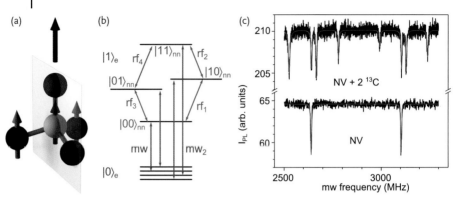

Figure 9.8 NV center with two ^{13}C nuclei in the first coordination shell. (a) NV center in the diamond lattice including the first coordination shell of carbon atoms, two of which are ^{13}C isotopes; (b) Spin energy level spectrum including electron spin $m_s = 0, -1$ ($|0\rangle_e$, $|1\rangle_e$) levels, times the four levels corresponding to the two nuclear spins with $I = ½$. For details of names, see the text; (c) Upper ODMR spectrum at 8.32 mT shows large hyperfine splitting due to the two nuclear spins. Both EPR transitions are shown, the red curve is a fit. The lower ODMR spectrum is a bare NV center at the same magnetic field, for comparison.

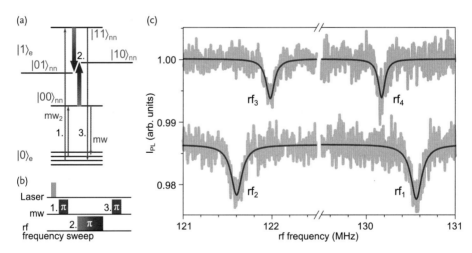

Figure 9.9 Electron-nuclear double resonance (ENDOR) spectra of ^{13}C nuclear spins. (a) Energy level scheme containing microwave (mw) and radiofrequency (rf) pulses to find rf resonance frequencies. The bold orange arrows with changing color indicate frequency sweeps; (b) Measurement sequence containing laser initialization, mw π-pulse (either mw or mw$_2$) and rf π-pulse with swept frequency. A spectrum is recorded as the rf frequency is swept while the sequence is running; (c) Actual spectra showing the correct transition frequencies for Figure 9.8b.

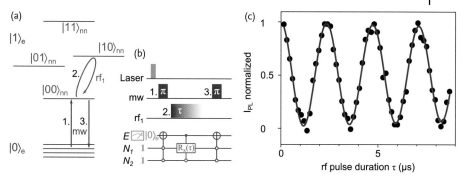

Figure 9.10 Coherent manipulation of ^{13}C nuclear spins. (a) Energy level scheme with arrows corresponding to microwave (mw) and radiofrequency (rf) transitions used for monitoring nuclear spin Rabi oscillations; (b) Pulse sequence for the experiment. A laser pulse initializes the electron spin into $|0\rangle_e$, after which a mw π-pulse flips the electron spin for nuclear spin state $|00\rangle_{nn}$ and a subsequent rf pulse resonant for the illustrated transition rotates the nuclear spin. Finally, spin state $|1\rangle_e |00\rangle_{nn}$ is probed by a second mw pulse and the laser pulse of the next sequence; (c) Measurement outcome. The fluorescence level oscillates while the nuclear spin is rotated. The vertical axis corresponds to the population that is found in state $|1\rangle_e \otimes |00\rangle_{nn}$.

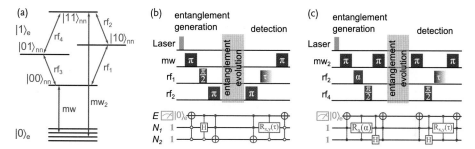

Figure 9.11 Sequence for entanglement generation and detection. (a) Energy level scheme showing contributing spin states and microwave (mw) and radiofrequency (rf) transitions used; (b) Exemplary sequence to generate and detect the Φ^- Bell state (see text). The other Bell states are created by changing the frequency of the rf$_2$ pulse and length and frequency of the rf$_1$ pulse. To record Ramsey oscillations, the free evolution time is varied and the final rf$_1$ pulse is a $\pi/2$-pulse. For state tomography, the final rf$_1$ pulse length is varied (Rabi oscillation) for two 90° phase-shifted rf fields; (c) Sequence to generate and detect the W state (analogous to Bell states; see text).

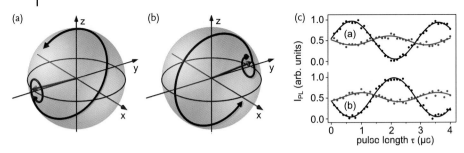

Figure 9.13 Quantum state tomography – Rabi signal. (a) Bloch sphere illustrating Rabi oscillations for quantum state tomography (see text). Here, the main density matrix entries of a Φ^- state (green arrow) have been transferred onto the working transition rf_1 (Bloch sphere), where the Rabi oscillations are performed around the x axis (black line) and the y axis (red line); (b) Tomography of Φ^+; (c) Actual measurement data of the Rabi oscillations (z-components of quantum states) for Φ^+ and Φ^- Bell states that lead to the state reconstruction shown in panel (a, upper part) and panel (b, lower part).

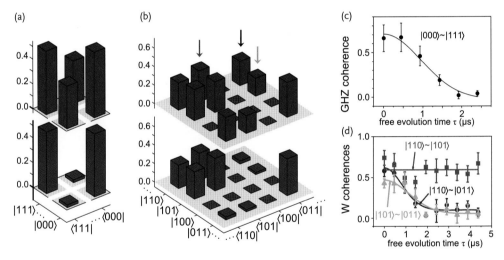

Figure 9.15 Results of quantum state tomography of GHZ and W states. (a) Partial tomography of the GHZ state, showing only the main density matrix entries. The upper part is for zero evolution time (see Figure 9.11), and the lower part shows the decayed coherences after 2.4 μs; (b) Partial tomography of the W state (gray entries have not been measured). The red, green, and black arrows mark the three different coherences. The upper topography is taken immediately after generation, and the lower tomography after an evolution time of 4.4 μs. The bare nuclear spin coherence (red arrow) survives longest; (c, d) Decay of the coherence of GHZ and W states, respectively, for increasing free evolution time. For the GHZ state, the decay of the single coherence (off-diagonal density matrix entry) is shown, whereas for the W state three coherences decay (for color coding, see panel b).

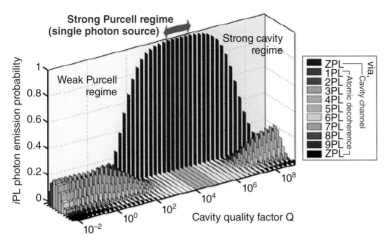

Figure 10.1 By ensuring that the ZPL emission is resonant with a cavity mode in the strong Purcell regime, all of the emission can be directed to be on the ZPL [38].

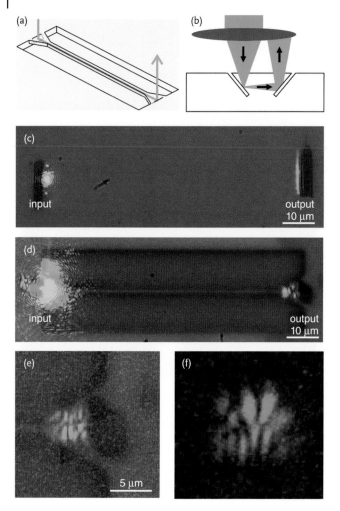

Figure 10.7 (a, b) Schematics of the (a) input/output light-collection system and (b) coupling to the waveguide structure; (c) Optical images of the incoming and outgoing 532 nm laser light between pairs of 45° mirrors facing each other in the bulk and (d) coupled through the bridge structure. In the former case, the outgoing light is spread along the output mirror, while in the latter case the propagating light is waveguided, which is indicated by the structured modal pattern at the output mirror; (e) Higher-magnification image of the output pattern; (f) same as (e), but without external illumination [86].

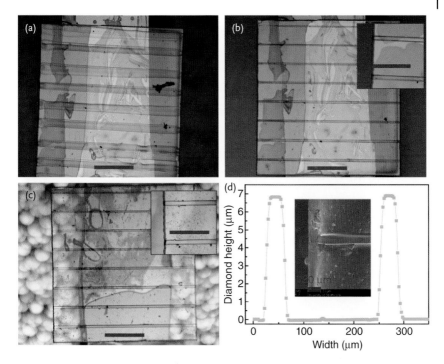

Figure 10.8 Optical microscopy images of the diamond sample during the process. (a) Inkjet printer-deposited photoresist stripes; (b) Photolithographic resist lines and inset with higher magnification; (c) Diamond waveguides etched by ICP and inset with higher magnification; (d) The surface profilometer measurements of the finished waveguide and the corresponding SEM image (inset). Scale bars: 500 μm for main figures; 200 μm for insets [92].

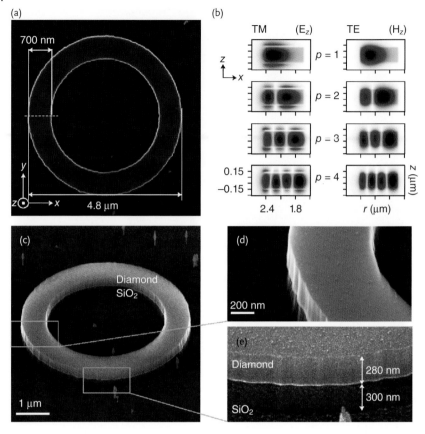

Figure 10.10 (a) Scanning electron microscopy image of a multimode ring resonator etched into a 280 nm-thick single crystal diamond membrane; (b) Simulated field profiles of 4 TM and 4 TE modes, with radial quantum numbers $p = 1$ to 4; (c) Side-view of the ring measuring 2.05 μm in radius and 700 nm in width, sitting on a SiO_2 pedestal that has been overetched by 300 nm; (d) and (e) Highlights of the surface roughness in the side and top walls of the structure [25].

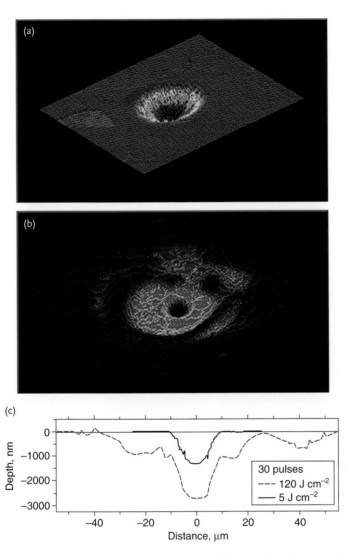

Figure 12.13 Influence of air ionization by 100 fs (λ = 800 nm) laser pulses on the quality and productivity of diamond ablation. (a) Image of the crater produced at E = 5 J cm^{-2} (low-intensity regime); (b) Image of the crater formed as a result of irradiation at E = 120 J cm^{-2} when air optical breakdown takes place; (c) Corresponding WLI crater profiles.

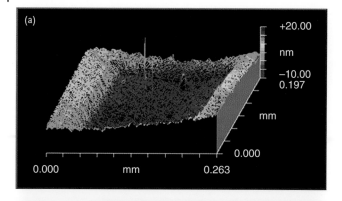

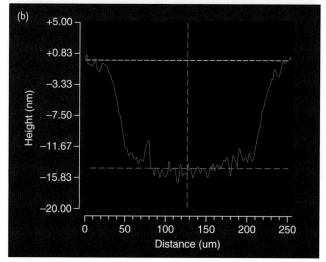

Figure 12.35 White light interferometric image (a) and profile (b) of the shallow crater formed as a result of KrF laser-induced nanoablation of IIa type diamond (irradiation in air by 3×10^5 laser pulses with $E = 15$ J cm^{-2}) [20].

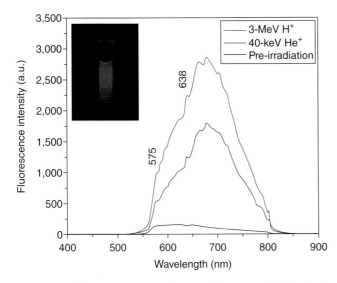

Figure 13.2 Fluorescence spectra of 35 nm FNDs prepared with either 40 keV He⁺ or 3 meV H⁺ irradiation. Inset: Fluorescence image of a cuvette containing FND suspension (1 mg ml^{-1}) excited by a 532 nm laser. Adapted from Ref. [15].

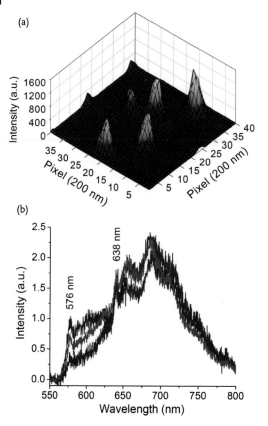

Figure 13.3 Fluorescence images and spectra of single 35 nm FNDs. (a) Confocal scanning image of 35 nm FNDs dispersed on a glass coverslip. The FWHM of each peak is 2–3 pixels, corresponding to a physical distance of 400–600 nm; (b) Fluorescence spectra of three different 35 nm FND particles. Adapted from Ref. [13].

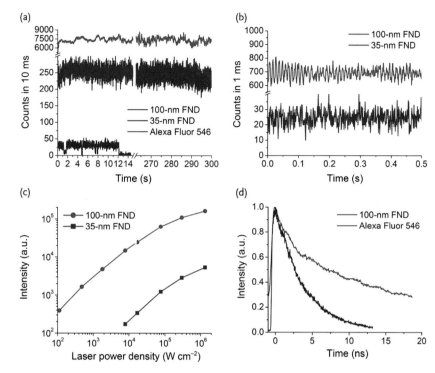

Figure 13.4 Spectroscopic characterization of 35 nm and 140 nm FNDs. (a) Typical time traces of fluorescence from a single 140 nm (green), 35 nm (red) and single Alexa Fluor 546 dye molecule attached to a single double-stranded DNA molecule (blue); (b) Time traces at a resolution of 1 ms, showing no photoblinking behavior; (c) Plot of fluorescence intensity as a function of laser power density; (d) Fluorescence lifetime measurements of 140 nm FNDs (green) and Alexa Fluor 546 dye molecule (blue). Fitting the time traces with two exponential decays for the FND revealed a fast component of 1.7 ns (4%) and a slow component of 17 ns (96%). The latter was fourfold longer than that (4 ns) of Alexa Fluor 546. Adapted from Ref. [13].

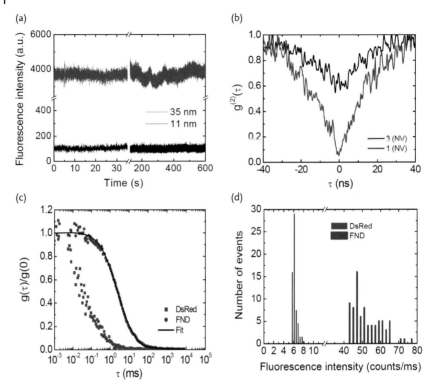

Figure 13.6 Optical characteristics of 11 nm FNDs produced by differential centrifugation. (a) Photostability tests of particles spin-coated on a bare glass substrate; (b) Typical photon correlation functions of the fluorescence intensities of two single FND particles containing one and three NV⁻ centers, respectively; (c) Spectroscopic characterization of DsRed-monomers and 11 nm FNDs by FCS; (d) Normalized autocorrelation functions of the intensity fluctuations and histograms of the fluorescence intensities of the particles. Adapted from Ref. [36].

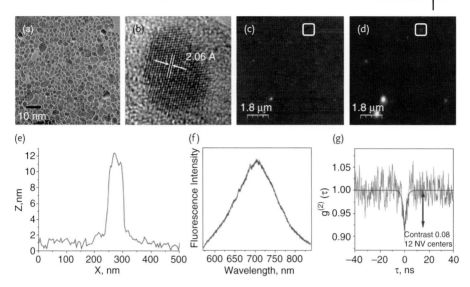

Figure 13.7 Particle size and optical characterization of 10 nm FNDs produced by milling. (a) TEM and (b) HRTEM images of the smallest particles obtained; (c) AFM and (d) fluorescence image of the same particles; (e) AFM dimensions of a 10 nm bright FND and (f) photoluminescence spectrum of the same particle; (g) Second-order autocorrelation function for the same particle. Adapted from Ref. [37].

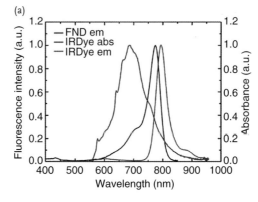

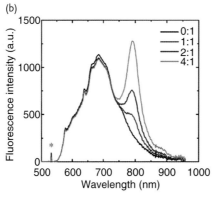
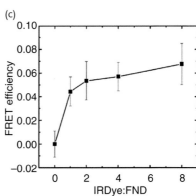

Figure 13.8 FRET between 19 nm FND and IRDye-800CW. (a) Overlap of the emission spectrum of FND and the absorption spectrum of IRDye; (b) Variation of the fluorescence spectra with the molar ratio of IRDye:FND; (c) FRET efficiency as a function of the IRDye:FND ratio. Adapted from Ref. [36].

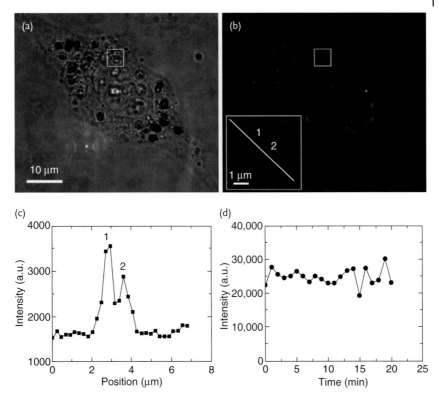

Figure 13.9 Observation of single 35 nm FNDs in HeLa cells. (a) Bright-field and epifluorescence merge images of a HeLa cells after uptake of 35 nm FNDs. Most of the particles are seen in the cytoplasm; (b) Epifluorescence image of a single HeLa cell after the FND uptake. Inset: An enlarged view of two fluorescence spots (denoted by **1** and **2**) with diffraction-limited size (FWHM = 500 nm). The separation between the two particles is 1 μm; (c) Intensity profile of the fluorescence image along the line drawn in the inset of panel (b); (d) Integrated fluorescence intensity as a function of time for particle **1**. The signal integration time was 0.1 s. Adapted from Ref. [13].

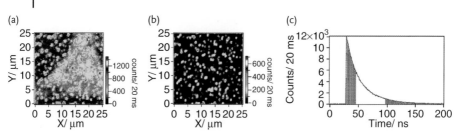

Figure 13.11 Time-resolved confocal raster scans of a fixed HeLa cell containing FND particles. (a) Raster scan obtained from the detection of all photons, displaying both FND and cell autofluorescence signals; (b) Time-gated raster scan constructed from photons detected at between 15 and 53 ns after the laser excitation pulses; (c) Fluorescence lifetime decay curve from the FNDs shown in panel (a). The gray region is the time-delay range, in which the photons are excluded to build the time-gated raster scan in panel (b). Adapted from Ref. [14].

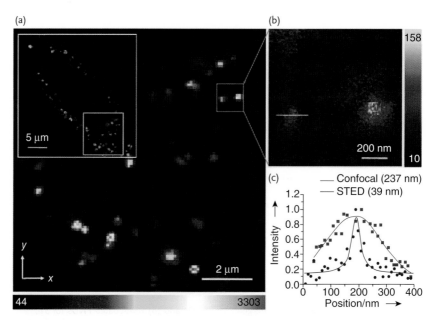

Figure 13.12 Confocal and STED imaging of HeLa cells labeled with BSA-conjugated FNDs by endocytosis. (a) Confocal image acquired by raster scanning of an FND-labeled cell. The fluorescence image of the entire cell is shown in the white box, and demonstrates fairly uniform cell labeling by BSA-conjugated FNDs; (b) STED image of single BSA-conjugated FND particles enclosed within the green box in panel (a); (c) Confocal and STED fluorescence intensity profiles of the particle indicated in panel (b) with a blue line. Solid curves are best fits to one-dimensional Gaussian (confocal) or Lorentzian (STED) functions. The corresponding full widths at half-maximum are given in parentheses. Adapted from Ref. [48].

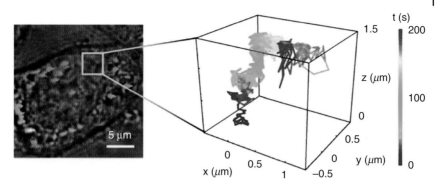

Figure 13.13 Three-dimensional tracking of a single 35 nm FND in a live cell. An overlay of bright-field and epifluorescence images (left) shows the existence of 35 nm FNDs in the cytoplasm of a live HeLa cell. The fluorescence intensity is sufficiently high and stable to allow tracking of a single FND in three dimensions over a time span of 200 s (right). Adapted from Ref. [15].

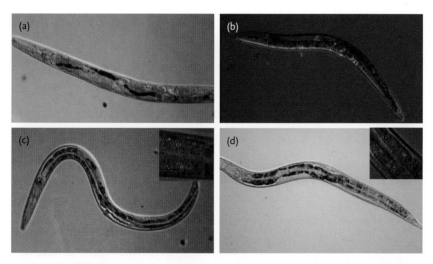

Figure 13.14 Epifluorescence/DIC-merged images of wild-type *C. elegans*. (a, b) Worms fed with bare FNDs for (a) 2 h and (b) 12 h; (c, d) Worms fed with (c) dextran-coated FNDs and (d) BSA-coated FNDs for 3 h. FNDs can be seen to be localized within in the intestinal cells. Adapted from Ref. [49].

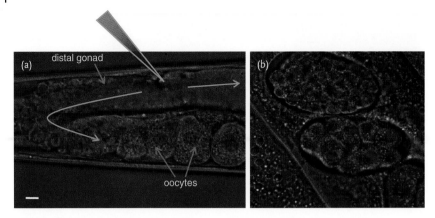

Figure 13.15 Microinjection of bare FNDs into *C. elegans*. Epifluorescence/DIC-merged images of an injected worm (a) and its progeny at the early (b) embryonic stage. The FNDs dispersed in the distal gonad and oocytes at approximately 30 min after injection (a). The green arrows indicate bulk streaming of FNDs with cytoplasmic materials, while the red triangle indicates the site of injection. Note that the injected FNDs are present in the cytoplasm of many cells in the early embryos (b). Scale bars = 10 μm. Adapted from Ref. [49].